Prebiotics and Probiotics Science and Technology

Dimitris Charalampopoulos, Robert A. Rastall (Eds.)

Prebiotics and Probiotics Science and Technology

Volume 1

With 67 figures and 97 tables

 Springer

Editors:
Dimitris Charalampopoulos
Department of Food Biosciences
University of Reading
Whiteknights, Reading
UK
d.charalampopoulos@reading.ac.uk

Robert A. Rastall
Department of Food Biosciences
University of Reading
Whiteknights, Reading
UK
r.a.rastall@reading.ac.uk

Library of Congress Control Number: 2009929416

ISBN: 978-0-387-79057-2

This publication is available also as:
Electronic version under ISBN 978-0-387-79058-9
Print and electronic bundle under ISBN 978-0-387-79059-6

springer.com

Printed on acid-free paper   SPIN: 12084159 – 5 4 3 2 1 0

To Elie Metchnikoff, Glenn Gibson and Marcel Roberfoid for originating
the concept of probiotics and prebiotics

Preface

With growing consumer and industrial interest in self-care and integrative medicine, our understanding of the relationship between health and diet has grown stronger. As a result, the market for functional foods, dietary supplements and nutraceuticals is continuing to expand rapidly. Among these products, probiotics and prebiotics have carved their own special niche because of their scientifically-supported health promoting properties, and have been in the forefront of research over the past twenty years or so. This is driven by the realisation that the gut microbiota can play a critical role in human health. Important functions of the gut microbiota include the inhibition of the colonisation of the gut by potentially pathogenic microorganisms, the microbial fermentation of substrates yielding metabolic products which can serve as sources of energy for the gut cell wall, and the modulation of the immune system. A substantial amount of research has shown that the human gut microbiota can be modulated using probiotics and prebiotics leading to various beneficial effects.

The prebiotics and probiotics area is a fast evolving field that attracts significant interest by both the academic and industrial communities. As a result, a substantial amount of research is generated every year. The use of post-genomics, encompassing transcriptomics, proteomics, metabolomics and meta-genomics has helped greatly in making significant advances in the field, as they provide the means to carry out in-depth studies of the mechanisms involved and of the beneficial effects likely to be exerted to the host. The target of many probiotics and prebiotics is the prevention and treatment of disorders associated with the gastrointestinal tract (GIT), including infectious, traveller's, and antibiotic-associated diarrhoeas, *Clostridium difficile* infections, and constipation. They have also been suggested as therapeutic agents against irritable bowel syndrome and inflammatory bowel diseases. An increasing amount of evidence from *in vitro* and *in vivo* studies suggests that they are effective in the prevention of atopic allergies and may have potential anti-carcinogenic effects. In addition to the above, there has been considerable interest in extra-intestinal application of probiotics and prebiotics, such as in urogenital infections and oral health, and in their applications in animals.

This book will give a detailed and up to date account of the advances in the prebiotics and probiotics field, covering their taxonomy, the potential beneficial effects to humans and animals, the proposed mechanisms of actions, the molecular techniques used, and the challenges faced in manufacturing such products. The book consists of thirty one chapters. It starts with an introduction into the gastrointestinal tract ecosystem and its interaction with prebiotics and probiotics (Chapter 1), it then describes the molecular techniques used for analysing the complex gut ecosystem (Chapters 2 and 3), and discusses the design of human trials for evaluating their efficacy (Chapter 4). It then moves into the area of prebiotics and starts off by discussing the mechanisms of prebiotic impact on health (Chapter 5); it then provides a systematic overview of established as well as potential prebiotic oligosaccharides (Chapters 6 to 11). Each of these chapters covers the processes used to manufacture these compounds, and critically discusses their efficacy based on published data from *in vitro* and *in vivo* studies. The rest of the chapters in the prebiotics part cover the application of prebiotics in animals (Chapter 12), the analysis of prebiotic oligosaccharides (Chapter 13), and their manufacture from biomass sources (Chapter 14). The book then focuses on probiotics and begins with a taxonomic study of probiotic microorganisms (Chapter 15). It then discusses their interaction in the human gut (Chapter 16), provides an overview of functional genomic studies of probiotics (Chapter 17), discusses the challenges in their manufacture (Chapter 18) and in their addition to foods (Chapter 19), and presents the encapsulation methodologies used to improve probiotic delivery (Chapter 20). Subsequent chapters cover the potential beneficial effects of probiotics focusing on antibiotic-associated and *Clostridium difficile* diarrhea (Chapter 21), infectious and traveller's diarrhea (Chapter 22), immune modulation (Chapter 23), chronic gastrointestinal infections (Chapter 24), allergic diseases (Chapter 25), potential anti-carcinogenic effects (Chapter 26), urogenital applications (Chapter 27), oral health (Chapter 28), and on the development of mucosal vaccines based on lactic acid bacteria (Chapter 29). Finally, the remaining chapters cover the applications of probiotics in livestock animals (Chapter 30) and safety aspects of probiotics (Chapter 31).

The aim of this book is to provide a comprehensive overview of the research in the field of prebiotics and probiotics covering the achievements, challenges and future needs. As such, we hope that this book will be a valuable reference to both researchers and industrialists working in the field.

Dimitris Charalampopoulos
Bob Rastall

Table of Contents

Volume 1

Volume 2

Editors

Dimitris Charalampopoulos
Department of Food Biosciences
University of Reading
Whiteknights, Reading
UK
d.charalampopoulos@reading.ac.uk

Robert Rastall
Department of Food Biosciences
University of Reading
Whiteknights, Reading
UK
r.a.rastall@reading.ac.uk

List of Contributors

Leticia Abecia
Food Microbial Sciences
Department of Food Biosciences
University of Reading
Reading, UK

A. Adam-Perrot
TATE & LYLE Innovation Centre
Parc Scientific de la Haute Borne
Villeneuve d'Ascq
France

Philip Allsopp
Northern Ireland Centre for Food
and Health
Centre for Molecular Biosciences
University of Ulster
Coleraine, UK

Jośe Luis Alonso
Department of Chemical Engineering
University of Vigo
As Lagoas, Ourense, Spain

Charlotte Atkinson
Lecturer in Nutrition
University of Bristol and United Bristol
Healthcare, NHS Trust
Bristol, Avon, UK

Kathleen A. Barry
Department of Animal Sciences
University of Illinois at
Urbana-Champaign
Champaign, IL, USA

Virender K. Batish
National Dairy Research Institute
Karnal (Haryana), India

Luis G. Bermúdez-Humarán
Unité d'Ecologie et de
Physiologie du Système Digestif
INRA 0910, Jouy-en-Josas, France

Douwina Bosscher
Laboratory of Functional Food Science
and Nutrition
Department of Pharmaceutical Sciences
University of Antwerp
Wilrijk, Belgium

S. Bouvier
TATE & LYLE Innovation Centre
Parc Scientific de la Haute Borne
Villeneuve d'Ascq, France

Robert J. Boyle
Department of Paediatrics
Imperial College, London, UK

Randal Buddington
Health and Sports Sciences
University of Memphis
Memphis, Tennessee, USA

Claude P. Champagne
Food R & D Centre
Agriculture and Agri-Food Canada
Casavant, St. Hyacinthe
QC, Canada

Jean-Marc Chatel
Unité d'Ecologie et de
Physiologie du Système Digestif
INRA 0910, Jouy-en-Josas, France

James W. Collins
Department of Bacterial Diseases
VLA (Veterinary Laboratories Agency)
Woodham Lane
New Haw, UK

C. Combe
TATE & LYLE Innovation Centre
Parc Scientific de la Haute Borne
Villeneuve d'Ascq, France

N. Corzo
Instituto de Fermentaciones Industriales
(C.S.I.C.) Juan de la Cierva
Madrid, Spain

Luc De Vuyst
Research Group of Industrial
Microbiology and Food Biotechnology
Department of Applied Biological
Sciences and Engineering
Vrije Universiteit Brussel
Pleinlaan, Brussels, Belgium

Eddie R. Deaville
Food Microbial Sciences
Department of Food Biosciences
University of Reading
Reading, UK

Franco Dellaglio
Dipartimento Scientifico e Technologico
Facoltà di Scienze MM. FF. NN.
Università degli Studi di Verona
Verona, Italy

Herminia Domínguez
Department of Chemical Engineering
University of Vigo
As Lagoas, Ourense, Spain

A. W. C. Einerhand
TATE & LYLE Innovation Centre
Parc Scientific de la Haute Borne
Villeneuve d'Ascq, France

George C. Fahey, Jr.
Department of Animal Sciences
University of Illinois at
Urbana-Champaign
Champaign, IL, USA

Gwen Falony
Research Group of Industrial
Microbiology and Food Biotechnology
Department of Applied Biological
Sciences and Engineering
Vrije Universiteit Brussel
Pleinlaan, Brussels, Belgium

Francesca Fava
Food Microbial Sciences
Department of Food Biosciences

University of Reading
Reading, UK

Giovanna E. Felis
Dipartimento di Scienze Tecnologie
e Mercati della Vite e del Vino
Facoltà di Scienze MM. FF. NN.
Università degli Studi di Verona
Verona, Italy

G. F. Fitzgerald
Department of Microbiology
University College Cork
Cork, Co. Cork, Ireland

Rafael Frias
Central Animal Laboratory
University of Turku
Finland

Harsharn S. Gill
Primary Industries Research Victoria
Department of Primary Industries
Australia and Victoria University
Werribee, Victoria, Australia

Preet Gill
School of Medicine
Griffith University
Southport, Australia

Yong Jun Goh
Department of Food
Bioprocessing and Nutrition Sciences
and
Southeast Dairy Foods
Research Center

North Carolina State University
Raleigh, NC, USA

Sunita Grover
National Dairy Research Institute
Karnal (Haryana), India

Francisco Guarner
Digestive System Research Unit
University Hospital Vall d'Hebron
Barcelona, Spain

Miguel Gueimonde
Department of Microbiology and
Biochemistry of Dairy Products
Instituto de Productos Lacteos de
Asturias. CSIC
Asturias, Spain

Beatriz Gullón
Department of Chemical Engineering
University of Vigo
As Lagoas, Ourense, Spain

Patricia Gullón
Department of Chemical Engineering
University of Vigo
As Lagoas, Ourense, Spain

L. Gutton
TATE & LYLE Innovation Centre
Parc Scientific de la Haute Borne
Villeneuve d'Ascq, France

O. Hasselwander
Technology & Business Development
Danisco (UK) Limited
Redhill, UK

Patricia L. Hibberd
Director, Center for Global Health
Research
and
Departments of Public Health and
Family Medicine,
Medicine, and Pediatrics
Tufts University School of Medicine
Boston, MA, USA

Silvia Innocentin
Unité d'Ecologie et de Physiologie du
Système Digestif INRA 0910
Jouy-en-Josas, France
and
Unité de Virologie et Immunologie
Moléculaires
Domaine de Vilvert
INRA, UR892, Jouy-en-Josas, France

Erika Isolauri
Department of Paediatrics
University of Turku
and
Turku University Central Hospital
Finland

M. Juntunen
Danisco Finland Health & Nutrition
Kantvik, Finland

Todd R. Klaenhammer
Department of Food
Bioprocessing, and Nutrition Sciences;
and
Southeast Dairy Foods Research Center
North Carolina State University
Raleigh, NC, USA

Annett Klinder
Food Microbial Sciences
Department of Food Biosciences
University of Reading
Reading, UK

Roberto M. La Ragione
Dept. of Food & Environmental Safety
Veterinary Laboratories Agency (Defra)
Woodham Lane, KT15 3NB, Addlestone
Surrey, UK

Sampo J. Lahtinen
Danisco Finland Health & Nutrition
Kantvik, Finland

Philippe Langella
Unité d'Ecologie et de Physiologie du
Système Digestif
INRA 0910, Jouy-en-Josas, France

Christophe Lay
Department of Microbiology and
Immunology
University of Otago
Dunedin, New Zealand

Francois Lefèvre
Unité de Virologie et Immunologie
Moléculaires
INRA, UR892, Domaine de Vilvert
Jouy-en-Josas, France

Stephen Lewis
Derriford Hospital
Plymouth, Devon, UK

S. Macfarlane
Microbiology and Gut Biology Group
Ninewells Hospital Medical School
Dundee, UK

Abelardo Margolles
Department of Microbiology and
Biochemistry of Dairy Products
Instituto de Productos Lacteos de
Asturias. CSIC
Asturias, Spain

I. Martínez-Castro
Instituto de Química Orgánica General
(C.S.I.C.)
Juan de la Cierva Madrid, Spain

Jean-Antoine Meiners
Meiners Commodity Consultants S.A
Colombier, Switzerland

J. H. Meurman
Institute of Dentistry of Helsinki
and
Department of Oral and Maxillofacial
Disease
Helsinki University Central Hospital
Finland

H. Mäkeläinen
Danisco Finland Health & Nutrition
Kantvik, Finland

Pierre F. Monsan
LISBP-INSA, UMR CNRS 5504
UMR INRA, Toulouse cedex, France

Andrés Moure
Department of Chemical Engineering
University of Vigo
As Lagoas, Ourense, Spain

J. A. Muller
Teagasc
Moorepark Food Research Centre
Fermoy, Co. Cork, Ireland

Sarah O'Flaherty
Department of Food
Bioprocessing and Nutrition Sciences;
and
Southeast Dairy Foods Research Center
North Carolina State University
Raleigh, NC, USA

Francois Ouarné
CRITT Bioindustries, INSA
Toulouse cedex, France

Juan Carlos Parajó
Department of Chemical Engineering
University of Vigo
As Lagoas, Ourense, Spain

S. Potter
TATE & LYLE Innovation Centre
Parc Scientific de la Haute Borne
Villeneuve d'Ascq, France

Gregor Reid
Canadian Research & Development
Centre for Probiotics
Lawson Health Research Institute
and

Department of Microbiology &
Immunology
University of Western Ontario
ON, Canada

R. P. Ross
Teagasc
Moorepark Food Research Centre
Fermoy, Co. Cork, Ireland

Ian Rowland
Hugh Sinclair Unit for Human Nutrition
Department of Food Biosciences
University of Reading
Whiteknights, Reading, UK

A. I. Ruiz-Matute
Instituto de Química Orgánica General
(C.S.I.C.)
Juan de la Cierva Madrid, Spain

Seppo Salminen
Functional Foods Forum
University of Turku
20014 Turku, Finland

L. Sanders
TATE & LYLE Innovation Centre
Parc Scientific de la Haute Borne
Villeneuve d'Ascq, France

M. L. Sanz
Instituto de Química Orgánica General
(C.S.I.C.)
Juan de la Cierva Madrid, Spain

Laura E. J. Searle
Department of Bacterial Diseases
VLA (Veterinary Laboratories Agency)
Woodham Lane
New Haw, UK

Qing Shen
Food Microbial Sciences
Department of Food Biosciences
University of Reading
Whiteknights, Reading, UK

C. Stanton
Teagasc
Moorepark Food Research Centre
Fermoy, Co. Cork, Ireland

H. Steed
Microbiology and Gut Biology Group
Ninewells Hospital Medical School
Dundee, UK

Julian D. Stowell
Danisco Sweeteners
Danisco (UK) Ltd
Redhill, Surrey, UK

Christina M. Surawicz
Division of Gastroenterology
Department of Medicine
School of Medicine
University of Washington
Seattle, WA, USA

Sandra Torriani
Dipartimento di Scienze Technologie
e Mercati della Vite e del Vino
Università degli Studi di Verona,
Verona, Italy

Kieran M. Tuohy
Food Microbial Sciences
Department of Food Biosciences
Food Biosciences and Pharmacy
University of Reading
Reading, UK

George Tzortzis
Clasado Ltd
Wolverton Mill
Milton Keynes, UK

R. Van Den Abbeele
TATE & LYLE Innovation Centre
Parc Scientific de la Haute Borne
Villeneuve d'Ascq, France

Brittany M. Vester
Department of Animal Sciences
University of Illinois at
Urbana-Champaign
Champaign, IL, USA

Jelena Vulevic
Department of Food Biosciences
University of Reading
Whiteknights, Reading, UK

Martin J. Woodward
Department of Bacterial Diseases
VLA (Veterinary Laboratories Agency)
Woodham Lane
New Haw, UK

1 Using Probiotics and Prebiotics to Manage the Gastrointestinal Tract Ecosystem

Randal Buddington

1.1 Introduction

Natural and man-made ecosystems are routinely managed to increase productivity and provide desired characteristics. The management approaches most commonly used include the addition of desired organisms, provision of fertilizers or feeds to encourage desired species, alteration of the physical or chemical features of the environment, and the selective removal of undesirable species. The selection of specific management strategies and their success are dependent on a thorough understanding of existing ecosystem characteristics and the short and long-term responses to the management strategy.

Ecosystems are generally recognized as consisting of structural elements that include living (biotic) and non-living (abiotic) components in a physically defined area and are maintained by functional elements that are involved in the cycling of materials and energy. The concept of the gastrointestinal tract (GIT) as an ecosystem is based on the interactions among the resident assemblages of microorganisms, the structural and functional characteristics of the GIT, and the responses to dietary inputs, and has proven useful for understanding the complexities of the interactions. Moreover, the application of ecological principles should assist in guiding "management" decisions to improve GIT characteristics and host health and nutrition.

The resident bacteria (the microbiome) are recognized to play a central role in GIT characteristics and host health (Falk et al., 1998; Martin et al., 2008; Tappenden and Deutsch, 2007). This has encouraged the development of approaches to manage the GIT bacteria to improve health and nutrition. Although the use of antibiotics for removal of undesired species has been a mainstay for managing the GIT bacteria, concerns about bacterial adaptation, development of antibiotic resistance, and destabilization of the commensal assemblages have stimulated the search for alternative strategies. At the forefront have been probiotic products, which include viable microorganisms that when administered in doses that reach the intestine in an active state are able to exert positive health effects. More recent is the introduction of prebiotic products, which are "selectively fermented food ingredients that elicit specific changes, both in the composition and/or metabolic activity of the gastrointestinal microbiota, and thereby confer health benefits to the host" (Roberfroid, 2007). The combination of pro- and prebiotics to obtain synergistic benefits has been termed synbiotics. The efficacy of probiotics, prebiotics, and synbiotics to improve host health and nutrition is dependent on the changes they elicit in the composition and metabolic activities of the resident assemblages of microorganisms.

This chapter describes the interactions among GIT structure and functions, the resident assemblages of bacteria, and dietary inputs, and how those interactions change from birth to senescence, are responsive to health status, and can be influenced by probiotics and prebiotics. Readers will first be introduced to the concept of the GIT as an ecosystem and provided with descriptions of the physical, chemical, functional, and biotic components. The objective is to encourage readers to consider the GIT from a different perspective. Examples and selected citations are then provided to demonstrate the application of probiotics and prebiotics as tools that can be used to manage the GIT ecosystem during development and in normal and disease states, and thereby improve the health and nutritional status of the host. The objective is to inform readers about the opportunities and challenges of using probiotics and prebiotics to manage the GIT ecosystem.

1.2 Gastrointestinal Tracts and River Ecosystems

The GIT ecosystem is in many ways similar to river systems that originate in a lake or reservoir (i.e., stomach) with an outflow into a fast flowing stream (i.e., small intestine) that receives inputs from other sources (pancreas and gall bladder) and eventually become slow moving, large rivers (colon) that empty into the ocean (◉ *Figure 1.1*). Many of the tenets of the "river continuum concept" (Vannote et al., 1980) apply to the GIT. For example, both rivers and GIT's are characterized by physical and chemical gradients, which include regional differences in size, velocity of flow, and lumenal composition. The gradients shape the patterns of species distributions along the GIT, with the resident assemblages of biota

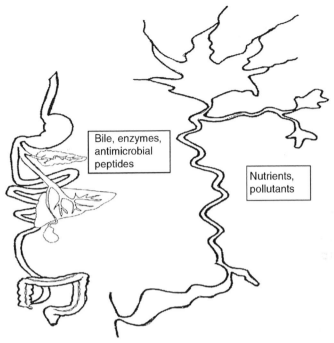

◙ Figure 1.1

The gastrointestinal tract shares several similarities with many river ecosystems.

adapted to specific regions. Sharp gradients, such as the transition between the stomach and small intestine, result in distinct regional differences in biota, whereas the gradual gradient along the length of the small intestine results in a less distinct distribution of the resident biota. Furthermore, in both types of ecosystems the downstream communities of organisms are largely dependent on upstream inefficiencies that allow for the availability of needed nutrients.

Each region of the GIT represents a habitat with unique structural, chemical, and biotic components. Moreover, the regional processes of digestion and secretion alter the composition of the lumenal contents as dietary inputs are processed and flow distally, thereby influencing the species composition and metabolic activities of the resident bacteria. Exemplary are the profound differences between the stomach, the small intestine, and colon. Even within each GIT region there are sub-habitats. Specifically, the physical, chemical, and biotic characteristics of the lumenal contents differ from the layer of material that is immediately adjacent to the epithelial lining. Similarly, the physical, chemical, and biotic characteristics found in the middle of a stream or river differ markedly from those at interface between the water and the land (riparian zone), resulting in different biotic communities.

Many of the ecological principles that apply to rivers are also relevant to the understanding of the GIT ecosystem. For example, the characteristics and functions of other ecosystems are closely related to the species composition, densities, and functions of the resident biota (Tilman et al., 1997). Of importance are the numbers of species and functional groups that constitute biodiversity (Hooper and Vitousek, 1997), with productivity and ecosystem stability increasing with greater species diversity. Similarly, the GIT and host health is related to the abundances, diversity, and metabolic activities of the commensal and pathogenic bacteria, and the responses of the bacteria to dietary inputs, including probiotics and prebiotics.

Differences do exist between rivers and GIT ecosystems. Unlike the abiotic substrates of rivers and streams, the living cells constituting the epithelium and underlying layers of the GIT are responsive to the chemical composition and bacterial assemblages of the lumenal contents. Exemplary are the changes in epithelial cell patterns of gene expression in response to different species of bacteria (Bry et al., 1996; Shirkey et al., 2006) and to changes in nutrient concentrations (Beaslas et al., 2008; Le Gall et al., 2007). Moreover, unlike river systems, signaling from distal to proximal regions of the GIT provides a mechanism whereby characteristics in the proximal regions of the GIT can be altered in response to events "downstream." This is exemplified by the regulatory peptides secreted by the ileum and colon (e.g., glucagon-like peptides 1 and 2, peptide YY) in response to the presence of nutrients and bacterial metabolites (e.g., short chain fatty acids) and the changes they elicit in the proximal small intestine (i.e., regulate digestion and stimulate growth and functions).

1.3   The Components of the GIT Ecosystem

The GIT ecosystem involves a dynamic balance between GIT structure and functions, the resident microbiota, and dietary inputs (● *Figure 1.2*). The ability of bacterial populations to adapt to different GIT characteristics and dietary inputs (Dunne, 2001) results in GIT ecosystems that are unique for each individuals. Even monozygotic twins have different assemblages of GIT bacteria, though the differences are of lower magnitude than those of non-related individuals (Stewart et al., 2005). The following sections acquaint readers with the major components of the GIT ecosystem.

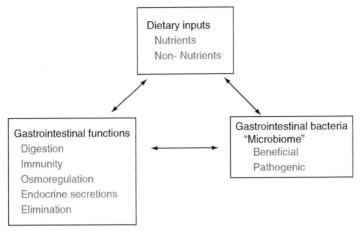

◻ Figure 1.2
There are complex interactions among dietary inputs, gastrointestinal tract functions, and the resident assemblages of bacteria that influence host health and nutrition.

The physical features. Anatomically, the GIT is the most variable organ system of vertebrates. This is obvious by comparing GIT anatomy among carnivores, herbivores, and omnivores (Stevens and Hume, 1995). The different functional demands associated with processing the wide diversity of natural diets among species has led to adaptive changes in the structural features of the different GIT regions. Moreover, although the basic GIT plan for a species is established during evolution, adaptive changes in GIT structure can and do occur during development and in response to different diets.

The variation in flow rates that occurs in the different regions and during the processing of foods has an influence on the biotic components of the GIT ecosystem. Specifically, rapid movement of digesta in the proximal small intestine contributes to the reduced population densities by shortening residence time and hindering persistence. In contrast, the slower movement of digesta in the distal ileum and colon is associated with higher abundances and diversity of species.

The chemical features. The chemical composition in the different regions of the GIT is determined by the combination of dietary inputs and GIT functions. This includes secretions from the different regions of the GIT (stomach, small and large intestine) and the associated organs (i.e., pancreas and gall bladder). Correspondingly, the lumenal contents in the different regions of the GIT have unique chemical compositions. During the processing of meals, the contents of the stomach are nutrient rich, with nutrient levels declining along the intestine, into the colon. Regional differences in pH, bile acids, electrolytes, antimicrobial peptides, and oxygen tension impose an additional selective pressure on the assemblages and metabolic activities of the resident bacteria. Notable is the presence of *Helicobacter* spp in the stomach, whereas populations of obligate anaerobes are highest in the colon.

The functional features. The functional features (physiology) of the GIT are key determinants of the chemical characteristics that influence the resident assemblages of bacteria. They also represent barriers to the introduction of species, such as probiotics as well as pathogens. The five basic functions of the GIT include (1) digestion of feedstuffs, (2) osmoregulation, (3) endocrine regulation of digestion and host metabolism, (4) immunity and defense against potential pathogens and harmful substances, and (5) the detoxification and elimination of toxic molecules originating from the environmental or the host.

Despite the diversity of anatomical structures, the functional characteristics of the vertebrate GIT are similar and are based on shared mechanisms. Hence, observed interspecies differences in GIT functions are related much more to adaptive responses of shared mechanisms to match dietary inputs and environmental influences rather than different mechanisms. For example, absorption of sugars and amino acids by the intestine is dependent on a diversity of apical membrane transporters that are shared among vertebrates. The higher rates of glucose absorption by omnivores compared with carnivores are caused by higher densities and activities of the apical membrane sodium glucose cotransporter SGLT-1, not because of a different type of transporter. Similarly, the exocrine pancreas of all vertebrates secretes a complex mixture of digestive enzymes and other proteins. The specific mixture of digestive enzymes secreted by the exocrine pancreas is modulated to match diet composition. Granted, over evolutionary time, individual proteins in pancreatic secretion have undergone changes in amino acid composition, thereby altering functional properties.

Following are brief reviews to acquaint readers with each of the GIT functions, with selected examples of the responses to the bacteria in the GIT. Extensive reviews of GIT physiology are available (e.g., Johnson, 2006).

Digestion. Feedstuffs are made available for metabolism by the host processes of secretion, motility, and absorption. Secretion of electrolytes, digestive enzymes, and bile by the GIT and associated organs is essential for the hydrolysis of the complex polymers in feedstuffs into the smaller molecules that are then absorbed and made available to the host. Although some food molecules are taken in by simple diffusion, the vast majority are absorbed from the GIT lumen by various specialized proteins (active and facilitative transporters) in the apical membrane. Additional transporters and ion channels are important for recovering the electrolytes, water, and other molecules (e.g., bile acids) present in the digestive secretions. Contractions by the smooth muscle that lines the alimentary canal provides the motility that is necessary to mix foodstuffs with digestive secretions, reduce unstirred layer influences that reduce absorption, and the peristalsis that propels the food distally and reduces stagnation between the processing of meals. The patterns of intestinal motility are responsive to the resident bacteria (Husebye et al., 2001), including probiotics (Lesniewska et al., 2006).

Microbial fermentation, though not a host process per se, plays an important role in the ability of the host acquiring energy and nutrients from undigested feedstuffs. The products of bacterial fermentation, and particularly the short chain fatty acids (SCFA) provide up to 10% of the total metabolic energy requirement of humans and even higher proportions among animals with larger hindguts or rumens (Rechkemmer et al., 1988). Moreover, SCFA produced by the bacteria in the ileum and colon stimulates the secretion of regulatory peptides that enhance growth and functions of the proximal small intestine (Bartholome et al., 2004). Also to be considered is the interesting, but poorly understood competition between the host and the resident bacteria for nutrients. Obviously, it is in the best interest of the host to rapidly digest and absorb dietary nutrients before they can be metabolized by bacteria. By maintaining lower densities of bacteria in the proximal small intestine by peristalsis and antibacterial secretions from the pancreas and intestine, the host has first access to readily available, digestible nutrients. This leaves undigestible or poorly digested feedstuffs for the bacteria. Prebiotics represent dietary components that are not digestible by the host, but can be metabolized by some, but not all of the resident bacteria. The resulting products of fermentation from prebiotics and undigested foodstuffs are then available to the host, and provide energy, nutrients and additional health benefits.

Osmoregulation. The combination of dietary inputs and digestive secretions introduce about 10 L of fluid into the human GIT. Yet, only 100–150 ml are lost per day in the feces of normal individuals. A large fraction of the fluid is absorbed passively through the "leaky" epithelium lining the small intestine. The remainder is absorbed in the colon by the combination of ion channels, transporters, and aquaporins (Itoh et al., 2003), and the "tight" epithelium that lines the colon restricts the passive movement of water back into the colon. Small disturbances of the osmoregulatory functions can have profound consequences (Ewe, 1988), and particularly if they occur in the colon (Rolfe, 1999). Notably pathogens trigger diarrhea by disturbing osmoregulation.

The amounts of water and electrolytes absorbed by the colon are related to residence time of digesta. This is evident by the high water content of stools produced during diarrhea and by the characteristic desiccated stools of people with constipation and long residence times for materials in the colon. The importance of the colon in osmoregulation is also evident from the ability of the colon to compensate for the additional volumes of fluid that enter the colon of short bowel syndrome patients (Nightingale, 1999). Conversely if the colon is absent or bypassed (e.g., colectomized and jejunostomy patients) the increased loss of fluid and electrolytes requires interventions.

Endocrine secretion. Collectively, the various regions of the GIT and the associated organs represent the largest endocrine organ in the vertebrate body. The vast diversity of peptides regulate local (GIT) functions (e.g., gastrin, secretin) and metabolic processes throughout the host body (e.g., insulin, glucagon), with some GIT hormones regulating other host processes (e.g., cholecystokinin and satiety). The endocrine functions of the GIT act in concert with the enteric and central components of the nervous system to regulate digestion and host physiology and metabolism. It has been recently recognized that enteroendocrine cells can directly respond to resident bacteria by the secretion of hormones (Palazzo et al., 2007). However, the responses of enteroendocrine cells to probiotic bacteria are uncertain.

Immunity and host defense. The epithelium lining the GIT represents a critical interface between the external environment and the host and the combined surface area ($\sim$300 m^2) greatly exceeds the exposed surface area of the skin or the lungs. When one considers the combination of dietary inputs and the resident microorganisms, no other portion of the body is exposed to such a diversity and concentration of antigens. Corresponding with this, the cells constituting the enteric immune system exceed in number the immune cells associated with the rest of the body. Furthermore, the immune cells in the GIT have an even greater challenge of being able to distinguish between potentially harmful and beneficial molecules in the lumenal contents (Magalhaes et al., 2007; Tlaskalova-Hogenova et al., 2004). It is no surprise that the interactions between the enteric immune system and the resident bacteria are vital to host health (Schaible and Kaufmann, 2005; Tannock, 2007).

When considering the relationship between GIT immune functions and the resident bacteria, one is reminded of being asked as a small child to help remove weeds from the garden. Without instruction, one was unable to distinguish between weeds and flowers? They looked the same and one started to pull indiscriminately. Obviously, the outcome was not beneficial for the garden.

Once one was instructed, one was more selective in what plants were to be removed and those to be ignored. The result of one's actions thereafter will be beneficial. In some ways, the enteric immune system faces the same dilemma. It must learn to distinguish between potentially dangerous antigens and microorganisms and from those that are either neutral or potentially beneficial to the host. This "learning" process has occurred during the co-evolution of hosts with the bacteria in the GIT and during the life history of individuals and has resulted in the ability of the GIT to modulate immune responses to match the types of food and bacterial antigens (Comstock and Kasper, 2006).

The ability of the enteric immune system to differentiate between antigens that represent a threat and those that should be tolerated is dependent on a diversity of extracellular Toll-like Receptors (TLR's) and intracellular nucleotide-binding oligomerization domain (NOD) receptors. These provide intestinal cells with the ability to recognize a variety of pathogen associated molecular patterns (PAMP's) (Cario and Podolsky, 2005; Shaw et al., 2008). Importantly, during the co-evolution process, the TLR's and NLR's have acquired the ability to distinguish between pathogens and commensals, including probiotic bacteria. Furthermore, the patterns of expression for the TLR's and NLR's are modulated in response to the composition of the resident bacteria (Lundin et al., 2008).

The co-evolution between the GIT and the bacteria has selected for a set of innate and adaptive defense functions that respond to particular antigens. The innate responses of the GIT respond to antigens without prior exposure and include the secretion of antimicrobial peptides and other antimicrobial molecules, and the activation of macrophages and other defense functions (Eckmann, 2006). The adaptive components of the enteric immune system have similarly responded to co-evolutionary pressures and the populations of associated cells (i.e., T, B, and dendritic cells) have acquired characteristics that differ from those of cells in the systemic circuit (Coombes and Maloy, 2007) and prevent what could be massive and counterproductive immune response to the abundant and diverse antigens in the GIT. The adaptive components interact with the innate components of the GIT immune system to provide the GIT and the host with a comprehensive, multi-layered defense (Winkler et al., 2007) that is capable of "cultivating" the commensal assemblages of bacteria. When either component of the enteric immune system is compromised, the assemblages of bacteria are altered, and host health is compromised.

By changing the species composition of the GIT bacteria using probiotics and prebiotics it is possible to beneficially modulate enteric immune functions, thereby improving resistance to GIT pathogens and other health challenges

(Dogi et al., 2008). Importantly, the responses of the enteric immune system can be transferred to the systemic circuit enhancing disease resistance to systemic infections (Buddington et al., 2002).

Elimination. The GIT plays an important role in the detoxification and elimination of ingested toxins, including drugs, and metabolic wastes. The elimination functions include various GIT tissues that express numerous xenobiotic converting enzymes (e.g., cytochrome P-450's and other Phase I and II enzymes) (Kaminsky and Zhang, 2003; Kato, 2008; Lampe, 2007; Paine and Oberlies, 2007) and export transporters (e.g., P-glycoproteins, Multidrug Resistance transporters, and various ATP-binding cassette (ABC) transporters) (Oude Elferink and de Waart, 2007; Takano et al., 2006). The enzymes and transporters act in concert to detoxify and export toxic molecules.

A limited number of studies suggest management of the GIT bacteria can reduce accumulation and increase elimination of some environmental contaminants (Gratz et al., 2007; Kimura et al., 2004). Corresponding with this, changing the assemblages of the resident bacteria by dietary inputs or antibiotic regimens can influence the bioavailability and efficacy of some therapeutic compounds. For example, administration of some antibiotics can affect the efficacy of oral contraceptives (Dickinson et al., 2001). At the present time the potential influences of commensal, probiotic, and pathogenic bacteria on expression of xenobiotic converting enzymes and export transporters are poorly understood.

The biotic component. The GIT harbors a diverse collection of bacteria, generally estimated to include 400–500 species, with some estimates of >800 species and >7,000 strains (O'Keefe, 2008). Most of the GIT bacteria are unculturable, remain to be characterized, and there is little known about their functional characteristics and influences on host health (Flint et al., 2007). Collectively the GIT bacteria exceed the number of host cells, have a vast metabolic potential, and the interactions with the GIT have a profound impact on host health (Norin and Midtvedt, 2000). Comparisons of germ-free and conventional animals have revealed how the commensal bacteria are essential for normal GIT characteristics. Notable are the profound differences between germ-free and conventional rodents with respect to villus architecture, enterocyte proliferation, differentiation, and patterns of gene expression (Hooper et al., 2001). Significantly, there are differences in the responses to commensal and pathogenic bacteria (Lu and Walker, 2001). The resident bacteria also play an important role in host nutrition. Gnotobiotic rodents require 30% more dietary energy and vitamin supplements compared with conventional rodents harboring commensal bacteria that ferment undigested feedstuffs.

The interactions between the GIT bacteria and the host have been shaped by co-evolution (Ley et al., 2008) and have led to the emergence of commensal and symbiotic relationships (Hooper and Gordon, 2001). Additional interactions occur during the life history of individuals and these contribute to determining the characteristics of the assemblages (densities, diversity, evenness, regional distribution, and functional attributes). The interactions include the ability of some bacteria to alter the patterns of host gene expression, such as glycosylation patterns of extracellular proteins (Freitas and Cayuela, 2002), in ways that benefit both the bacteria and the host (Bry et al., 1996).

The regional distribution of the GIT bacteria. The densities, diversity, and evenness of bacterial species vary among the different regions of the GIT and over time (Franks et al., 1998). The persistence of a bacterial species in a specific region of the GIT is dependent on its being able to tolerate local environmental conditions, or exist in a refuge that provides shelter from adverse conditions (e.g., crypt regions), or adhere to the epithelium, or proliferate rapidly enough to avoid wash out from that region. The different attributes and environmental requirements of the various species and strains of GIT bacteria has resulted in an unequal distribution of species composition and densities along the length of the GIT.

Despite the high input of nutrients into the stomach of monogastric vertebrates, the density of viable bacteria in the stomach is typically low ($<10^4$ cfu/g or/ml). The diversity of species in the stomach is also low compared to the small intestine and colon. This can be attributed to the acidic environment. Although the assemblages of bacteria in the rumen of digastric animals (e.g., cows, sheep, goats) reach much higher densities and are more diverse (Larue et al., 2005; Nelson et al., 2003; Rodriguez et al., 2003) than in the monogastric stomach, bacterial densities and diversity are low in the acid secreting portion of the ruminant stomach (abomasum).

Bacterial densities and diversity in the intestine increase distally, with the highest densities (10^{11-12} cfu/gm) and greatest diversity in the colon. The relative distributions and proportions of aerotolerant and anaerobic species correspond with the declining gradient for oxygen from the proximal small intestine to the colon. Moreover, the rapid movement of the digesta in combination with the introduction of bile acids from the gall bladder and antibacterial peptides secreted by the pancreas and the intestine (e.g., defensins) result in comparatively low bacterial densities and diversity in the duodenum and proximal jejunum. Even within the colon, there is a proximal to distal distribution of species and metabolic activities. Notable is the higher proportion of saccharolytic bacteria and SCFA production in the proximal colon, whereas proteolytic bacteria and the production of putrefactive metabolites are more prevalent in the distal colon. The higher incidence of colon cancer in the developed world has been attributed to a low fiber diet, causing lower abundances of saccharolytic bacteria and an increase in the proportion of proteolytic forms. The consequence is a higher production of ammonia, phenols, indoles, amines, and other toxic and carcinogenic metabolites.

Disturbances of the environment also play an important role in determining the densities, diversity, and relative proportions of resident species in ecosystems. Densities, diversity and evenness are reduced in environments that are harsh or highly disturbed as well as those that are highly stable. Typically, harsh or highly stable environments are dominated by relatively few species with a higher proportion of rare species. In contrast, the densities, diversity, and evenness of species are higher in environments characterized by intermediate frequencies and magnitudes of disturbances as this prevents one or a few species from establishing dominance. This "intermediate disturbance hypothesis" of species diversity (Connell, 1978) can be applied to the GIT ecosystem. Corresponding with the hypothesis, the harsh environments of the stomach and upper small intestine with the wide swings in chemical conditions and intermittently high flow rates, correspond with lower observed densities, diversities, and evenness of resident bacterial species. In contrast, the greatest densities, diversities, and evenness of bacterial species are found in the distal ileum and proximal colon where the environment is more benign, yet the conditions are subject to moderate fluctuations.

In addition to the longitudinal distribution of species in the GIT, there are vertical gradients in each region that extend from the epithelium into the lumen (Kleessen and Blaut, 2005). Although the populations of bacteria adherent or adjacent to the epithelium have a profound impact on the GIT and the host, they are less understood.

Adaptive responses of the GIT bacteria. The biotic components of ecosystems respond to environmental changes. The GIT ecosystem and the resident bacteria are somewhat unique in a two-way adaptation; bacteria adapt to changes in the characteristics of the GIT, and the GIT adapts to changes in the assemblages of bacteria. Of particular importance are how shifts in bacterial populations elicit responses by the GIT. The most obvious contrasts are the divergent responses to pathogenic versus the health promoting commensal and probiotic bacteria. The underlying concept of probiotics and prebiotics is that by increasing the beneficial and decreasing the pathogenic bacteria resident in the GIT, it is possible to improve GIT characteristics and have a beneficial impact on host health and nutrition.

1.4   Dietary Inputs and the GIT Ecosystem

The exclusion of terrestrial inputs into streams reduces the abundances, diversity, and production of the resident organisms. Similarly, during prolonged total parenteral nutrition (Alverdy et al., 2005), starvation (Nettelbladt et al., 1997), and hibernation (Banas et al., 1988) there are analogous changes in the assemblages of GIT bacteria. Elemental diets that are rapidly and efficiently absorbed by the proximal small intestine result in reduced amounts of nutrients entering the colon and thereby also disturb the resident bacteria (Alverdy et al., 2005). However, restriction (not exclusion) of complex feedstuffs is considered to have only a minor impact on the GIT bacteria (Henderson et al., 1998). Of particular concern during parenteral nutrition are the increases in pathogens and the elevated risk of secondary diseases (Deplancke et al., 2002; Harvey et al., 2006). These findings highlight the dependence of the GIT on dietary inputs for normal functions and health. Moreover, management approaches that change the species composition and metabolic activities can be expected to alter GIT characteristics (Tilman et al., 1997).

The dependence of the colonic bacteria on dietary inputs has led to the concept of "colonic foods." Simply increasing the amounts and types of feedstuffs that reach the colon increases bacterial diversity. Correspondingly, adding fermentable fibers, including prebiotics, to elemental diets partially restores the assemblages and metabolic activities of the colonic bacteria and improves GIT and host health (Schneider et al., 2006; Whelan et al., 2005). Importantly, the responses of the GIT bacteria and the production of SCFA and other metabolites vary among different types of dietary carbohydrate (Rabiu and Gibson, 2002; Silvi et al., 1999) and digestion-resistant proteins (Morita et al., 1998), and this impacts GIT functions and health (Topping and Clifton, 2001). Supplementing elemental diets with probiotics also reduces the magnitude of decline in GIT characteristics (Shen et al., 2006).

1.5   Applications of Probiotics and Prebiotics

There are complex interactions among the bacteria resident in the GIT that establish the commensal assemblages and exclude invading species by out competing for limiting resources (food, attachment sites, etc.) and by the production of metabolites (bacteriocins and SCFA) that inhibit the growth or kill outright invading bacteria. Additionally, the interactions between the commensal bacteria and immune functions of the GIT can enhance the defenses against invading pathogens. Of particular interest is having an understanding of how probiotics and prebiotics can be used to manage the diverse and dynamic assemblages of bacteria such that pathogens are excluded or are unable to become established.

Even when probiotics are administered at relatively high doses (i.e., 10^{11} cfu/d), they represent <0.01% of the total number of bacteria in the GIT (10^{14-15}). In reality, the percentage is even less because a fraction of the probiotic bacteria do not remain viable. Moreover, competitive exclusion by the commensal assemblages can prevent probiotics from becoming established in the GIT and in most cases probiotics are transients. As a consequence, densities of probiotic bacteria in fecal samples are typically low. Yet, there are numerous examples that demonstrate the administration of probiotics provides health benefits. This emphasizes how "rare" species can affect the GIT ecosystem and provide health benefits. The success of a probiotic is dependent on the ability to survive in the GIT ecosystem and to interact with the other components of the ecosystem in a manner that fosters improved health. Corresponding with this, the responses to some probiotics are neutral.

Prebiotics encourage the proliferation of beneficial bacteria already resident in the GIT and can elicit larger changes than probiotics in the abundances of commensal bacteria that confer health benefits. Specifically, if the total mass of colon contents is 1 kg and a prebiotic increases the density of a bacterium from 10^7 to 10^9 cfu/g, the result is an almost 10^{12} increase in the total abundance (cfu) of that species of bacterium. Since several species can respond to prebiotics, the influence on the GIT ecosystem can be even more pronounced. Additionally, prebiotics may affects the assemblages of bacteria associated with the epithelial "biofilm" and the lumen (Kleessen and Blaut, 2005). However, the species that are responsive to prebiotics may not provide the same magnitude of benefits as some probiotics.

The following are examples of the application of probiotics and prebiotics to manage the GIT ecosystems of infants and the elderly, for prevention and treatment of acute and chronic GIT diseases, and for emerging health issues. Readers are referred to later chapters for more detailed descriptions.

GIT development. The two most dramatic periods of postnatal development of the GIT occur at birth with the onset of feeding and colonization with bacteria and again at weaning when dietary inputs changing from milk to the adult diet, with corresponding changes in GIT characteristics and assemblages of bacteria. At birth the sterile GIT of neonates is colonized by bacteria originating from the maternal GIT, vagina, skin, and the surrounding environment (Mackie et al., 1999).

Since the species composition and the relative abundances of the colonizers is not consistent, it is not surprising there is variation in the GIT bacterial assemblages that colonize the GITs among littermates and twins. After the initial period of colonization, the densities, diversity and distribution of species within the GIT ecosystems are determined by a combination of dietary inputs, GIT functions, and inter-bacterial interactions, including the processes of competition and facilitation. The process of facilitation, which can be described as cooperative relationships among bacteria, may not be considered enough in understanding the influences of probiotics and prebiotics, and particularly during development. An example is the concept of "metabolic cross-feeding" whereby metabolites produced by one group of bacteria are used or converted by other groups of bacteria (Flint et al., 2007). One of the better examples is the conversion of lactate produced by bifidobacteria into butyrate and other SCFA by anaerobic members of the GIT bacteria.

The interactions between dietary inputs, the resident bacteria, and health are particularly evident during infancy (Gil and Rueda, 2000) and can have profound health consequences. Necrotizing enterocolitis (NEC) is an inflammatory reaction that is the most common GIT disorder of neonates, and particularly those born premature (Schnabl et al., 2008). The NEC disease process is multifactorial and the protection provided by colostrum versus the increased risk associated with formula indicates dietary inputs play a central role. Moreover, the absence of NEC among germ-free animals, regardless of diet demonstrates the essential role of the resident bacteria. A review of trials that evaluated the efficacy of probiotics as a prophylactic for NEC suggest the disease risk is reduced (Barclay et al., 2007). However, the use of different strains and administration regimes confounds interpretations. Although the potential benefits of including prebiotics in formula fed to preterm infants are uncertain, animal models suggest including prebiotics may reduce the risk of NEC (Butel et al., 2002).

Dietary inputs can and do influence the successional changes that occur in the developing bacterial assemblages. By doing so, they influence postnatal GIT development as well as neonatal and infant health (Amarri et al., 2006; Moreau and Gaboriau-Routhiau, 2001; Walker, 2000). Of particular interest are the responses of the GIT bacteria to diet and the associated influences on the developing immune system of infants (Cebra, 1999; Gaskins et al., 2008; Haverson et al., 2007). This includes the relations with development of tolerance and allergy (Kukkonen et al., 2007, 2008). Coinciding with these findings, there is understandable interest in increasing the proportion of beneficial bacteria in the GIT of infants to improve health and disease resistance. There are three principle approaches that have been used to date. The first is the administration of probiotics to infants. This has proven effective. However, the benefits are generally transient and do not persist after the probiotic is no longer administered. The second approach is to provide prebiotics to encourage the growth of beneficial bacteria. Human breast milk, which is high in oligosaccharides, encourages higher densities of health promoting lactobacilli and bifidobacteria. The addition of prebiotics to formulas has similarly increased the densities of lactobacilli and bifidobacteria (Veereman, 2007), thereby providing health benefits (Arslanoglu et al., 2007, 2008). The third approach has been to provide probiotic bacteria to pregnant women to facilitate colonization of infants born vaginally (Schultz et al., 2004). Since probiotics given to pregnant women can be detected in infants delivered by caesarian section (Schultz et al., 2004), colonization can be independent of exposure to the maternal GIT bacteria at the time of birth.

Elderly. The diversity, evenness, and relative proportions of bacterial species in the GIT typically decline with advanced age (Mitsuoka, 1992; Woodmansey, 2007). Of concern are the reciprocal declines in health promoting bacteria (e.g., *Lactobacillus* spp. and *Bifidobacterium* spp.) and increases in bacterial groups that are considered to be opportunistic pathogens or to be putrefactive. Although there is evidence that changes in diet during senescence contribute to the adverse shifts in the bacterial assemblages (Benno et al., 1989), the same patterns of changes during increasing age are known for animals that have consumed the same diet throughout life (Benno et al., 1992). Therefore, the changes in GIT structure and functions that occur during senescence must contribute to the changes of the bacterial assemblages.

The numbers and proportions of elderly subjects is increasing as is interest in alternative medicines to improve GIT health. Administering probiotics and prebiotics to elderly subjects has helped to restore the bacterial assemblages and apparently provide benefits (Hebuterne, 2003; Woodmansey, 2007). There is a need to expand these efforts to gain a better understanding of how probiotics and prebiotics can be used to manage the aging GIT ecosystem.

Diarrheal diseases. Diarrhea can be caused by food infections and toxins, by administration of antibiotics that disturb the commensal bacteria, and by dysfunctions of digestive functions (Huebner and Surawicz, 2006). To understand the application of probiotics and prebiotics for treatment of the various types of diarrhea it is important to first consider the impact of diarrhea on the GIT ecosystem. In many ways diarrhea is analogous to how floods affect river systems. Specifically, minor increases in the flow of rivers mainly disturb the center of the river, with little impact on the resident organisms and ecosystem functioning.

As flow rates increase, the disturbance radiates out toward the interface between the stream and substrate until at high flow rates (floods) the interface itself is disturbed, exchange between the substrate and stream is disrupted, and the residents are displaced. The disturbances associated with floods alter biodiversity, enhances the invasion of exotic species, and increases the abundances of "weed species."

Before the onset of diarrhea, anaerobes dominate the GIT bacteria, whereas the disturbance caused by diarrhea elicits an increase in aerotolerant forms, which includes the majority of pathogens, throughout the GIT (Balamurugan et al., 2008; Bhan et al., 1989; Neto et al., 1976). During diarrhea there are also concurrent declines in bacterial diversity and evenness of species (Mai et al., 2006), with adverse changes in bacterial metabolism and immune status (Khalif et al., 2005), thereby increasing the risk of secondary infections.

The length of time required for the GIT bacteria to recover from diarrhea is dependent on the severity and duration. It is of health importance to promote a rapid recovery of commensal assemblages of bacteria after diarrhea. Unfortunately, undesirable species (i.e., pathogens) usually recover faster and achieve higher densities. Corresponding with this, some of the strongest claims for the benefits of probiotics and prebiotics are for the treatment and prevention of diarrhea. Administration of probiotics during and after diarrhea are known to accelerate the recovery of the commensal bacteria, reduce the abundances of pathogens, and enhance GIT characteristics (Huebner and Surawicz, 2006; Sullivan and Nord, 2005). The adverse impacts of diarrhea on the GIT and the benefits of probiotic therapy are even more pronounced for infants with diarrhea (Johnston et al., 2007). However, the efficacy of probiotics is not universal and the benefits are widely variable, with some studies showing little if any benefit (de Vrese and Marteau, 2007). These findings highlight the need to identify strains of probiotics and establish doses that are most effective in accelerating recovery. There is evidence that probiotics can be used prophylactically to provide protection against diarrhea (Sullivan and Nord, 2005).

The benefits of using prebiotics, prophylactically and therapeutically, for diarrhea are not well established. Yet, supplementing an oral electrolyte solution with prebiotics accelerates recovery of the commensal bacteria and improves GIT structure and functions in an animal model of secretory diarrhea (Oli et al., 1998). However, caution must be exercised to not provide doses of prebiotics that elicit osmotic diarrhea.

Constipation. At the other end of the spectrum from diarrhea is constipation caused by one or more GIT dysfunctions that result in prolonged intervals between defecations. The longer retention of digesta in the colon is associated with declines in commensal bacteria that promote health, increased abundances of potential pathogens, and alterations of digestive functions (Khalif et al., 2005; Zoppi et al., 1998). Although chemical laxatives are often effective and help to restore the assemblages of colonic bacteria, there is growing interest in identifying and using "natural" products.

The incidence of constipation is as high as 30% in developed countries with diets lower in fiber, and afflicts an even greater proportion of the elderly and individuals on hypocaloric diets. Increasing the consumption of fiber is often recommended to reduce constipation. Although including prebiotics as a fiber source for constipation can be predicted to provide benefits, studies directed at this possibility are limited. Those that are reported provide evidence of improved bowel habits and colonic assemblages of bacteria when the diets of constipated individuals are supplemented with prebiotics (Kleessen et al., 1997).

Studies of the efficacy of probiotics for treatment of constipation are actually more common than for prebiotics. There is evidence that probiotics can increase stool frequency (Drouault-Holowacz et al., 2008). However, these findings are not universal, with efficacy varying among strains of probiotic bacteria.

Chronic diseases of the GIT. The most common chronic diseases of the GIT include inflammatory bowel disease (IBD), irritable bowel syndrome with diarrhea (IBS-D) or constipation (IBS-C), celiac sprue, and colorectal cancer (CRC). These are reviewed elsewhere in this volume. Animal models have demonstrated the presence of bacteria is essential for triggering the damaging inflammatory reactions and for converting environmental contaminants into carcinogens (Bohn et al., 2006; Clavel and Haller, 2007; Geier et al., 2007). Hence, effective management of the resident bacteria can have a profound influence on the risk of chronic GIT diseases.

IBD involves a dysregulation of the immune responses to the GIT bacteria (Sartor et al., 2007) in conjunction with a genetic predisposition (Torres and Rios, 2008). The disease is characterized by chronic relapsing inflammation of the GIT, and includes Crohn's disease, ulcerative colitis, and the pouchitis associated with the formation of a "neo-colon" using segments of ileum after colectomy. Of concern is the increased risk of colorectal cancer among individuals with IBD (Geier et al., 2007). The assemblages of GIT bacteria are altered with IBD, including a decline in diversity (Dicksved et al., 2008; Xenoulis et al., 2008). The majority of evidence supporting the use of probiotics for treatment of IBD is largely based on various animals models that showed probiotics trigger signaling pathways that result in the secretion of anti-inflammatory cytokines and thereby decrease the severity of IBD (Winkler et al., 2007). The findings from these studies have fostered the rationale and rapidly growing interest in the application of probiotics and prebiotics to restore the commensal assemblages for prevention and treatment of IBD (Quigley and Flourie, 2007). This is evident by more than 90% of the published reports having been published since 2000. Although clinical trials suggest probiotic preparations can be used for treatment of IBD, not all probiotic strains are effective (Heilpern and Szilagyi, 2008). The most effective probiotic preparations tend to be multi-species mixtures (Madsen et al., 2001). Still, the evidence for the efficacy of probiotics for treatment of IBD is not conclusive (Geier et al., 2007) and additional well-controlled studies are necessary to obtain definitive proof. Similarly, although there is evidence probiotics may reduce the incidence of IBD remission, the findings are inconclusive, largely because of the lack of adequately controlled and designed studies (Butterworth et al., 2008).

In light of the essential role of the resident bacteria in triggering IBD, the *a priori* expectation is that increasing the proportions of beneficial commensal bacteria by administration of prebiotics should stimulate anti-inflammatory responses and reduce the risk and severity of IBD. Yet, the application of pre-biotics for prevention and treatment of IBD has received far less investigation. Still, the available evidence from animal models indicate that prebiotic-induced changes in the GIT bacteria can provide benefits (Leenen and Dieleman, 2007), and the limited, but promising results from clinical trials (Guarner, 2007). Although synbiotic combinations may provide the most effective approach, there are too few clinical trials to draw conclusions.

IBS is a growing health concern that afflicts up to 25% of the population in developed countries. Patients with IBS have chronic and recurring abdominal pain, altered bowel habits (diarrhea or constipation), and other GIT dysfunc-tions. There are multiple underlying causes of IBS, including the resident bacteria (Clavel and Haller, 2007), and curative therapies do not yet exist. A consistent finding is that the assemblages of GIT bacteria are disrupted with IBS, and there is evidence that suggests prior enteric infections increase the risk of developing IBS (Cuomo et al., 2007). These findings suggest therapeutic approaches that restore or improve the GIT bacteria may reduce the incidence or severity of IBS. A metanalysis of the limited number of studies that have been conducted suggests probiotics may be effective for IBS, but additional studies are needed (McFarland and Dublin, 2008). Again, there are fewer studies of the efficacy of prebiotics for treatment of IBS. One of the more promising applications of prebiotics is for treatment of IBS-C due to the laxative and bacterial normalization effects.

Celiac disease is an allergic response of the GIT to the gliadin component of wheat gluten that triggers an inflammatory reaction. Similar to IBD and IBS, there are shifts in the species composition of the resident bacteria of individuals with celiac disease (Sanz et al., 2007). This has fostered speculation that restoring the commensal assemblages of bacteria using probiotics and prebiotics should reduce the incidence and severity of celiac disease and other food allergies. Although cell culture studies have demonstrated probiotics abrogate the responses to the components of gliadin that elicit inflammation (Lindfors et al., 2008), clinical evaluations of efficacy are lacking.

There is also interest in exploring the efficacy of probiotics for preventing or reducing the severity of other food allergies (Mengheri, 2008). However, it can be expected that not all probiotic preparations will provide protection for enteric or systemic allergic reactions (Brouwer et al., 2006; Hol et al., 2008). Obviously, there is a need to identify the strains of probiotics that are effective.

There is increasing awareness of the relationship among diet composition, the GIT bacteria, and the risk of colorectal cancer (Juste, 2005). Of interest are the metabolic activities of the colonic bacteria that can activate procarcinogens or deactivate carcinogens. The majority of evidence demonstrating probiotics and prebiotics reduce the risk of colorectal cancer is based on animal models exposed to chemical carcinogens. A comprehensive description of the evidence is provided elsewhere in this volume. The best clinical evidence of a protective effect of probiotics and prebiotics is from the SYNCAN project (Pool-Zobel and Sauer, 2007).

Surgical interventions and dysfunctions that alter the GIT ecosystem. Fragmentation and reductions in the sizes of ecosystems, and changes in inputs are associated with shifts in the abundances and distributions of the resident organisms, and these can cause dramatic changes in the functions and productivity of the affected ecosystems. It is not surprising that surgical interventions and GIT dysfunctions similarly affect the species composition and metabolic activities of the resident bacteria. Exemplary is how resection of portions of the small bowel (e.g., short bowel syndrome) alters the species composition and metabolic activities of the GIT bacteria (Kaneko et al., 1997). Similarly, the bacteria resident in neo-colons (pouches) prepared using segments of ileum after colectomy differ from those in the normal colon and this has been associated with inflammation ("pouchitis"; Welters et al., 2002). In both cases, probiotics and prebiotics have been successfully administered to address the changes in the bacterial assemblages and to decrease the risk and severity of mucosal inflammation.

The popularity of gastric bypass surgery with the Roux-en-Y procedure for treatment of obesity has increased the number of individuals who have a GIT with vastly altered structural and functional attributes. This surgical procedure changes the resident bacteria throughout the GIT, including the stomach (Bjorneklett et al., 1984; Ishida et al., 2007). At the present time, there is very little known about whether providing probiotics or prebiotics to patients following bariatric procedures will provide health benefits.

Cholecystectomy and the loss of regulated bile acid secretion into the small intestine change the chemical composition of the GIT contents. There are corresponding changes in the metabolic activities of the colonic bacteria and the production of secondary bile acids and other metabolites. Although there is concern that these changes can increase the risk of colorectal cancer (Zuccato et al., 1993), the potential health benefits of probiotics and prebiotics after cholecystectomy have not been explored.

Exocrine pancreatic insufficiency and pancreatitis and the decline in secretion are associated with overgrowth of bacteria in the proximal small intestine and increase the risk of bacterial translocation and septicemia (Besselink et al., 2004; Simpson et al., 1990; Trespi and Ferrieri, 1999). Despite the decline in digestive efficiency, a large fraction of the malabsorbed feedstuffs are "salvaged" by bacterial metabolism and are available to the host as SCFA. Although probiotics are considered for prophylactic and therapeutic interventions with pancreatic dysfunctions, the benefits are not yet established and there is a need to identify strains that provide benefit without risk for such patients (Bengmark, 2008).

1.6   Considerations, Limitations, and Opportunities

The ability to manage ecosystems is dependent on being able to analyze, measure, and manage the biotic components. Important characteristics of the biotic component of any ecosystem include the abundances, diversity, and evenness (relative proportions) of the resident species. However, investigators and clinicians seeking to manage the GIT of human subjects face challenges. For obvious reasons, people know much more about fecal microbiology, yet many of the most important interactions among diet, the resident bacteria, and the host occur in the "upstream" habitats of the GIT. The difficulty of obtaining samples from throughout the GIT has impeded the understanding of this ecosystem and how it responds to dietary inputs, including probiotics and prebiotics, and other interventions. Similarly, if river ecologists were restricted to studying only water samples collected at the beginning and termination of large rivers, their understanding of the entire river system would be limited, and the design of management strategies would be severely compromised. Moreover, even when other regions are accessible, the majority of information about the GIT bacteria is based on samples of the lumenal contents. Yet, the interactions between the host and the bacteria at the boundary layer have an impact on health, and this critical habitat should be considered as an important target of management strategies. Obviously, a better understanding of the entire GIT ecosystem is essential for developing and implementing management approaches based on probiotics, prebiotics, and other dietary interventions.

One of the foundations of community ecology is the relationship between the number of species and the area sampled. Determining the absolute and relative abundances of species, not just their presence or absence, is essential to understanding ecosystems. These principles highlight how appropriate sampling protocols (i.e., frequency, size, methods of identification and enumeration) are critical for investigating the biotic components of ecosystems. Also to be considered are the several orders of magnitude that separate the abundances of rare ($<10^3$ cfu/g) and common species ($>10^{11}$ cfu/g) of GIT bacteria. Rare species can be missed by both culture dependent and independent methods, particularly when the amount of material used for detection and identification is small. Yet, the ability to detect rare species is essential for characterizing and understanding ecosystems, including the GIT. Many studies of the GIT bacteria are dependent on a single sample, often less than 1 g. Because rare species may be detected only infrequently, they can be missed. Moreover, even with multiple samples the inconsistent detection may lead to the misinterpretation that such rare species are transient, even when they are continuously present.

The variation in bacterial populations (species composition and metabolic activities) among species, individuals, even among identical twins (Dicksved et al., 2008), and at different ages emphasize how GIT structure and functions and dietary inputs are important determinants of the characteristics and health of the GIT ecosystem, Although probiotics and prebiotics have proven to be effective tools to manage the GIT ecosystem, the responses and benefits are not universal. Coinciding with this, the patterns of gene expression by epithelial cells lining the small intestine and colon differ when exposed to a diversity of probiotic bacteria (Shima et al., 2008). These findings explain the divergent responses of host species and individuals to probiotics and prebiotics (Fujiwara et al., 2001; Simpson et al., 2000; Sullivan and Nord, 2005) and why a specific strain of probiotic or type of prebiotic does not elicit consistent responses among individuals.

As a consequence, there are difficulties associated with developing "one-for-all" management strategies. A better understanding of host-microbiome interactions and the responses to probiotics and prebiotics will greatly enhance efforts to improve management of the GIT ecosystem and thereby the health of individuals.

List of Abbreviations

ABC	ATP-binding cassette
CRC	Eolorectal cancer
GIT	Gastrointestinal Tract
IBD	Inflammatory bowel disease
IBS-C	irritable bowel syndrome with constipation
IBS-D	irritable bowel syndrome with diarrhea
NEC	Necrotizing enterocolitis
NLR	NOD-like receptor
NOD	Nucleotide-binding oligomerization domain
PAMP	Pathogen associated molecular pattern
SCFA	Short Chain Fatty Acids
SGLT-1	Sodium Glucose cotransporter-1
TLR	Toll-like Receptor

References

Alverdy J, Zaborina O, Wu L (2005) The impact of stress and nutrition on bacterial-host interactions at the intestinal epithelial surface. Curr Opin Clin Nutr Metab Care 8:205–209

Amarri S, Benatti F, Callegari ML, Shahkhalili Y, Chauffard F, Rochat F, Acheson KJ, Hager C, Benyacoub J, Galli E, Rebecchi A, Morelli L (2006) Changes of gut microbiota and immune markers during the complementary feeding period in healthy breast-fed infants. J Pediatr Gastroenterol Nutr 42:488–495

Arslanoglu S, Moro GE, Boehm G (2007) Early supplementation of prebiotic oligosaccharides protects formula-fed infants against infections during the first 6 months of life. J Nutr 137:2420–2424

Arslanoglu S, Moro GE, Schmitt J, Tandoi L, Rizzardi S, Boehm G (2008) Early dietary intervention with a mixture of prebiotic oligosaccharides reduces the incidence of allergic manifestations and infections during the first two years of life. J Nutr 138:1091–1095

Balamurugan R, Janardhan HP, George S, Raghava MV, Muliyil J, Ramakrishna BS (2008) Molecular studies of fecal anaerobic commensal bacteria in acute diarrhea in children. J Pediatr Gastroenterol Nutr 46:514–519

Banas JA, Loesche WJ, Nace GW (1988) Classification and distribution of large intestinal bacteria in nonhibernating and hibernating leopard frogs (Rana pipiens). Appl Environ Microbiol 54:2305–2310

Barclay AR, Stenson B, Simpson JH, Weaver LT, Wilson DC (2007) Probiotics for necrotizing enterocolitis: a systematic review. J Pediatr Gastroenterol Nutr 45:569–576

Bartholome AL, Albin DM, Baker DH, Holst JJ, Tappenden KA (2004) Supplementation of total parenteral nutrition with butyrate acutely increases structural aspects of intestinal adaptation after an 80% jejunoileal resection in neonatal piglets. JPEN J Parenter Enteral Nutr 28:210–222

Beaslas O, Torreilles F, Casellas P, Simon D, Fabre G, Lacasa M, Delers F, Chambaz J, Rousset M, Carriere V (2008) Transcriptome response of enterocytes to dietary lipids: impact on cell architecture, signaling and metabolism genes. Am J Physiol Gastrointest Liver Physiol 295:G942–G952

Bengmark S (2008) Is probiotic prophylaxis worthwhile in patients with predicted severe acute pancreatitis? Nat Clin Pract Gastroenterol Hepatol 5:602–603

Benno Y, Endo K, Mizutani T, Namba Y, Komori T, Mitsuoka T (1989) Comparison of fecal microflora of elderly persons in rural and urban areas of Japan. Appl Environ Microbiol 55:1100–1105

Benno Y, Nakao H, Uchida K, Mitsuoka T (1992) Impact of the advances in age on the gastrointestinal microflora of beagle dogs. J Vet Med Sci 54:703–706

Besselink MG, Timmerman HM, Buskens E, Nieuwenhuijs VB, Akkermans LM, Gooszen HG (2004) Dutch Acute Pancreatitis Study Group. Probiotic prophylaxis in patients with predicted severe acute pancreatitis (PROPATRIA): design and rationale of a double-blind, placebo-controlled randomised multicenter trial [ISRCTN38327949]. BMC Surg 4:12

Bhan MK, Raj P, Khoshoo V, Bhandari N, Sazawal S, Kumar R, Srivastava R, Arora NK (1989) Quantitation and properties of fecal and upper small intestinal aerobic microflora in infants and young children with persistent diarrhea. J Pediatr Gastroenterol Nutr 9:40–45

Bjørneklett A, Viddal KO, Midtvedt T, Nygaard K (1981) Intestinal and gastric bypass. Changes in intestinal microecology after surgical treatment of morbid obesity in man. Scand J Gastroenterol 16:681–687

Bohn E, Bechtold O, Zahir N, Frick JS, Reimann J, Jilge B, Autenrieth IB (2006) Host gene expression in the colon of gnotobiotic interleukin-2-deficient mice colonized with commensal colitogenic or noncolitogenic bacterial strains: common patterns and bacteria strain specific signatures. Inflamm Bowel Dis 12:853–862

Brouwer ML, Wolt-Plompen SA, Dubois AE, van der Heide S, Jansen DF, Hoijer MA, Kauffman HF, Duiverman EJ (2006) No effects of probiotics on atopic dermatitis in infancy: a randomized placebo-controlled trial. Clin Exp Allergy 36:899–906

Bry L, Falk PG, Midtvedt T, Gordon JI (1996) A model of host-microbial interactions in an open mammalian ecosystem. Science 273:1380–1383

Buddington KK, Donahoo JB, Buddington RK (2002) Dietary oligofructose and inulin protect mice from enteric and systemic pathogens and tumor inducers. J Nutr 132:472–477

Butel MJ, Waligora-Dupriet AJ, Szylit O (2002) Oligofructose and experimental model of neonatal necrotising enterocolitis. Br J Nutr 87(Suppl. 2):S213–S219

Butterworth AD, Thomas AG, Akobeng AK (2008) Probiotics for induction of remission in Crohnâs disease. Cochrane Database Syst Rev (3):CD006634

Cario E, Podolsky DK (2005) Intestinal epithelial TOLLerance versus in TOLLerance of commensals. Mol Immunol 42:887–893

Cebra JJ (1999) Influences of microbiota on intestinal immune system development. Am J Clin Nutr 69:1046S–1051S

Clavel T, Haller D (2007) Molecular interactions between bacteria, the epithelium, and the mucosal immune system in the intestinal tract: implications for chronic inflammation. Curr Issues Intest Microbiol 8:25–43

Comstock LE, Kasper DL (2006) Bacterial glycans: key mediators of diverse host immune responses. Cell 126:847–850

Connell JH (1978) Diversity in tropical forests and coral reefs. Science 199:1302–1310

Coombes JL, Maloy KJ (2007) Control of intestinal homeostasis by regulatory T cells and dendritic cells. Semin Immunol 19:116–126

Cuomo R, Savarese MF, Gargano R (2007) Almost all irritable bowel syndromes are post-infectious and respond to probiotics: consensus issues. Dig Dis 25:241–244

de Vrese M, Marteau PR (2007) Probiotics and prebiotics: effects on diarrhea. J Nutr 137:803S–811S

Deplancke B, Vidal O, Ganessunker D, Donovan SM, Mackie RI, Gaskins HR (2002) Selective growth of mucolytic bacteria including Clostridium perfringens in a neonatal piglet model of total parenteral nutrition. Am J Clin Nutr 76: 1117–1125

Dickinson BD, Altman RD, Nielsen NH, Sterling ML (2001) Council on Scientific Affairs, American Medical Association. Drug interactions between oral contraceptives and antibiotics. Obstet Gynecol 98:853–860

Dicksved J, Halfvarson J, Rosenquist M, Järnerot G, Tysk C, Apajalahti J, Engstrand L, Jansson JK (2008) Molecular analysis of the gut microbiota of identical twins with Crohn's disease. ISME J 2:716–727

Dogi CA, Galdeano CM, Perdigón G (2008) Gut immune stimulation by non pathogenic Gram(+) and Gram(−) bacteria. Comparison with a probiotic strain. Cytokine 41:223–231

Drouault-Holowacz S, Bieuvelet S, Burckel A, Cazaubiel M, Dray X, Marteau P (2008) A double blind randomized controlled trial of a probiotic combination in 100 patients with irritable bowel syndrome. Gastroenterol Clin Biol 32:147–152

Dunne C (2001) Adaptation of bacteria to the intestinal niche: probiotics and gut disorder. Inflamm Bowel Dis 7:136–145

Eckmann L (2006) Innate immunity. In: Johnson LR, Barrett KE, Ghishan FK, Merchant JL, Said HM, Wood J (eds) Physiology of the gastrointestinal tract, 4th ed, Elsevier, Amsterdam, pp 1033–1066

Ewe K (1988) Intestinal transport in constipation and diarrhoea. Pharmacology 36 (Suppl. 1):73–84

Falk PG, Hooper LV, Midtvedt T, Gordon JI (1998) Creating and maintaining the gastrointestinal ecosystem: what we know and need to know from gnotobiology. Microbiol Mol Biol Rev 62:1157–1170

Flint HJ, Duncan SH, Scott KP, Louis P (2007) Interactions and competition within the microbial community of the human colon: links between diet and health. Environ Microbiol 9:1101–1111

Franks AH, Harmsen HJ, Raangs GC, Jansen GJ, Schut F, Welling GW (1998) Variations of bacterial populations in human feces measured by fluorescent in situ hybridization with group-specific 16S rRNA-targeted oligonucleotide probes. Appl Environ Microbiol 64:3336–3345

Freitas M, Axelsson LG, Cayuela C, Midtvedt T, Trugnan G (2002) Microbial-host interactions specifically control the glycosylation pattern in intestinal mouse mucosa. Histochem Cell Biol 118:149–161

Fujiwara S, Seto Y, Kimura A, Hashiba H (2001) Establishment of orally-administered Lactobacillus gasseri SBT2055SR in the gastrointestinal tract of humans and its influence on intestinal microflora and metabolism. J Appl Microbiol 90: 343–352

Gaskins HR, Croix JA, Nakamura N, Nava GM (2008) Impact of the intestinal microbiota on the development of mucosal defense. Clin Infect Dis 46(Suppl. 2):S80–S86

Geier MS, Butler RN, Howarth GS (2006) Probiotics, prebiotics and synbiotics: a role in chemoprevention for colorectal cancer? Cancer Biol Ther 5:1265–1269

Geier MS, Butler RN, Howarth GS (2007) Inflammatory bowel disease: current insights into pathogenesis and new therapeutic options; probiotics, prebiotics and synbiotics. Int J Food Microbiol 115:1–11

Gil A, Rueda R (2000) Modulation of intestinal microflora by specific dietary components. Microb Ecol Health Dis (Suppl. 2): 31–39

Gratz S, Wu QK, El-Nezami H, Juvonen RO, Mykkänen H, Turner PC (2007) Lactobacillus rhamnosus strain GG reduces aflatoxin B1 transport, metabolism, and toxicity in Caco-2 Cells. Appl Environ Microbiol 73:3958–3964

Guarner F (2007) Prebiotics in inflammatory bowel diseases. Br J Nutr 98(Suppl. 1): S85–S89

Harvey RB, Andrews K, Droleskey RE, Kansagra KV, Stoll B, Burrin DG, Sheffield CL, Anderson RC, Nisbet DJ (2006) Qualitative and quantitative comparison of gut bacterial colonization in enterally and parenterally fed neonatal pigs. Curr Issues Intest Microbiol 7:61–64

Haverson K, Rehakova Z, Sinkora J, Sver L, Bailey M (2007) Immune development in jejunal mucosa after colonization with selected commensal gut bacteria: a study in germ-free pigs. Vet Immunol Immunopathol 119:243–253

Hébuterne X (2003) Gut changes attributed to ageing: effects on intestinal microflora. Curr Opin Clin Nutr Metab Care 6:49–54

Heilpern D, Szilagyi A (2008) Manipulation of intestinal microbial flora for therapeutic benefit in inflammatory bowel diseases: review of clinical trials of probiotics, pre-biotics and synbiotics. Rev Recent Clin Trials 3:167–184

Henderson AL, Cao WW, Wang RF, Lu MH, Cerniglia CE (1998) The effect of food restriction on the composition of intestinal microflora in rats. Exp Gerontol 33:239–247

Hol J, van Leer EH, Elink Schuurman BE, de Ruiter LF, Samsom JN, Hop W, Neijens HJ, de Jongste JC, Nieuwenhuis EE (2008) Cow's milk allergy modified by elimination and Lactobacilli study group. The acquisition of tolerance toward cow's milk through probiotic supplementation: a randomized, controlled trial. J Allergy Clin Immunol 121: 1448–1454

Hooper DU, Vitousek PM (1997) The effects ofplant composition and diversity on ecosystem processes. Science 177:1302–1305

Hooper LV, Gordon JI (2001) Commensal host-bacterial relationships in the gut. Science 292:1115–1118

Hooper LV, Wong MH, Thelin A, Hansson L, Falk PG, Gordon JI (2001) Molecular analysis of commensal host-microbial relationships in the intestine. Science 291:881–884

Huebner ES, Surawicz CM (2006) Probiotics in the prevention and treatment of gastrointestinal infections. Gastroenterol Clin North Am 35:355–365

Husebye E, Hellström PM, Sundler F, Chen J, Midtvedt T (2001) Influence of microbial species on small intestinal myoelectric activity and transit in germ-free rats. Am J Physiol Gastrointest Liver Physiol 280: G368–G380

Ishida RK, Faintuch J, Paula AM, Risttori CA, Silva SN, Gomes ES, Mattar R, Kuga R, Ribeiro AS, Sakai P, Barbeiro HV, Barbeiro DF, Soriano FG, Cecconello I (2007) Microbial flora of the stomach after gastric bypass for morbid obesity. Obes Surg 17:752–758

Itoh A, Tsujikawa T, Fujiyama Y, Bamba T (2003) Enhancement of aquaporin-3 by vasoactive intestinal polypeptide in a human colonic epithelial cell line. J Gastroenterol Hepatol 18:203–210

Johnson LR (Ed) (2006) Physiology of the Gastrointestinal Tract, 4th edn. Elsevier, New York

Johnston BC, Supina AL, Ospina M, Vohra S (2007) Probiotics for the prevention of pediatric antibiotic-associated diarrhea. Cochrane Database Syst Rev CD004827

Juste C (2005) Dietary fatty acids, intestinal microbiota and cancer. Bull Cancer 92:708–721

Kaminsky LS, Zhang QY (2003) The small intestine as a xenobiotic-metabolizing organ. Drug Metab Dispos 31:1520–1525

Kaneko T, Bando Y, Kurihara H, Satomi K, Nonoyama K, Matsuura N (1997) Fecal microflora in a patient with short-bowel syndrome and identification of dominant lactobacilli. J Clin Microbiol 35:3181–3185

Kato M (2008) Intestinal first-pass metabolism of CYP3A4 substrates. Drug Metab Pharmacokinet 23(2):87–94

Khalif IL, Quigley EM, Konovitch EA, Maximova ID (2005) Alterations in the colonic flora and intestinal permeability and evidence of immune activation in chronic constipation. Dig Liver Dis 37:838–849

Kimura Y, Nagata Y, Buddington RK (2004) Some dietary fibers increase elimination of orally administered polychlorinated biphenyls but not that of retinol in mice. J Nutr 134:135–142

Kleessen B, Blaut M (2005) Modulation of gut mucosal biofilms. Br J Nutr 93(Suppl. 1): S35–S40

Kleessen B, Sykura B, Zunft HJ, Blaut M (1997) Effects of inulin and lactose on fecal microflora, microbial activity, and bowel habit in elderly constipated persons. Am J Clin Nutr 65:1397–1402

Kukkonen K, Savilahti E, Haahtela T, Juntunen-Backman K, Korpela R, Poussa T, Tuure T, Kuitunen M (2007) Probiotics and prebiotic galacto-oligosaccharides in the prevention of allergic diseases: a randomized, double-blind, placebo-controlled trial. J Allergy Clin Immunol 119:192–198

Kukkonen K, Savilahti E, Haahtela T, Juntunen-Backman K, Korpela R, Poussa T, Tuure T, Kuitunen M (2008) Long-term safety and impact on infection rates of postnatal probiotic and prebiotic (synbiotic) treatment: randomized, double-blind, placebo-controlled trial. Pediatrics 122:8–12

Lampe JW (2007) Diet, genetic polymorphisms, detoxification, and health risks. Altern Ther Health Med 13:S108–S111

Larue R, Yu Z, Parisi VA, Egan AR, Morrison M (2005) Novel microbial diversity adherent to plant biomass in the herbivore gastrointestinal tract, as revealed by ribosomal intergenic spacer analysis and rrs gene sequencing. Environ Microbiol 7:530–543

Le Gall M, Tobin V, Stolarczyk E, Dalet V, Leturque A, Brot-Laroche E (2007) Sugar sensing by enterocytes combines polarity, membrane bound detectors and sugar metabolism. J Cell Physiol 213:834–843

Leenen CH, Dieleman LA (2007) Inulin and oligofructose in chronic inflammatory bowel disease. J Nutr 137(Suppl. 11):2572S–2575S

Lesniewska V, Rowland I, Laerke HN, Grant G, Naughton PJ (2006) Relationship between dietary-induced changes in intestinal commensal microflora and duodenojejunal myoelectric activity monitored by radiotelemetry in the rat in vivo. Exp Physiol 91:229–237

Ley RE, Hamady M, Lozupone C, Turnbaugh PJ, Ramey RR, Bircher JS, Schlegel ML, Tucker TA, Schrenzel MD, Knight R, Gordon JI (2008) Evolution of mammals and their gut microbes. Science 320:1647–1651

Lindfors K, Blomqvist T, Juuti-Uusitalo K, Stenman S, Venäläinen J, Mäki M, Kaukinen K (2008) Live probiotic Bifidobacterium lactis bacteria inhibit the toxic effects induced by wheat gliadin in epithelial cell culture. Clin Exp Immunol 152:552–558

Lu L, Walker WA (2001) Pathologic and physiologic interactions of bacteria with the gastrointestinal epithelium. Am J Clin Nutr 73:1124S–1130S

Lundin A, Bok CM, Aronsson L, Björkholm B, Gustafsson JA, Pott S, Arulampalam V, Hibberd M, Rafter J, Pettersson S (2008) Gut flora, Toll-like receptors and nuclear receptors: a tripartite communication that tunes innate immunity in large intestine. Cell Microbiol 10:1093–1103.

Mackie RI, Sghir A, Gaskins HR (1999) Developmental microbial ecology of the neonatal gastrointestinal tract. Am J Clin Nutr 69:1035S–1045S

Madsen K, Cornish A, Soper P, McKaigney C, Jijon H, Yachimec C, Doyle J, Jewell L, De Simone C (2001) Probiotic bacteria enhance murine and human intestinal epithelial barrier function. Gastroenterology 121:580–591

Magalhaes JG, Tattoli I, Girardin SE (2007) The intestinal epithelial barrier: how to distinguish between the microbial flora and pathogens. Semin Immunol 19:106–115

Mai V, Braden CR, Heckendorf J, Pironis B, Hirshon JM (2006) Monitoring of stool microbiota in subjects with diarrhea indicates distortions in composition. J Clin Microbiol 44:4550–4552

Martin FP, Wang Y, Sprenger N, Yap IK, Rezzi S, Ramadan Z, Peré-Trepat E, Rochat F, Cherbut C, van Bladeren P, Fay LB, Kochhar S, Lindon JC, Holmes E, Nicholson JK (2008) Top-down systems biology integration of conditional prebiotic modulated transgenomic interactions in a humanized microbiome mouse model. Mol Syst Biol 4:205

McFarland LV, Dublin S (2008) Meta-analysis of probiotics for the treatment of irritable bowel syndrome. World J Gastroenterol 14:2650–2661

Mengheri E (2008) Health, probiotics, and inflammation. J Clin Gastroenterol 42 (Suppl. 3) Pt 2:S177–S178

Mitsuoka T (1992) Intestinal flora and aging. Nutr Rev 50:438–446

Moreau C-M, Gaboriau-Routhiau V (2001) Influence of resident intestinal microflora on the development and functions of the gut-associated lymphoid tissue. Microb Ecol Health Dis 13:65–86

Morita T, Kasaoka S, Ohhashi A, Ikai M, Numasaki Y, Kiriyama S (1998) Resistant proteins alter cecal short-chain fatty acid profiles in rats fed high amylose cornstarch. J Nutr 128:1156–1164

Nelson KE, Zinder SH, Hance I, Burr P, Odongo D, Wasawo D, Odenyo A, Bishop R (2003) Phylogenetic analysis of the microbial populations in the wild herbivore gastrointestinal tract: insights into an unexplored niche. Environ Microbiol 5:1212–1220

Neto UF, Toccalino H, Dujovney F (1976) Stool bacterial aerobic overgrowth in the small intestine of children with acute diarrhoea. Acta Paediatr Scand 65:609–615

Nettelbladt CG, Katouli M, Volpe A, Bark T, Muratov V, Svenberg T, Möllby R, Ljungqvist O (1997) Starvation increases the number of coliform bacteria in the caecum and induces bacterial adherence to caecal epithelium in rats. Eur J Surg 163:135–142

Nightingale JM (1999) Management of patients with a short bowel. Nutrition 15:633–637

Norin E, Midtvedt T (2000) Interactions of bacteria with the host: alteration of microflora-associated characteristics of the host; non-immune functions. Microb Ecol Health Dis (Suppl. 2):186–193

O'Keefe SJ (2008) Nutrition and colonic health: the critical role of the microbiota. Curr Opin Gastroenterol 24:51–58

Oli MW, Petschow BW, Buddington RK (1998) Evaluation of fructooligosaccharide supplementation of oral electrolyte solutions for treatment of diarrhea: recovery of the intestinal bacteria. Dig Dis Sci 43:138–147

Oude Elferink RP, de Waart R (2007) Transporters in the intestine limiting drug and toxin absorption. J Physiol Biochem 63:75–81

Paine MF, Oberlies NH (2007) Clinical relevance of the small intestine as an organ of drug elimination: drug-fruit juice interactions. Expert Opin Drug Metab Toxicol 3:67–80

Palazzo M, Balsari A, Rossini A, Selleri S, Calcaterra C, Gariboldi S, Zanobbio L, Arnaboldi F, Shirai YF, Serrao G, Rumio

C (2007) Activation of enteroendocrine cells via TLRs induces hormone, chemokine, and defensin secretion. J Immunol 178:4296–4303

Physiology of the Gastrointestinal Tract (2006) In: Johnson LR, Barrett KE, Ghishan FK, Merchant JL, Said HM, Wood J (eds), 4th edn. Elsevier, Amsterdam

Pool-Zobel BL, Sauer J (2007) Overview of experimental data on reduction of colorectal cancer risk by inulin-type fructans. J Nutr 137(Suppl. 11):2580S–2584S

Quigley EM, Flourie B (2007) Probiotics and irritable bowel syndrome: a rationale for their use and an assessment of the evidence to date. Neurogastroenterol Motil 19:166–172

Rabiu BA, Gibson GR (2002) Carbohydrates: a limit on bacterial diversity within the colon. Biol Rev Camb Philos Soc 77:443–453

Rechkemmer G, Rönnau K, von Engelhardt W (1988) Fermentation of polysaccharides and absorption of short chain fatty acids in the mammalian hindgut. Comp Biochem Physiol A 90(4):563–568

Roberfroid MB (2007) Inulin-type fructans: functional food ingredients. J Nutr 137 (Suppl. 11):2493S–2502S

Rodríguez CA, González J, Alvir MR, Redondo R, Cajarville C (2003) Effects of feed intake on composition of sheep rumen contents and their microbial population size. Br J Nutr 89:97–103

Rolfe V (1999) Colonic fluid and electrolyte transport in health and disease. Vet Clin North Am Small Anim Pract 29:577–588

Sanz Y, Sánchez E, Marzotto M, Calabuig M, Torriani S, Dellaglio F (2007) Differences in faecal bacterial communities in coeliac and healthy children as detected by PCR and denaturing gradient gel electrophoresis. FEMS Immunol Med Microbiol 51:562–568

Sartor RB, Blumberg RS, Braun J, Elson CO, Mayer LF (2007) CCFA microbial-host interactions workshop: highlights and key observations. Inflamm Bowel Dis 13:600–619

Schaible UE, Kaufmann SH (2005) A nutritive view on the host-pathogen interplay. Trends Microbiol 13:373–380

Schnabl KL, Van Aerde JE, Thomson AB, Clandinin MT (2008) Necrotizing enterocolitis: a multifactorial disease with no cure. World J Gastroenterol 14:2142–2161

Schneider SM, Girard-Pipau F, Anty R, van der Linde EG, Philipsen-Geerling BJ, Knol J, Filippi J, Arab K, Hébuterne X (2006) Effects of total enteral nutrition supplemented with a multi-fibre mix on faecal short-chain fatty acids and microbiota. Clin Nutr 25:82–90

Schultz M, Göttl C, Young RJ, Iwen P, Vanderhoof JA (2004) Administration of oral probiotic bacteria to pregnant women causes temporary infantile colonization. J Pediatr Gastroenterol Nutr 38:293–297

Shaw MH, Reimer T, Kim YG, Nuñez G (2008) NOD-like receptors (NLRs): bona fide intracellular microbial sensors. Curr Opin Immunol 20:377–382

Shen TY, Qin HL, Gao ZG, Fan XB, Hang XM, Jiang YQ (2006) Influences of enteral nutrition combined with probiotics on gut microflora and barrier function of rats with abdominal infection. World J Gastroenterol 12:4352–4358

Shima T, Fukushima K, Setoyama H, Imaoka A, Matsumoto S, Hara T, Suda K, Umesaki Y (2008) Differential effects of two probiotic strains with different bacteriological properties on intestinal gene expression, with special reference to indigenous bacteria. FEMS Immunol Med Microbiol 52:69–77

Shirkey TW, Siggers RH, Goldade BG, Marshall JK, Drew MD, Laarveld B, Van Kessel AG (2006) Effects of commensal bacteria on intestinal morphology and expression of proinflammatory cytokines in the gnotobiotic pig. Exp Biol Med 231: 1333–1345

Silvi S, Rumney CJ, Cresci A, Rowland IR (1999) Resistant starch modifies gut microflora and microbial metabolism in human flora-associated rats inoculated with faeces from Italian and UK donors. J Appl Microbiol 86:521–530

Simpson JM, McCracken VJ, Gaskins HR, Mackie RI (2000) Denaturing gradient gel electrophoresis analysis of 16S ribosomal DNA amplicons to monitor changes in fecal bacterial populations of weaning pigs after introduction of Lactobacillus reuteri strain MM53. Appl Environ Microbiol 66:4705–4714

Simpson KW, Batt RM, Jones D, Morton DB (1990) Effects of exocrine pancreatic insufficiency and replacement therapy on the bacterial flora of the duodenum in dogs. Am J Vet Res 51:203–206

Stevens CE, Hume ID (1995) Comparative physiology of the vertebrate digestive system, 2nd ed. Cambridge University Press, UK

Stewart JA, Chadwick VS, Murray A (2005) Investigations into the influence of host genetics on the predominant eubacteria in the faecal microflora of children. J Med Microbiol 54(Pt 12):1239–1242

Sullivan A, Nord CE (2005) Probiotics and gastrointestinal diseases. J Intern Med 257:78–92

Takano M, Yumoto R, Murakami T (2006) Expression and function of efflux drug transporters in the intestine. Pharmacol Ther 109:137–161

Tannock GW (2007) What immunologists should know about bacterial communities of the human bowel. Semin Immunol 19:94–105

Tappenden KA, Deutsch AS (2007) The physiological relevance of the intestinal microbiota–contributions to human health. J Am Coll Nutr 26:679S–683S

Tilman D, Knops J, Wedin D, Reich P, Mitchie M, Siemann E (1997) The influence of functional diversity and composition of ecosystem processes. Science 277:1300–1302

Tlaskalová-Hogenová H, Stepánková R, Hudcovic T, Tucková L, Cukrowska B, Lodinová-Zádníková R, Kozáková H, Rossmann P, Bártová J, Sokol D, Funda DP, Borovská D, Reháková Z, Sinkora J, Hofman J, Drastich P, Kokesová A (2004) Commensal bacteria (normal microflora), mucosal immunity and chronic inflammatory and autoimmune diseases. Immunol Lett 93:97–108

Topping DL, Clifton PM (2001) Short-chain fatty acids and human colonic function: roles of resistant starch and nonstarch polysaccharides. Physiol Rev 81:1031–1064

Torres MI, Rios A (2008) Current view of the immunopathogenesis in inflammatory bowel disease and its implications for therapy. World J Gastroenterol 14:1972–1980

Trespi E, Ferrieri A (1999) Intestinal bacterial overgrowth during chronic pancreatitis. Curr Med Res Opin 15(1):47–52

Vannote RL, Minshall GW, Cummins KW, Sedell JR, Cushing CE (1980) The river continuum concept. Can J Fish Aquat Sci 37:130–137

Veereman G (2007) Pediatric applications of inulin and oligofructose. J Nutr 137 (Suppl. 11):2585S–2589S

Walker WA (2000) Role of nutrients and bacterial colonization in the development of intestinal host defense. J Pediatr Gastroenterol Nutr 30(Suppl. 2):S2–S7

Welters CF, Heineman E, Thunnissen FB, van den Bogaard AE, Soeters PB, Baeten CG (2002) Effect of dietary inulin supplementation on inflammation of pouch mucosa in patients with an ileal pouch-anal anastomosis. Dis Colon Rectum 45:621–627

Whelan K, Judd PA, Preedy VR, Simmering R, Jann A, Taylor MA (2005) Fructooligosaccharides and fiber partially prevent the alterations in fecal microbiota and short-chain fatty acid concentrations caused by standard enteral formula in healthy humans. J Nutr 135:1896–1902

Winkler P, Ghadimi D, Schrezenmeir J, Kraehenbuhl JP (2007) Molecular and cellular basis of microflora-host interactions. J Nutr 137(3 Suppl. 2):756S–772S

Woodmansey EJ (2007) Intestinal bacteria and ageing. J Appl Microbiol 102:1178–1186

Xenoulis PG, Palculict B, Allenspach K, Steiner JM, Van House AM, Suchodolski JS (2008) Molecular-phylogenetic characterization of microbial communities imbalances in the small intestine of dogs with inflammatory bowel disease. FEMS Microbiol Ecol 66:579–589

Zoppi G, Cinquetti M, Luciano A, Benini A, Muner A, Bertazzoni Minelli E (1998) The intestinal ecosystem in chronic functional constipation. Acta Paediatr 87: 836–841

Zuccato E, Venturi M, Di Leo G, Colombo L, Bertolo C, Doldi SB, Mussini E (1993) Role of bile acids and metabolic activity of colonic bacteria in increased risk of colon cancer after cholecystectomy. Dig Dis Sci 38:514–519

2 Molecular Tools for Investigating the Gut Microbiota

Christophe Lay

2.1 Introduction

The "microbial world within us" (Zoetendal et al., 2006) is populated by a complex society of indigenous microorganisms that feature different "ethnic" populations. Those microbial cells thriving within us are estimated to outnumber human body cells by a factor of ten to one. Insights into the relation between the intestinal microbial community and its host have been gained through gnotobiology. Indeed, the influence of the gut microbiota upon human development, physiology, immunity, and nutrition has been inferred by comparing gnotoxenic and axenic murine models (Hooper et al., 1998, 2002, 2003; Hooper and Gordon, 2001).

First interventional human studies on the structure of the gut microbiota derived from a common initiative aiming to seek a link between bacterial inhabitants of the fecal microbiota and their implication in colon cancer (Finegold et al., 1983; Moore and Holdeman, 1974; Moore and Moore, 1995). Culture based methods have been applied to characterize the bacterial species diversity in fecal samples. According to Moore and Holdeman (1974), the number of species could be estimated at 400–500. Nevertheless, this extrapolated number may not reflect the reality. Indeed an inherent bias of classical culture techniques concerns the cultivability potential of bacteria. Several studies have indicated differential counts resulting from the comparison between enumeration of colony-forming units and bacterial concentrations determined by microscopy. This phenomenon has been called "the great plate count anomaly" (Staley and Konopka, 1985). According to studies, 15–54% of the fecal bacterial community is cultivable (Langendijk et al., 1995; Matsuki et al., 2002; Suau et al., 1999; Tannock et al., 2000). Apajalahti et al. (2003) estimated that one-third of the fecal bacterial community is not viable, and this mortality rate could partially explain "the great plate count anomaly."

Nucleic acid based methods have been developed to circumvent biases inherent to bacterial culture. Those approaches based on the analysis and comparison of ribosomal nucleic acid sequences, have permitted to unravel the hidden diversity of the gut microbiota. Microbial classification is mainly based on the 16S rRNA taxonomic gene marker. Currently, more than 700,000 16S rRNA gene sequences are available in the RDP or ribosomal database project (http://rdp.cme.msu.edu/). This valuable resource gathers sequences isolated from diverse microbial communities and comprises two types of sequences, those originated from cultivated bacterial strains and those generated from clone libraries. Using 16S rRNA gene sequences based approaches, an unprecedented microbial diversity has been described within the human colonic microbiota. The structure of the human gut bacterial community features mainly four phyla, Firmicutes, Bacteroidetes, Actinobacteria and Proteobacteria. Consistent data regarding bacterial members belonging to the "healthy gut" population have been gained. Members of the *Eubacterium rectale-Clostridium coccoides*, *Clostridium leptum* (both belonging to Firmicutes), *Bacteroides* (Bacteroidetes), *Bifidobacterium* and *Atopobium* (both belonging to Actinobacteria) groups form the main core representative of the bacterial collection resident in the human bowel (Eckburg et al., 2005; Harmsen et al., 2002; Hayashi et al., 2002; Lay et al., 2005a; Suau et al., 1999). Advances in genome sequencing technologies have opened up new perspectives in studying the gut microbiota. Beyond the structural parameters, an insight into the genetic potential of the human microbiome, a term referring to the total number of genes encoded by the collective genomes of the gut bacterial members, is being revealed through metagenomic approaches (Gill et al., 2006; Kurokawa et al., 2007; Manichanh et al., 2006).

Interest of the research community in the study of the intestinal microbial ecosystem is expanding. The Human Microbiome Project (HMP) has recently been launched by the National Institutes of Health (NIH) with the mission of generating resources enabling comprehensive characterization of the human microbiota and analysis of its role in human health and disease (Turnbaugh et al., 2007). In Europe, the MetaHIT (Metagenomics of the Human Intestinal Tract) project funded by the European Commission is investigating the role of the gut microbiota in obesity and inflammatory bowel diseases (Mullard, 2008). Topics surrounding the gut microbiota and its intrinsic relations with health and disease also interest the food and pharmaceutical industries. Nowadays, the concept of functional foods to maintain and enhance the bacterial balance of the gut microbiota through probiotic and prebiotic administration is anchored in

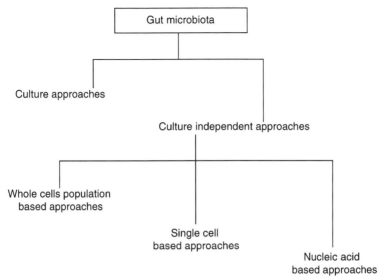

⬛ Figure 2.1
Molecular tools to investigate the gut microbiota.

our nutrition habits. This bacterial balance is dynamic and reflects the homeosta-sis of the intestinal microbial ecosystem, a concept to characterize this heteroge-neous community of microorganisms, interacting with its host (Hooper et al., 1998, 2002, 2003; Hooper and Gordon, 2001), capable of self-regulation, even after allochtonous invasion (Tannock et al., 2000) or temporary antibacterial disturbance (De La Cochetiere et al., 2005).

In this chapter, a description of the molecular tools for investigating the gut microbiota and their applications in human intervention studies will be sum-marized. A review of what has been revealed on the structure and genetic potential of the human gut microbiota through the applications of these molec-ular tools will be conducted (⊙ *Figure 2.1*).

2.2   Culture-Dependent Molecular Approaches

2.2.1   PCR Screening and 16S rRNA Sequence Analysis

Identification of bacterial colonies using phenotypic approaches is laborious and time-consuming. Molecular methods based on 16S rDNA have been developed to circumvent these constraints.

Matsuki et al. (2002) designed and validated specific 16S rDNA PCR primers for the detection of *Bacteroides fragilis*, *Bifidobacterium*, *Clostridium coccoides*, and *Prevotella* groups, in order to characterize the cultivable fraction of the fecal bacterial community of six healthy adult volunteers. Fifty bacterial colonies were randomly selected on non-selective media from each individual. The 300 bacterial isolates were then tested with the group-specific primers, 117 isolates were affiliated to the *Bacteroides fragilis* group, 22 to the *Bifidobacterium* group, 65 to the *Clostridium coccoides* group, and 17 isolates to the *Prevotella* group. These results indicated that 74% of the bacterial colonies were identified with the four pairs of primers. The remaining 79 isolates were identified using 16S rRNA gene sequence analysis. Forty colonies were affiliated to *Colinsella aerofaciens*, 24 to the *Clostridium leptum* subgroup, and 15 isolates to disparate clusters. In this study the authors compared bacterial concentrations determined on non-selective media with microscopic counting. On average 46% of the total fecal bacterial population corresponding to the non cultivable fraction could not be characterized (● *Figure 2.2*).

2.2.2   Ribotyping and Pulse-Field Gel Electrophoresis

Both techniques are based on RFLP (Restriction Fragment Length Polymorphism) and allow the discrimination of bacterial strains affiliated to the same species. Both molecular typing methods are culture-based dependent.

2.2.2.1   Ribotyping

Genomic DNA extracted from bacterial colonies randomly selected on selective medium is digested with appropriate restriction endonucleases. The resulting

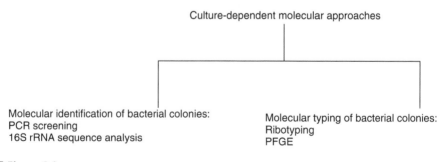

■ Figure 2.2
Culture-dependent molecular approaches.

restriction fragments are separated by agarose gel electrophoresis and transferred to nylon membranes. A radioactively labeled 16S or 23S rDNA group-specific probe is used to detect the restriction fragments in the digests that contains rRNA sequences. Hybridization is detected by autoradiography. Most bacteria carry multiple rRNA operons, therefore several restriction fragments of different sizes are detected after hybridization with the DNA probe. The hybridization pattern displays a genetic fingerprint or ribotype for the bacterial strain analyzed.

2.2.2.2 Pulsed-Field Gel Electrophoresis

Genomic DNA extracted from bacterial colonies randomly selected on selective medium is digested with an appropriate restriction endonuclease. The resulting restriction fragments are separated by PFGE (pulsed-field gel electrophoresis). Compared to conventional electrophoresis, PFGE separates large DNA fragments in the megabase range such as chromosomal DNA. After electrophoresis, gel is stained with ethidium bromide and examined by UV illumination. The gel obtained displays a genetic fingerprint or pulsotype for the bacterial strain analyzed.

2.2.2.3 Application in Human Intervention Studies

McCartney et al. (1996) applied ribotyping and PFGE to describe and monitor the *Bifidobacterium* and *Lactobacillus* populations in two adult volunteers over a period of one year. Five strains of *Bifidobacterium* were identified in one subject throughout the 12-month period. The *Bifidobacterium* population was relatively simple and stable over time, and among the isolates detected one strain was numerically predominant. In contrast, the other subject harbored a more complex and unstable *Bifidobacterium* composition, 36 sporadic or transient strains were identified during the 12-month period. A unique and distinctive collection of *Bifidobacterium* featuring its level of diversity and strain specificity was a characteristic of each subject. The authors also observed that both volunteers harbored a distinctive and predominant strain of *Lactobacillus*. McCartney and colleagues highlighted the complementarity of both molecular typing methods. Indeed the same ribotype was observed for both *Lactobacillus* isolates; however PFGE permitted to discriminate them in two distinctive strains. Kimura et al. (1997) corroborated the results of McCartney and colleagues in a similar study involving ten volunteers.

2.2.2.4 Application in Probiotic Intervention Studies

The composition of the *Lactobacillus* population of ten healthy subjects was monitored before (6 months control period), during (6 months test period), and after (3 months post test period) the consumption of a dairy product containing *Lactobacillus rhamnosus* DR20 (daily dose, 1.6 × 10⁹ lactobacilli) (Tannock et al., 2000). PFGE was used to differentiate DR20 from the other strains present in the samples. Inter-individual comparisons of pulsotypes showed that each subject harbored a unique and distinctive collection of lacto-bacilli. Intra-individual comparisons revealed that DR20 was temporary detected in dominance among the subjects whose *Lactobacillus* populations fluctuated in terms of size or composition. In contrast DR20 did not predominate among subjects whose *Lactobacillus* composition was stable. Tannock and colleagues defined two types of bacterial strains characteristics of the *Bifidobacterium* and *Lactobacillus* populations. Those which permanently resided in the fecal bacterial community were defined as autochtonous or indigenous, and those which were detected occasionally were defined as allochtonous or transient.

2.2.3 Limits

The methods described can only be applied to the cultivable fraction of the fecal bacterial community.

Those 16S rRNA based approaches rely on the availability of group-specific PCR primers and probes.

2.3 Culture-Independent Molecular Approaches Based on 16S rRNA

2.3.1 Electrophoresis of PCR-Amplified 16S rDNA Amplicons

2.3.1.1 Principle

The technique allows profiling the genetic diversity and complexity of a bacterial community. The principle relies on: (1) the isolation of total genomic DNA or RNA (cDNA) from the sample; (2) amplification of 16S rRNA gene with specific

PCR primers; and (3) electrophoretic separation of 16S rDNA amplicons of equal length but differing in base-pair, and according to their sequence-dependent melting temperature. Separation occurs within a polyacrylamide gel where a denaturing gradient is generated either with a linearly increasing gradient of temperature over time (TTGE: temporal temperature gradient gel electrophoresis and TGGE: temporal gradient gel electrophoresis) or a linearly increasing gradient of chemical denaturants (urea, formamide) (DGGE: denaturing gradient gel electrophoresis). Once the amplicon reaches its melting temperature at a particular position in the denaturing gel, a conformational change from helical structure to partially melted molecule occurs, and the migration of the DNA fragment will stop. A GC-clamp attached to the 5' end of one of the PCR primers prevents the amplicons from complete denaturation. The gel is then stained (ethidium bromide, silver staining or SYBR green), digitally captured and analyzed using specific software. The electrophoresis profile generated represents a 16S rDNA genetic fingerprint displaying a banding pattern. Differences in band intensities reflect a relative number of specific amplicons. The technique is considered as a semi-quantitative method rather than a quantitative one. Such approach has been applied to directly visualize and evaluate the heterogeneity or genetic diversity of the gut bacterial community. Zoetendal et al. (1998) estimated that PCR-TGGE profile generated with Eubacterial PCR primers displayed 90–99% of the amplicons. Therefore only dominant bacterial members are detected, nevertheless the use of group-specific PCR primers allows for analysis of specific bacterial sub-populations (❂ *Figure 2.3*).

2.3.1.2 Application in Human Intervention Studies

Dynamic Characterization of the Complexity and Genetic Diversity of the Adult Fecal Microbiota

Zoetendal et al. (1998) applied RT-PCR-TGGE and PCR-TGGE to describe and characterize the species diversity of the dominant fecal bacterial community. The authors compared TGGE profiles derived from fecal RNA and DNA, generated from 16 healthy and unrelated adult volunteers. TGGE patterns derived from RNA and DNA extracted from the same fecal sample were similar. A closer comparison between RNA and DNA derived profiles showed the presence of predominant bacterial species but expressing a low metabolic activity and reciprocally. Inter-individual comparisons showed differences in the position of specific bands, the intensity and number of bands, demonstrating that

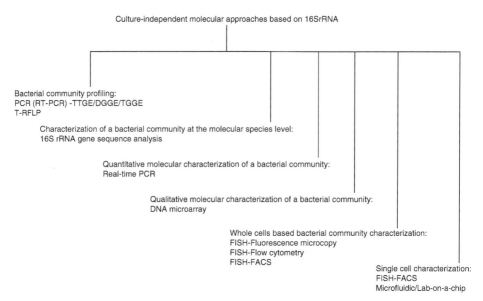

■ Figure 2.3
Culture-independent molecular approaches based on 16S rRNA.

each individual harbored its unique fecal bacterial community. The dominant fecal bacterial diversity of two subjects was monitored and showed a remarkable stability over time. These observations were corroborated by Seksik et al. (2003) and Vanhoutte et al. (2004). The uniqueness of the fecal bacterial community is not absolute since some distinct and dominant bands are found in all the subjects. Those bands were identified using a cloning and sequencing approach, and were affiliated to *Ruminococcus obeum*, *Faecalibacterium prausnitzii* and *Eubacteriuum hallii.*

According to Zoetendal et al. (2001) reasons of the uniqueness of the fecal bacterial community are likely to be found in host-related factors. The authors observed that PCR-DGGE profiles generated from monozygotic twins were significantly more similar than those derived from marital couples or unrelated individuals. Diversity, uniqueness and stability of the fecal microbiota were also demonstrated within specific bacterial groups. The *Bifidobacterium*, *Bacteroides* and *Clostridium leptum* groups were characterized by their genetic diversity, uniqueness and stability over time (Lay et al., 2007; Satokari et al., 2001a; Vanhoutte et al., 2004). In comparison, fluctuation of bacterial populations within the lactobacilli group was observed over time (Heilig et al., 2002; Walter et al., 2001).

Temporal stability of the fecal bacterial community is disrupted upon anti-biotic administration. However, the fecal microbiota is able to cope with this challenge. De La Cochetière et al. (2005) investigated the ability of the dominant fecal bacterial community to return to its initial structure after temporary anti-bacterial treatment. The dominant species diversity of six healthy subjects was monitored before, during a 5-day course of amoxicillin, and after antibiotic therapy. Alterations of PCR-TTGE profiles were observed under antimicrobial chemotherapy. A return to the initial profile or resilience of the dominant fecal bacterial community was described in five out of the six volunteers within 2 months after the experimental period.

Zoetendal et al. (2002b) compared the distribution of the bacterial community adhering to the colonic mucosa to the luminal fecal bacteria. Fecal samples and biopsies from the ascending, transverse, and descending colons of ten individuals were analyzed by using PCR-DGGE. Stool samples were collected before the colonoscopy, and during the clinical examination three patients were diagnosed with an ulcer and one with a colon polyp. Intra-individual comparisons revealed that the predominant mucosa-associated bacterial community was uniformly distributed along the colon but significantly different from the fecal community. Inter-individual comparisons showed that the colonic mucosa-associated bacterial community is host specific. According to Zoetendal and colleagues these observations support the hypothesis that host-related factors are involved in the determination of the structure of the gut microbiota. No significant differences were observed between healthy individuals and those diagnosed with a disease. Group-specific PCR-DGGE on the lactobacilli showed low species diversity, and for 6 out of the 10 volunteers the *Lactobacillus* community adhering to the intestinal mucosa was similar to the luminal one. Profiles generated from biopsies taken from the descending colon were compared between them; one common band was found in nine individuals, and was affiliated to *Lactobacillus gasseri*. Results from this study were corroborated by Lepage et al. (2005) and Nielsen et al. (2003).

Dynamic Characterization of the Complexity and Genetic Diversity of the Infant Fecal Microbiota

Favier et al. (2002) combined PCR-DGGE and 16S rDNA sequence analysis to describe and characterize the establishment of the fecal bacterial community in two healthy infants during the first ten months of life. DGGE patterns revealed low species diversity during the first weeks following birth. Appearance and disappearance of bands occurred during the first two months. A certain number of bands were identified by constructing and sequencing 16S rDNA clone libraries generated from both infant fecal samples collected at different ages. Electrophoretic mobility of the clones was also analyzed by PCR-DGGE and compared to the mobility of the amplicons separated in the former denaturing gel. Both infants showed similar bacterial colonization pattern featuring the detection of *Bifidobacterium*, *Streptococcus*, *Enterococcus*, *Ruminococcus*, *Clostridium*, and *Enterobacter*. The authors observed an increase in the complexity of PCR-DGGE profiles during the first ten months of life. These fluctuations of bacterial populations reflect a diversification in the species composition of the dominant infant bacterial community over time.

The species diversity of the fecal bacterial community of 13 infants was monitored at one and seven months following birth (Satokari et al., 2002). Seven babies were breast-fed and the others received an infant formula. Intra-individual comparisons of PCR-DGGE profiles at one and seven months showed fluctuations of the dominant fecal bacterial community. No significant differences were observed between the two experimental groups. Satokari and colleagues identified the species composition by constructing and sequencing 16S rDNA clone libraries. Common DGGE bands found in the majority of the infant profiles were affiliated to *Ruminococcus gnavus*, *Escherichia coli* and the genus *Bifidobacterium*. Group-specific PCR-DGGE on the genus *Bifidobacterium* revealed low species diversity and relative stability in half of the newborns. *Bifidobacterium infantis* was frequently found in both groups. Regarding the *Lactobacillus* group, a similar pattern was observed with however a variable stability. *Lactobacillus acidophilus* was the most detected.

Evidence of Vertical Transmission of Maternal Gut Bacteria to the Infant via Breast Feeding

The neonatal colonization of the infant gut may be promoted by the transmission of maternal gut bacteria through breast feeding. Perez et al. (2007) used eubacterial PCR primers and TTGE to detect and monitor the presence of bacteria in samples of maternal origin (feces, breast milk and blood) and in infant fecal samples. All samples were collected during lactation and over a 4-week period following birth. Control blood samples from non pregnant women were also analyzed. PCR-TTGE profiles generated from fecal samples were host specific, and species diversity was greater in maternal than in infant feces. Bacterial community detected in breast milk leukocytes was less complex than the maternal fecal microbiota. The authors built 16S rRNA clone libraries from breast milk leukocytes and revealed in addition of human milk bacterial members, the presence of bacterial signatures from the colonic microbiota (*Bacteroides*, *Clostridium* and *Eubacterium*). Detection of bacteria in peripheral blood mononuclear cells was also observed and species diversity was greater in lactating women than in control women. Moreover PCR-TTGE profiles from lactating women were host specific whereas those generated from control women were similar. Common bacterial species were detected in infant feces and in samples of maternal origin. Those TTGE bands were identified by excision, cloning and sequencing, and were affiliated to *Bifidobacterium longum*, *Streptococcus thermophilus and Streptococcus epidermidis*. Perez and colleagues described the translocation of gut bacteria signatures from the mother to the infant via blood circulation and breast milk feeding. A clue of vertical transmission of maternal gut bacteria to the infant via breast feeding was also shown by Martin et al. (2007). The authors demonstrated with PCR-DGGE that the *Lactobacillus* populations detected in infant feces resembled those from the breast milk of the respective mothers.

2.3.1.3   Application in Probiotic and Prebiotic Intervention Studies

Tannock et al. (2000) applied PCR-DGGE to determine whether the daily consumption of a probiotic had an impact or not on the structure of the dominant fecal bacterial community. The bacterial species diversity of ten healthy subjects was monitored before (6 months control period), during (6 months test period), and after (3 months post test period) the administration of a dairy product containing *Lactobacillus rhamnosus* DR20 (daily dose, 1.6×10^9 lactobacilli). Long-term consumption of *Lactobacillus rhamnosus* DR20 did not affect the dominant members of the fecal microbiota regarding the PCR-DGGE profiles generated with universal PCR primers. Using *Lactobacillus* group-specific primers, Walter et al. (2001) investigated in 2 out of the 10 volunteers the impact of DR20 ingestion on the *Lactobacillus* populations. The authors visualized the transient passage of the probiotic strain within the *Lactobacillus* populations during the test period.

Satokari et al. (2001b) observed that the consumption of *Bifidobacterium lactis* Bb-12 and/or galacto-oligosaccharides did not affect the endogenous *Bifidobacterium* populations. Only a transient implantation of the probiotic strain was detected during the ingestion period.

Modulation of RT-PCR-DGGE profiles targeting the Eubacterial domain was observed following daily consumption of biscuits containing prebiotic (Tannock et al., 2004). Appearance and intensification of bands in RNA-DGGE patterns was a response to galacto-oligosaccharides or fructo-oligosaccharides consumption. Conversely, DGGE profiles generated from bacterial DNA did not show any prebiotic effect. The authors identified the bacterial species incriminated in the alteration of RT-PCR-DGGE profiles. *Bifidobacterium adolescentis* and *Colinsella aerofaciens* were found to be metabolically stimulated when biscuits were consumed.

2.3.1.4 Limits

Heterogeneity between multiple 16S ribosomal RNA operons is characterized by the detection of multiple bands. 16S rRNA sequences are highly conserved among the genus *Bifidobacterium*, and there are multiple copies of the 16S rRNA gene per chromosome. In order to avoid those biases, Requena and colleagues developed a method that enabled the identification and detection of *Bifidobacterium* species by PCR targeting a conserved region of the transaldolase gene common to the *Bifidobacterium* genus (Requena et al., 2002).

Approximately, one band corresponds to one species, however amplicons with variable sequences can also migrate at the same position in the gel. Therefore, the bacterial diversity might be underestimated.

Methodological biases are encountered in all the experimental steps: nucleic acid extraction; PCR; cloning; sequencing and sequences analysis (von Wintzingerode et al., 1997).

2.3.2   T-RFLP

2.3.2.1 Principle

Terminal restriction fragment length polymorphism (T-RFLP) permits profiling the genetic diversity and complexity of a microbial community by analyzing the polymorphism of 16S rRNA gene. The approach is based on: (1) the isolation of total genomic DNA or RNA (cDNA) from the sample; (2) amplification of 16S rRNA gene using fluorescent labeled group-specific PCR primers; (3) restriction enzyme digestion of amplicons; and (4) separation of the fluorescent terminal restriction fragments through either capillary or poly-acrylamide electrophoresis in a DNA sequencer. Only the sizes of the different terminal fluorescently labeled fragments are determined by the fluorescence detector. T-RFLP profile displays a graph or electropherogram where the X axis represents the sizes of the fragment and the Y axis represents their fluorescence intensity. Putative phylogenetic identification of the terminal restriction fragments (T-RFs) generated can be predicted in silico using T-RFLP analysis program which assigns the T-RFs to existing sequences listed in database.

2.3.2.2 Application in Human Intervention Studies

Sakamoto et al. (2003) applied T-RFLP to describe and compare the fecal bacterial community of three healthy adult volunteers. T-RFLP patterns showed that each individual harbored a specific bacterial composition; this host specificity was seen at the dominant level and within the *Bifidobacterium* group.

Dicksved et al. (2007) investigated the impact of life styles on the composition of the fecal microbiota. A cohort of 90 children between 5 and 13 years-old from three European countries were involved in this study. Children were recruited and grouped in two main groups according to their life style. The anthroposophic lifestyle group with its reference group was compared to the farm lifestyle with its reference group. Inter-individual comparisons showed that each child had a unique fecal bacterial community. However, some terminal restriction fragments were observed in almost all the profiles, and those fragments were affiliated to the genus *Eubacterium* and *Clostridium*. Comparison between the experimental groups showed that anthroposophic children had significantly higher species diversity than farm children.

The species diversity of the dominant fecal microbiota was characterized in two groups of infants with and without atopic eczema (Wang et al., 2008). Selected from a cohort, fecal samples from 20 healthy newborns and 15 atopic infants diagnosed with atopic eczema at the age of 18 months were analyzed by T-RFLP and PCR-TTGE. Stool samples had been collected one week after birth, before tests for atopy were performed at the age of 18 months. Both molecular approaches showed a reduced diversity in infant fecal microbiota who later were diagnosed with atopic eczema.

2.3.2.3 Application in Probiotic and Prebiotic Intervention Studies

Jernberg et al. (2005) monitored the impact of antibiotic and probiotic administration on the fecal microbiota in healthy adults. Eight volunteers participated in this study, and were separated into two groups. The probiotic group ingested a dairy product containing *Lactobacillus acidophilus* NCFB 1748, *Lactobacillus paracasei* F19, and *Bifidobacterium lactis* Bb12 (10^8 CFU of each strain per ml) (250 ml) twice daily for 14 days. The placebo group received a yogurt without any of the probiotic strains. Both experimental groups were under antibiotic therapy (clindamycin) for 7 days, and the administration of clindamycin and dairy products were initiated on the same day. Fecal samples were collected before administration (day 0), on the last day of clindamycin administration (day 7), and one week after the end of the experimental diet on day 21. Terminal restriction fragment length polymorphism was used to describe and monitor the dominant fecal bacterial community during the experimental period. In both groups, the authors observed that the consumption of antibiotic induced alterations of the fecal bacterial community characterized by a marked reduction or complete eradication of the genus *Eubacterium*, and appearance of bacterial species belonging to the *Bacteroides* and *Prevotella* group, and the *Enterobacteriaceae*. By day 21, a reestablishment of the fecal microbiota to its initial composition was observed. Finally, a transit of both *Lactobacillus* probiotic strains was observed using *Lactobacillus* group-specific PCR primers.

The effect of prebiotic consumption on the mouse gut microbiota was monitored using T-RFLP approach (Nakanishi et al., 2006). Four experimental groups of mice were designed: a control group and three groups of mice (n = 7 per group) fed with short-chain fructo-oligosaccharide (scFOS) for 5 weeks. The composition of the prebiotic diet was different for each group of mice. Fecal and caecal samples were collected before and at 5 weeks after starting the experimental diet. Compared to the control group, T-RFLP profiles of the dominant microbiota generated from the prebiotic groups showed an increase in a particular group of terminal restriction fragments that were affiliated to the *Bacteroidetes* phylum.

2.3.2.4 Limits

Methodological biases are encountered in all the experimental steps: nucleic acid extraction; PCR; restriction enzyme digestion of amplicons.

The bacterial diversity may be underestimated because same T-RFs can be generated by different species of bacteria.

2.3.3 16S rRNA Gene Sequence Analysis

2.3.3.1 Principle

16S rRNA gene sequence analysis permits characterizing the members of a microbial community at the molecular species level. The method is based on the amplification and cloning of 16S rRNA gene from genomic DNA or RNA (cDNA) extracted from the total microbial community. Selected clones are then sequenced and analyzed in silico. Bioinformatic tools and phylogenetic approaches are utilized to align and affiliate the clones to their closest relatives. The aligned sequences are finally grouped in "an entity" named molecular species or OTUs (Operational Taxonomic Unit) or phylotypes and visualized in a phylogenetic tree. Stackebrandt and Goebel (1994) established that microorganisms having a 16S rRNA gene sequence similarity >97% belong to the same species. This threshold value delimitating two species is a reference in microbial ecology. Therefore a group of 16S rRNA sequences or clones having a sequence similarity >97% is affiliated to the same molecular species. One OTU or molecular species can be either affiliated to a cultivated bacterial species, or a not yet cultivated bacterial species. 16S rRNA gene sequences are a valuable resource because they constitute a database, that allows designing in silico oligonucleotide probes and primers.

2.3.3.2 Application in Human Intervention Studies

This molecular approach has been applied to explore in depth the bacterial diversity of the gut microbiota in healthy adults (Eckburg et al., 2005; Hayashi et al., 2002; Suau et al., 1999; Wilson and Blitchington, 1996). Large inter-individual variations were observed reflecting the uniqueness of the gut microbiota in each subject. Phylogenetic distribution of the clones generated from those studies showed that three main bacterial groups, *Clostridium coccoides*, *Clostridium leptum* and the *Bacteroides* constituted the core of the colonic microbiota. A striking observation was that large numbers of the determined phylotypes were affiliated to novel and uncultivated species. Suau et al. (1999)

estimated the phylogenetic diversity of the human fecal microbiota by using the "Good's coverage" as a statistical model. The authors analyzed a 16S rRNA gene clone library generated from one healthy adult volunteer. The 284 clones isolated from the library were gathered into 82 OTUs for which 76% were affiliated to uncultivated species. The majority of the clones (95%) were distributed within the *Clostridium coccoides, Clostridium leptum* and *Bacteroides* groups. Proportions of OTUs corresponding to novel species were 85% for the *Clostridium leptum* group, 77.4% for the *Clostridium coccoides* group, and 62% for the *Bacteroides* group. Suau and colleagues statistically measured the coverage of their clone library in terms of species diversity. From the 284 clones analyzed, a global coverage of 85% was determined meaning that the isolation of 100 supplementary clones would have permitted to detect potentially 15 novel species. In other terms, wider species diversity would have been described.

Large-scale phylogenetic analyses of the human gut microbiota in terms of number of 16S rDNA clones generated, were undertaken in two studies. Bacterial and archaeal 16S rRNA gene libraries were generated from feces and biopsies taken at different colonic sites from three healthy adult subjects (Eckburg et al., 2005). From the pool of sequences generated (13,355) 395 bacterial phylotypes, and one single archaeal phylotype were determined. The bacterial phylotypes were mainly distributed within the Firmicutes and the Bacteroidetes phyla, and most of the sequences were affiliated to novel OTUs and derived from uncultivated species. Intra-individual comparisons showed differences in bacterial composition between the mucosal and luminal microbiota. Eckburg and colleagues observed large inter-individual variations, and those differences between individuals were greater than the differences observed between different sampling sites collected from one subject. Ley et al. (2006) monitored the fecal bacterial diversity of 12 obese volunteers randomly assigned to either a fat-restricted or to a carbohydrate-restricted low calorie-diet over a period of one year. Stool samples were collected before and at 12, 26, and 52 weeks after starting diet therapy. Two control lean humans also participated in this study. The authors built a 16S rRNA gene clone library to monitor the composition of the fecal microbiota. The large pool of 16S rDNA sequences generated (18,348) were classified into 4,074 OTUs, and were mainly distributed within the Bacteroidetes and the Firmicutes. Most of the phylotypes identified (70%) were unique to each individual. Intra-individual comparisons showed a stability of the fecal bacterial community over time. The authors observed that compared to the lean controls, obese subjects had fewer Bacteroidetes and more Firmicutes before starting low calorie-diet. In the obese group regardless the type of diet, the relative abundance of Bacteroidetes increased and the proportion of Firmicutes decreased during the experimental period. Statistical analysis showed that the proportion of Bacteroidetes increased with weight loss on the two types of low-calorie diet.

2.3.3.3 Limits

The threshold limit for delimitating an OTU is variable according to studies (Martin, 2002). Therefore, determining the number of molecular species constitutive of the gut microbial community is rather subjective. The coverage is a biodiversity indices and its determination reveals that the total number of species constitutive of the gut microbiota is difficult to assess.

Methodological biases are encountered in all the experimental steps: nucleic acid extraction; PCR; cloning; sequencing and sequences analysis (von Wintzingerode et al., 1997).

2.3.4   Whole Cell Fluorescence *In Situ* Hybridization

2.3.4.1 Principle

Fluorescence *in situ* hybridization is based on the complementary binding of labeled nucleic acid probes to complementary sequences in cells or tissue sections. In microbial ecology, this technique is mainly applied to detect and quantify specific bacterial populations within a complex microbial community. Using specific 16S or 23S rRNA oligonucleotide probes (DNA probes) whole bacterial cells can be targeted *in situ* within an assemblage of mixed microbial populations. This strategy has the advantage of taking into account the cultivable and uncultivable fractions of a bacterial community. The DNA probes target the single strand rRNA molecules localized within the ribosomes. Probes are 5' labeled with a fluorescent dye permitting detection and quantification of the specific bacterial population targeted. Two systems of fluorescence detection are combined with the technology of whole cell Fluorescence *in situ* Hybridization (FISH): Fluorescence Microscopy and Flow Cytometry.

2.3.4.2   Fluorescence *In Situ* Hybridization Combined with Microscopic Detection

Principle

According to Welling et al. (1997), the detection limit of whole cell fluorescence *in situ* hybridization combined with microscopic detection is estimated at 10^6 bacteria per gram of stool sample. Laborious and time consuming, visual and manual microscopic counting has been automated by coupling fluorescence microscopy to image analysis (Jansen et al., 1999). Using this automated counting system, the threshold is estimated at 10^7 bacteria per gram of stool sample. Therefore, only the dominant fecal bacteria can be detected. In addition to detection and quantification, *in situ* identification, morphological and topographical information are others advantages offered by fluorescence microscopy.

Application in Human Intervention Studies

DNA probes targeting diverse bacterial populations have been developed and applied to quantify and monitor the composition of the fecal microbiota in healthy volunteers (Franks et al., 1998; Harmsen et al., 1999, 2000, 2002; Langendijk et al., 1995; Suau et al., 2001). Large inter-individual variations in term of bacterial composition were observed; however three bacterial groups, *Clostridium coccoides*, *Clostridium leptum* and *Bacteroides* constituted the main core of the adult fecal bacterial community (Franks et al., 1998; Harmsen et al., 2002).

Application in Probiotic Intervention Studies

The composition of the fecal microbiota of ten healthy subjects was monitored before (6 months control period), during (6 months test period) and after (3 months post test period) the administration of a probiotic product containing *Lactobacillus rhamnosus* DR20 (daily dose, 1.6×10^9 lactobacilli) (Tannock et al., 2000). FISH combined with automated microscopy showed that long-term consumption of *Lactobacillus rhamnosus* DR20 did not affect the dominant members of the fecal microbiota, regarding the bacterial groups investigated: *Bacteroides*; *Clostridium coccoides-Eubacterium rectale*; *Atopobium*; *Bifidobacterium* and the gram-positive low-G + C content group 2 bacteria.

The establishment of the fecal bacterial community of breastfed infants at risk of allergy (n = 132) was monitored throughout the first 2 years of life (Rinne et al., 2006). Following birth, infants received either a probiotic *Lactobacillus rhamnosus* GG (daily dose, 1.0×10^{10} lactobacilli) or a placebo throughout the first 6 months of life. Fecal bacterial community were analyzed and monitored at 6, 12, 18, and

24 months. Results showed that administration of *Lactobacillus rhamnosus* GG in the first months of life did not significantly interfere with the establishment of the *Bifidobacterium*, *Lactobacillus-Enterococcus*, *Bacteroides-Prevotella* and *Clostridium histolyticum* groups.

In a preclinical study, Dinoto et al. (2006) investigated the effects of administration of raffinose and encapsulated *Bifidobacterium breve* JCM 1192^T on the rat cecal microbiota. Twenty-four rats divided in four groups, were fed for 3 weeks with four different diets: basal diet (group BD); basal diet supplemented with raffinose (group RAF); basal diet supplemented with encapsulated *Bifidobacterium breve* JCM 1192^T (group CB) and basal diet supplemented with both raffinose and *Bifidobacterium breve* JCM 1192^T (group RCB). The combination of raffinose and *Bifidobacterium breve* JCM 1192^T was referred as a synbiotic, an association of prebiotic and probiotic. A *Bifidobacterium breve* specific oligonucleotide probe, PBR2, combined with helper probes to increase its *in situ* accessibility, was developed in order to monitor the influence of the probiotic consumption on the rat cecal microbiota. Interestingly, the probiotic strain administrated was only detected in the RCB group (7.3% of the total bacterial population) compared to the CB group where no proliferation was observed. These results indicated that raffinose influences on the cecal proliferation of *Bifidobacterium breve* JCM 1192^T and supports the concept of combining a probiotic strain with its preferred substrate.

2.3.4.3 Fluorescence *In Situ* Hybridization Combined with Flow Cytometry Detection

Principle
Combined with flow cytometry, whole cell fluorescence *in situ* hybridization represents a high throughput quantitative and qualitative method of analysis. Rapid and easy to set up, flow cytometry associates quantitative and multiparametrics analysis system (size, internal granularity and fluorescence signal). According to Lay (2004), a threshold of 0.4% relative to the total number of bacteria determined with the probe EUB338 was shown. A limit of detection around 10^4 CFU/ml was demonstrated using FISH and serial dilutions of *Escherichia coli* cells suspension (unpublished data).

Application in Human Intervention Studies
Flow cytometry has been used as a powerful tool to determine the specificity and labeling efficiency of DNA probes (Fallani et al., 2006; Fuchs et al., 1998; Lay et al.,

2005b; Rigottier-Gois et al., 2003b; Saunier et al., 2005). 16S rRNA group and species-specific probes validated with fluorescence microscopy, and newly developed probes, have been applied to quantify and monitor the composition of the fecal microbiota with flow cytometry (Fallani et al., 2006; Lay et al., 2005b; Rigottier-Gois et al., 2003a; Zoetendal et al., 2002a). Rigottier-Gois et al. (2003a) applied a panel of 6 DNA probes to investigate the fecal bacterial community in 23 healthy adult volunteers. Lay et al. (2005a) expanded the panel of probes to 18, and analyzed the bacterial composition of fecal samples (n = 91) collected from five northern European countries (Denmark, France, Germany, the Netherlands, and the United Kingdom). On average 75% of fecal bacteria could be identified with the extended panel of probes, whereas a proportion of 51% of fecal bacteria were recognized in the study of Rigottier-Gois and colleagues. Large inter-individual variations were observed. Each individual harbors a specific fecal bacterial community or signature in term of bacterial composition. The *Clostridium coccoides* and *Clostridum leptum* groups were co-dominant and constituted more than 50% of the total bacterial population, followed by the *Bacteroides, Bifidobacterium* and *Atopobium*. No significant correlations between the composition of the colonic microbiota and the parameters; age, gender, and geographical origin were found. Conversely, in a study involving a cohort of 230 healthy volunteers (85 adults and 145 elderlies (>60 years)) recruited from four European countries (France, Germany, Italy, and Sweden) some associations were identified (Mueller et al., 2006). The Italian cohort was characterized by a higher proportion of *Bifodobacterium*, two- to threefold higher than in the three others groups of European volunteers. Independently of the geographical location higher proportions of enterobacteria were found in elderly subjects. Male volunteers harbored a higher proportion of *Bacteroides* than female.

Mah et al. (2007) monitored the establishment of the fecal bacterial community of 37 infants at risk of developing allergic disease, from birth to one year. FISH combined with flow cytometry (FISH-FC) showed that the colonic bacterial colonization followed a pattern featuring an expansion of the bifidobacterial population during the first three months of life. Whereas enterobacterial and *Bacteroides-Prevotella* populations decreased over time, *Eubacterium rectale-Clostridium coccoides* and *Atopobium* groups gradually increased. Members of the *Clostridium leptum* subgroup formed only a small fraction of the total fecal microbiota at one year of age.

Application in Probiotic Intervention Studies

Garrido et al. (2005) monitored the composition of the fecal bacterial community in ten volunteers, who ingested daily different amounts of the probiotic preparation *Lactobacillus johnsonii* La1 (La1): 100 ml of 10^8 CFU/ml of La1 during the first week; 200 ml of 10^8 CFU/ml of La1 during the second week and 500 ml of 10^8 CFU/ml of La1 during the third week. A baseline period (before administration, 5 days) and a post-ingestion period (7 weeks) were also monitored during the trial. FISH-FC showed that La1 intake increased the populations of *Clostridium histolyticum*, *Lactobacillus-Enterococcus* and *Bifidobacterium*, and decreased those of *Faecalibacterium prausnitzii*. These bacterial groups returned to their basal levels during the post-ingestion period. No control group was defined in this study.

The influence of consumption of the fermented milk containing *Lactobacillus casei* DN-114001 (daily dose of 3.0×10^{10} CFU) on the fecal bacterial community was monitored in 12 healthy subjects, before (control period of 1 week), during (10 days supplementation) and after (10 days post-ingestion) the probiotic administration (Rochet et al., 2006). Compared to the study of Garrido et al. (2005), the probiotic supplementation did not affect the dominant members of the fecal microbiota regarding the seven bacterial groups investigated.

Mah et al. (2007) monitored the fecal bacterial community of 37 infants at risk of developing atopy, with (n = 20) and without probiotic administration (n = 17) during the first year of life. The probiotic group received daily an infant formula containing *Bifidobacterium longum* BB536 (1.0×10^7 CFU/g) and *Lactobacillus rhamnosus* GG (2.0×10^7 CFU/g) during the first six months. The authors observed that the intestinal colonization followed a pattern regardless of probiotic administration.

2.3.4.4 Phylogenic Gap

The determination of the composition of the fecal bacterial community using FISH has shown that the panel of existing phylogenetic probes does not cover the totality of the colonic bacterial population. This cellular fraction, phylogenetically unaffiliated has been defined as the phylogenetic gap (Lay et al., 2005a). Twelve DNA probes were applied in the study of Harmsen et al. (2002) permitting to take into account more than 56% of the total cell population. Rigottier-Gois et al. (2003a) observed that 49% of the total fecal bacterial population was not recognized by any of the six phylogenetic probes selected. The characterization of the fecal microbiota composition of 91 individuals from five northern European countries showed that on average, a proportion of 25% of the fecal bacteria still remained unidentified (Lay et al., 2005a). Such proportion of untargeted bacteria also exists within specific bacterial groups. Rigottier-Gois et al. (2003b) determined 34% of unidentified species within the *Bacteroides* group. Lay et al. (2007) observed a phylogenetic gap of 9% within the *Clostridium leptum* subgroup.

2.3.4.5 Fluorescence Activated Cell Sorting

Flow cytometry equipped with a cell sorting function allows sorting specific bacterial cell populations from a heterogeneous community of bacteria. This approach was applied to identify the phylogenetic gap observed within the *Clostridum leptum* subgroup (Lay et al., 2007). Using FISH-Fluorescence activated cell sorting (FISH-FACS) combined with 16S rRNA gene cloning strategy the authors identified new molecular species within this Gram-positive bacterial group.

Amor et al. (2002) used FACS and fluorescent viability probes to sort viable, injured, and dead bacteria from fecal microbiota. Three bacterial sub populations were sorted and characterized by PCR-DGGE, and cloning of PCR-DGGE dominant bands. Members of *Clostridium coccoides, Clostridium leptum*, and *Bacteroides* were found in the three physiological groups of sorted bacterial populations.

2.3.4.6 Limits

Parameters Affecting the Fluorescence Detection

Cell Wall Permeability In particular, the cell wall of Gram-positive bacteria represents a protective structure to crossover. Therefore, a standardized hybridization procedure including a permeabilization step is needed (Lay et al., 2005b). Once within the cell, probes face with another physical hurdle, the secondary structure of rRNA molecules and their molecular interactions within the ribosome, which may hinder the access of the probes to their target sites.

In Situ **Accessibility** The *in situ* accessibility of a probe to its target site determines the probe-conferred fluorescence or probe-mediated fluorescence (Fuchs et al., 1998). The concept relies on the principle that high *in situ* accessibility would facilitate the binding of the probe to its target site, and hence allowing the probe to emit a bright fluorescence signal. The determination of the brightness of fluorescence, or probe relative fluorescence conferred by a probe is a means to evaluate its *in situ* accessibility (Fuchs et al., 1998; Lay et al., 2005b). Modeling secondary structure of the 16S rRNA has permitted an in silico diagnostic of the *in situ* accessibility of the entire molecule (Fuchs et al., 1998; Kumar et al., 2005; Saunier et al., 2005). Therefore, the target site of a probe can be assessed in terms of *in situ* accessibility. If the target region is determined as a lower or non-accessible site, a strategy using helpers or unlabeled oligonucleotide probes can be developed in order to promote the binding of the probe to its target and hence, amplifying the fluorescence signal emission (Fuchs et al., 2000). Helpers are designed in silico to open inaccessible regions and are complementary to regions adjacent to the probe's target site (Dinoto et al., 2006; Fuchs et al., 2000; Saunier et al., 2005).

Ribosomal Content or Metabolic Activity of the Cell   The ribosomal content is linked to the physiological state or metabolic activity of the cell. Therefore, bacterial cells in a quiescent state i.e., harboring a low number of ribosomes have weak fluorescence emission and may escape detection.

Hybridization Conditions   The stringency of hybridization depends on three parameters: temperature; salt concentration and formamide concentration. Manipulation of these factors influences the specificity of hybridization and hence the fluorescence signal detection.

Others Parameters Affecting the Detection and Quantification
Beside the parameters influencing the fluorescence signal detection, specificity and coverage of the probes may also impact on the outcome. In gastrointestinal microbial ecology, 16S rRNA probes mainly have been developed to quantify and monitor the composition of the fecal microbiota. Cross-hybridization with non-target cells and partial coverage of a bacterial group are the main criteria of exclusion in probes selection. So far, more than 700,000 16S rRNA sequences are available from the Ribosomal Database Project website (http://rdp.cme.msu.edu), permitting the in silico development and validation of a large panel of phylogenetic probes. Several probes targeting members of the gut microbiota were designed using the previous version of the Ribosomal Database Project website (Release 8.1). With the increasing number of 16S rRNA sequences available, reassessment of the specificity and coverage of those oligonucleotide probes is essential in order to update and reconfirm their reliability. The use of competitor strategy has enhanced the specificity of several DNA probes, previously and newly designed (Lay et al., 2005b). The approach consists to

combine the "specific" probe with unlabeled oligonucleotides targeting the non-specific targets.

2.3.5 Quantitative Real-Time PCR

2.3.5.1 Principle

Quantitative real-time PCR (qPCR) enables to amplify, detect and quantify a target DNA or RNA. Frequently, real-time PCR is combined with reverse transcription to quantify RNA. A system of detection based on fluorescence signal emission is applied to monitor and quantify the accumulation of amplicons during the reaction. The fluorescence signal emission is proportional to the accumulation of amplicons, therefore the initial amount of DNA or RNA present in a sample can be indirectly quantified by determining the cycle threshold (Ct). The cycle threshold corresponds to the point at which the reaction reaches the fluorescence intensity above the background. The Ct is conversely proportional to the initial amount of DNA or RNA present in the sample. The comparison of the Ct value to a standard curve determined from samples of known concentrations enables to extrapolate the initial quantity of target DNA or RNA. Two fluorescence based technology of amplicons detection are used in quantitative real-time PCR, the fluorescent dye SYBR Green that intercalates with double-stranded DNA and fluorescent reporter probes that fluoresce when hybridized with a complementary DNA.

2.3.5.2 Application in Human Intervention Studies

SYBR Green-based real-time quantitative PCR has been used to detect and quantify different fecal bacterial populations. Matsuki et al. (2004) described the distribution of *Bifidobacterium* in 46 healthy adult volunteers. The *Bifidobacterium catenulatum* group, *Bifidobacterium adolescentis* group, and *Bifidobacterium longum* species were the most predominant populations detected. The authors validated their results with those derived from culture, and FISH combined with fluorescence microscopy. Real-time quantitative PCR was 10–100 times more sensitive than plate-count and FISH methods. The composition of the elderly-like fecal bacterial community was described in healthy volunteers (n = 35), hospitalized patients (n = 38), and hospitalized patients receiving antibiotic treatment for non-intestinal affections (n = 31) (Bartosch et al., 2004).

Compared to the healthy control group, both cohorts of hospitalized patients were characterized by a marked reduction in the *Bacteroides-Prevotella* group. Antibiotic treatment was correlated with an increase of *Enterococcus faecalis*.

Another alternative to the fluorescent dye SYBR Green is the TaqMan hydrolysis probe. TaqMan probes are oligonucleotides designed to hybridize specifically between the forward and reverse primer to an internal region of the target sequence, and are labeled with a reporter fluorophore at the 5' end, and a quencher at the 3' end. A phosphate group is also added at the 3' end to prevent extension of the reporter probe by the Taq polymerase. During the amplification the 5' exonuclease activity of the Taq polymerase degrades the hybridized probe, releasing the reporter from the quencher, and as a result the fluorescence intensity of the reporter dye increases. The sensitivity and specificity of TaqMan probes have been increased by combining at the 3' end conjugated minor groove binders (MGBs), which form extremely stable duplexes with the target DNA. Compared with unmodified DNA probes, MGB probes have higher melting temperature and increased specificity (Ott et al., 2004). PCR primers in combination with TaqMan probes targeting diverse bacterial groups and species have been developed and applied to quantify and monitor the composition of the gut microbiota (Gueimonde et al., 2004; Huijsdens et al., 2002; Ott et al., 2004; Requena et al., 2002).

In a large scale epidemiologic study the composition of the fecal samples of 1,032 infants collected at 1 month of age were analyzed by quantitative real time PCR (Penders et al., 2006). Penders and colleagues examined external factors influencing the composition of the fecal bacterial community in infants, and particularly regarding the *Bifidobacterium* group, *Escherichia coli*, *Clostridium difficile*, *Bacteroides fragilis*, and lactobacilli. Factors influencing the gut microbiota development include mode of delivery, hospitalization, prematurity, antibiotic intake, breast feeding versus infant formula and presence of siblings.

2.3.5.3 Application in Prebiotic Intervention Studies

Haarman and Knol (2005) designed and validated a panel of primers and probes targeting several *Bifidobacterium* species in order to monitor their distribution in infant fecal samples. Three groups of infants participated in this study, formula-fed (n = 10), formula-fed supplemented with galacto-oligosaccharides and fructo-oligosaccharides (n = 10), and a breast-fed control group (n = 10). Fecal samples were collected at the beginning of the study and after six weeks of intervention. An increase in the total number of *Bifidobacterium* was observed in the prebiotic group, and the species diversity was similar to that of the breast fed group. *Bifidobacterium infantis*, *Bifidobacterium breve*, and *Bifidobacterium longum* were the predominant species detected in both groups. At the end of the study, the standard formula group was characterized by a decrease in *Bifidobacterium breve* and an increase in *Bifidobacterium catenulatum* and *Bifidobacterium adolescentis*.

2.3.5.4 Limits

Methodological biases are encountered in all the experimental steps: nucleic acid extraction; PCR (von Wintzingerode et al., 1997).

Biases are also introduced into the analysis with the different methods to express qPCR results using the standard curve.

2.3.6 DNA Microarray

2.3.6.1 Principle

DNA microarray analysis is based on nucleic acid hybridization and relies on the complementary binding of labeled amplicons to complementary nucleic acid probes pre-immobilized onto a surface. Multiple 16S rDNA probes targeting a panel of dominant fecal bacterial species are pre-immobilized onto a surface, and those probes are potentially revealed after hybridization with labeled polynucleotides amplified by PCR from fecal DNA or RNA (cDNA). Hybridization is detected either with a colorimetric- enzymatic reaction or a physical reaction based on the excitation of fluorochrome. The approach is rather qualitative than quantitative, indeed according to the signal intensity generated from the hybridization reaction, only positive or negative detection is revealed.

2.3.6.2 Application in Human Intervention Studies

Wang et al. (2002) developed and validated a nitrocellulose-based membrane-array featuring a panel of 20 triplets of 16S rDNA probes targeting 20 predominant bacterial species of the fecal microbiota. The full length 16S rRNA gene was amplified from fecal samples collected from three volunteers. Amplicons labeled with digoxigenin were hybridized to the membrane-array and colorimetric-enzymatic reaction was used to reveal the hybridization. Results showed that some species were common to all subjects; however each volunteer had its unique bacterial species composition. Following the development of this membrane-array, Wang et al. (2004) designed an oligonucleotide-microarray for the detection of 40 bacterial species. Fecal samples collected from 11 volunteers were screened. 16S rDNA amplicons labeled with a fluorochrome were hybridized to the array of probes pre-immobilized on an epoxy-based matrix. A microarray laser scanner was used to read the fluorescence emission corresponding to a positive signal of hybridization. Inter-individual comparisons revealed that the bacterial species composition is host-specific. The authors observed that 25–37 of the 40 bacterial species could be detected in each fecal sample, and that 33 of the species were found in a majority of the samples.

Palmer et al. (2007) developed and validated a 16S rDNA microarray to describe and monitor qualitatively the bacterial colonization of the infant gut. The authors also investigated the provenance of the colonizers by looking at the bacterial composition of samples (stool, vaginal and breast milk) collected from parents and siblings. The microarray featured 9,121 group and species-level taxonomic probes. Palmer and colleagues tested the performance of their microarray with a set of biological samples. For each individual sample, they compared the 16S rDNA profile generated from the microarray with sequence analysis generated from a clone library. Results obtained from both strategies of 16S rDNA analysis were similar and showed that Bacteroidetes, Firmicutes, Proteobacteria, and Actinobacteria were the main phyla represented in the samples. Fecal samples from 14 healthy babies were collected from birth to the age of one year. 16S rDNA microarray approach was used to profile and monitor the fecal bacterial community of each infant over a 1 year period. On average 26 fecal samples per infant were analyzed. The majority of the bacterial species detected was distributed within the Bacteroidetes, Firmicutes, Proteobacteria, and Actinobacteria. Inter-individual comparisons showed that each infant had its unique collection of fecal bacteria. Interestingly, similar bacterial profiles were observed for a pair of twins included in the study. Intra-individual comparisons revealed a dynamic fecal bacterial community characterized by periods of relative stability punctuated with the occurrence of marked fluctuations in bacterial populations. Similarities in bacterial composition were found between some infant samples and samples of their respective mothers. At the age of one year, the composition of the infant fecal microbiota is predominated by the Bacteroidetes and Firmicutes, a characteristic of the adult-like fecal microbiota.

2.3.6.3   Limits

The approach is qualitative rather than quantitative.

Methodological biases are encountered in all the experimental steps: nucleic acid extraction; PCR; hybridization.

Using universal PCR primers, only dominant bacterial species can be detected.

The coverage is a biodiversity index and its determination reveals that the total number of species constitutive of the gut microbiota is difficult to assess. Therefore, the development of a microarray that allows covering the species diversity of the colonic microbiota is difficult to consider.

2.4   Culture-Independent Genomic Approaches

2.4.1   Metagenomic Approach

2.4.1.1   Principle

The study of the genome of one single organism is called genomics. Metagenomics (also called environmental genomics or community genomics) is a culture-independent approach allowing the analysis of multiple bacterial genomes extracted from environmental samples. Accessing the collection of bacterial genomes belonging to a particular environmental niche (also termed the metagenome) through metagenomic methods allows investigation and characterisation of the genetic potential of the microbial community, and hence inferring its potential activity within the ecosystem. In gastro-intestinal microbial ecology, the term microbiome refers to the total number of genes encoded by the collective genomes of the gut bacterial members. Metagenomics analysis of the gut microbiota has been investigated in several studies. The approach relies on: (1) the isolation of the metagenomic DNA from the bowel sample; (2) cloning of large fragments of bacterial genomic DNA into BAC (bacterial artificial chromosome) or fosmids; and (3) analysis of the metagenomic library. Several approaches can be applied to analyze a metagenomic library. A genetic and functional screening is used when the gene of interest is known. A random approach based on shotgun sequencing is applied when the genetic information encoded by the metagenome is deciphered. Recently, metagenomics has been combined to an emerging high-throughput DNA sequencing strategy termed as pyrosequencing (Margulies

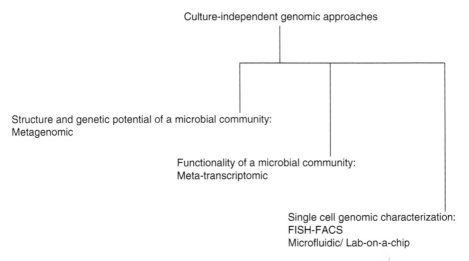

Culture-independent genomic approaches

Structure and genetic potential of a microbial community:
Metagenomic

Functionality of a microbial community:
Meta-transcriptomic

Single cell genomic characterization:
FISH-FACS
Microfluidic/ Lab-on-a-chip

■ Figure 2.4
Culture-independent genomic approaches.

et al., 2005). Pyrosequencing is a clone-free sequencing-by-synthesis technology, which allows sequencing directly circumventing the need for clone library construction (❯ *Figure 2.4*).

2.4.1.2   Application in Human Intervention Studies

Manichanh et al. (2006) developed a metagenomic approach to investigate the fecal bacterial diversity in Crohn's disease patients in remission (n = 6) and in healthy donors (n = 6). Two metagenomic libraries were generated from these two groups of volunteers. The clone libraries were screened for 16S rRNA gene using a DNA hybridization approach. Positives clones carrying 16S rDNA insert were then sequenced and analyzed. Four bacterial phyla, Bacteroidetes, Firmicutes, Actinobacteria and Proteobacteria were described in the fecal microbiome of both groups of volunteers. The authors observed a reduced bacterial diversity within the Firmicutes phylum and particularly within the *Clostridium leptum* subgroup in patients with Crohn's disease. Following this phylogenetic characterization, those metagenomic clones carrying a taxonomic signature were screened *in vitro* for the modulation of epithelial cell growth (Gloux et al., 2007). Gloux and colleagues developed a high throughput method to detect the presence of bacterial molecular signals involved in the stimulatory or inhibitory effects of eukaryote cell proliferation. The authors tested *in vitro* the cell growth using bacterial lysates extracted from the metagenomic clones. Putative clones displaying an effect were then screened by transposon mutagenesis and subcloning in order to map and identify the candidate loci. Ubiquitous genes were a common feature of those modulatory clones.

Gill et al. (2006) described the bacterial and functional genetic diversity of two healthy human fecal microbiomes. Phylogenetic analysis of the metagenomic libraries showed that both human gut microbiomes were dominated by the Firmicutes and the Actinobacteria. Functional annotation of the metagenomic DNA sequences showed that the human gut microbiome is a reservoir of genes encoding different metabolic pathways: metabolism of glycans, amino acids, and xenobiotics, methanogenesis, biosynthesis of vitamins.

Kurokawa et al. (2007) performed a comparative metagenomic analysis of fecal samples collected from 13 healthy individuals of various ages including infants. The phylogenetic distribution of the adult-type microbiome showed three distinct phyla, Bacteroidetes, Firmicutes and Actinobacteria. In the infant-type microbiome, members of the Actinobacteria and Proteobacteria phyla were predominant. Compared to the adult-type one, large inter-individual variations were described in the infant-type microbiome. The predicted functional genes identified in both metagenomic libraries showed that the colonic microbiota uses different adaptive strategies to thrive in the intestinal ecosystem, and establish symbiosis with its host. Interestingly, the authors described a rich reservoir of mobile genetic elements in the microbiome suggesting that the gut is a "hot spot" for horizontal gene transfer between intestinal bacterial members. This metagenomic analysis also revealed that 25% of the total genes identified were orphan.

2.4.1.3 Application in Animal Studies

Walter et al. (2005) prepared a metagenomic library from the large bowel microbiota of mice. The clone library was screened for the expression of β-glucanases. Three positive clones were detected and sequenced. Besides the genes encoding glucanolytic enzymes, other putative genes were annotated on the metagenomic DNA inserts. Predictive functions in nutrient acquisition, host-bacterial interactions and bacterial coaggregation were described. Among the three metagenomic clones, two originated from uncultivated bacteria and one featured sequence similarity with *Bacteroides* species.

In a preliminary study based on 16S rRNA gene sequence analysis, Ley et al. (2005) revealed that genetically obese mice compared to their lean counterparts, had a 50% reduction in the abundance of Bacteroidetes and a proportional increase in Firmicutes. Using the same analytical approach, the authors observed the same microbiota structure in obese versus human volunteers (Ley et al., 2006). In this study, Ley and colleagues monitored the fecal bacterial diversity of obese human volunteers randomly assigned to either a fat-restricted or to a carbohydrate-restricted low calorie-diet over a period of one year. In the obese group regardless the type of diet, the relative abundance of Bacteroidetes increased and the proportion of Firmicutes decreased during the experimental period. Also the increase proportion of Bacteroidetes was correlated to weight loss on the two types of low-calorie diet. In order to determine if the collection of bacterial genomes present in obese subjects had an association with obesity, Turnbaugh et al. (2006) used metagenomic to characterize the genetic potential of the bacterial community present in genetically obese mice. Shotgun sequencing of metagenomic DNA prepared from the caecal contents of obese mice and their lean counterparts, revealed that compared to the lean microbiome, the obese microbiome harbored a higher proportion of Firmicutes and had an increase genetic capacity for fermenting polysaccharides. This genetic potential of the gut microbiome to harvest more energy from the diet was demonstrated in another elegant study. Turnbaugh et al. (2008) created a mouse model of obesity by raising conventional mice on a western diet high in saturated fats, unsaturated fats, and carbohydrates. Using 16S rRNA gene sequencing and metagenomic approaches, the authors observed that conventionalized mice initially on a low-fat diet, when switched to a western diet harbored a distinctive cecal microbiota characterized by a higher relative abundance of Firmicutes and a lower proportion of Bacteroidetes. The shift towards the Firmicutes was due to the expansion of an uncultivated Mollicute lineage. Moreover the authors revealed that the western diet-associated gut microbiome was enriched in the import and fermentation of carbohydrates.

2.4.1.4   Limits

Methodological biases are encountered in all the experimental steps: nucleic acid extraction; cloning; genetic (sequencing) or/and functional screening of the metagenomic library.

Linking a metagenomic sequence to its microbial origin is a difficult task.

2.4.2 Meta-Transcriptomic Approach

2.4.2.1 Principle and Recent Applications in Studying Complex Microbial Communities

An inventory of the microbial genes content is performed when metagenomic is applied to investigate the genetic potential of a microbial community. However, this approach does not reflect what genes are actually being expressed within the community. Meta-transcriptomic is the logical continuation of metagenomics. The approach to study the expression of genes in environmental samples relies on: (1) the isolation of the environmental transcripts; (2) reverse transcription of the transcripts to cDNA; (3) synthesis of double-stranded cDNA; (4) cloning the cDNA; and (5) sequencing and analysis of the cDNA clone library. Such method has been applied to investigate gene expression in marine and fresh water bacterioplankton communities (Poretsky et al., 2005). A recent application of meta-transcriptomic approach to investigate the activity of the gut microbiota was performed by Turnbaugh et al. (2008). The authors built a cDNA clone library from enriched mRNA isolated from the cecum of a mouse model of obesity fed with a western diet high in saturated fats, unsaturated fats, and carbohydrates. Phosphotransferase systems involved in the transport of sugars were expressed in the diet-induced obesity cecal microbiome's transcriptome.

Recently, meta-transcriptomic analysis has been combined to pyrosequencing, a clone-free sequencing technology (Margulies et al., 2005). So far, there have been few studies on the application of meta-transcriptomic combined with pyrosequencing to analyze gene expression in complex microbial communities. Frias-Lopez et al. (2008) developed a method to analyze bacterial gene expression in seawater. The authors used an approach based on polyadenylation-dependent RNA amplification to isolate and convert bacterial messenger RNA into cDNA. The bacterial community cDNA was then sequenced by pyrosequencing and analyzed. Genes involved in photosynthesis, carbon fixation, and nitrogen acquisition, were highly expressed in the ocean bacterial community transcriptome. Frias-Lopez and colleagues observed that half of the bacterial messengers identified were novel.

Urich et al. (2008) applied a meta-transcriptomic approach to simultaneously gain insight on both structure and function of a soil microbial community. Total RNA extracted from a soil sample was used as a template for random-hexamer primed reverse transcription. The generated cDNA was directly subjected to pyrosequencing resulting in 258,411 RNA-tags with an average size of 98 bp. The authors set up a two-step analysis process to identify and differentiate rRNA-tags from mRNA-tags. All RNA-tags were first compared against a compiled rRNA database gathering a small subunit and a large subunit rRNA reference database. In the second step, all unassigned RNA-tags were compared against the Genbank non-redundant protein database to identify mRNA-tags. Seventy five percent of the RNA-tags derived from rRNA, 8% from mRNA, and 17% could not be assigned. The rRNA-tags or ribo-tags were distributed within the three domains of life, Eukaryota, Bacteria and Archaea. Bacterial ribo-tags were the most abundant (85.5%) and were characterized by their wide range of diversity. Among the phyla detected, Actinobacteria, Proteobacteria, Firmicutes, Plancto-mycetes and Acidobacteria were the most encountered prokaryotic groups. The Crenarchaoeta phylum was the predominant archaeal ribo-tag (<2% of the total ribo-tags) identified in the soil sample. Eukaryotal rRNA-tags represented 11% of the total ribo-tags and were affiliated to the Fungi, Viridiplantae, Metazoa and Protist kingdoms. Global functional analysis of the mRNA-tags was compared to metagenomic data derived from the same soil habitat and a farm soil microbial community. Differences in RNA, protein and carbohydrate metabolism were observed between the meta-transcriptome and the metagenomes.

2.4.2.2 Limits

Methodological biases are encountered in all the experimental steps: nucleic acid extraction; reverse transcription; cloning and sequencing or pyrosequencing of the metatranscriptomic library.

2.4.3 From Population Based Analysis to Single Cell Analysis

Metagenomic studies look at the collection of bacterial genomes present in the microbial community. It involves deciphering the genome of an entire commu-nity rather than that of an individual bacterial species. A major issue encountered in metagenomic approach concerns the genome assembly which raises the diffi-culty to assign a functional gene to a bacterial species. Nevertheless, the assign-ment of a metabolic gene of interest is deconvoluted when a phylogenetic marker is located on the same large environmental DNA insert. Sorting and reassembling the environmental bacterial chromosomes which were in first instance truncated and then displayed in a metagenomic library comes to resemble searching for a needle in a haystack. However, the ability to obtain a complete genome sequence from an individual bacterial cell directly isolated from the environment has been demonstrated using single-cell techniques. Fluorescence activated cell sorting and innovative technology based on microfluidics have recently been used to sort or isolate single bacteria cells from complex microbial communities. Once isolated, genome analysis at single-cell level can be carried out by combining whole-genome multiple displacement amplification (MDA) to PCR screening or genome sequencing.

2.4.3.1   Fluorescence Activated Cell Sorting

Stepanauskas and Sieracki (2007) combined high-speed droplet based FACS, whole-genome multiple displacement amplification (MDA), and PCR screening to investigate the marine bacterioplankton. Individual bacterioplankton cells fluorescently labeled and displaying high nucleic acid content were sorted from coastal water sample. Single bacterial cells sorted into 96-well plates were then lysed before performing single-cell whole-genome MDA. Large quantity of genomic DNA was generated from single-MDA reaction, which allowed the authors to use PCR approach to screen the presence or absence of specific phylogenetic and functional markers. Using this approach Stepanauskas and Sieracki were able to detect, identify and analyze multiple metabolic genes in not yet cultivated marine bacteria.

Podar et al. (2007) combined FISH, FACS, MDA and cloning-sequencing to characterize an uncultivated bacterial group of the soil microbial community. Present in low abundance among the soil bacterial community, bacteria belonging to the TM7 phylum were detected by hybridization with a fluorescently labeled DNA specific probe, then analyzed and sorted with a fluorescence activated cell sorter. The authors determined empirically that with five sorted cells, an efficient MDA genomic DNA amplification could be achieved with a low level of non-target bacterial cells contamination. 16S rRNA clone library and TM7 genomic DNA library were constructed from the MDA-amplified genomic DNA. With this approach Podar and colleagues inferred some phylogenetic and metabolism features of bacterial members belonging to the TM7 phylum that have not yet been cultivated.

2.4.3.2   Microfluidic Based Devices

Microfluidics is a multidisciplinary field of research requiring competencies from engineering, physics, chemistry, microtechnology and biology. Microfluidic based technology deals with the behavior, precise control and manipulation of fluids that are geometrically constrained to a small volume, typically sub-millimeter scale. The physical dimensions of bacteria matches with such micron-scale size and therefore rendering possible the experimental manipulation of individual bacterial cells. Recently, microfluidic based devices have been developed to analyze directly complex microbial community at single-cell level without any cultivation step.

Ottesen et al. (2006) used an innovative approach based on microfluidic digital PCR technology to separate and interrogate individual bacterial cells harvested from the hindguts of wood-feeding termites. Diluted gut content samples from termites were mixed with PCR reagents and then loaded onto the microfluidic PCR chips. The microfluidic device displayed an array of thousands of interconnected nanoliter reaction chambers. Micromechanical valves were used to randomly partition single PCR mixture into thousands of independent nanoliter reaction chambers. A flat-block thermocycler was used to carry out the PCR, and the amplification reaction was monitored with TaqMan probes. Fluorescent signal generated during the reaction was detected with a customized microarray laser scanner. Amplicons were then retrieved from the reaction chambers and reamplified off-chip, cloned, sequenced and analyzed. Using this microfluidic PCR chip the authors demonstrated the co-amplification of eubacterial 16S rRNA gene and a metabolic gene of interest from single bacterial cells partitioned in nanoliter reaction chambers. Looking for the identity of bacteria encoding homoacetogenesis, Ottesen and colleagues identified not yet cultivated *Treponema* species as carrier of this fermentation activity.

Marcy et al. (2007) designed a microfluidic device allowing to select, isolate and amplify the whole genome of single bacteria from the human oral microbiota. Biofilm samples were collected from periodontal pockets by scraping subgingival tooth surfaces of a healthy volunteer. Samples were loaded onto the chip which was placed on an optical microscope. The single-cell genome amplification device includes series of interconnected nanoliter reaction chambers. The authors targeted an unexplored bacterial phylum of the oral mouth bacterial community, TM7, and particularly a subset of this phylum represented by long filamentous bacteria with a rod-like morphotype. Using this criterion of selection, the single bacterial cells of interest were optically selected and isolated before undergoing the whole genome amplification process. Sequentially after the on chip optical trapping, the single cell is conveyed to the sorting, lysis, neutralization, and reaction chambers allowing the whole genome amplification reaction to be performed. The chip was placed on a hot plate to carry out the isothermal MDA reaction. The amplified genomic DNA was then recovered from the microfluidic chip and re-amplified off-chip in order to increase the amount of genomic material. 16S rRNA amplification and direct sequencing were performed from the amplified genomic DNA in order to identify the filamentous isolated rod-like cells. Several bacterial cells were affiliated to the TM7 phylum. In order to gain insight into the genome of TM7 bacterial members, high-throughput pyrosequencing and genome assembly were performed on the amplified genomic DNA. Using this approach the authors identified several metabolic pathways within this particular group of bacteria from the human mouth, which has no cultivated members.

2.4.3.3   Limits

Methodological biases are encountered in all the experimental steps: single bacterial cell isolation; whole genome amplification; cloning and sequencing or pyrosequencing.

2.5   Conclusion

This chapter describes the molecular tools that have been applied to investigate the structure and genetic potential of the human gut microbiota, and their applications in human intervention studies. Knowledge on the gut bacterial residents has been gained through a variety of techniques based on 16S rRNA. Beyond the structure of the human gut microbiota, insight into its genetic potential is being deciphered through metagenomic. Once a comprehensive human microbiome metagenomic database is available, the global analysis of transcribed genes expressed by the colonic microbiota will be achievable through meta-transcriptomic. Finally, the possibility to identify which proteins are expressed through meta-proteomic approach, will pave the way to characterize the functionality of the human microbiome (Klaassens et al., 2007). The determination of which genes the gut microbiota is using in its daily activities, or when and in which circumstances those genes are expressed, will allow profiling the functional signatures of the human microbiome in relation to nutrition, health and disease. Understanding how the microbiota and the microbiome respond to disease or specific diet, and how both entities are regulated will bring clues on how we could improve health by manipulating our intestinal microbiota. The bacterial balance of the gut microbiota is not altered under probiotic supplementation, only a transient intestinal passage of the probiotic is detected. Application of "omics" approaches may certainly elucidate the functional effect of probiotic and prebiotic supplementation on the gut bacterial balance.

What could we expect in gastro-intestinal microbial ecology in the future?

The human fecal viral community is largely unexplored. The implication of bacteriophages in the regulation of the structure of the gut bacterial community, and their suspected role in inflammatory bowel diseases interests the research community (Breitbart et al., 2003, 2008; Lepage et al., 2008). Recently, Finkbeiner and colleagues applied metagenomic to characterize the diversity of the viral gut structure in pediatric patients suffering from diarrhea (Finkbeiner et al., 2008).

From bacterial colonies isolation to total population based approaches, that has been the way microbiologists are used to study complex microbial community. We are now moving to a new dimension, towards single cell technology. Deciphering or interrogating the genome of single bacteria directly isolated from the environmental sample without the need for culture will be the new tendency. The possibility to separate and interrogate hundreds of individual bacteria in parallel to investigate for instance the heterogeneity of clonal bacterial population, or perform a metabolic mapping of uncultivated microorganisms will revolutionize microbiology. With the growing interest in lab-on-a-chip technologies development, on chip cultivation of not yet cultivated bacteria might not be unfeasible, as well as the study on chip of simplified model of bacterial community (Balagadde et al., 2005, 2008).

2.6   Summary

- The bacterial diversity of the human intestinal microbiota has been investigated using a battery of techniques, which has highlighted its complexity, uniqueness and stability.
- Consistent data regarding bacterial members belonging to the "healthy gut" population have been gained. Members of the *Eubacterium rectale-Clostridium coccoides*, *Clostridium leptum* are predominant and represent more than half of the total

bacterial population, followed by the *Bacteroides*, *Bifidobacterium* and *Atopobium* groups. These bacterial groups form the main core representative of the microbial collection resident in the human bowel.

● Each individual harbors a specific gut microbiota signature in terms of bacterial composition reflecting a bacterial balance unique to each individual. Not yet identified host-related factors are involved in the determination of the structure of the gut microbiota.

● The bacterial community adhering to the intestinal mucosa is uniformly distributed along the colon but significantly different from the luminal fecal community.

● The dynamic and idiosyncratic colonization of the infant gut is characterized by substantial demographic change occurring during the first year of life. Clues of vertical transmission of maternal gut bacteria to the infant via breast feeding have been demonstrated.

● Temporal stability of the fecal bacterial community is disrupted upon antibiotic administration, nevertheless the fecal microbiota is able to recover its bacterial balance.

● A probiotic supplementation does not alter the dominant bacterial members of the fecal microbiota. However, modulation at the subdominant level is detected through the transient passage of the probiotic strain during the ingestion period. The bacterial balance of the fecal microbiota withstands the long-term establishment of exogenous invaders. This ecological barrier fulfils its entire function from birth.

● Prebiotic effect is characterized by either the proliferation of a particular bacterial population or the stimulation of the metabolic activity of a particular bacterial population.

● The human gut microbiome is a reservoir of genes encoding different metabolic pathways. A "hot spot" for horizontal gene transfer between intestinal bacterial members has also been described within the microbiome.

● Studies with animal models of obesity have shown that the gut microbiome plays important roles in nutrient and energy harvest, moreover a bacterial signature of obese microbiota has been observed.

● The study of complex microbial communities can be performed at single cell level.

List of Abbreviations

BAC	bacterial artificial chromosome
BD	basal diet
Ct	cycle threshold

DGGE	denaturing gradient gel electrophoresis
FISH	Fluorescence *in situ* hybridization
FISH-FC	FISH combined with flow cytometry
FISH-FACS	FISH-Fluorescence activated cell sorting
HMP	human microbiome project
MDA	multiple displacement amplification
MGBs	minor groove binders
NIH	National Institutes of Health
OTUs	operational taxonomic units
PFGE	pulsed-field gel electrophoresis
qPCR	quantitative real-time PCR
RDP	ribosomal database project
RFLP	restriction fragment length polymorphism
scFOS	short-chain fructo-oligosaccharide
TGGE	temporal gradient gel electrophoresis
T-RFLP	terminal restriction fragment length polymorphism
T-RFs	terminal restriction fragments
TTGE	temporal temperature gradient gel electrophoresis

References

Amor KB, Breeuwer P, Verbaarschot P, Rombouts FM, Akkermans ADL, De Vos WM, Abee T (2002) Multiparametric flow cytometry and cell sorting for the assessment of viable, injured, and dead bifidobacterium cells during bile salt stress. Appl Environ Microbiol 68:5209–5216

Apajalahti JH, Kettunen A, Nurminen PH, Jatila H, Holben WE (2003) Selective plating underestimates abundance and shows differential recovery of bifidobacterial species from human feces. Appl Environ Microbiol 69:5731–5735

Balagadde FK, Song H, Ozaki J, Collins CH, Barnet M, Arnold FH, Quake SR, You L (2008) A synthetic Escherichia coli predator-prey ecosystem. Mol Syst Biol 4:187

Balagadde FK, You L, Hansen CL, Arnold FH, Quake SR (2005) Long-term monitoring of bacteria undergoing programmed population control in a microchemostat. Science 309:137–140

Bartosch S, Fite A, Macfarlane GT, Mcmurdo ME (2004) Characterization of bacterial communities in feces from healthy elderly volunteers and hospitalized elderly patients by using real-time PCR and effects of antibiotic treatment on the fecal microbiota. Appl Environ Microbiol 70:3575–3581

Breitbart M, Haynes M, Kelley S, Angly F, Edwards RA, Felts B, Mahaffy JM, Mueller J, Nulton J, Rayhawk S, Rodriguez-Brito B, Salamon P, Rohwer F (2008) Viral diversity and dynamics in an infant gut. Res Microbiol 159:367–373

Breitbart M, Hewson I, Felts B, Mahaffy JM, Nulton J, Salamon P, Rohwer F (2003) Metagenomic analyses of an uncultured viral community from human feces. J Bacteriol 185:6220–6223

De La Cochetiere MF, Durand T, Lepage P, Bourreille A, Galmiche JP, Dore J (2005) Resilience of the dominant human fecal microbiota upon short-course antibiotic challenge. J Clin Microbiol 43:5588–5592

Dicksved J, Floistrup H, Bergstrom A, Rosenquist M, Pershagen G, Scheynius A, Roos S, Alm JS, Engstrand L, Braun-Fahrlander C, Von Mutius E, Jansson JK (2007) Molecular fingerprinting of the fecal microbiota of children raised according to different lifestyles. Appl Environ Microbiol 73:2284–2289

Dinoto A, Suksomcheep A, Ishizuka S, Kimura H, Hanada S, Kamagata Y, Asano K, Tomita F, Yokota A (2006) Modulation of rat cecal microbiota by administration of raffinose and encapsulated Bifidobacterium breve. Appl Environ Microbiol 72:784–792

Eckburg PB, Bik EM, Bernstein CN, Purdom E, Dethlefsen L, Sargent M, Gill SR, Nelson KE, Relman DA (2005) Diversity of the human intestinal microbial flora. Science 308:1635–1638

Fallani M, Rigottier-Gois L, Aguilera M, Bridonneau C, Collignon A, Edwards CA, Corthier G, Dore J (2006) Clostridium difficile and Clostridium perfringens species detected in infant faecal microbiota using 16S rrna targeted probes. J Microbiol Methods 67:150–161

Favier CF, Vaughan EE, De Vos WM, Akkermans AD (2002) Molecular monitoring of succession of bacterial communities in human neonates. Appl Environ Microbiol 68:219–226

Finegold SM, Sutter VL, Mathisen GE (1983) Normal indigenous intestinal flora. In: Hentges DJ (ed) Human intestinal microflora in health and disease. Academic Press, New York.

Finkbeiner SR, Allred AF, Tarr PI, Klein EJ, Kirkwood CD, Wang D (2008) Metagenomic analysis of human diarrhea: viral detection and discovery. PLoS Pathog 4(2):e1000011

Franks AH, Harmsen HJ, Raangs GC, Jansen GJ, Schut F, Welling GW (1998) Variations of bacterial populations in human feces measured by fluorescent in situ hybridization with group-specific 16S rrna-targeted oligonucleotide probes. Appl Environ Microbiol 64:3336–3345

Frias-Lopez J, Shi Y, Tyson GW, Coleman ML, Schuster SC, Chisholm SW, Delong EF (2008) Microbial community gene expression in ocean surface waters. Proc Natl Acad Sci USA 105:3805–3810

Fuchs BM, Glockner FO, Wulf J, Amann R (2000) Unlabeled helper oligonucleotides increase the in situ accessibility to 16S rrna of fluorescently labeled oligonucleotide probes. Appl Environ Microbiol 66:3603–3607

Fuchs BM, Wallner G, Beisker W, Schwippl I, Ludwig W, Amann R (1998) Flow cytometric analysis of the in situ accessibility of Escherichia coli 16S rrna for fluorescently labeled oligonucleotide probes. Appl Environ Microbiol 64:4973–4982

Garrido D, Suau A, Pochart P, Cruchet S, Gotteland M (2005) Modulation of the fecal microbiota by the intake of a Lactobacillus johnsonii La1-containing product in human volunteers. FEMS Microbiol Lett 248:249–256

Gill SR, Pop M, Deboy RT, Eckburg PB, Turnbaugh PJ, Samuel BS, Gordon JI, Relman DA, Fraser-Liggett CM, Nelson KE (2006) Metagenomic analysis of the human distal gut microbiome. Science 312:1355–1359

Gloux K, Leclerc M, Iliozer H, L'haridon R, Manichanh C, Corthier G, Nalin R, Blottiere HM, Dore J (2007) Development of high-throughput phenotyping of metagenomic clones from the human gut microbiome for modulation of eukaryotic cell growth. Appl Environ Microbiol 73:3734–3737

Gueimonde M, Tolkko S, Korpimaki T, Salminen S (2004) New real-time quantitative PCR procedure for quantification of bifidobacteria in human fecal samples. Appl Environ Microbiol 70: 4165–4169

Haarman M, Knol J (2005) Quantitative real-time PCR assays to identify and quantify fecal bifidobacterium species in infants receiving a prebiotic infant formula. Appl Environ Microbiol 71:2318–2324

Harmsen HJM, Elfferich P, Schut F, Welling GW (1999) A 16S rRNA-targeted probe for detection of lactobacilli and enterococci in faecal samples by fluorescent in situ hybridization. Microb Ecol Health Dis 11:3–12

Harmsen HJ, Raangs GC, He T, Degener JE, Welling GW (2002) Extensive set of 16S rrna-based probes for detection of bacteria in human feces. Appl Environ Microbiol 68:2982–2990

Harmsen HJM, Wildeboer-Veloo ACM, Grijpstra J, Knol J, Degener JE, Welling GW (2000) Development of 16S rrna-based probes for the *Coriobacterium* group and the *Atopobium* cluster and their application for enumeration of *Coriobacteriaceae* in human feces from volunteers of different age groups. Appl Environ Microbiol 66:4523–4527

Hayashi H, Sakamoto M, Benno Y (2002) Phylogenetic analysis of the human gut microbiota using 16S rdna clone libraries and strictly anaerobic culture-based methods. Microbiol Immunol 46:535–548

Heilig HG, Zoetendal EG, Vaughan EE, Marteau P, Akkermans AD, De Vos WM (2002) Molecular diversity of Lactobacillus spp. and other lactic acid bacteria in the human intestine as determined by specific amplification of 16S ribosomal DNA. Appl Environ Microbiol 68:114–123

Hooper LV, Bry L, Falk PG, Gordon JI (1998) Host-microbial symbiosis in the mammalian intestine: exploring an internal ecosystem. Bioessays 20:336–343

Hooper LV Gordon JI (2001) Commensal host-bacterial relationships in the gut. Science 292:1115–1118

Hooper LV, Midtvedt T, Gordon JI (2002) How host-microbial interactions shape the nutrient environment of the mammalian intestine. Annu Rev Nutr 22:283–307

Hooper LV, Stappenbeck TS, Hong CV, Gordon JI (2003) Angiogenins: a new class of microbicidal proteins involved in innate immunity. Nat Immunol 4:269–273

Huijsdens XW, Linskens RK, Mak M, Meuwissen SG, Vandenbroucke-Grauls CM, Savelkoul PH (2002) Quantification of bacteria adherent to gastrointestinal mucosa by real-time PCR. J Clin Microbiol 40:4423–4427

Jansen GJ, Wildeboer-Veloo AC, Tonk RH, Franks AH, Welling GW (1999) Development and validation of an automated, microscopy-based method for enumeration of groups of intestinal bacteria. J Microbiol Methods 37:215–221

Jernberg C, Sullivan A, Edlund C, Jansson JK (2005) Monitoring of antibiotic-induced alterations in the human intestinal microflora and detection of probiotic strains by use of terminal restriction fragment length polymorphism. Appl Environ Microbiol 71:501–506

Kimura K, Mccartney AL, Mcconnell MA, Tannock GW (1997) Analysis of fecal populations of bifidobacteria and lactobacilli and investigation of the immunological responses of their human hosts to the predominant strains. Appl Environ Microbiol 63:3394–3398

Klaassens ES, De Vos WM, Vaughan EE (2007) Metaproteomics approach to study the functionality of the microbiota in the human infant gastrointestinal tract. Appl Environ Microbiol 73:1388–1392

Kumar Y, Westram R, Behrens S, Fuchs B, Glockner FO, Amann R, Meier H, Ludwig W (2005) Graphical representation of ribosomal RNA probe accessibility data using ARB software package. BMC Bioinformatics 6:61

Kurokawa K, Itoh T, Kuwahara T, Oshima K, Toh H, Toyoda A, Takami H, Morita H, Sharma VK, Srivastava TP, Taylor TD, Noguchi H, Mori H, Ogura Y, Ehrlich DS,

Itoh K, Takagi T, Sakaki Y, Hayashi T, Hattori M (2007) Comparative metagenomics revealed commonly enriched gene sets in human gut microbiomes. DNA Res 14:169–181

Langendijk PS, Schut F, Jansen GJ, Raangs GC, Kamphuis GR, Wilkinson MH, Welling GW (1995) Quantitative fluorescence in situ hybridization of *Bifidobacterium* spp. with genus-specific 16S rrna-targeted probes and its application in fecal samples. Appl Environ Microbiol 61:3069–3075

Lay C (2004) Caractérisation moléculaire à haut débit de la diversité phylogénétique de la microflore digestive humaine. Ph.D. Thesis. Faculté de Pharmacie, Université Paris XI, Châtenay-Malabry, France

Lay C, Dore J, Rigottier-Gois L (2007) Separation of bacteria of the Clostridium leptum subgroup from the human colonic microbiota by fluorescence-activated cell sorting or group-specific PCR using 16S rrna gene oligonucleotides. FEMS Microbiol Ecol 60:513–520

Lay C, Rigottier-Gois L, Holmstrom K, Rajilic M, Vaughan EE, De Vos WM, Collins MD, Thiel R, Namsolleck P, Blaut M, Dore J (2005a) Colonic microbiota signatures across five northern European countries. Appl Environ Microbiol 71:4153–4155

Lay C, Sutren M, Rochet V, Saunier K, Dore J, Rigottier-Gois L (2005b) Design and validation of 16S rrna probes to enumerate members of the Clostridium leptum subgroup in human faecal microbiota. Environ Microbiol 7:933–946

Lepage P, Colombet J, Marteau P, Sime-Ngando T, Dore J, Leclerc M (2008) Dysbiosis in inflammatory bowel disease: a role for bacteriophages? Gut 57:424–425

Lepage P, Seksik P, Sutren M, De La Cochetiere MF, Jian R, Marteau P, Dore J (2005) Biodiversity of the mucosa-associated microbiota is stable along the distal digestive tract in healthy individuals and patients with IBD. Inflamm Bowel Dis 11:473–480

Ley RE, Backhed F, Turnbaugh P, Lozupone CA, Knight RD, Gordon JI (2005) Obesity alters gut microbial ecology. Proc Natl Acad Sci USA 102:11070–11075

Ley RE, Turnbaugh PJ, Klein S, Gordon JI (2006) Microbial ecology: human gut microbes associated with obesity. Nature 444:1022–1023

Mah KW, Chin VI, Wong WS, Lay C, Tannock GW, Shek LP, Aw MM, Chua KY, Wong HB, Panchalingham A, Lee BW (2007) Effect of a milk formula containing probiotics on the fecal microbiota of asian infants at risk of atopic diseases. Pediatr Res 62:674–679

Manichanh C, Rigottier-Gois L, Bonnaud E, Gloux K, Pelletier E, Frangeul L, Nalin R, Jarrin C, Chardon P, Marteau P, Roca J, Dore J (2006) Reduced diversity of faecal microbiota in Crohn's disease revealed by a metagenomic approach. Gut 55:205–211

Marcy Y, Ouverney C, Bik EM, Losekann T, Ivanova N, Martin HG, Szeto E, Platt D, Hugenholtz P, Relman DA, Quake SR (2007) Dissecting biological "dark matter" with single-cell genetic analysis of rare and uncultivated TM7 microbes from the human mouth. Proc Natl Acad Sci USA 104:11889–11894

Margulies M, Egholm M, Altman WE, Attiya S, Bader JS, Bemben LA, Berka J, Braverman MS, Chen YJ, Chen Z, Dewell SB, Du L, Fierro JM, Gomes XV, Godwin BC, He W, Helgesen S, Ho CH, Irzyk GP, Jando SC, Alenquer ML, Jarvie TP, Jirage KB, Kim JB, Knight JR, Lanza JR, Leamon JH, Lefkowitz SM, Lei M, Li J, Lohman KL, Lu H, Makhijani VB, Mcdade KE, Mckenna MP, Myers EW, Nickerson E, Nobile JR, Plant R, Puc BP, Ronan MT, Roth GT, Sarkis GJ, Simons JF, Simpson JW, Srinivasan M, Tartaro KR, Tomasz A, Vogt KA, Volkmer GA, Wang SH, Wang Y, Weiner MP, Yu P, Begley RF, Rothberg JM (2005) Genome sequencing in microfabricated high-density picolitre reactors. Nature 437:376–380

Martin AP (2002) Phylogenetic approaches for describing and comparing the diversity of microbial communities. Appl Environ Microbiol 68:3673–3682

Martin R, Heilig GH, Zoetendal EG, Smidt H, Rodriguez JM (2007) Diversity of the Lactobacillus group in breast milk and vagina of healthy women and potential role in the colonization of the infant gut. J Appl Microbiol 103: 2638–2644

Matsuki T, Watanabe K, Fujimoto J, Kado Y, Takada T, Matsumoto K, Tanaka R (2004) Quantitative PCR with 16S rrna-gene-targeted species-specific primers for analysis of human intestinal bifidobacteria. Appl Environ Microbiol 70:167–173

Matsuki T, Watanabe K, Fujimoto J, Miyamoto Y, Takada T, Matsumoto K, Oyaizu H, Tanaka R (2002) Development of 16S rrna-gene-targeted group-specific primers for the detection and identification of predominant bacteria in human feces. Appl Environ Microbiol 68:5445–5451

Mccartney AL, Wenzhi W, Tannock GW (1996) Molecular analysis of the composition of the bifidobacterial and lactobacillus microflora of humans. Appl Environ Microbiol 62:4608–4613

Moore WE, Holdeman LV (1974) Human fecal flora: the normal flora of 20 Japanese-Hawaiians. Appl Environ Microbiol 27: 961–979

Moore WE, Moore LH (1995) Intestinal floras of populations that have a high risk of colon cancer. Appl Environ Microbiol 61:3202–3207

Mueller S, Saunier K, Hanisch C, Norin E, Alm L, Midtvedt T, Cresci A, Silvi S, Orpianesi C, Verdenelli MC, Clavel T, Koebnick C, Zunft HJ, Dore J, Blaut M (2006) Differences in fecal microbiota in different European study populations in relation to age, gender, and country: a cross-sectional study. Appl Environ Microbiol 72:1027–1033

Mullard A (2008) Microbiology: the inside story. Nature 453:578–580

Nakanishi Y, Murashima K, Ohara H, Suzuki T, Hayashi H, Sakamoto M, Fukasawa T, Kubota H, Hosono A, Kono T, Kaminogawa S, Benno Y (2006) Increase in terminal restriction fragments of Bacteroidetes-derived 16S rrna genes after administration of short-chain fructooligosaccharides. Appl Environ Microbiol 72:6271–6276

Nielsen DS, Moller PL, Rosenfeldt V, Paerregaard A, Michaelsen KF, Jakobsen M (2003) Case study of the distribution of mucosa-associated Bifidobacterium species, Lactobacillus species, and other lactic acid bacteria in the human colon. Appl Environ Microbiol 69:7545–7548

Ott SJ, Musfeldt M, Ullmann U, Hampe J, Schreiber S (2004) Quantification of intestinal bacterial populations by real-time PCR with a universal primer set and minor groove binder probes: a global approach to the enteric flora. J Clin Microbiol 42:2566–2572

Ottesen EA, Hong JW, Quake SR, Leadbetter JR (2006) Microfluidic digital PCR enables multigene analysis of individual environmental bacteria. Science 314:1464–1467

Palmer C, Bik EM, Digiulio DB, Relman DA, Brown PO (2007) Development of the Human Infant Intestinal Microbiota. PLoS Biol 5(7):e177, 1556–1573

Penders J, Thijs C, Vink C, Stelma FF, Snijders B, Kummeling I, Van Den Brandt PA, Stobberingh EE (2006) Factors influencing the composition of the intestinal microbiota in early infancy. Pediatrics 118:511–521

Perez PF, Dore J, Leclerc M, Levenez F, Benyacoub J, Serrant P, Segura-Roggero I, Schiffrin EJ, Donnet-Hughes A (2007) Bacterial imprinting of the neonatal immune system: lessons from maternal cells? Pediatrics 119:e724–e732

Podar M, Abulencia CB, Walcher M, Hutchison D, Zengler K, Garcia JA, Holland T, Cotton D, Hauser L, Keller M (2007) Targeted access to the genomes of low-abundance organisms in complex microbial communities. Appl Environ Microbiol 73:3205–3214

Poretsky RS, Bano N, Buchan A, Lecleir G, Kleikemper J, Pickering M, Pate WM, Moran MA, Hollibaugh JT (2005) Analysis of microbial gene transcripts in environmental samples. Appl Environ Microbiol 71:4121–4126

Requena T, Burton J, Matsuki T, Munro K, Simon MA, Tanaka R, Watanabe K, Tannock GW (2002) Identification, detection, and enumeration of human bifidobacterium species by PCR targeting the transaldolase gene. Appl Environ Microbiol 68:2420–2427

Rigottier-Gois L, Le Bourhis A-G, Gramet G, Rochet V, Dore J (2003a) Fluorescent hybridisation combined with flow cytometry and hybridisation of total RNA to analyse the composition of microbial communities in human faeces using 16S rrna probes. FEMS Microbiol Ecol 43:237–245

Rigottier-Gois L, Rochet V, Garrec N, Suau A, Dore J (2003b) Enumeration of Bacteroides species in human faeces by fluorescent in situ hybridisation combined with flow cytometry using 16S rrna probes. Syst Appl Microbiol 26:110–118

Rinne M, Kalliomaki M, Salminen S, Isolauri E (2006) Probiotic intervention in the first months of life: short-term effects on gastrointestinal symptoms and long-term effects on gut microbiota. J Pediatr Gastroenterol Nutr 43:200–205

Rochet V, Rigottier-Gois L, Sutren M, Krementscki MN, Andrieux C, Furet JP, Tailliez P, Levenez F, Mogenet A, Bresson JL, Meance S, Cayuela C, Leplingard A, Dore J (2006) Effects of orally administered Lactobacillus casei DN-114 001 on the composition or activities of the dominant faecal microbiota in healthy humans. Br J Nutr 95:421–429

Sakamoto M, Hayashi H, Benno Y (2003) Terminal restriction fragment length polymorphism analysis for human fecal microbiota and its application for analysis of complex bifidobacterial communities. Microbiol Immunol 47:133–142

Satokari RM, Vaughan EE, Akkermans AD, Saarela M, De Vos WM (2001a) Bifidobacterial diversity in human feces detected by genus-specific PCR and denaturing gradient gel electrophoresis. Appl Environ Microbiol 67:504–513

Satokari RM, Vaughan EE, Akkermans AD, Saarela M, De Vos WM (2001b) Polymerase chain reaction and denaturing gradient gel electrophoresis monitoring of fecal bifidobacterium populations in a prebiotic and probiotic feeding trial. Syst Appl Microbiol 24:227–231

Satokari RM, Vaughan EE, Favier CF, Doré J, Edwards C, De Vos WM (2002) Diversity of bifidobacterium and lactobacillus spp. in breast-fed and formula-fed infants as assessed by 16S rdna sequence differences. Microb Ecol Health Dis 14:97–105

Saunier K, Rouge C, Lay C, Rigottier-Gois L, Dore J (2005) Enumeration of bacteria from the Clostridium leptum subgroup in human faecal microbiota using Clep 1156 16S rrna probe in combination with helper and competitor oligonucleotides. Syst Appl Microbiol 28:454–464

Seksik P, Rigottier-Gois L, Gramet G, Sutren M, Pochart P, Marteau P, Jian R, Dore J (2003) Alterations of the dominant faecal bacterial groups in patients with Crohn's disease of the colon. Gut 52:237–242

Stackebrandt E, Goebel BM (1994) Taxonomic note: a place for DNA-DNA reassociation and 16S rrna sequence analysis in the present species definition in bacteriology. Int J Syst Bacteriol 44:846–849

Staley JT, Konopka A (1985) Measurement of in situ activities of nonphotosynthetic microorganisms in aquatic and terrestrial habitats. Annu Rev Microbiol 39:321–346

Stepanauskas R, Sieracki ME (2007) Matching phylogeny and metabolism in the uncultured marine bacteria, one cell at a time. Proc Natl Acad Sci USA 104:9052–9057

Suau A, Bonnet R, Sutren M, Godon JJ, Gibson GR, Collins MD, Doré J (1999) Direct analysis of genes encoding 16S rrna from complex communities reveals many novel molecular species within the human gut. Appl Environ Microbiol 65:4799–4807

Suau A, Rochet V, Sghir A, Gramet G, Brewaeys S, Sutren M, Rigottier-Gois L, Dore J (2001) *Fusobacterium prausnitzii* and related species represent a dominant group within the human fecal flora. Syst Appl Microbiol 24:139–145

Tannock GW, Munro K, Bibiloni R, Simon MA, Hargreaves P, Gopal P, Harmsen H, Welling G (2004) Impact of consumption of oligosaccharide-containing biscuits on the fecal microbiota of humans. Appl Environ Microbiol 70:2129–2136

Tannock GW, Munro K, Harmsen HJ, Welling GW, Smart J, Gopal PK (2000) Analysis of the fecal microflora of human subjects consuming a probiotic product containing *Lactobacillus rhamnosus* DR20. Appl Environ Microbiol 66:2578–2588

Turnbaugh PJ, Backhed F, Fulton L, Gordon JI (2008) Diet-induced obesity is linked to marked but reversible alterations in the mouse distal gut microbiome. Cell Host Microbe 3:213–223

Turnbaugh PJ, Ley RE, Hamady M, Fraser-Liggett CM, Knight R, Gordon JI (2007) The human microbiome project. Nature 449:804–810

Turnbaugh PJ, Ley RE, Mahowald MA, Magrini V, Mardis ER, Gordon JI (2006) An obesity-associated gut microbiome with increased capacity for energy harvest. Nature 444:1027–1031

Urich T, Lanzen A, Qi J, Huson DH, Schleper C, Schuster SC (2008) Simultaneous assessment of soil microbial community structure and function through analysis of the meta-transcriptome. PLoS ONE 3(6): e2527

Vanhoutte T, Huys G, De Brandt E, Swings J (2004) Temporal stability analysis of the microbiota in human feces by denaturing gradient gel electrophoresis using universal and group-specific 16S rrna gene primers. FEMS Microbiol Ecol 48:437–446

Von Wintzingerode F, Gobel UB, Stackebrandt E (1997) Determination of microbial diversity in environmental samples: pitfalls of PCR-based rrna analysis. FEMS Microbiol Rev 21:213–229

Walter J, Hertel C, Tannock GW, Lis CM, Munro K, Hammes WP (2001) Detection of Lactobacillus, Pediococcus, Leuconostoc, and Weissella species in human feces by using group-specific PCR primers and denaturing gradient gel electrophoresis. Appl Environ Microbiol 67:2578–2585

Walter J, Mangold M, Tannock GW (2005) Construction, analysis, and beta-glucanase screening of a bacterial artificial chromosome library from the large-bowel microbiota of mice. Appl Environ Microbiol 71:2347–2354

Wang RF, Beggs ML, Erickson BD, Cerniglia CE (2004) DNA microarray analysis of predominant human intestinal bacteria in fecal samples. Mol Cell Probes 18:223–234

Wang M, Karlsson C, Olsson C, Adlerberth I, Wold AE, Strachan DP, Martricardi PM, Aberg N, Perkin MR, Tripodi S, Coates AR, Hesselmar B, Saalman R, Molin G, Ahrne S (2008) Reduced diversity in the early fecal microbiota of infants with atopic eczema. J Allergy Clin Immunol 121:129–134

Wang RF, Kim SJ, Robertson LH, Cerniglia CE (2002) Development of a membrane-array method for the detection of human intestinal bacteria in fecal samples. Mol Cell Probes 16:341–350

Welling GW, Elfferich P, Raangs GC, Wildeboer-Veloo AC, Jansen GJ, Degener JE (1997) 16S ribosomal RNA-targeted oligonucleotide probes for monitoring of intestinal tract bacteria. Scand J Gastroenterol Suppl 222:17–19

Wilson KH, Blitchington RB (1996) Human colonic biota studied by ribosomal DNA sequence analysis. Appl Environ Microbiol 62:2273–2278

Zoetendal EG, Akkermans AD, De Vos WM (1998) Temperature gradient gel electrophoresis analysis of 16S rrna from human fecal samples reveals stable and host-specific communities of active bacteria. Appl Environ Microbiol 64:3854–3859

Zoetendal EG, Akkermans ADL, Akkermans-Van Vliet WM, De Visser JAGM, De Vos WM (2001) The host genotype affects the bacterial community in the human gastrointestinal tract. Microb Ecol Health Dis 13:129–134

Zoetendal EG, Ben-Amor K, Harmsen HJ, Schut F, Akkermans AD, De Vos WM (2002a) Quantification of uncultured Ruminococcus obeum-like bacteria in human fecal samples by fluorescent in situ hybridization and flow cytometry using 16S rrna-targeted probes. Appl Environ Microbiol 68:4225–4232

Zoetendal EG, Vaughan EE, De Vos WM (2006) A microbial world within us. Mol Microbiol 59:1639–1650

Zoetendal EG, Von Wright A, Vilpponen-Salmela T, Ben-Amor K, Akkermans AD, De Vos WM (2002b) Mucosa-associated bacteria in the human gastrointestinal tract are uniformly distributed along the colon and differ from the community recovered from feces. Appl Environ Microbiol 68:3401–3407

3 Post-Genomics Approaches towards Monitoring Changes within the Microbial Ecology of the Gut

Kieran M. Tuohy · Leticia Abecia · Eddie R. Deaville · Francesca Fava · Annett Klinder · Qing Shen

3.1 Introduction

The human gut microbiota, comprising many hundreds of different microbial species, has closely co-evolved with its human host over the millennia. Diet has been a major driver of this co-evolution, in particular dietary non-digestible carbohydrates. This dietary fraction reaches the colon and becomes available for microbial fermentation, and it is in the colon that the great diversity of gut microorganisms resides. For the vast majority of our evolutionary history humans followed hunter-gatherer life-styles and consumed diets with many times more non-digestible carbohydrates, fiber and whole plant polyphenol rich foods than typical Western style diets today. Adaption of the Western-style diet over the past 50 years may thus have resulted in a significant down-shift or adverse modulation in the fermentative and metabolic output of the colonic microbiota. Short chain fatty acids (SCFA), the major end-products of bacterial carbohydrate fermentation, impact on many critical physiological processes including energy metabolism, lipid and cholesterol levels, carcinogenesis and gene expression. Additionally, the close inter-relationship between bacteria within our gut and the gastrointestinal lymphatic tissue is critical to the optimal functioning of our immune system. Modulation of these microbiota related processes as a result of our modern dietary environment may have played a role in the growing incidence of allergic diseases and chronic diseases like cardiovascular disease, obesity and cancer, often called the diseases of affluence. Recent advances in

© Springer Science+Business Media, LLC 2009

functional foods, particularly prebiotics, offer a means of modulating the gut microbiota and potentially redressing some of the deleterious consequences of modern diet on our gut microbiota and health, both locally within the gut and systemically. There is a growing body of evidence, mainly from animal studies, but also from notable human interventions, that prebiotic dietary supplementation can improve host health. Similarly, prebiotics have been repeatedly shown in human and animal interventions to modulate the relative abundance of bacteria within the gut microbiota, particularly bringing about an increased abundance of bifidobacteria, seen as beneficial microbiota constituents. However, in many cases the underlying mechanisms of effect remain to be determined. Until recently, studies examining the impact of the gut microbiota on human health or the efficacy of microbiota targeted functional foods were greatly limited by the fact that up to 70% of the gut microbiota are resistant to culture under laboratory conditions. Additionally, many gut bacteria are new to science, lacking even closely related representatives within microbial culture collections. This made assigning function to particular intestinal bacteria difficult if not impossible in most cases. Recent advances in post-genomics technologies such as genomics, metagenomics and metabonomics, are allowing researchers to derive the metabolic and microbiological consequences of microbiota modulation in a culture independent manner and link these to physiological biomarkers. These culture independent approaches allow the determination of not just the metabolic potential of the gut microbiota, in terms of the genes comprising the microbiota metagenome, but the metabolic kinetics of the gut microbiota in terms of metabolic fluxes or changes in metabolite profiles upon microbiota modulation. In this chapter, we will discuss how metagenomics and metabonomics are being applied to study the consequences of diet:microbe interactions in the gut and particularly, the microbiological and metabolic consequences of prebiotic ingestion. These technologies hold great promise in elucidating the underlying molecular basis of prebiotic induced health effects.

3.2   From Genomics to Metagenomics

The Human Genome Project (HGP) (Abdellah et al., 2004; Venter et al., 2001) provided a human genetic blueprint which has proved an extremely useful tool in identifying determinants of heritable diseases and providing the possibility of linking particular genotypes to disease risk (Desiere, 2004). During its completion the HGP also gave rise to many of the bioinformatic tools necessary for the emergence of other high resolution, data rich molecular technologies such as transcriptomics, proteomics and metabolomics. Additionally, it created the necessary scientific and commercial interest which has lead to technological leaps in terms of DNA sequencing capabilities. However, it has become apparent that this genetic blueprint does not tell the whole story of our individuality and other factors, environmental factors, play an important role in human health and disease susceptibility. These extra-genomic environmental factors which include diet, xenobiotic compounds (e.g., cooked food mutagens, drugs, environmental chemicals) and our intestinal microbiota closely interact with an individuals genome to regulate gene expression, metabolic pathways and epigenetics, and consequently, impact on an individuals health and disease susceptibility (Johnson et al., 2008; Martin et al., 2007). The advent of high through-put sequencing has enabled the rapid genetic characterization of microbial genomes and whole microbial community genomes or "metagenomes" derived directly from micro-bial consortia in a culture independent manner (Goldberg et al., 2006). Metage-nomics is defined as "the application of modern genomics techniques to the study of communities of microbial organisms directly in their natural environments, bypassing the need for isolation and lab cultivation of individual species" and has provided unparalleled insight into the composition and metabolic potential of microorganisms comprising important communities in the terrestrial and marine environments (Chen and Pachter, 2005; McHardy and Rigoutsos, 2007; Rusch et al., 2007). This approach has now been adopted by a large scale international sequencing project, the Human Microbiome Project, run by the NIH in the United States (see Turnbaugh et al., 2007). This multi-disciplinary project aims to sequence and characterize the gut microbiota giving a global view of the metabolic potential encoded by the gut microbiota. The collaborative sequencing effort will result in the completion of an estimated 1,000 microbial genome sequences as well as metagenomic analysis of microbial communities resident in different regions of the body including the gastrointestinal and urogenital tract, oral and nasal cavities and the skin. This genomic encyclopaedia will revolutionize both our understanding of the human associated microbiota and the way in which human microbiology is carried out in the future, given the enormous reference database of sequence information on both culturable and un-culturable microbiota members that will be generated.

3.3 The Human Gut Microbiota as an Extension of the Human Genome

The human gut microbiota is estimated to comprise up to 1,000 different microbial species which colonize to various degrees and in different ecological niches along the length of the alimentary canal, from mouth to anus. The composition and activity of the gut microbiota differs between the different geographical regions of the gut. Host physiology plays an important role in nurturing these gastrointestinal microbial communities and can also play a regulatory role in bacterial colonization. Gastric acid, rapid flow of digesta, enzymatic and bile secretions all limit the growth of bacteria in the upper gut and it is not until the colon that sizable bacterial populations develop (Conway, 1995; Hayashi et al., 2005; Wang et al., 2005). In the healthy adult, a climax microbial consortium resides within the colon and this microbiota displays a remarkable compositional stability and a high degree of homeostasis and self-regulation which greatly impedes colonization by allochthonous microorganisms ingested by the host (Marchesi and Shanahan, 2007). Additionally, many bacteria, even distantly related species share phenotypic traits and there is a degree of functional redundancy shared between some microbiota components, ensuring for the host a steady supply of metabolites of microbial origin including vitamins and SCFAs for example. This homeostasis also ensures essential functions within the gut microbiota are maintained irrespective of fluctuations in the relative abundance of individual bacterial strains e.g., upon bacteriophage attack (Kurokawa et al., 2007). These activities include fermentative and enzymatic transformation of plant storage and structural carbohydrates, many of which are recalcitrant to degradation and digestion in the upper gut, allowing the host to derive energy from otherwise inaccessible substrates. The gut microbiota also enables host access to plant derived polyphenolic compounds most of which escape digestion in the upper gut of which some possess strong biological activities including antioxidant, phytoeostrogenic, vaso-dilation activities. Important co-metabolic processes occur between host and microbiota, explemified by the entero-hepatic circulation of bile acids (Ley et al., 2008; Nicholson et al., 2005). The gut microbiota is largely fermentative, converting un-digested carbohydrate and protein into short chain fatty acids and there is also a complex array of mutualistic, symbiotic and cross-feeding interactions between bacteria within the gut microbiota in SCFA transformation and metabolism (Duncan et al., 2004; Morrison et al., 2006).

3.4   Limitations of the Microbiological Culture Based Approaches

The extent to which the gut microbiota contributes towards human health is only becoming clear now and until relatively recently the colon was viewed solely as a retention tank for faeces, where water absorption was considered its most impor- tant contribution to systemic biological function. A major contributor towards this lack of understanding and a major hurdle to studying the gut microbiota is the fact that the vast majority of intestinal bacteria are recalcitrant to growth under laboratory conditions and up to 30% of bacterial species within an individual's gut microbiota may be novel phylotypes and new to science (Eckburg et al., 2005; Suau et al., 1999). Therefore there are major limitations to the traditional culture based microbiological approach for studying the gut micro- biota. These include;

- Traditional culture based approaches are limited to monitoring only the culturable minority of the gut microbiota. About 70–80% of intestinal bacteria are considered recalcitrant to cultivation in pure culture, many requiring thus far unidentified growth factors or co-culture with other intestinal microorganisms to fulfil their mutualistic or symbiotic life strategies.

- Many phylogenetically diverse bacteria share phenotypic traits and can grow under the same growth conditions and on the same nutrient media, even in the presence of selective supplements (Tuohy and McCartney, 2006). This necessitates sub-culturing followed by biochemical and genetic characterization before the microorganisms may be phylogenetically positioned; a time consuming and costly process.

- Bacteria show a high degree of genetic plasticity, acquiring genetic determinants through horizontal DNA transfer from even distantly related species and genome duplication followed by divergent evolution within bacterial species is common. This genomic plasticity is driven by environmental selective pressure and often involves transfer of environmentally important genetic traits. Repeated passage of bacterial cultures under the selective pressure of pure culture growth conditions in the laboratory will thus result in loss of traits relevant to the natural environment from which these organisms were isolated (Lee et al., 2008). Environmentally driven genomic instability in bacteria should thus be taken into consideration when con- ducting functional studies on bacteria, such as probiotics, isolated from gastrointes- tinal ecosystems.

Thus, the reasons why we can not culture the majority of gut bacteria include the following factors; the unknown growth requirements of the bacteria, the selectiv- ity of the media that are used, the stress imposed by the cultivation procedures, the necessity of strictly anoxic conditions, and difficulties with simulating the interactions of bacteria with other microbes and host cells. The traditional microbiological culture approach is therefore is not well suited to the enumera- tion of intestinal bacteria or species diversity measures especially for complex microbial consortia, such as the human gut microbiota. Advances in microbial genetics, and the adoption of the 16S rRNA gene as a universal microbial molecular chronometer, have allowed the development of tools which greatly facilitate culture independent microbial ecology studies (Frank and Pace, 2008; Handelsman, 2004; Tuohy and McCartney, 2006). Ribosomal RNAs are excellent molecules for measuring evolutionary relationships among organisms (Olsen et al., 1986). In contrast to traditional taxonomy, which is based on phenotypic traits, this kind of taxonomy reflects natural evolutionary relationships among organisms (Woese et al., 1990). Prokaryotic ribosomes contain two subunits, the sizes being 50S and 30S. The 50S subunit contains about thirty-four proteins as well as 5S rRNA (120 bases), and 23S rRNA (about 2,900 bases), and the 30S subunit contains about twenty-one proteins and 16S rRNA (about 1,500 bases). The 16S rRNA has been the most widely employed molecule to develop the phylogeny of prokaryotes. Within the various regions of 16S rRNA, the degree of conservation differs considerably. Analyses of rRNA sequences have revealed signature sequences, short stretches of rRNA, that are unique to a certain group or groups of organisms enabling the phylogenetic placement and identification of bacteria (Blaut et al., 2002).

Recent large scale sequencing projects have significantly increased the number of 16S rRNA genes archived in public databases such as GenBank, enhancing our understanding of how bacteria are related to one another at the phylogenetic level (Frank and Pace, 2008). However, many of these 16S rRNA gene sequences were derived from direct cloning experiments where environmental DNA is extracted, purified and sequenced either directly or upon generation of clone libraries and as such do not correspond to previously cultured bacteria. Indeed, many phylotypes identified within the gut microbiota are only distantly related to previously cultured bacteria. Where close relatives have been cultivated, comparisons have been drawn between these known and cultured bacteria and close unculturable novel relatives in terms of putative physiology and ecological function. However, such functional inferences are limited since many core metabolic functions appear to be shared between distantly related bacteria whilst traits enabling occupation of a particular ecological niche may be strain specific and may not be shared with close relatives. The gut microbiota in particular appears to be a hot bed of bacterial promiscuity, with many intestinal bacteria showing evidence of high-frequency heterologous DNA transfer. Many of the genes carried by highly transmissible genetic elements within the gut microbiota encode ecologically important functional genes, involved in substrate metabolism, drug resistance or antimicrobial production (Salyers et al., 2004; Tuohy et al., 2002, 2004).

Recent comparative genomic studies on the intestinal microorganism *Bifi-dobacterium longum*, clearly illustrated the significance of DNA transfer within the gut microbiota. This important member of what may be considered the beneficial gut microbiota, and close relative of common probiotic microorganisms, proved to be highly susceptible to deletion and loss of genetic traits upon repeated passage through pure culture. Upon comparative genomic analysis, several DNA regions, encoding intestinally important traits including oligosaccharide and polyol utilization, arsenic resistance and lantibiotic production, unique to the intestinally isolated *B. longum* strain DJO10A appeared to have been deleted from the culture collection strain, *B. longum* NCC2705. The authors were able to demonstrate this loss of intestinally relevant traits upon culturing *B. lonugm* for more than 1,000 generations under laboratory conditions (Lee et al., 2008). Similarly, comparative genomics studies with the important gastrointestinal and nosocomial pathogen, *Clostridium difficile* showed that a core of only 19.7% of genes were shared between 75 clinically relevant strains compared to a *C. difficile* 630 whole genome DNA microarray (Stabler et al., 2006). Thus despite the growing number of novel 16S rRNA species identified, we know little about how intestinal bacteria behave in their own specific ecological niche within the gut.

3.5   Gut Microbiota Community Level Phylogenetic Analysis

Studies on the composition and species richness of the human gut microbiota have been greatly facilitated by the development of molecular techniques and tools based around the phylogenetic information encoded by the bacterial 16S rRNA gene (Tuohy and McCartney, 2006; Zoetendal et al., 2004). The gut microbiota species composition may be determined by directly isolating bacterial DNA from intestinal samples followed by separation and DNA sequencing of the 16S rRNA genes present. Upon phylogenetic positioning, these 16S rRNA gene sequences can be used to generate a picture of the relative abundance of different bacterial species or phylotypes present in the original environmental sample. Using this and similar approaches a detailed picture of the composition of the human gut microbiota has now emerged. Suau et al. (1999) showed that 95% of 16S rRNA gene sequences recovered from a single fecal sample of a healthy adult volunteer fell within one of three phylogenetic groupings: the *Bacteroides* group, the *C. coccoides-E. rectale* group, and the *C. leptum* group. The *Clostridium* groups, *C. leptum* and *C. coccoides-E. rectale* include many bacteria previously described as important members of the gut microbiota including species of *Eubacterium, Ruminococcus, Butyrivibrio* and *Faecalibacterium prausnitzii*. More recently the divisional structure of the gut microbiota at the phylum level has been described using a similar approach with the two most abundant bacterial phyla being the Gram negative *Bacteroidetes* and the Gram positive, low GC% *Firmicutes* (Eckburg et al., 2005; Louis et al., 2007; Wang et al., 2005). Seventy two per cent of the 395 phylotypes detected by Eckburg et al. (2005) belonged to the clostridial group of Firmicute bacteria, most in clusters XIVa (also referred to as the *Clostridium coccoides* group) or IV (*Clostridium leptum*) and the majority of these were either novel, or unrelated to species held in culture collections. Other important though less abundant phyla include the *Proteobacteria, Actinobacteria, Fusobacteria* and the *Verrucomicrobia* phyla (Eckburg et al., 2005; Wang et al., 2005). These lesser phyla none-the-less contain many bacteria important for human health and disease including the enterobacteria and bifidobacteria. More-over, the bacterial community in the stomach and jejunum has been shown to differ from that in the distal ileum, ascending colon and rectum, with the major phyloge-netic groups being similar in the distal ileum and rectum (Bik et al., 2006; Suau et al., 1999; Wang et al., 2005) also showed that a major proportion, up to 70% of the 16S rRNA species present in this fecal microbiota belonged to novel phyloge-netic lineages all be it within the three dominant groupings. Only 24% of clones corresponded to previously identified bacterial phylotypes. More recently, Gill et al. (2006) using a direct high throughput sequencing of metagenomic DNA present in fecal samples collected from two American individuals found that 22.2% of phy-lotypes were novel and 83.3% of phylotypes corresponded to previously uncultured bacterial species. It is likely that as more metagenomic-level studies add 16S rRNA gene sequences onto the databases, the proportion of un-recognized phylotypes within an individual's gut microbiota will fall considerably. However, as discussed above, full community phylogeny does not necessarily translate into an understand-ing of bacterial function or community function and other approaches are needed to link bacterial identity and ecological functioning.

3.6   Community Finger-Printing Techniques (e.g., DGGE)

Fingerprinting techniques have been used to study bacterial communities and appear to be ideal for monitoring community shifts and comparing communities between GI sites and among animals.

Denaturing gradient gel electrophoresis (DGGE) and similar techniques (e.g., temperature gradient gel electrophoresis, TGGE and temporal temperature gradient gel electrophoresis, TTGE) profile or finger-print of the gut microbiota by generating snap-shots of bacterial richness encoded by 16S rRNA gene fragments within environmental samples (Muyzer et al., 1993). Two additional microbial community fingerprinting techniques are single strand conformation polymorphism (SSCP) and terminal-restriction fragment length polymorphism (TRFLP) analyses. DGGE, TGGE, and TTGE are based on sequence-specific melting behavior of amplicons, SSCP on the secondary structure of single stranded DNA, and T-RFLP on specific target sites for restriction enzymes. Double stranded DNA (ds-DNA) melts into single-stranded DNA at different rates across a gradient of denaturant (e.g., urea, or temperature) depending on its DNA sequence. DGGE takes advantage of this differential ds-DNA melting to separate 16S rRNA gene fragments of equal length amplified from microbial communities across a denaturing urea and formamide gradient on a polyacrylamide gel. The resultant pattern of DNA bands on the DGGE gel constitutes a fingerprint of the different 16S rRNA gene fragments amplified from the original sample and thus the species make-up of the sampled microbial community. This approach generates a snap-shot that can then be used to assess changes in species richness over time or in response to different environmental stimuli, such as the presence of growth substrate, antibiotics or xenobiotics. The phylogenetic identity of differential bands can be determined following band excision and sequencing. It has been reported that DGGE or TGGE are sensitive enough to detect bacteria that constitute up to 1% of the total bacterial community. This means that only the most dominant bacteria will be represented. Theoretically, each DGGE band corresponds to a single operational taxonomic unit (OTU), where the total banding pattern is reflective of a community's species richness and diversity (Muyzer et al., 1993). Earlier workers have excised and directly PCR-amplified and sequenced (Ampe et al., 1999; Ovreas et al., 1997) or PCR cloned and sequenced DGGE bands to successfully identify the taxonomic units of interest (Iwamoto et al., 2000; Zwart et al., 1998). Conversely, recent investigators (Abecia et al., 2007; Ercolini, 2004; Ercolini et al., 2003; Kowalchuk et al., 1997; van Beek and Priest, 2002) have reported that band excision and sequencing of DGGE bands might not provide unequivocal identifications as a result of the co-migration of DNA fragments from different taxa to the same positions within DGGE gels.

This approach has been employed to show that individuals harbor unique collections of intestinal bacteria, and that the species composition of the adult gut microbiota is remarkably stable over time (Zoetendal et al., 1998). DGGE has recently been employed to follow the compositional changes that occur within the human gut microbiota upon ingestion of prebiotic functional foods (Tannock et al., 2004); to characterize the mucosa associated microbiota in Crohn's disease patients (Martinez-Medina et al., 2006); to monitor microbiota changes within the infant gut upon dietary intervention (Nielsen et al., 2007); to monitor the impact of antibiotics and probiotic intervention on the gut microbiota (Saarela et al., 2007; Yap et al., 2008) and has shown that the gut microbiota profile presented by the patient before pharmaceutical intervention can predict individual susceptibility to diarrhoea (De La Cochetière et al., 2008).

3.7   Limitations of PCR Based Techniques

Many of molecular microbiology tools rely on efficient extraction of community DNA, PCR and cloning to characterize community 16S rRNA. This approach has a number of limitations, and bias towards recovery of particular 16S rRNA species may be introduced at different stages. Firstly, the method used to isolate bacterial DNA from faeces or mucosal specimens may select for a certain bacteria based on cell wall conformations; Gram positive bacteria by and large are more difficult to disrupt than Gram negative species. PCR is a competitive reaction governed by melting and renaturation efficiencies of the target DNA sequences and therefore PCR based approaches may select for particular pools of 16S rRNA based on PCR amplification efficiency rather than the relative abundance of bacteria and their 16S rRNA genes in the test sample. High GC, Gram positive bacteria, such as the bifidobacteria, consequently are often under-represented in 16S rRNA gene libraries derived from competitive PCR reactions (Farris and Olson, 2007). Primer sequences to amplify bacterial DNA are based on highly conserved regions within the 16S rRNA gene. Even slight differences in the DNA sequence in these primer binding regions might result in bacteria not being detected and information on the highly conserved regions within the 16S rRNA gene derives only from already sequenced species. Finally, cloning of 16S rRNA gene fragments may introduce bias since different cloning vectors and host strains favor certain DNA sequences (Bonnet et al., 2002; Sipos et al., 2007). The limitations of these approaches must be recognized particularly when considering the relative abundance of bacteria within an ecosystem and methods which enable direct visualization and enumeration of phylogenetically related groups of bacteria within environmental samples may be more suited to this task. Hongoh et al. (2003) detected a significant increase in the expected number of phylotypes by lowering the annealing temperature, and a significant decrease in the proportion of clones belonging to the predominant group by raising the number of PCR cycles.

3.8   Molecular Characterization of the Gut Microbiota *in situ*

Fluorescence *in situ* hybridization (FISH) using 16S rRNA targeted oligonucleotide probes provides a direct means of enumerating specific bacterial populations in environmental samples without the need for microbiological culture and in a phylogenetically relevant manner (Amann et al., 1992). Here, bacterial cells are visualized either by epifluorescence microscopy or flow cytometry. The advantage of this approach is that phylogenetically related bacteria ranging from kingdom or phylum levels to species level, depending on probe design, may be enumerated directly in situ. This approach is now commonly applied to enumerate changes in relative abundance of bacteria within the human gut microbiota. As with other molecular approaches it is important to realize the limitations of FISH, which include: variability in permeabilization of target cells and the need for prior 16S rRNA gene information for probe design. FISH does however, allow the estimation of relative bacterial population levels within mixed microbial consortia without the bias introduced by DNA recovery and subsequent PCR and cloning (Amann and Fuchs, 2008). FISH gives an accurate picture of relative abundance of bacteria in mixed consortia because the bacteria are enumerated directly without selective amplification or culture (Amann and Fuchs, 2008). ⊘ *Table 3.1* shows the relative abundances of bacteria within the human gut microbiota as determined by FISH.

Use of flow cytometry allows fluorescence activated cell sorting of bacterial cells. Where 16S rRNA targeted FISH is employed to label bacteria, FACS can be used for the physical separation or sorting of microorganisms according to their phylogenetic groupings from mixed microbial consortia (Ben-Amor et al., 2005; Kalyuzhnaya et al., 2006; Lay et al., 2007). The cell sorted bacterial groups can then be subjected to other molecular procedures to identify bacteria enumerated by the particular FISH probe employed or to investigate the genetic potential of cell sorted bacteria following cloning and DNA sequencing in a culture independent manner. This approach has been used to identify the species make up of viable, injured and dead fractions in human faeces (Ben-Amor et al., 2005; Lay et al., 2007) also used this approach to monitor which bacterial species were

◻ Table 3.1

Adapted from Flint (2006) and Stewart et al. (2006)

Bacterial group	Abundance (typical % of total bacteria)	Fermentation end-products
Phylum Firmicutes		
Clostridial clusters XIV a + b		
Eubacterium rectale – Clostridium coccoides	14.5–33.0	Butyrate, formate, lactate
Eubacterium hallii	0.6–3.8	Butyrate, formate, acetate
Ruminococcus obeum	2.5	Acetate
Lachnospira spp.	3.6	Formate, acetate, lactate, succinate
Clostridial cluster IV		
Clostridium leptum	21.7–26.8	
Faecalibacterium prausnitzii	4.9–20.4	Butyrate, formate, lactate
Ruminococcus bromii, Ruminococcus flavefaciens	1.8–10.2 0.4–1.3	Acetate, formate, lactate, succinate
Clostridium viride	0.5–2.6	Acetate, propionate, butyrate, valerate, ammonia
Eubacterium desmolans	0.1–0.4	Acetate, butyrate
Clostridial cluster IX		
Veillonella spp.	0.9–2.5	Propionate, various minor acids
Clostridial cluster XVI		
Eubacterium cylindroides	0.3–1.7	Butyrate, acetate, lactate, succinate, formate
Phylum Firmicutes		
Lactobacillus/Enterococcus	0.2–2.7	Lactate, acetate
Phylum Actinobacteria		
Bifidobacterium spp.	1.1–5.8	Lactate, acetate, formate
Atopobium spp.	0.8–6.3	Acetate, formate, Lactate
Phylum Bacteroidetes		
Bacteroides-Prevotella gp.	3.9–13.6	Acetate, propionate, succinate
Bacteroides putredinis	0.1–0.8	Acetate, succinate
Bacteroides fragilis	0.4–4.2	Acetate, propionate
Phylum Enterobacteriaceae		
Escherichia coli	0.1–0.2	Lactate, acetate, succinate, formate
Salmonella		
Klebsiella		
Desufovibrio	<5.5	SCFA cross-feeding, H2S production
Kingdom Archaea		
Methanobrevibacter smithii	<0.2	SCFA cross-feeding, CH4 production

Relative abundance of dominant human gut bacterial groups and corresponding main acidic fermentation end-products. The abundance is expressed as mean values of the percentage of total bacteria using FISH, based on data from Lay et al. (2005) and Mueller et al. (2006). The reported fermentation end-products are indicative of cultured representatives enumerated by *C. leptum* group specific FISH probes by carrying out PCR-TTGE (a DGGE type method) on cell sorts of fecal bacteria labeled with the *C. leptum* probes (Clep866-CY5/cp or Fprau645-CY5). The *C. leptum* group, which include important butyrate producing bacteria, constitute one of the dominant groupings within the gut microbiota and covers many different phylogenetic clades. Since the majority of bacteria within the gut are unculturable such culture independent techniques which have the potential for separating out bacterial species without the need for microbiological culture may prove invaluable in assigning ecological function to these dominant and prevalent unculturable moieties.

The prebiotic concept, defined as "non-digestible (by the host) food ingredients that have a beneficial effect through their selective metabolism in the intestine" requires that certain dietary fibers are selectively metabolized by the gut microbiota leading to a concomitant improvement in host health (Gibson and Roberfroid, 1995). The FISH technique has recently been employed to demonstrate that the ability of various foodstuffs and ingredients to bring about a change within the relative abundance of gut bacterial populations seen as beneficial towards human health. The fructans, inulin and oligofructose, galactooligosaccharides, lactulose, resistant starch, whole grain wheat have all been shown to elicit a significant relative increase in numbers of bifidobacteria within the human gastrointestinal tract upon dietary intervention using FISH (Costabile et al., 2008; Depeint et al., 2008). Indeed, true prebiotics can reproducibly be shown to bring about a selective modulation of the gut microbiota, whereby relative abundance of what are termed the beneficial microbiota (currently including bifidobacteria and/or lactobacilli) using either FISH or more traditional culture based approaches. In parallel, as outlined elsewhere in this volume, there is a growing body of evidence from both animal and human studies that prebiotics can bring about improvements in mineral absorption and bone health, improved immune function and relief of allergic and inflammatory disease; protection from gastrointestinal infections including infectious diarrhea and travellers diarrhea; reduced risk of colon cancer and through modulation of blood lipid profiles and energy metabolism, reduce the risk of the metabolic syndrome, type 2 diabetes and coronary vascular disease. However, there is a lack of mechanistic data linking changes within the gut microbiota with biological mechanisms underpinning the observed health effects. The challenge for the prebiotics field of research in the coming years is to bridge this gap and link prebiotic induced modulations within the gut microbiota with specific physiological processes responsible for improved health or protection from disease.

3.9   Culture-Independent Functional Characterization of the Gut Microbiota's Metabolic Potential

The 16S rRNA based molecular tools mentioned above have contributed to a picture of the gut microbiota as being made up of mainly unculturable bacteria, with many novel species. This approach has also contributed greatly to the extensive 16S rRNA gene database used for bacterial phylogeny and taxonomy and has proved invaluable in characterizing species make up and composition of the gut microbiota. However, knowing the phylogeny of a microorganism may not necessarily shed light on its ecological function. These issues highlight an important problem, how to assign ecological function to novel or unculturable bacteria? Recent advances in high throughput sequencing allow the rapid genome sequencing and more over rapid and accurate compilation of community meta-genomes (Committee on Metagenomics, 2007). Sequence based metagenomics applies shot-gun sequencing of total DNA extracted from an environmental sample and was originally pioneered in marine and terrestrial environments (Gill et al., 2006; Kurokawa et al., 2007; Turnbaugh et al., 2007). This approach has also recently been applied to study the human gut microbiota, e.g., the Human Microbiome Project described above. The functional or metabolic potential of the human gut microbiota may thus be accessed through these large scale sequencing projects targeting not just community 16S rRNA genes, but potentially all bacterial genes present.

Gill et al. (2006) examined the fecal metagenome of two unrelated healthy American adults. They found that these metagenomes were enriched for genes involved in the metabolism of carbohydrates, amino acids and xenobiotics, methanogenesis and the biosynthesis of vitamins and isoprenoids, compared to the human genome itself and other bacterial communities. They also pointed out that the fecal metagenome encoded many functions not represented in the human genome, in particular, genes involved in the metabolism of major plant structural and storage polysaccharides such as xylans, pectins, arabino-containing carbohydrates and fructans. The abundance of such genes within the gut micro-biome and their rarity within the human genome is of key importance to the overall metabolic capacity encoded in the combined human-microbiota metagenome. Some 81 different glycosyl hydrolase families as well as key genes involved in the production of short chain fatty acids (SCFA, acetate, butyrate, propionate and succinate) could be assigned to the gut microbiota high-lighting the fact that carbohydrate fermentation appears to be a major energy source in the colon, driving the energy economy within the microbiota, building cross-feeding interrelationships and providing about 50% of the daily energy requirements of the gut wall. In a second metagenomics study, Kurokawa et al. (2007) compared the metagenomes of 13 healthy Japanese individuals of different age, with two American fecal metagenomes (Gill et al., 2006). They found that genes involved in carbohydrate metabolism and transport were enriched in all 13 metagenomes, compared to a database constructed from COG (Cluster of Orthologous Groups) assigned genome sequences of 243 different bacteria not commonly found within human gut microbiota. Genes involved in lipid metabolism however, were under-represented in the gut metagenomes compared to this reference database. These observations support the notion that the human gut microbiota has evolved along-side its human host to complement human encoded functions and allows the host access to dietary nutrients not digested or absorbed in the upper gut like complex plant storage polysaccharides. SCFA derived from microbial fermentation in the colon contribute about 10% of our daily energy intake, and play important physiological roles including regulating cellular differentiation and proliferation in the gut wall, cholesterol synthesis and *de novo* lipogenesis in the liver, and may act as an energy source systemically (Macfarlane and Gibson, 1997). Similarly, 90% of plant derived polyphenolic compounds reach the colon where they are transformed into biologically active and available intermediates by the resident microbiota (Spencer et al., 2001). This interaction between our intestinal microbiota and plant foods, both carbohydrate and polyphenolic moieties has important implications for human health and disease and in devising optimal human diet as well as functional food design. For much of our evolutionary history, humans followed a hunter-gatherer life strategy and consumed diets rich in whole plant foods particularly high in dietary fiber and polyphenols. Whole plant foods (fruits, vegetables, whole grain cereals) and dietary fiber are recognized as beneficial for human health and in epidemiological studies these foods are often inversely related to risk of chronic human diseases including coronary vascular disease and cancer. Both for present day humans and for our prehistoric ancestors the maintenance of a stable, fermentative gut microbiota would have been essential to maximize energy and nutrient recovery as well as non-nutrient functional benefit from our diet (Cassidy, 2006; Pompei et al., 2007). However, our modern, Western style diets, low in fiber and whole plant foods, and high in refined sugars, protein and saturated fat, appear to be out of step with our co-evolved hunter-gatherer intestinal microbiota. It is not surprising that there appears to be a strong link between adoption of this Western style diet and increased incidence of certain cancers, coronary vascular disease, obesity and the metabolic syndrome, and immunological diseases, like inflammatory bowel disease and allergy, in which the gut microbiota have been proposed to play an aetiological role and for which certain whole plant foods are proving protective. Modulating our modern gut microbiota through dietary supplementation with Prebiotics in particular can mediate positive health effects on a number of important physiological functions including *de novo* lipogenesis Beylot (2005), mineral absorption (Abrams et al., 2005; Holloway et al., 2007), regulation of satiety and body fat deposition (Delzenne et al., 2007; So et al., 2007) and importantly, providing butyrate as an energy supply to the colonic mucosa (Pool-Zobel and Sauer, 2007).

One of the more surprising insights provided by recent metagonomic studies has been the observation that the gut microbiota may be involved in the aetiology of obesity and that the composition of the gut microbiota is fundamentally different in the obese compared to the lean individuals at the phylum level. There is currently an epidemic of obesity which appears to develop at the population level upon adoption of a Western-style diet. In the UK there has been a sharp increase in the incidence of obesity over the past 15 years and it has been estimated that by 2,050 60% men in the UK will be obese Foresight (2007). There is a genetic component to obesity with over 600 genes reported to play a role in energy metabolism and body weight, and there is a sub-population genetically predisposed to obesity (Wardle et al., 2008). However, the sudden increase in the incidence of obesity since the 1980s has occurred at a rate which far outstrips that of human genomic evolution, showing that the obesogenic environment has a major role to play. This obesogenic environment impacts on the quantity of food we eat, the types of foods we eat, satiety, energy recovery from the diet, epigenetic programming, mental state and exercise, all of which play important roles in determining the risk of obesity. There are strong epidemiological data linking diets high in fat and refined carbohydrates with obesity (Johnson et al., 2008). Conversely, diets rich in fiber and whole plant foods are inversely associated with obesity (Astrup et al., 2008). Gordon and co-workers have recently shown that the gut microbiota differs between obese and lean individuals at the phylum level in both animal models of obesity and in human subjects (Ley et al., 2005, 2006; Turnbaugh et al., 2006). These authors report that obesity is associated with a reduced abundance of intestinal Bacteroidetes and increased abundance of Firmicutes in the genetic *ob/ob* mouse model of obesity and in humans. This altered microbiota is associated with increased fermentation end products in caecal and fecal samples, and differences in the metabolic potential of the gut microbiota with an increased energy harvest from the diet (Ley et al., 2005, 2006). Interestingly, obesity appears to be transferred concomitantly with this obese type microbiota and when germ-free animals are associated with caecal contents of obese mice they too become obese despite despite eating less food (Turnbaugh et al., 2006, 2008). However, it also appears that diet has a strong influence on the composition and activity of the obese-associated microbiota. It appears that the gut microbiota of obese humans who lose weight after one year on either a low carbohydrate or a low fat diet returns to a lean type profile (Ley et al., 2006). Similarly, there is some evidence from animal studies that certain fibers, particularly prebiotics may reduce the risk of obesity itself and its associated pathologies. Cani et al. (2007a) showed that when conventional mice are placed on a high fat diet obesity and insulin resistance may be induced by increased plasma levels of the highly inflammatory Gram negative bacterial wall fragment, lipopolysaccharide. They observed a die-off in the intestinal microbiota in high fat fed animals. This high-fat diet with low fermentable carbohydrate may have contributed towards increased availability or though depleted mucosal barrier function have contributed to increased uptake of intestinal LPS thus providing the immunological trigger for the chronic low grade inflammation characteristic of obesity and predisposing to insulin resistance. The same authors later found that dietary supplementation with the prebiotic oligofructose, in these same high-fat fed animals, resulted in reduced plasma LPS, reduced inflammation and improved insulin sensitivity. These physiological changes were strongly associated with an increased abundance of intestinal bifidobacteria in these animals (Cani et al., 2007b). Further studies are required to investigate in more depth any apparent aetiological role of the obese-type microbiota in body weight gain and conversely, in the ability of different diets to induce an obese-type microbiota or reduce the risk of becoming obese by modulating gut microbiota composition and activity through dietary interventions for example with prebiotics.

The sequence based metagonomics is a powerful tool for measuring the metabolic potential of the gut microbiota and recent high level investment in metagenomic studies of the gut microbiota will generate a wealth of sequence data for data-mining and the design of future gut microbial ecology studies. Identifying the genetically encoded functions which are enriched or under-represented within a microbial community can shed light on the ecological role of that community as a whole (Turnbaugh et al., 2007). Additionally, identifying ecologically sensitive genes or sets of genes which enable a particular strain to successfully colonize an ecological niche within the gut will greatly enrich our understanding of how bacteria interact with each other, our diet and their human host to mediate either beneficial health effects or induce disease. A good place to start may be to characterize the composition and commonality of ecologically important genes constituting the gut microbiota mobile metagenome, encoded by transmissible plasmids, transposons and bacteriophages, which is becoming recognized as an important determinant of strain identity and ecological function within the gut (Lee et al., 2008). However, as observed by Kurokawa et al. (2007) not all open reading frames (ORFs) identified in metagenomic studies can be annotated to known function by comparisons with reference databases. In the existing gut microbiota metagenomes between 45% and 80% of protein-coding genes observed in the 13 Japanese (Kurokawa et al., 2007) and two American (Gill et al., 2006) fecal metagenomes respectively could not be assigned a metabolic function at the 90% threshold identity upon BLASTP analysis against their reference database of 243 non-gut bacterial genomes. Another limitation of the metagonomic approach is that it generates data on the metabolic potential of a microbial ecosystem, the potential encoded by the genotype of the organisms present which is only translated into metabolic kinetics or phenotype in response to particular environmental stimuli. Another post-genomics approach has recently emerged with the potential to measure the metabolic kinetics or community phenotype of the gut microbiota and possibly relate these changes in microbial derived metabolites to particular intestinal bacteria irrespective of whether they can be cultured under laboratory conditions or whether they are new to science.

3.10   Measuring the Metabolic Kinetics of the Human Gut Microbiota Through Metabonomics

Metabonomics is defined as "a systems approach to examining the changes in hundreds or thousands of low-molecular-weight metabolites in an intact tissue or biofluid" (Nicholson et al., 2005). The human metabonome thus comprises; the metabolites derived from human genome encoded determinants, metabolites of derived from the human microbiota, and the flux in these combined metabolite profiles under different environmental perturbations e.g., interactions with diet, pharmaceuticals, carriage of parasites, and chronic disorders like cardiovascular disease and cancer. This technology employs ^{1}H-nuclear magnetic resonance (NMR) spectrometry and mass spectrometry-based techniques to profile metabolites in biofluids like urine, plasma and fecal water, generating a picture of the metabolic kinetics of an organism at a particular point in time. By applying image analysis followed by multivariate statistics, metabonomics is currently being used to track changes in these metabolite profiles in response to dietary modulation, including prebiotic, or pharmaceutical interventions, and to generate distinctive metabolite profiles in disease states such as inflammatory bowel disease, bowel cancer and cardiovascular disease. Such studies offer the possibility of developing new diagnostic tools or novel therapeutic targets and importantly deliver tools with the breath and resolution to derive functional data on the behavior of the gut microbiota in situ and in response to dietary change (Marchesi et al., 2007; Martin et al., 2008; Holmes et al., 2008; Solanky et al., 2005). The metagenomic studies described above illustrate clearly the degree of co-operation between the human genome and the gut microbiota and that the gut microbiota has co-evolved alongside mammals over time (Ley et al., 2008; Nicholson et al., 2005) also put forward the concept of the gut microbiota co-evolving and metabolically complementing the human genome resulting in a close symbiotic relationship involved in co-metabolism of a range of dietary and xenobiotic compounds. These authors suggested that gut microbiota structure and composition reflects this symbiotic relationship with only the bacterial populations beneficial to the host predominating in the human microbiota in health and contributing to co-metabolic activities. Of course this symbiotic relationship is impacted by other life-style influences such as diet, exercise, stress and drug or xenobiotic intake. Recognizing the contribution of microbiota derived compounds observable in metabolite profiles of human biofluids to metabolic processes at the whole organism level, Nicholson et al. (2005) combined high resolution analytical techniques with image analysis and multivariate statistics to establish the meta-bonomics concept.

Recent metabonomics studies have highlighted the potential contribution of microbiotal derived metabolites or co-metabolites in the aetiology of chronic human diseases. Dumas et al. (2006) found that in mice genetically predisposed to impaired glucose homeostasis and non-alcoholic fatty liver disease (NAFLD) maintained on a high fat diet, microbial activities within the gut lead to reduced choline bioavailability. This mimics choline-deficient diets already known to induce NAFLD and insulin resistance (IR), key initial steps in the development of the metabolic syndrome and obesity. ^{1}H-NMR metabolite profiling using principle component analysis was able to identify low circulating plasma phospha-tidylcholine and high urinary excretion levels of methylamines (dimethylamine, trimethylamine and trimethylamine-N-oxide), gut microbe:host co-metabolites, as characteristic of high-fat fed animals with NAFLD and impaired glucose metab-olism. Holmes et al. (2008) later illustrated the power of the metabonomics approach in grouping individuals according to their urine metabolite profiles at the population level, and relating these profiles to geography, diet and disease risk. These authors showed that urinary formate (and, to a lesser extent, hippurate) was inversely related to blood pressure (BP) and coronary vascular disease (CVD) risk. Again ^{1}H-NMR was used to generate urinary metabolite profiles of 4,630 human volunteers in the UK, the USA, China and Japan. Individuals could be separated in a blind manner according to geography with East Asians separating from UK/USA populations; Japanese living in Japan separating from Japanese living in the USA, and populations of northern (Guangxi) and Southern (Beijing and Shanxi) China showing distinct urinary metabolite profiles. The authors reported that urinary profiles of people in the UK and USA were similar. The main differentiating metabolites between the populations were of dietary origin, including amino acids, creatine and trimethylamine-N-oxide; acetylcarnitine, tricarboxylic acid cycle intermediates involved in energy metabolism and dicarboxylic acids like suberate. Holmes et al. (2008) also identified a group of microbial compounds or which derive from the gut microbiota:host co-metabolic processing including hippurate, phenylacetylglutamine, methylamines and formate. Formate can either be formed from endogenous one-carbon metabolism or upon fermentation of non-digestible carbohydrates in the colon by diverse bacteria including certain clostridia (mainly belonging to clusters XVI, IV, XIV), the *Actinobacteria* including *Bifidobacterium* species, and to a lesser extent, the *Proteobacteria*, see ❷ *Table 3.1*. Hippurate, an end product of aromatic (benzoic acid) co-metabolism and a likely end product of polyphenolic metabolism by the host and gut microbiota, was also inversely related to BP and positively correlated with fiber intake, while high BP was associated with diets high in animal protein and the urinary metabolite, aniline (❷ *Table 3.2*).

The same group have also recently shown that the close relationship between mammals and their intestinal microbiota extends through-out the life-span. Distinct ^{1}H-HMR urinary metabolite profiles were found in dogs in response to dietary change (a calorie restricted diet compared to normal chow) and at different ages in this longitudinal study. Urinary metabolite profiles shifted rapidly before age 1 year (early life) after which the metabolic signature stabilized between 1 and 2 years of age. This change in metabolite profiles in to first 12 months of life corresponds to the emergence and successive development of the gut microbiota and the dietary change upon weaning. A second metabolic shift was observed in middle-age (years 5–9) before profiles again underwent a metabolic transformation in old age after about 10 years. Many of the differentiative metabolites had their origins in the microbiota:host co-metabolism highlighting the role of

☐ Table 3.2

Estimated daily fiber intake in palaeolithic diet and modern diet

Dietary pattern	Fiber content (g)	Reference
Palaeolithic diet modified in 1997 (50% meat, 50% vegetables)	104	Eaton et al. (1997)
Rural Chinese diet	77	Campbell and Chen (1994)
Rural African diet	60–120	Dunitz (1983)
Current US diet	12–18	Institute of Medicine (2002)
Recommended fiber content in US	20–35	Institute of Medicine (2002)
Current UK diet	12	British Nutrition Foundation (2004)
Recommended fiber content in UK	18 (minimum)	British Nutrition Foundation (2004)

the gut microbiota in the ageing process, from successive development of the gut microbiota in puppies to modulation of the gut microbiota in middle and old age, times often associated with the onset of chronic disease (Wang et al., 2007).

3.11 Metabonomics and Disease States (IBD and Colon Cancer)

Metabonomics has been applied to investigate changes in metabolic profiles in order to identify mechanisms involved in certain diseases. Metabonomic analysis of either fecal extracts (Marchesi et al., 2007) or colonic mucosa (Balasubramanian et al., 2008) in patients with active inflammatory bowel disease (IBD) both showed a reduction in SCFA in these patients compared to control individuals, in particular in acetate but also in butyrate. Balasubramanian et al. found however, that in patients in remission the values for these metabolites were similar to control. They also reported that the concentration of formate was significantly lower in patients with active ulcerative colitis (UC) compared to patients with active Crohn's disease (CD) and that this difference may serve as a biomarker for the distinction between active UC and CD. It is important to accurately diagnose IBD at an early stage as a correct differentiation between CD and UC defines treatment and prognosis.

The above metabolites differentiating UC and CD include microbial meta-bolites originating from the fermentation of carbohydrates by the gut microbiota highlighting the importance of the microbiota in the aetiology of IBD. Similar findings have been observed in studies looking at blood metabolites in different mouse models of colitis (Chen et al., 2008). The metabonomics analysis revealed increased levels of stearoyl lysophosphatidylcholine and lower levels of oleoyl lysophosphatidylcholine in blood which the authors traced to an inhibition of stearoyl-CoA desaturase 1 (SCD1) expression in the liver. As this inhibition did not only occur in a dextran sulphate sodium (DSS)-induced colitis model but also in *Citrobacter rodentium*-induced colitis Chen et al. concluded that the observed inhibition of SCD1 is highly likely to be due to the disruption of the intestinal microbiota and the resulting inflammation. Furthermore Marchesi et al. reported higher quantities of amino acids lysine, leucine, isoleucine, valine and alanine – products of bacterial protein metabolism – in faeces of CD patients compared to controls.

Although these studies identified metabolites of microbial origin as playing important physiological roles at the whole body or system level, they did not attempt to link these metabolite profiles with specific bacteria within the gut microbiota.

A study in colon cancer and polypectomized patients attempted to attribute changes in metabolic profiles to changes in bacterial diversity by combining DGGE and metabonomics analysis of fecal water (Scanlan et al., 2008). The authors reported a significantly increased diversity in the *Clostridium leptum* and the *Clostridium coccoides* subgroups as well as relatively higher levels of amino acids such as valine, leucine, isoleucine, glutamate and tyrosine and lower levels of methylamine in fecal water of colon cancer and polypectomized patients compared to control individuals. The altered amino acid profile together with the increased diversity may suggest a higher incidence of potentially detri-mental species of clostridia.

The modulating effect of microorganisms on systemic metabolite profiles (blood, jejunal wall and longitudinal mysenteric muscle tissue) was also confirmed by infection with *Trichinella spiralis* in NIH Swiss mice which subse-quently caused post-infective irritable bowel syndrome (IBS) – an intestinal disorder characterized by abdominal pain, vomiting and either diarrhoea or constipation (Martin et al., 2006). The metabonomic signature of the *T. spira-lis*-infected mice revealed an increased energy metabolism, fat mobilisation and a disruption of amino acid metabolism as well as muscular hypertrophy. The treatment of the infected mice with probiotic *Lactobacillus paracasei* resulted in metabolic profiles closer to those of uninfected mice indicating a partial normalisation of the muscular activity and the disordered energy metabolism.

3.12   Measuring the Impact of Microbiota Modulation Using Metabonomics

As described above, the gut microbiota interacts intimately with host metabolism to mediate health and disease. These interactions are complex, often involving multiple and interconnected metabolic pathways, which makes a classical approach whereby one or a few metabolites are monitored impractical for investigating microbe:host metabolic interactions. Recently, metabonomics has been employed to measure the consequences of microbiota modulation. Yap et al. (2008) investigated the impact of the broad spectrum glycopeptide antibiotic, vancomycin, on the gut microbiota and the metabonome of female mice. Vancomycin was chosen because it is active against Gram positive bacteria and is poorly absorbed across the gut wall which means it will reach the large bowel. Vancomycin induced changes in the composition of the gut microbiota as determined by 16S rRNA targeted PCR-DGGE was reflected in changes in metabolite profiles in faeces and urine. Vancomycin intervention had a dramatic impact on phenolic regions of the NMR spectrum, with reduced levels of urinary hippurate and phenylacetylglycine which are produced through microbiota:host co-metabolic pathways. Although these changes in urine metabolites appeared to be transitory, particular metabolites took longer to return to pre-treatment levels, with hippurate in particular only returning to pre-vancomycin levels 19 days after the vancomycin intervention. Microbiotal choline metabolism also appeared to be disrupted by vancomycin treatment. Reduced concentrations of trimethylamine (TMA) and trimethylamine-N-oxide (TMAO), gut microbial and hepatic detoxification end products of choline metabolism respectively, were observed in urine post-vancomycin treatment. Vancomycin intervention had a dramatic effect on carbohydrate fermentation, a key functional activity of the gut microbiota, with reduced levels of acetate, propionate and n-butyrate and elevated fecal oligosaccharide concentrations. Considering the important and diverse biological roles of these SCFA in the host, antibiotic disruption of this key microbiotal function may have a significant impact on host health. Such an impact on carbohydrate fermentation and SCFA production may have been expected considering the key roles played by Gram positive bacteria like the bifidobacteria and species belonging to the *C. leptum* and *C. coccoides* groups in polysaccharide and oligosaccharide fermentation and in cross-feeding on SCFA. Interestingly, although alterations in fermentation end products lasted 13 days post-vancomycin treatment, populations of *C. leptum* and *C. coccoides* appeared to recover from day 2 onwards, indicating that disruption of other bacterial groups present may have had a more dramatic impact on SCFA production than bacteria belonging to these clostridial groupings. Vancomycin intervention also disrupted protein handling by the gut microbiota with elevated levels of amino acids in faeces and reduced levels of creatine and α-ketoisocaproate in urine.

Using a defined microbiota animal model of the infant gut microbiota, Martin et al. (2008) recently described the impact of probiotics (*Lactobacillus paracasei* and *L. rhamnosus*) and prebiotics (two different GOS preparations) on the gut microbiota and host metabonome. This simplified animal model of the human infant gut microbota comprised ex-germ-free animals colonized with strains of *E. coli, B. breve, B. longum, Staphylococcus epidermis, S. aureus, C. perfringens* and *Bacteroides distasonis* isolated from a healthy twenty day old breast fed human infant. The authors showed distinct metabolite profiles in urine, plasma, fecal extracts and intact liver tissue upon prebiotic induced microbiota modulation using NMR based metabonomics. Supplementation with either prebiotic resulted in elevated population levels of the bifidobacterial strains present, *B. breve* and *B. longum* and reduced levels of *C. perfringens*. When given in combination with *L. paracasei*, reduced numbers of *Bacteroides distasonis* were observed, while reduced numbers of fecal *E. coli* were observed in animals dosed with prebiotic and *L. rhamnosus*. Prebiotic treatment appeared to reduce bacterial proteolysis, with lower concentrations of lysine observed in faeces, isobutyrate in the caecum and *N*-acetyl-glycoproteins in urine. Changes were also observed in choline metabolism upon prebiotic intervention. Co-metabolic processes in the metabolism of dietary choline have been previously shown to impact on insulin resistance, non-alcoholic fatty liver disease and type 2 diabetes in animal models. Both TMA and TMAO concentrations were altered in the urine and liver respectively, indicating prebiotic induced changes in choline metabolism by the gut microbiota. Prebiotic intervention impacted significantly on levels of lipids stored in the liver, with reduced triglycerides and increased concentrations of polyunsaturated fatty acids. Similarly, Prebiotic intervention resulted in increased hepatic glutamate, gutamine, branched-chain amine acids and alanine, and when mice were dosed with prebiotic plus *L. paracasei*, increased hepatic glycogen was observed indicating a stimulation of gluconeogenesis and glycogenesis. Prebiotic intervention also appeared to stimulate animal energy expenditure as indicated by increased levels of taurine and creatine in urine post-prebiotic supplementation which derive from increased muscular activity. Changes in SCFA within the gastrointestinal tract were measured by gas chromatography and correlated with prebiotic induced alterations in microbiota population levels. In general there was a negative association between SCFA concentrations and population levels of *C. perfringens* and *E. coli* and a positive association between SCFA and bifidobacteria, the lactobacilli and *Bacteroides diastasonis*, corresponding with perceived prebiotic modes of action within the gut microbiota. However, it is difficult to extrapolate these data to the human situation since there are a number of major differences between these model microbiota systems and humans. Principally, in the lack of microbiota complexity, differences in gut physiology (rodents are copiophagious, and their upper gut are colonized by large populations of bacteria unlike healthy humans) and bacterial species may have different biological roles in different animals.

Recently, Li et al. (2008) in an attempt facilitate direct human microbial ecology studies at the "omics" level, combined both metagenomics (16S rRNA targeted PCR-DGGE) and metabonomics (metabolite profiling by NMR) to generate a matrix of differential urinary metabolites and unique bacterial genotypes present in fecal samples collected from four generations of a single Chinese family. These authors were thus able to correlate individual bacterial species identified within the fecal microbiota of the human volunteers by 16S rRNA gene targeted PCR-DGGE with particular profiles of metabolites present in urine. This powerful approach offers for the first time a real insight into the *in vivo* functioning of even unculturable and previously uncharacterized members of the gut microbiota directly without the need for bacterial cultivation and in biological samples which can be collected in a non-invasive manner.

3.13   Conclusion

Recent insights into the composition and make up of the human gut microbiota and the evolution of powerful and high resolution data rich analytical techniques are revolutionizing the way we view the human intestinal microbiota. It is clear that our resident microbiota, which has co-evolved with us over the millennia, impacts on a range of human metabolic processes and appears to be particularly effected by recent population level changes in human diet, particularly the reduction in fiber and whole plant food ingestion and adoption of the Western-style diet which as occurred with growing affluence over the past 50 years. Nowhere is this more clearly illustrated than in the fact that the gut microbiota of the obese people appears to be different to lean people and that this obese-type microbiota returns to a lean type profile upon weight loss. The application of metagenomics approaches, particularly the recent Human Microbiome Project which aims to genome sequence up to 1,000 gastrointestinal bacteria as well as directly sequence functional communities in different body sites will provide a valuable encyclopaedia of genetic information mapping out the metabolic potential of bacteria residing on or in the human body. Metabonomics, on the other hand offers the possibility of tracking changes in metabolite profiles at the systems level allowing direct measurement of metabolic kinetic or metabolite flux over experimental time courses such as before and after dietary intervention or in the presence or absence of disease. A key recent development has been the combining of these two approaches (Li et al., 2008) offering a powerful tool for direct study of the human gut microbiota *in vivo* and upon dietary modulating. These omics based approaches are thus providing tools of sufficient resolution to allow researchers to realistically address one of the most fundamental and tantalizing questions in the area of functional foods research, "how do probiotics and prebiotics really work."

References

Abdellah Z, Ahmadi A, Ahmed S, Aimable M, Ainscough R, Almeida J et al. (2004) International human genome sequencing consortium. Nature 409:860–921

Abecia L, Fondevila M, Balcells J, Edwards JE, Newbold CJ McEwan NR (2007) Effect of antibiotics on the bacterial population of the rabbit caecum FEMS Microbiol. Lett 272:144–153

Abrams SA, Griffin IJ, Hawthorne KM, Liang L, Gunn SK, Darlington G et al. (2005) A combination of prebiotic short- and long-chain inulin-type fructans enhances calcium absorption and bone mineralization in young adolescents. Am J Clin Nutr 82:471–476

Amann R, Fuchs BM (2008) Single-cell identification in microbial communities by improved fluorescence in situ hybridization techniques. Nat Rev Microbiol 6:339–348

Amann RI, Zarda B, Stahl DA, Schleifer KH (1992) Identification of individual prokaryotic cells by using enzyme-labeled, rRNA-targeted oligonucleotide probes. Appl Environ Microbiol 58:3007–3011

Ampe F, ben Omar N, Moizan C, Wacher C, Guyot JP (1999) Polyphasic study of the spatial distribution of microorganisms in Mexican pozol, a fermented maize dough, demonstrates the need for cultivation-independent methods to investigate traditional fermentations. Appl Environ Microbiol 65:5464–5473

Astrup A, Dyerberg J, Selleck M, Stender S (2008) Nutrition transition and its relationship to the development of obesity and related chronic diseases. Obes Rev 9: Suppl 1:48–52

Balasubramanian K, Kumar S, Singh RR, Sharma U, Ahuja V, Makharia GK, Jagannathan NR (2008) Metabolism of the colonic mucosa in patients with inflammatory bowel diseases: an in vitro proton magnetic resonance spectroscopy study. Magnon Reson Imaging 2. [Epub ahead of print]

Ben-Amor K, Heilig H, Smidt H, Vaughan EE, Abee T, de Vos WM (2005) Genetic diversity of viable, injured, and dead fecal bacteria assessed by fluorescence-activated cell sorting and 16S rRNA gene analysis. Appl Environ Microbiol 71:4679–4689

Beylot M (2005) Effects of inulin-type fructans on lipid metabolism in man and in animal models. Br J Nutr 93:S163–S168

Bik EM, Eckburg PB, Gill SR, Nelson KE, Purdom EA, Francois F et al. (2006) Molecular analysis of the bacterial microbiota in the human stomach. Proc Natl Acad Sci USA 103:732–737

Blaut M, Collins MD, Welling GW, Doré J, van Loo J, de Vos W (2002) Molecular biological methods for studying the gut microbiota: the EU human gut flora project. Br J Nutr 87(Suppl. 2):S203–S211

Bonnet R, Suau A, Doré J, Gibson GR, Collins MD (2002) Differences in rDNA libraries of faecal bacteria derived from 10- and 25-cycle PCRs. Int J Syst Evol Microbiol 52:757–763

British Nutrition Foundation (2004) http://www.nutrition.org.uk/home.asp?siteId=43&sectionId=609&parentSection=324&which=1

Campbell TC, Chen J (1994) Diet and chronic degenerative diseases: perspectives from China. Am J Clin Nutr 59:1153S–1161S

Cani PD, Amar J, Iglesias MA, Poggi M, Knauf C, Bastelica D, Neyrinck AM, FavaF, Tuohy KM, Chabo C, Waget A, Delmee E, Cousin B, Sulpice T, Chamontin B, Ferrieres J, Tanti JF, Gibson GR, Casteilla L, Delzenne NM, Alessi MC, Burcelin R (2007a) Metabolic endotoxemia initiates obesity and insulin resistance. Diabetes 56(7):1761–1772

Cani PD, Neyrinck AM, Fava F, Knauf C, Burcelin RG, Tuohy KM et al. (2007b) Selective increases of bifidobacteria in

gut microflora improve high-fat-diet-induced diabetes in mice through a mechanism associated with endotoxaemia. Diabetologia 50:2374–2383

Cassidy A (2006) Factors affecting the bioavailability of soy isoflavones in humans. J AOAC Int 89:1182–1188

Chen K, Pachter L (2005) Bioinformatics for whole-genome shotgun sequencing of microbial communities. PLoS Comp Biol 1:24

Chen C, Shah YM, Morimura K, Krausz KW, Miyazaki M, Richardson TA, Morgan ET, Ntambi JM, Idle JR, Gonzalez FJ (2008) Metabolomics reveals that hepatic stearoyl-CoA desaturase 1 downregulation exacerbates inflammation and acute colitis. Cell Metab 7:135–147

Committee on Metagenomics (2007) Challenges and Functional Applications The new science of metagenomics: revealing the secrets of our microbial planet. National Research Council, The National Academies Press, Washington, DC, USA http://www.nap.edu/catalog/11902.html accessed on 27 March 2007.

Conway PL (1995) Human colonic bacteria: role in nutrition. In: Gibson GR, Macfarlane GT (eds) Physiology and Pathology, CRC Press, Boca Raton, FL, pp. 1–18

Costabile A, Klinder A, Fava F, Napolitano A, Fogliano V, Leonard C et al. (2008) Whole-grain wheat breakfast cereal has a prebiotic effect on the human gut microbiota: a double-blind, placebo-controlled, crossover study. Br J Nutr 99:110–120

De La Cochetière MF, Durand T, Lalande V, Petit JC, Potel G, Beaugerie L (2008) Effect of antibiotic therapy on human fecal microbiota and the relation to the development of Clostridium difficile. Microb Ecol 56:395–402

Delzenne NM, Cani PD, Neyrinck AM (2007) Modulation of glucagon-like peptide 1 and energy metabolism by inulin and oligofructose: experimental data. J Nutr 137:2547S–2551S

Depeint F, Tzortzis G, Vulevic J, I'anson K, Gibson GR (2008) Prebiotic evaluation of a novel galactooligosaccharide mixture produced by the enzymatic activity of Bifidobacterium bifidum NCIMB 41171: in healthy humans: a randomized, double-blind, crossover, placebo-controlled intervention study. Am J Clin Nutr 87:785–789

Desiere F (2004) Towards a systems biology understanding of human health: interplay between genotype, environment and nutrition. Biotechnol Annu Rev 10:51–84

Duncan SH, Louis P, Flint HJ (2004) Lactate-utilizing bacteria, isolated from human feces, that produce butyrate as a major fermentation product. Appl Environ Microbiol 70:5810–5817

Dumas ME, Barton RH, Toye A, Cloarec O, Blancher C, Rothwell A et al. (2006) Metabolic profiling reveals a contribution of gut microbiota to fatty liver phenotype in insulin-resistant mice. Proc Natl Acad Sci USA 103:12511–12516

Dunitz M (1983) In: Bukitt D (ed) Don't forget fiber in your diet. Singapore. Arco, p. 32

Eaton SB, Eaton SB III, Konner MJ (1997) Paleolithic nutrition revisited: a twelve year retrospective. Euro J Clin Nut 51:207–216

Eckburg PB, Bik EM, Bernstein CN, Purdom E, Dethlefsen L, Sargent M et al. (2005) Diversity of the human intestinal microbial flora. Science 308:1635–1638

Ercolini D (2004) PCR-DGGE fingerprinting: novel strategies for detection of microbes in food. J Microbiol Methods 56:297–314

Ercolini D, Hill PJ, Dodd CE (2003) Bacterial community structure and location in Stilton cheese. Appl Environ Microbiol 69:3540–3548

Farris MH, Olson JB (2007) Detection of Actinobacteria cultivated from environmental samples reveals bias in universal primers. Lett Appl Microbiol 45:376–381

Frank DN, Pace NR (2008) Gastrointestinal microbiology enters the metagenomics era. Curr Opin Gastroenterol 24:4–10

"Forsight: tackling obesity" document www.foresight.gov.uk (2007)

Flint HJ (2006) In: Logan NA, Lappin-Scott HM, Oyston PCF (eds) Prokaryote diversity: mechanisms and significance. The significance of prokaryote diversity in the human gastrointestinal tract. SGM Symposium, UK, pp. 65–90

Goldberg SM, Johnson J, Busam D, Feldblyum T, Ferriera S, Friedman R et al. (2006) A Sanger/pyrosequencing hybrid approach for the generation of high-quality draft assemblies of marine microbial genomes. Proc Natl Acad Sci USA 103: 11240–11245

Gill SR, Pop M, Deboy RT, Eckburg PB, Turnbaugh PJ, Samuel BS et al. (2006) Metagenomic analysis of the human distal gut microbiome. Science 312:1355–1359

Gibson GR, Roberfroid MB (1995) Dietary modulation of the human colonic microbiota: introducing the concept of prebiotics. J Nutr 125:1401–1412

Handelsman J (2004) Metagenomics: application of genomics to uncultured microorganisms. Microbiol Mol Biol Rev 68: 669–685

Hayashi H, Takahashi R, Nishi T, Sakamoto M, Benno Y (2005) Molecular analysis of jejunal, ileal, caecal and recto-sigmoidal human colonic microbiota using 16S rRNA gene libraries and terminal restriction fragment length polymorphism. J Med Microbiol 54:1093–1101

Hongoh Y, Yuzawa H, Ohkuma M, Kudo T (2003) Evaluation of primers and PCR conditions for the analysis of 16S rRNA genes from a natural environment. FEMS Microbiol Lett 221(2):299–304

Holloway L, Moynihan S, Abrams SA, Kent K, Hsu AR, Friedlander AL (2007) Effects of oligofructose-enriched inulin on intestinal absorption of calcium and magnesium and bone turnover markers in postmenopausal women. Br J Nutr 97:365–372

Holmes E, Loo RL, Stamler J, Bictash M, Yap IK, Chan Q et al. (2008) Human metabolic phenotype diversity and its association with diet and blood pressure. Nature 453:396–400

Institute of Medicine (2002) Dietary reference intakes. energy, carbohydrate, fiber, fat, fatty acids, cholesterol, protein, and amino acids. Washington, DC: National Academy Press

Iwamoto T, Tani K, Nakamura K, Suzuki Y, Kitagawa M, Eguchi M, Nasu M (2000) Monitoring impact of in situ biostimulation treatment on groundwater bacterial community by DGGE. FEMS Microbiol Ecol 32:129–141

Johnson IT, Belshaw NJ (2008) Environment, diet and CpG island methylation: epigenetic signals in gastrointestinal neoplasia. Food Chem Toxicol 46: 1346–1359

Johnson L, Mander AP, Jones LR, Emmett PM, Jebb SA (2008) Energy-dense, low-fiber, high-fat dietary pattern is associated with increased fatness in childhood. Am J Clin Nutr 87:846–854

Kalyuzhnaya MG, Zabinsky R, Bowerman S, Baker DR, Lidstrom ME, Chistoserdova L (2006) Fluorescence in situ hybridization-flow cytometry-cell sorting-based method for separation and enrichment of type I and type II methanotroph populations. Appl Environ Microbiol 72:4293–4301

Kurokawa K, Itoh T, Kuwahara T, Oshima K, Toh H, Toyoda A (2007) Comparative metagenomics revealed commonly enriched gene sets in human gut microbiomes. DNA Res 14:169–181

Kowalchuk GA, Stephen JR, De Boer W, Prosser JI, Embley TM, Woldendorp JW (1997) Analysis of ammonia-oxidizing bacteria of the beta subdivision of the class Proteobacteria in coastal sand dunes by denaturing gradient gel electrophoresis and sequencing of PCR-amplified 16S ribosomal DNA fragments. Appl Environ Microbiol 63:1489–1497

Lay C, Doré J, Rigottier-Gois L (2007) Separation of bacteria of the Clostridium leptum subgroup from the human colonic microbiota by fluorescence-activated cell sorting or group-specific PCR using 16S rRNA gene oligonucleotides. FEMS Microbiol Ecol 60:513–520

Lay C, Rigottier-Gois L, Holmstrøm K, Rajilic M, Vaughan EE, de Vos WM et al. (2005) Colonic microbiota signatures across five northern European countries. Appl Environ Microbiol 71:4153–4155

Lee JH, Karamychev VN, Kozyavkin SA, Mills D, Pavlov NV, Polpuchine NN, Richardson PM, Shakhova VV, Slesarev AI, Weimer B, O'Sullivan DJ (2008) Comparative gneomic analysis of the gut bacterium Bifidobacterium longum reveals loci susceptible to deletion during pure culture growth. BMC Genomics 27:9:247

Ley RE, Bäckhed F, Turnbaugh P, Lozupone CA, Knight RD, Gordon JI (2005) Obesity alters gut microbial ecology. Proc Natl Acad Sci USA 102:11070–11075

Ley RE, Hamady M, Lozupone C, Turnbaugh PJ, Ramey RR, Bircher JS et al. (2008) Evolution of Mammals and Their Gut Microbes. Science 320:1647–1651

Ley RE, Turnbaugh PJ, Klein S, Gordon JI (2006) Microbial ecology: human gut microbes associated with obesity. Nature 444:1022–1023

Li M, Wang B, Zhang M, Rantalainen M, Wang S, Zhou H et al. (2008) Symbiotic gut microbes modulate human metabolic phenotypes. Proc Natl Acad Sci USA 105:2117–2122

Louis P, Scott KP, Duncan SH, Flint HJ (2007) Understanding the effects of diet on bacterial metabolism in the large intestine. J Appl Microbiol. 102:1197–208

Macfarlane GT, Gibson GR (1997) In Mackei RI, White BA (eds) Gastrointestinal microbiology: gastrointestinal ecosystems and fermentations, Carbohydrate Fermentation, Energy Transduction and Gas Metabolism in the Human Large Intestine. vol 1. International Thomson Publishing, UK p. 269

Marchesi J, Shanahan F (2007) The normal intestinal microbiota. Curr Opin Infect Dis 20:508–513

Marchesi JR, Holmes E, Khan F, Kochhar S, Scanlan P, Shanahan F et al. (2007) Rapid and noninvasive metabonomic characterization of inflammatory bowel disease. Proteome Res 6:546–551

Martin FP, Dumas ME, Wang Y, Legido-Quigley C, Yap IK, Tang H et al. (2007) A top-down systems biology view of microbiome-mammalian metabolic interactions in a mouse model. Mol Syst Biol 3:112

Martin FP, Wang Y, Sprenger N, Yap IK, Lundstedt T, Lek P et al. (2008) Probiotic modulation of symbiotic gut microbial-host metabolic interactions in a humanized microbiome mouse model. Mol Syst Biol 4:157

Martinez-Medina M, Aldeguer X, Gonzalez-Huix F, Acero D, Garcia-Gil LJ (2006) Abnormal microbiota composition in the ileocolonic mucosa of Crohn's disease patients as revealed by polymerase chain reaction-denaturing gradient gel electrophoresis. Inflamm Bowel Dis 12:1136–1145

McHardy AC, Rigoutsos I (2007) What's in the mix: phylogenetic classification of metagenome sequence samples. Curr Opin Microbiol 10:499–503

Morrison DJ, Mackay WG, Edwards CA, Preston T, Dodson B, Weaver LT (2006) Butyrate production from oligofructose fermentation by the human faecal flora: what is the contribution of extracellular acetate and lactate? Br J Nutr 96:570–577

Mueller S, Saunier K, Hanisch C, Norin E, Alm L, Midtvedt T et al. (2006) Differences in fecal microbiota in different European study populations in relation to age, gender, and country: a cross-sectional study. Appl Environ Microbiol 72:1027–1033

Muyzer G, de Waal EC, Uitterlinden AG (1993) Profiling of complex microbial populations by denaturing gradient gel electrophoresis analysis of polymerase chain reaction-amplified genes coding for 16S rRNA. Appl Environ Microbiol 59:695–700

Nielsen S, Nielsen DS, Lauritzen L, Jakobsen M, Michaelsen KF (2007) Impact of diet on the intestinal microbiota in 10-month-old infants. J Pediatr Gastroenteorl Nutr 44:613–618

Nicholson JK, Holmes E, Wilson ID (2005) Gut microorganisms, mammalian metabolism and personalized health care. Nat Rev Microbiol 3:431–438

Olsen GJ, Lane DJ, Giovannoni SJ, Pace NR & Stahl DA (1986) Microbial ecology and evolution: a ribosomal RNA approach. Annual Reviews of Microbiology 40:337–365

Ovreas L, Forney L, Daae FL, Torsvik V (1997) Distribution of bacterioplankton in meromictic Lake Saelenvannet, as determined by denaturing gradient gel electrophoresis of PCR-amplified gene fragments coding for 16S rRNA. Appl Environ Microbiol 63:3367–3373

Pool-Zobel BL, Sauer J (2007) Overview of experimental data on reduction of colorectal cancer risk by inulin-type fructans. J Nutr 137:2580S–2584S

Pompei A, Cordisco L, Amaretti A, Zanoni S, Matteuzzi D, Rossi M (2007) Folate production by bifidobacteria as a potential probiotic property. Appl Environ Microbiol 73:179–185

Rusch DB, Halpern AL, Sutton G, Heidelberg KB, Williamson S, Yooseph S et al. (2007) The Sorcerer II Global Ocean Sampling expedition: northwest Atlantic through eastern tropical Pacific. PLoS Biol 53:77

Saarela M, Maukonen J, von Wright A, Vilpponen-Salmela T, Patterson AJ, Scott KP, Hämynen H, Mättö J (2007) Tetracycline susceptibility of the ingested Lactobacillus acidophilus LaCH-5 and Bifidobacterium animalis subsp. lactis Bb-12 strains during antibiotic/probiotic intervention. Int J Antimicor Agenet 29 (3):271–280

Salyers AA, Gupta A, Wang Y (2004) Human intestinal bacteria as reservoirs for antibiotic resistance genes. Trends Microbiol 12:412–416

Sipos R, Székely AJ, Palatinszky M, Révész S, Márialigeti K, Nikolausz M (2007) Effect of primer mismatch, annealing temperature and PCR cycle number on 16S rRNA gene-targetting bacterial community analysis. FEMS Microbiol Ecol 60:341–350

So PW, Yu WS, Kuo YT, Wasserfall C, Goldstone AP, Bell JD et al. (2007) Impact of resistant starch on body fat patterning and central appetite regulation. PLoS ONE 2:e1309

Suau A, Bonnet R, Sutren M, Godon JJ, Gibson GR, Collins MD et al. (1999) Direct analysis of genes encoding 16S rRNA from complex communities reveals many novel molecular species within the human gut. Appl Environ Microbiol 65:4799–4807

Stabler RA, Gerding DN, Songer JG, Drudy D, Brazier JS, Trinh HT, Witney AA, Hinds J, Wren BW (2006) Comparative phylogenomics of Clostridium difficile reveals clade specificity and microevolution of hypervirulent strains. J Bacteroil 188: 7297–7305

Spencer JP, Schroeter H, Rechner AR, Rice-Evans C (2001) Bioavailability of flavan-3-ols and procyanidins: gastrointestinal tract influences and their relevance to bioactive forms in vivo. Antioxid Redox Signal 3:1023–1039

Solanky KS, Bailey NJ, Beckwith-Hall BM, Bingham S, Davis A, Holmes E et al. (2005) Biofluid 1H NMR-based metabonomic techniques in nutrition research - metabolic effects of dietary isoflavones in humans. J Nutr Biochem 16:236–244

Stewart JA, Chadwick VS, Murray A (2006) Carriage, quantification, and predominance of methanogens and sulfate-reducing bacteria in faecal samples. Lett Appl Microbiol 43:58–63

Tannock GW, Munro K, Bibiloni R, Simon MA, Hargreaves P, Gopal P et al. (2004) Impact of consumption of oligosaccharide-containing biscuits on the fecal microbiota of humans. Appl Environ Microbiol 70:2129–2136

Turnbaugh PJ, Bäckhed F, Fulton L, Gordon JI (2008) Diet-induced obesity is linked to marked but reversible alterations in the mouse distal gut microbiome. Cell Host Microbe 3:213–223

Turnbaugh PJ, Ley RE, Hamady M, Fraser-Liggett CM, Knight R, Gordon JI (2007) The human microbiome project. Nature 449:804–810

Turnbaugh PJ, Ley RE, Mahowald MA, Magrini V, Mardis ER, Gordon JI (2006) An obesity-associated gut microbiome with increased capacity for energy harvest. Nature 444:1027–1031

Tuohy K, Rowland IR, Rumsby PC (2002) In: Atherton KT (ed) Genetically modified crops assessing safety. Taylor and Francis, London, 110–137

Tuohy KM, McCartney AL (2006) Molecular microbial ecology of the human gut. In: Gibson GR, Rastall RA (eds) Prebiotics: development and application. Wiley, London, 135–155

van Beek S, Priest FG (2002) Evolution of the lactic acid bacterial community during malt whisky fermentation: a polyphasic study. Appl Environ Microbiol 68: 297–305

Venter JC, Adams MD, Myers EW, Li PW, Mural RJ, Sutton GG et al. (2001) The sequence of the human genome. Science 291:1304–1351

Wardle J, Carnell S, Haworth CM, Plomin R (2008) Evidence for a strong genetic influence on childhood adiposity despite the force of the obesogenic environment. Am J Clin Nutr 87:398–404

Wang M, Ahrné S, Jeppsson B, Molin G (2005) Comparison of bacterial diversity along the human intestinal tract by direct cloning and sequencing of 16S rRNA genes. FEMS Microbiol Ecol 54:219–231

Wang Y, Lawler D, Larson B, Ramadan Z, Kochhar S, Holmes E et al. (2007) Metabonomic investigations of aging and caloric restriction in a life-long dog study. J Proteome Res 6:1846–1854

Woese CR, Kandler O, Wheelis ML (1990) Towards a natural system of organisms: proposal for the domains Archaea, Bacteria and Eucarya. Proc Natl Acad Sci USA 87:4576–4579

Yap IK, Li JV, Saric J, Martin FP, Davies H, Wang Y, Wilson ID, Nicholson JK, Utzinger J, Marchesi JR, Holmes E (2008) Metabonomic and microbiological analysis of the dynamic effect of vancomycin-induced gut microbiota modification in the mouse. J Proteome Res 7:3718–3728

Zoetendal EG, Cheng B, Koike S, Mackie RI (2004) Molecular microbial ecology of the gastrointestinal tract: from phylogeny to function. Curr Issues Intest Microbiol 5:31–47

Zoetendal EG, Akkermans AD, De Vos WM (1998) Temperature gradient gel electrophoresis analysis of 16S rRNA from human fecal samples reveals stable and host-specific communities of active bacteria. Appl Environ Microbiol 64:3854–3859

Zwart G, Huismans R, van Agterveld MP, Van de Peer Y, De Rijk P, Eenhoorn H, Muyzer G, van Hannen EJ, Gons HJ, Laanbroek HJ (1998) Divergent members of the bacterial division Verrucomicrobiales in a temperate freshwater lake1. FEMS Microbiol Ecol 25:159–169

4 Designing Trials for Testing the Efficacy of Pre- Pro- and Synbiotics

Stephen Lewis · Charlotte Atkinson

4.1 Introduction

Providing an evidence base for the rational delivery of medicines and treatments is the cornerstone of modern health care delivery. Much of this evidence base is gained through conducting clinical trials.

Superficially, designing a clinical trial seems straightforward. However, in practice many unforeseen difficulties arise with long setting up times, poor recruitment rates and patients or interventions not behaving in the way expected. Unfortunately, clinical trials examining the efficacy of pre-, pro- and synbiotics have developed a reputation for being published in low impact journals and reaching unconvincing conclusions. As a generalization, the reason for this poor reputation is that the trials have tended to be too small and have not used meaningful clinical endpoints. The level of evidence required to alter clinical practice is expected to be high and robust. Trials of drugs such as those used to treat hypertension are often very large with hundreds (if not thousands) of patients and have hard clinical end points, such as stroke, myocardial infarction or death. Many clinical trials involving pre-, pro- or synbiotics have less than 200 patients and often use surrogate markers of health benefit as main outcome measures. This chapter sets out to give an overview of how to design and run a clinical trial highlighting examples and problems related to studies using pre-, pro- and synbiotics.

4.2 Before You Begin

A thorough literature search is required to ensure that any prospective trialist has an in-depth knowledge of the subject area to be researched. Traditionally a trialist would start by looking in electronic databases (e.g., PUBMED, COCHRANE), then searching through the references of relevant published papers. Writing to the manufacturers of a given pre-, pro- or synbiotic can be helpful as can writing to other trialists with experience in the field. In particular, it is worth checking to see if any other researchers are studying the chosen or a similar subject such that any new study can complement, rather than duplicate, existing trials. A number of web sites provide up-to-date information on currently registered clinical trials (e.g., www.controlled-trials.com, http://clinicaltrials.gov and http://eudract.emea.europa.eu/index.html).

Much attention is now paid to health claims for functional foods, especially those containing pre or probiotics. The Process for the Assessment of Scientific Support for Claims on Foods (PASSCLAIM) project provides a scheme by which health claims for functional foods could be justified in a scientific manner. The project was initiated by the International Life Science Institute (ILSI) Europe (http://europe.ilsi.org/), which is a European Union backed multiprofessional organization, principally funded by industry. The published criteria (Aggett et al., 2005) provide an excellent source of information for researchers designing both non-clinical and clinical trials. In particular the authors promote the importance of high scientific standards and the requirement to examine physiologically relevant end-points.

Collaboration with other researchers may be desirable. Indeed grant awarding committees are often impressed by a multidisciplinary team approach, which will draw on a breadth of expertise, e.g., pharmacologist, health economist, statistician as well as an experienced trialist.

The International Conference for Harmonization of Good Clinical Practice in Research (ICH-GCP in research) has produced a combined framework for research conducted in Europe, Japan and the United States of America. Researchers are obliged to be compliant with this framework and thus should ensure they are familiar with its requirements. It is also important to ensure that the appropriate resources and motivation are available, as most studies will take a minimum of 3–4 years from conception to publication.

4.3   Hypothesis

The hypothesis is the main focus of the study. It cannot be emphasized how important it is to define clearly the question you would like to answer. The hypothesis is usually based on previous observations or assumptions, and the goal of the study is to either prove or disprove the hypothesis. Keeping the question simple and focused will greatly increase the study's success in proving or disproving the stated hypothesis. It is often desirable also to examine any underlying mechanisms that support the hypothesis, to explain why the trial either produced the desired result or to give an insight into the reasons for any unexpected results. Generally, researchers test the validity of the "null" hypothesis. The "null" hypothesis is where the assumption is that no clinically important difference of interest in the outcome exists between groups. For example, a trialist may set out to show that an interventional agent such as a probiotic yogurt when taken by patients does not alter the risk of developing endocarditis when compared with patients not taking the probiotic yogurt. An "alternative" hypothesis is that patients taking the probiotic yogurt will have a lower or higher risk of developing endocarditis when compared with patients not taking the probiotic yogurt.

4.4 Choosing an Interventional Agent, Placebo and Packaging

When choosing a pre-, pro- or synbiotic as an interventional agent, justification is required for the choice of agent and the proposed dose to be given. A trialist may be guided by factors such as faecal recovery rates as an indicator of intestinal viability, the demonstration of, such as: immune stimulation in previous studies, or other studies suggesting a benefit in terms of health outcomes for the target population of patients.

If a trialist chooses to use live microorganisms as the interventional agent, quality control and storage are important considerations as inappropriate bacteria may be present (e.g., through contamination) or viable counts of organisms may be less than desired if incorrectly stored (Gilliland and Speck, 1977; Hamilton-Miller et al., 1996). If a clinical trial requires multiple batches of microorganisms to be prepared or a long period of storage then repeated quality control is required.

In most countries pre-, pro- and synbiotic preparations are considered as foods or food additives. It is possible that some preparations – especially, if genetically modified to have specific characteristics (e.g., produce cytokines) would be classified as drugs. In such cases further approvals would be required (in the UK by the Medicines and Healthcare products Regulatory Agency (MHRA)) and considerably more stringent monitoring throughout the trial, especially for complications.

The power of suggestion should not be underestimated. It has been clearly shown that patients do better if they feel that they are receiving an active intervention rather than placebo. Likewise, if the treatment arm is known to the trialist then there may be a temptation for them to interpret the data according to their personal views of the expected outcome for those taking the active intervention. As such it is desirable that the patient and the treating physician are both unaware of which intervention the patient is receiving (e.g., a "double-blind" trial). This can be achieved by using a placebo. The placebo, where possible, should be indistinguishable in sight (including packaging), smell and taste from any interventional agent. Trials that do not use a placebo are usually much diminished in terms of credibility.

When choosing or developing a placebo, it is important to avoid the use of any materials that could affect the outcome measures or potentially affect recruitment rates. For example, "carrier" substances such as lactose in capsules or milk products in yogurts may cause diarrhea in some people, and therefore must be provided in similar amounts (if at all) to participants taking either the active or placebo intervention. Using products like gelatin or additives (for coloring, taste etc) may exclude some potential volunteers to a study because of vegetarianism or allergy to certain additives. Such products should be avoided where possible.

Where an intervention is compared with a standard treatment, e.g., an anti-motility agent against a probiotic to treat diarrhea, the principles of blinding still apply and both products should be indistinguishable from one another. Alternatively placebos can be given for both products. However, care should be taken when considering such a trial, as it is not always wise to assume that a standard treatment is effective particularly for disorders such as irritable bowel syndrome. Any trial that involves the use of a licensed drug will have to comply with further sets of regulations (see above). Indeed there will be a requirement for the drug, placebo and packaging to be prepared by a licensed facility, which will increase the cost of the trial and also increase the administrative burden on the trial due to the requirement for increased monitoring.

4.5   Choosing the Primary Study End-Point

Choosing the principal outcome measure of any given study, e.g., pneumonia, diarrhea etc is critical. Above all the principal end-point being studied has to be clinically relevant and represent an improvement, which if achieved is likely to alter clinical practice. Generally clinical trials are designed to examine mortality, morbidity, quality of life and economic benefits. Ideally, a clinical trial should be able to isolate the effects of a treatment on a study outcome and provide results that are free from bias. Surrogate markers of benefit, such as inflammatory or immune function markers are good for exploring mechanisms of action and for planning large, more definitive trials. However, surrogate markers of benefit, even if improved by the intervention, may not imply clinical benefit, especially if seen in healthy people and thus unlikely to alter clinical practice. A chosen primary end-point should be practical to measure and occur frequently enough in the target population to be statistically viable. If more than one principal end-point is chosen, then this usually will require the study to be larger, and thus increases its cost and duration and may reduce its chances of success. End-points can be binary, i.e., Yes/No, or continuous, e.g., blood pressure. A chosen endpoint must be clearly defined using a recognized and relevant definition. This can be difficult; for example what constitutes a wound infection? Trials examining the benefit of postoperative enteral nutrition revealed wound infection rates between 0–33% (Lewis et al., 2001), the large variation being due primarily to the different definitions used to describe a wound infection. A clinician may feel that a relevant definition of wound infection would include the need (or not) for treatment, increased cost, or patient discomfort, but the degree of erythema may be irrelevant, whereas another clinician may place more emphasis on the degree of erythema than other factors. It is also critical that any definition of an endpoint is reproducible; this subject has its own literature, which will not be covered in detail here but should be researched. For example what is a clinically meaningful and reproducible definition of diarrhea? Definitions range from one to three loose bowel motions over 1–3 days. Even trying to define what a loose bowel motion is can be fraught with difficulty, as different observers will interpret definitions such as semi-solid, loose and watery differently. Graphical representations of stool form have been developed such as the Bristol stool form scale, but even this scale has an intra and inter-observer error (Lewis and Heaton, 1997). Creating your own definitions of end-points will be open to criticism unless they have been substantiated. Again, any end point has to be clinically meaningful. If a clinical trial was to show that an intervention led to the resolution of diarrhea say one day earlier than the control group, would this change clinical practice? Probably not, but the answer to this would depend on a number of factors including the cost and ease of taking the intervention. It can be meaningful for trialists to adapt existing definitions or scales to complement their trial. Using the example of diarrhea, as well as collecting data on stool form it would also be relevant to know what a patients degree of urgency to defecate was or whether the presence of faecal incontinence was altered by an interventional agent.

In some clinical trials, assessing the economic benefits may be relevant, though this is difficult to do well. In order to do meaningful healthcare economic analysis it is usual to involve a heath economist early on within the trial design process. The most frequently used health economic measure is quality-adjusted life-years (QALYs) gained; this measurement takes into account the patient's duration of life, and heath related quality of life. QALYs can be used to calculate a cost-effectiveness ratio (the cost of gaining an extra year of good quality life). However, collecting data on costs can be problematic as identifying every health care cost may not be technically possible in many healthcare systems.

Many disease-specific as well as generic scoring systems are available to assess improvement in quality of life, or degree of dependence on carers, severity of illness and predicting outcome from illness. These can usually be easily incorporated into most trial designs. Indeed, it is becoming fashionable to use "composite" measures as the main end-point of clinical trials. Composite end-points may include several relevant outcome measures (e.g., for a study looking at outcomes of patients after a myocardial infarction the following measurements may be relevant: quality of life, heart rate, blood pressure and exercise tolerance tests) grouped together to give an overall score.

Secondary end-points may ask other relevant questions. When considering the study size, it may be appropriate to power the study to look at secondary end points if they are of sufficient interest to justify a larger study.

If possible, data on endpoints should be recorded prospectively rather than retrospectively. If the trial involves collecting data from a patient's notes going through their notes after the end of a study is inadequate. Data may have been either poorly recorded or not recorded at all, and it may not be possible to obtain the data at a later stage. Collecting data as the trial progresses will also allow interim analysis to be conducted if appropriate (see below). Furthermore, it could enable the identification of problems in data recording (e.g., patients may not understand a particular questionnaire and thus fill it in inadequately) or obtaining test results, which can be rectified quickly and with minimal loss of useable data.

4.6   Independent Variables

Independent variables are factors that may influence the main study outcome measure. A thorough review of previous literature to identify independent variables is important, and similar to primary end-points they need to be robustly defined. With regards to studies looking at *Clostridium difficile* related diarrhea, patient age, degree of disability, immune suppression, taking of proton pump inhibitors and recent antibiotic use need to be recorded, as they can all influence the development of the disease. It is likely that any study will have to be larger to be suitably powered to examine the influence of independent variables on the main outcome end-point, e.g., the use of immune suppressing drugs. The most commonly used approach to remove the potential effects of an independent variable on the outcome of interest is to use a randomized trial design where patients are randomly allocated to an intervention. The independent variables are recorded and if recruited numbers are large enough, then these independent variables will be equally distributed between study groups and their influence on the final outcome (e.g., *C difficile* related diarrhea) should also be equally distributed. If there is concern over the influence of a variable on the main trial outcome measure, e.g., immunosuppressive drugs on the development of diarrhea due to *C difficile*, then the patient's random allocation to different groups can be stratified by the variable. Where stratification is used, this may result in the requirement for greater numbers of patients in the trial to ensure reasonable numbers in each subgroup.

4.7 Clinical Trials

Any new medical drug has to go through various levels of clinical trials, often described as first, second, third and fourth phase trials. Pre-, pro- and synbiotics have not gone through these hurdles because in most countries they are considered to be nutritional supplements and are not subject to the same rules and regulations as drugs. Trialists are, however, interested in substantiating health claims, including benefit and lack of harm, for a given pre-, pro- or synbiotic. Whilst most pre-, pro- or synbiotics are considered safe, it must not be assumed that they are free from potential side effects. Indeed, there is a considerable literature on septicemias caused by many probiotic bacteria and yeasts. Furthermore, prebiotics may cause diarrhea. Thus, it would be wise to pay considerable attention to possible detrimental effects of an intervention when given to certain groups of patients such as those who are immunosuppressed or who are prone to diarrhea (e.g., those with ulcerative colitis).

4.7.1 Phase I Trials (Clinical Pharmacology and Toxicity, Typically 20–80 People)

These are the first experiments in humans (usually healthy volunteers or patients in whom usual treatments have failed), and are primarily concerned with drug safety and drug pharmacokinetics. Trials often involve small numbers of people and dose escalation studies using predetermined criteria can be conducted.

4.7.2 Phase II Trials (Initial Clinical Investigation of Treatment Effect, Typically 40–100 People)

These are small-scale investigations (although usually larger than phase I trials) into the effectiveness and safety of a drug, and require close monitoring of patients. Usually, but not always, there is no randomization process.

4.7.3 Phase III Trials (Evaluation of Intervention, Typically Greater than 200 People)

After a drug has been shown to be potentially effective in a phase II trial, it is essential to compare this drug against the current standard treatment or a placebo within a real life environment in order to provide definitive data on the effectiveness of an intervention. Phase III trials may require relatively large sample sizes and lengthy follow-up of study participants.

4.7.4 Phase IV Trials (Post Marketing Surveillance)

Large, long-term follow-up studies looking at morbidity and mortality are termed phase IV trials, and may detect uncommon problems, which were not picked up in phase III trials.

4.8   Trial Design

This chapter is principally concerned with phase III trials that are designed to show clinical benefit. The simplest and most widely used design involves an intervention and placebo randomly allocated to subjects or group of subjects, these are known as randomized placebo controlled trials. Where both the trialist and trial subject are blinded to the intervention these trials are called, "double blinded, placebo controlled" (DBPC) and are considered the "gold standard" of clinical trial design. Other ways of designing a clinical trial include studying cohorts of subjects or using a crossover design so that all subjects receive both

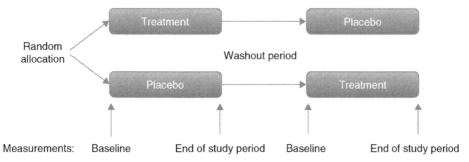

◻ Figure 4.1

Basic cross-over design for a clinical trial Patients are randomly allocated to one of two groups (treatment and placebo groups), they are assessed at the beginning and end of a study period. After a washout period the patients undergo another study period with beginning and end assessment. If there has been no "carry over of effect" then the two baseline assessments for each individual should be similar.

the interventional agent and placebo. The best trial design will depend on the study question, the study population and the nature of the end-points being studied.

If the end-points of a trial can be measured repeatedly over time (e.g., serum cholesterol) then using more complex trial designs such as crossover trial (◉ *Figure 4.1*) can be more economic in terms of numbers of subjects required for adequate statistical power. Any carry-over of effect between different phases of a trial can be assessed by collecting further "baseline" data after the washout period (but prior to the start of the next period) for comparison with the initial baseline data collected at the start of the trial. Using a crossover design any number of study periods can occur, though increasing the length of the time subjects are in the trial may increase the likelihood that they may withdraw from the trial due to the additional burden. If a subject withdraws from a crossover trial the loss of statistical power may be greater than if a simpler trial design had been used, because all of that subjects data from each of the study periods may then be unavailable for analysis.

There are many other design options and the exact format of any trial will depend exactly on what is being studied.

4.9 Protocol and Other Study Documents

Often, trial protocols go through many versions prior to being finalized, which enables ideas to be developed and refined (◉ *Table 4.1*). It is important to discuss

■ Table 4.1

Ideal contents of a protocol

A clear statement of the clinical question being asked, then how the trial will answer the question
Scientific background, and why the trial results will be meaningful
Trial design and methodology compliant with Good Clinical Practice (GCP) guidelines in research
Study population and how they will be recruited
How the trial will be analyzed then presented

trial design with people who may have done similar trials, in order to identify potential problems - many of which may not be initially obvious.

Protocols are commonly structured along the lines of: Title, Hypothesis, Background/Rationale, Objective/Value of research, Design, Methods, Timetable, Statistics, References, Costings. Some medical journals, e.g., The Lancet require that clinical trials conform to the CONSORT Statement (www.consort-statement.org, Altman et al., 2001; Moher et al., 2001), which is a flow diagram of how a clinical trial should be structured and is worth studying (❂ Table 4.2).

The protocol is used as the base document for submission of the study to any "ethics" committee for approval (see below) and, in many centres, for peer review by local research and development (R&D) committees. In most countries the conduct of clinical trials is regulated and audited. Within the European Union (EU), clinical trials are governed by the EU Clinical Trials Directive (http://www.wctn.org.uk/downloads/EU_Directive/Directive.pdf). Particular attention must be paid to data protection and reporting of adverse events. A protocol will be required to state how these requirements will be met.

The final version of the protocol needs to be well written, clear and be in compliance with ethical and regulatory requirements. Many trialists involve patients or members of the public in the design of protocols, as obtaining the view of potential participants may highlight difficulties not appreciated by the trialists; for example, procedures that cause patient discomfort (e.g., colonoscopy), may be off-putting to potential volunteers but may seem perfectly acceptable to a gastroenterologist!

Patient information leaflets and consent forms need to be prepared, which highlight any potential risks or inconveniences to the patient. The text needs to be easily understandable to a lay person. The more involved and complex the study,

◘ Table 4.2

CONSORT Statement 2001 – Checklist Items to include when reporting a randomized trial (*Cont'd* p. 122)

Paper section and topic	Item	Descriptor	Reported on page #
Title & Abstract	1	How participants were allocated to interventions (e.g., "random allocation," "randomised," or "randomly assigned")	
Introduction Background	2	Scientific background and explanation of rationale	
Methods Participants	3	Eligibility criteria for participants and the settings and locations where the data were collected	
Interventions	4	Precise details of the interventions intended for each group and how and when they were actually administered	
Objectives	5	Specific objectives and hypotheses	
Outcomes	6	Clearly defined primary and secondary outcome measures and, when applicable, any methods used to enhance the quality of measurements (e.g., multiple observations, training of assessors)	
Sample size	7	How sample size was determined and, when applicable, explanation of any interim analyses and stopping rules	
Randomization – sequence generation	8	Method used to generate the random allocation sequence, including details of any restrictions (e.g., blocking, stratification)	
Randomization – allocation concealment	9	Method used to implement the random allocation sequence (e.g., numbered containers or central telephone), clarifying whether the sequence was concealed until interventions were assigned	
Randomization – implementation	10	Who generated the allocation sequence, who enrolled participants, and who assigned participants to their groups	
Blinding (masking)	11	Whether or not participants, those administering the interventions, and those assessing the outcomes were blinded to group assignment. If done, how the success of blinding was evaluated	
Statistical methods	12	Statistical methods used to compare groups for primary outcome(s); Methods for additional analyses, such as subgroup analyses and adjusted analyses	

⊙ Table 4.2

Paper section and topic	Item	Descriptor	Reported on page #
Results Participant flow	13	Flow of participants through each stage (a diagram is strongly recommended). Specifically, for each group report the numbers of participants randomly assigned, receiving intended treatment, completing the study protocol, and analyzed for the primary outcome. Describe protocol deviations from study as planned, together with reasons	
Recruitment	14	Dates defining the periods of recruitment and follow-up	
Baseline data	15	Baseline demographic and clinical characteristics of each group	
Numbers analysed	16	Number of participants (denominator) in each group included in each analysis and whether the analysis was by "intention-to-treat." State the results in absolute numbers when feasible (e.g., 10/20, not 50%)	
Outcomes and estimation	17	For each primary and secondary outcome, a summary of results for each group, and the estimated effect size and its precision (e.g., 95% confidence interval)	
Ancillary analyses	18	Address multiplicity by reporting any other analyses performed, including subgroup analyses and adjusted analyses, indicating those pre-specified and those exploratory	
Adverse events	19	All important adverse events or side effects in each intervention group	
Discussion Interpretation	20	Interpretation of the results, taking into account study hypotheses, sources of potential bias or imprecision and the dangers associated with multiplicity of analyses and outcomes	
Generalisability	21	Generalisability (external validity) of the trial findings	
Overall evidence	22	General interpretation of the results in the context of current evidence	

the less likely patients will volunteer to be part of it. However, the information leaflet and consent form must provide a complete and balanced view of the study to enable the potential study participant to make an informed choice as to whether or not to participate.

4.10 Selection of Target Study Population

For clinical trials to be conducted for the prevention of a given problem, such as *C. difficile* associated diarrhea, then clearly it is clearly important to identify a target population where this is relevant. It may not be appropriate to exclude high-risk populations such as those receiving immunosuppressive drugs or suffering memory impairment, as these are the patients who suffer the majority of the morbidity and mortality associated with this disease. Excluding these patients will require the study to be much larger, be more likely to produce a non-significant result and would not be clinically relevant. Conversely, it may be harder for a clinical trial to show benefit in patients with more severe or end stage disease where there may be less ability to influence the disease process. As such, it may be more appropriate to study patients with milder disease in order to assess any benefits of an intervention. Thus, although it may be more convenient to study for example patients attending hospital clinics, for many conditions such as irritable bowel or asthma, these patients may have more severe problems than patients in the community and demonstration of benefit may be more difficult, and the results of the study may be not be generalisable. However, involving patients in the community rather than for example in a hospital setting, may bring its own set of difficulties that need to be addressed, e.g., low attendance at clinics if they live far away or have poor access to transport.

For the study to be clinically relevant, then the recruited patients should be as similar as possible to those on which the intervention will ultimately be used. Exclusion criteria should be as narrow as possible as the more people that are excluded the harder it will be to recruit and the less generally applicable the results will be. Furthermore, there may be ethical implications of excluding certain populations or groups of people. Thought should be given as to how volunteers will be identified and recruited. This is not always easy, especially if disease-specific databases are not available, and even then potential volunteers may be difficult to recruit if they are not prepared to come to a hospital or clinic, if that is necessary for recruitment purposes. The method of approach may also influence the characteristics of volunteers, e.g., advertisements in magazines or newspapers will only target the readership of those publications. For example, if trying to test whether a probiotic preparation will prevent travellers diarrhea, recruiting volunteers via high street travel agents may produce results that are not relevant to an intended "high-risk" target population (i.e., young backpackers roughing it in hostels with poor sanitation and food hygiene), as the recruited population may consist primarily of individuals who will be staying in expensive hotels (with hopefully good sanitation and food hygiene).

Clearly considerable thought needs to be spent on identifying a relevant target population of subjects to study, who are easy to recruit and who represent a population to whom the results of the study will be clinically relevant.

4.11 Pilot Studies

Obtaining local data pertinent to the area(s) in which the trial will be run may be informative (e.g., the prevalence of a disease within the local population) as local situations may not reflect national or previously published data. Such information may be useful in terms of assessing the feasibility of running a trial in a particular area.

Where there is little previous data on which to base a trial design a pilot trial will usually be very helpful. The results can provide information that can be used in the calculation of the sample size needed for a definitive study (see below) and the experience gained in running a pilot study may help to iron out any major study design problems prior to starting the main trial. Furthermore, the pilot study may identify any unforeseen ethical problems. A pilot study can quantify likely recruitment rates and enables examination of inclusion and exclusion criteria. They can also give an indication of likely compliance with the protocol. However, whilst doing a pilot study is generally a valuable exercise, many trialists decide not to do them because of the increased effort and expense of doing two studies.

4.12 Statistical Considerations: Power and Sample Size

Obtaining statistical advice when designing a trial is almost an obligatory prerequisite and usually well worth while. Determining the sample size of a study is not an exact science, as one has to make realistic assumptions before doing calculations to decide on an appropriate sample size. It is essential that trials be designed to recruit sufficient numbers of volunteers to avoid Type 2 errors (i.e., where there is a difference between treatments, but the study has failed to detect it), but clearly avoid recruiting too many patients. Type 2 errors occur where the natural variation for the outcome being measured is wider than expected, and the sample size was insufficient to detect any difference. When doing sample size calculations, consideration needs to be given to the likely "drop-out" rates and potential for poor compliance. Many studies using pre-, pro- or synbiotics are too small to detect benefit even if it occurs. Determining the number of patients required for recruitment will depend on the frequency of occurrence (and standard deviation) of the primary end-point in the study population, and the predicted degree of improvement likely to be seen with the interventional agent. This data may be available from previously published trials or pilot studies, but if not, educated guesses are required. Often trialists overestimate the degree of benefit likely to be seen which may result in an unrealistically small sample size estimate. The smaller the expected difference between treatment groups, the more people will be needed for the trial to be definitive. Another approach is to assume the degree of improvement of the primary outcome measure required from an intervention to be of clinical significance (likely to change clinical practice), e.g., a 20% reduction in length of hospital stay. When planning a large or expensive study, if there are no robust data on which to base sample size calculations it may be desirable to do a pilot study or even schedule an interim analysis to provide information, which will help determine the final size of the study.

Given a fixed sample size, it is nearly always true that simple one to one random allocation (i.e., one person assigned to the intervention for every one person assigned to the placebo/comparison treatment) is statistically the most efficient approach to the randomization process. Placing more participants in one group relative to the other reduces the chance of observing a difference if the sample size is fixed, although the power of the statistical test does not greatly decline unless the ratio exceeds 3–1. If the sample sizes can be increased then unequal distribution of subjects between groups may be beneficial if there are resource constraints or costs (i.e., if the intervention is very expensive or labor intensive), or if a high dropout rate is expected from an intervention because of poor tolerability.

To determine the sample size required for a clinical trial the "power" of a trial needs to be chosen. A trials "power," is the ability to show a significant difference between groups if it exists. The power of a study is calculated as 100%- β, where β is the Type 2 error (chance of arriving at a false negative result). Traditionally β values of 20–10% are used which equates to a power of 80 or 90%. The higher the study's "power" the increased chance the study has of detecting any difference, if a difference between groups exists. The increased confidence in detecting difference between groups is gained with increased study power and requires a larger sample size. Sample size calculations also require assumptions on the Type 1 error rate (detection of a false positive result); this is usually taken as a

1 in 20 chance (5%) and is denoted by α. The motivation behind most clinical trials is to show superiority for an interventional agent over a placebo or other treatment option. Occasionally, trials deliberately set out to show that two interventions are equally effective (called non-inferiority for clinical studies and equivalence for pharmacokinetic studies). In such cases a larger number of patients are usually required than for trials that are designed to show superiority of one agent over another. Calculating the sample size is best done with the aid of a statistician and suitable computer software (e.g., SPSS or STATA).

After obtaining a sample size estimate for the trial size the next step is to assess the likely recruitment rate of patients into a trial. Commonly, trialists overestimate how many patients would be eligible and willing to consent to participate. Once the likely recruitment rate has been established, an estimated time frame for the trial can be calculated. Long accrual periods are associated with failure to complete the trial, as the trialists may lose motivation. The best way to get around this is by getting more investigators and centers to participate. However, the organization of multicentre trials is considerably more complicated and expensive than a single centre trial, which perhaps explains why most studies of pre, pro and synbiotics are done in single centres.

4.13 Randomization Process and Labeling of Packaging

If the trial involves random allocation of one group of patients to a particular intervention (treatment) and another group to a placebo or current/standard treatment, computers are usually used to generate random allocations so that the investigators cannot influence who is given which treatment. Ideally, to avoid bias, the allocations need to be kept secret from the people running the trial and also from the trial subjects themselves (i.e., double-blind). The best way to achieve this is through the use of a separate randomization coordinator or service that can be contacted by telephone or email each time a new randomization code is required. The use of sealed envelopes, randomization lists, etc, are more open to abuse. A master code is often kept locally to enable decoding should there be a need due to, for example, a complication from a trial intervention.

Labeling of the packaging of trial products individually, i.e., 1,2,3,4,5..., rather than as "A" or "B" makes it harder for the trialist to predict the content. If all the placebos are labeled as "A" and the active interventions as "B," it is possible that the trialists could eventually work out which was which and potentially introduce bias. Ideally, a person unconnected with data collection and analysis should supervise the administration of the trial intervention to the study participant. However, this would not completely exclude the possibility of bias if they are not blinded to which treatment the study volunteer is getting. As an estimate of the quality of blinding, when each study participant has completed the study, they and the trialists can be asked whether they thought that they had received the active or placebo treatment.

4.14 Interim Analysis

Interim analysis is generally not encouraged as it may increase the opportunity for bias within the trial, and there are several instances in which early results were published which were later not substantiated by the completion of the whole trial. However, in some cases interim analysis may be appropriate (e.g., where there was little pre-existing data on which to base power calculations prior to the start of the study or if it is important to know if the intervention is providing a beneficial effect or if it is increasing morbidity). If an interim analysis is needed, it should be conducted by people independent from the trialists and if possible, the trialists should be kept blinded from the results if it does not impact on the subsequent running of the trial. Strict criteria for any interim analysis (end-points, safety data, patient accrual rates, quality issues or complaints) should be established in the design stage. The original protocol should contain details of how the trial will be analysed and have predefined criteria for either stopping the study early (either because the effect being studied is too small and extending the study will not detect it, or the effect is so obvious that a larger study is not needed), or allowing it to continue and perhaps increasing the sample size. Interim analysis may also look at recruitment rates as well as safety and compliance issues. Often, after the interim analysis is complete protocol changes are made after discussion with the trialists, interim monitors, trial sponsor and trial financer. Any substantial changes to the protocol would require ethical committee approval.

4.15 Data Analysis

The exact statistical analysis used will depend on the trial design and type of data collected (e.g., quantitative or qualitative). Collecting accurate data and entering it into a spreadsheet in a timely fashion is essential to avoid any potential for later confusion (e.g., from incorrectly recorded data or if original hard copies of data are lost). Analysis of data from a randomized trial is usually done on an intention to treat basis, where the analysis is based on the initial assigned treatment, not the treatment received. In real life, patients do not take all their medications for a variety of reasons (e.g., side effects), thus it is important to know how well an intervention would be expected to perform and be tolerated if used in a clinical situation. Indeed, one of the "problems" in conducting a clinical trial is that participation is voluntary; thus, the subject may be more or less likely to be compliant with an intervention, than if given to a patient as part of their standard clinical care.

Baseline data particularly on covariates can be used to demonstrate that the random allocation of groups was successful and that both treatment groups were similar at the start of the trial.

4.16 Ethical Considerations

There is a responsibility of the trialist to ensure that the trial design is of the highest quality and that the question being asked is meaningful. It is important for the researcher to be familiar with the ICH-GCP guidelines and any local guidelines that may be in place.

It is vitally important that serious consideration is given before subjecting patients to any risk or inconvenience such that the potential benefits are (1) proportionate to the value of the clinical question being asked, and (2) essential to the correct interpretation of the trial outcome. All clinical research requires approval by a research and ethics committee. These committees will demand that due attention has been paid to the ethical involvement of patients/volunteers, and especially to the involvement of vulnerable patients, such as those with learning difficulties or psychiatric problems. The research trial should be compliant with "The declaration of Helsinki" and subsequent amendments (http://www.wma. net/e/policy/b3.htm). Participants should be informed of what the trial involves and be fully aware of their rights whilst they are taking part. Patients should be given a suitable amount of time to think about their involvement and to ask questions before consenting to take part. They should be aware of the right for them to withdraw from a study at any time, without having to give a reason (and without it affecting their usual care). The involvement of minors or patients who are unable to comprehend the trial design for whatever reason is clearly even more complex. Informed consent needs to be obtained and if dealing with a patient population who are unable to give this then an ability to obtain appropriate consent from carers should be part of the trial design.

Many trials rely on the motivation of a few individuals and the goodwill of patients. The larger and more complex the study, the more difficult it will be to manage any trial to the required standard without providing incentives for trialists and patients. Financial incentives can be beneficial with respect to improved recruitment rates and motivation of trialists, but ethically these can be problematic (Lexchin et al., 2003). If financial incentives are available to trialists, it is possible that they may also be detrimental, as a trialist may ignore recruitment criteria, or recruit patients without the appropriate motivation, for financial gain. Similarly, if financial incentives are provided to study participants, it is possible that they may be interested only in financial gain and not the collection of reliable data.

All studies should have adequate indemnity insurance. Often the trial sponsor, whether a hospital, university or commercial company will be responsible for arranging this.

Any conflicts of interest must be declared when publishing the findings of a trial and many journals will require a declaration of not only where the monies for the trial came from, but also the sources of any relevant past contributions the trialists may have received.

4.17 Misconduct

Trialists have an ethical and legal responsibility to conduct research and present the results honestly. ICH-GCP framework highlights that all clinical trials have to be conducted to a high standard and that all data are made available for audit, thus ensuring compliance with the approved protocol and honesty in data collection. It is essential to have someone knowledgeable and up to date with this framework who is responsible for collecting trial related data in the appropriate way. Commercially sponsored trials often include continuous audit to ensure completeness in data collection and to highlight any potential problems early. It is important to ensure that data is clear and not open to miss-interpretation.

Data should be kept in a secure environment and if stored on a computer, it should be at least password protected. It is a usual requirement that data is not personalised and is linked only to a named patient/volunteer via a trial specific identification number, with the master key being held elsewhere.

4.18 Recruitment, Consent and Data Collection

Trialists should aim not only gain a legal knowledge of the consenting procedure but also improve their practical skills in this important aspect of a trial. How potential volunteers are approached and encouraged to participate in a trial can have a major influence on the way they are recruited and their subsequent compliance with the trial protocol (Donovan et al., 2002). Potential volunteers should be given background information, a trial specific information sheet, and be encouraged to ask questions. As noted above, potential volunteers should be given sufficient time to make up their minds and ask friends or relatives for advice before signing a consent form and should be aware that they can withdrawal from a trial without having to give a reason. Whilst a trialist should be enthusiastic about recruiting to a trial, a degree of realism is required when enrolling volunteers who are not completely comfortable with being in the study: if a volunteer later withdrawals as a result of, for example not having had a full understanding of the requirements of the trial from the outset, then this is counter productive. Conversely, being too negative about the demands of the trial will result in poor recruitment rates.

It is usual to keep a log not only of those who agree to participate in the trial, but also of those who do not and their given reason(s) for declining participation. Potentially this information can be used to help future recruitment.

Patients in clinical trials are often managed more closely than patients not in a trial. Occasionally this could lead to an overall better outcome in these patients. Conversely the improved data collection and patient management may identify higher than predicted complication rates. In order to reduce bias, individuals blinded to which intervention the patient received should collect and interpret the data.

Compliance with a trial should be recorded if practical (e.g., by recording the amount of remaining tablets at the end of the study, concentrations of an agent in stool or urine, or by direct observation). Information on side effects should also be recorded along with comments on the likelihood that the side effect is due to the interventional agent or not.

Trial data should be collected accurately and anonymously whether on paper or computer, and should always be compliant with any data protection laws. There is a requirement for the original data generated to be stored securely and be available for audit. The exact length of time data is required to be stored will depend on the type of trial and local advice should be taken.

4.19 Monitoring Trial Progress and Protocol Deviations

There is a requirement for trials to have a monitoring committee. It is clearly important to identify and deal with any unpredicted problems that may arise during a trial. This ensures compliance with the protocol and that adverse effects are identified and data is processed appropriately. There is a requirement for adverse event reporting to be done throughout a trial. If serious adverse events occur frequently or unexpectedly they can be investigated in a timely fashion and the trial stopped if it meets the pre-defined criteria for stoppage due to such events.

Maintaining the enthusiasm of the trialists, especially over long recruitment periods can be a challenge, especially if the trial is being conducted in more than one centre. Regular contact, meetings, newsletters and feedback on recruitment rates and problems are essential.

There are numerous reasons why trials do not run according to plan. Planning how protocol deviations, non-compliance, and withdrawal of patients are handled, is ideally done in the trial planning stage. Amendments to the trial protocols are often made after the trial has begun based on unforeseen problems that may have been encountered, newly available information from other published studies, or the results of interim analyses. Clearly identifying any issues with design or recruitment early on in the trial is imperative.

4.20 Dissemination of Research Findings

Even before starting a research project the trialist should ideally have given thought as to how the results will be written up and distributed. If the intention is to publish the trial results in a peer-reviewed journal, then the design and implementation will have to conform to the journal's requirements. Thought should also be given as to how publishable the trial will be if the results were not as expected or "null." Although null results are scientifically important, it is often more difficult to publish the results of a trial if it shows no difference between treatments. Anything that can be done to make a study more attractive even if null should be considered, e.g., producing data on the natural history of a disease process, looking at the underlying mechanism of the disease process, or the mode of action of the intervention. Any trial will be considerably enhanced if it not only answers the question as to whether or not the intervention provided benefit, but also if it sheds light on the mechanism of benefit.

The order in which the names of the trialist are presented on reports or publications should reflect their degree of contribution to the study. It is well to agree this order upfront to avoid later disagreements. It is inappropriate to include the names of people who have not contributed significantly to a trial (Hwang et al., 2003; Slone, 1996). At the time of submission for publication some journals require authors to state their contributions to the trial.

Most trials are written up using a standard format of Abstract, Introduction, Methods, Results and Discussion. If you are submitting your findings to a journal for publication, it is advisable to check the "information for authors" section as individual journals may have slightly different requirements. Generally, the best approach is to keep things as clear and succinct as possible. The title should indicate what the trial is about and possibly its results. The abstract should be brief and give the reader a quick overview of what was done, the main results and their implications. Trialists need to be realistic about their findings and it is advisable to avoid extravagant claims.

Introduction: The introduction should give the reader background information on the topic area, the experimental hypothesis and the importance of the research question being asked.

Methods: This section should include what was done in the trial in enough detail so that others could replicate the study if they so wished. Results should not be presented in this section. This section can be subdivided as relevant, e.g., Design, Participants, Apparatus, Procedures, Laboratory analysis and Data analysis.

Results: Traditionally, this section begins with descriptive statistics such as the age of participants in each group and distributions of other baseline variables, such as sex. The text then goes on to list the main findings of the study in an organized manner. The results should be intelligible to most readers. If results are presented in tables and/or as graphs, they should be understandable without reference to the text. Data should be presented without speculating.

Discussion: This should start with a brief statement of the main results. It is then usual to discuss the implications of the results and how they relate to previous studies and the provisional hypothesis. Comment should be made on any problems that may have occurred during the study, the strength and weaknesses of the study, and other possible interpretations of the results. Comment can then be made on the other findings and ideas for further work.

4.21 Summary

- The results of well-conducted clinical trials provide the evidence base for present day medical practice.
- Setting up and seeing a well-designed clinical trial through to completion and publication requires a considerable amount of energy and enthusiasm from both trialists and study participants.
- Trialists should ensure that they are asking a clear and useful question and seek appropriate advice and support where needed.
- The strict requirements to comply with ICH-GCP guidelines can place a significant burden on the trialists, and may require the employment of persons to help run the trials and ensure compliance with the guidelines.
- The choice of interventional agent and outcome measures needs careful thought.
- Obtaining appropriate advice from other trialists and statisticians is invaluable.
- A multidisciplinary team approach is often required for the successful completion of a clinical trial.
- Ensure an appropriate placebo is used and both trialists and subjects are blinded as to the interventional groups.
- Seeing a clinical trial through to completion and publication is a rewarding experience for those involved and the results may directly improve patient care.
- For further reading consult Field and Hole, 2006; Institute of Clinical Research, 2008; Torgerson and Torgerson, 2008; Wang and Bakhai, 2006.

List of Abbreviations

DBPC   Double Blinded, Placebo Controlled
ICH-GCP   International Conference for Harmonization of Good Clinical Practice

References

Aggett PJ, Antoine JM, Asp NG, Bellisle F, Contor L, Cummings JH, Howlett J, Muller DJ, Persin C, Pijls LT, Rechkemmer G, Tuijtelaars S, verhagen H (2005) PASSCLAIM: consensus on criteria. Eur J Nutr 44(Suppl 1):i5–i30

Donovan J, Mills N, Smith M, Brindle L, Jacoby A, Peters T, Frankel S, Neal D,

Hamdy F (2002) Quality improvement report: improving design and conduct of randomised trials by embedding them in qualitative research: Protect (prostate testing for cancer and treatment) study. Commentary: presenting unbiased information to patients can be difficult. Br Med J 325:766–770

Field A, Hole G (2006) How to design and report experiments. Sage Publications, London.

Gilliland SE, Speck ML (1977) Enumeration and identity of lactobacilli in dietary products. J Food Prot 40:760–762

Hamilton-Miller JMT, Shah S, Smith CT (1996) "Probiotic" remedies are not what they seem. Br Med J 312:55–56

Hwang SS, Song HH, Baik JH, Jung SL, Park SH, Choi KH, Park YH (2003) Researcher contributions and fulfillment of ICMJE authorship criteria: analysis of author contribution lists in research articles with multiple authors published in radiology. International Committee of Medical Journal Editors. Radiology 226:16–23

Institute of Clinical Research (2008) ICH harmonised tripartite guidelines for good clinical practice. Institute of clinical research, Marlow

Lewis SJ, Egger M, Sylvester PA, Thomas S (2001) Early enteral feeding versus "nil by mouth" after gastrointestinal surgery: systematic review and meta-analysis of controlled trials. Br Med J 323:773–776

Lewis SJ, Heaton KW (1997) Stool form scale as a useful guide to intestinal transit-time. Scand J Gastroenterol 32:920–924

Lexchin J, Bero LA, Djulbegovic B, Clark O (2003) Pharmaceutical industry sponsorship and research outcome and quality: systematic review. BMJ 326:1167–1170

Slone RM (1996) Coauthors' contributions to major papers published in the AJR: frequency of undeserved coauthorship. AJR Am J Roentgenol 167:571–579

Torgerson DJ, Torgerson CJ (2008) Designing randomised trials in health, education and the social sciences Palgrave macmillan, Basingstoke

Wang D, Bakhai A (2006) Clinical trials: practical guide to design, analysis, and reporting Remedica, London

5 Mechanisms of Prebiotic Impact on Health

H. Steed · S. Macfarlane

5.1 Introduction

Prebiotics were originally defined as non-digestible food ingredients that beneficially affect the host by selectively stimulating the growth and/or activities of one or a limited number of bacteria in the colon, thereby improving host health (Gibson and Roberfroid, 1995). However, a more recent definition is that "A prebiotic is a selectively fermented ingredient that allows specific changes, both in the composition and/or activity in the gastrointestinal microbiota that confers benefits upon host wellbeing and health" (Gibson et al., 2004). The principal concept associated with both of these definitions is that the prebiotic has a selective effect on the microbiota that results in an improvement in the health of the host. Common prebiotics in use include inulins, fructo-oligosaccharides (FOS), galacto-oligosaccharides (GOS), soya-oligosaccharides, xylo-oligosaccharides, pyrodextrins, isomalto-oligosaccharides and lactulose. The majority of studies carried out to date have focused on inulin, FOS and GOS (Macfarlane et al., 2008).

To be effective, prebiotics need to reach the large bowel with their chemical and structural properties essentially unchanged. Experimental evidence shows that feeding inulin to ileostomy subjects allows recovery of between 86 and 89% of what is fed. Breath tests after intake of prebiotics show an increase in hydrogen excretion, and no increase in blood glucose or insulin, after feeding healthy individuals 25 g of neosugar (fructo-oligosaccharides). Incubation of fructans in homogenized rat intestinal mucosa show a hydrolysis rate less than 1% of that of sucrose. Prebiotics are able to escape the rigors of digestive processes in the upper gut due to their molecular and structural composition, which renders them essentially resistant to mammalian digestive enzymes. In the main, prebiotics have been considered to be short-chain carbohydrates that have a degree of polymerization of two or more, and which are not susceptible to digestion by pancreatic and brush border enzymes.

5.2   Local and Physiological Effects of Prebiotics

5.2.1   Mucosal Structure

Prebiotics can have direct trophic effects on the structure of the colonic mucosa. Studies on rats fed chicory inulin and oligofructose, or pectin for 4 weeks caused an increase in numbers of epithelial cells, and the intensities of their secretory functions (Poldbeltsev et al., 2006). The prebiotic diet also increased the length and width of colonic crypts, which enlarged the available area of nutrient absorption, as well as micronutrients and trace elements in the gut. Studies on the rat caecal and colonic mucosa found that a concentration of 10 g/l of a FOS/inulin mixture (Raftilose® Synergy 1™) increased the macroscopic surface area 2.2-fold, and tissue wall weight increased 2.4-fold (Raschka and Deniel, 2005). Analysis of male Sprague-Dawley rat mucosa in animals fed the indigestible saccharide difructose anhydride III showed significant increases in the outer and inner circumferences of caecal epithelial tissue ($P < 0.005$ and <0.001, respectively), a significant increase in crypt depth ($P < 0.001$), and a significant increase in total cell numbers ($P = 0.001$), particularly goblet cells (Mineo et al., 2006).

The exact mechanisms whereby these morphological changes occur are not known, but butyrate, a major product of bacterial carbohydrate fermentation in the gut, is known to have trophic actions that stimulate mucosal proliferation (Blottiere et al., 2003), while lactic acid stimulates mitosis in the rat caecal epithelium (Ichikawa and Sakata, 1997). It is therefore possible that SCFA are responsible for these adaptive changes. This notion is supported by the fact that these trophic changes do not occur in prebiotic fed germ-free animals, implicating that bacterial metabolic processes are responsible for inducing major architectural alterations in the gut wall.

Administering prebiotics appears to increase the permeability of the paracellular pathway, which is dependent on tight junctions that link adjacent epithelial cells and large multi-protein complexes. Using the Caco-2 intestinal cell line model, the administration of non-digestible oligosaccharides (NDO) was found to rapidly increase paracellular ion transport. Studies are underway to characterize the molecular and physiological changes that occur in these important structures, to see if prebiotics can locally alter epithelial cell junction bonding structures.

5.2.2   Intestinal Mucus

Intestinal mucus is a cysteine-rich glycoprotein made up of high molecular weight mucin (m.w. $> 10^6$) and trefoil factor peptides, produced by goblet cells lining the gut. It is one of the first lines of defense in the intestine, and forms a thick slimy layer on mucosal surfaces. The mucus layer is a dynamic barrier, and the density and number of goblet cells varies throughout the digestive tract, being thickest in the most distal part of the colon. It is also more highly sulfated in the distal bowel, where bacteria are at their highest numbers, giving it a more negative charge to make it less sensitive to bacterial enzyme attack. Mucus is a source of nitrogen and carbon for bacteria; its continuous production by the host makes the gut an inviting niche for microorganisms, despite the fact that very few (culturable) bacteria are able to produce the complete panel of enzymes required to degrade the macromolecule. These organisms include certain species belonging to the genera *Bifidobacterium*, *Bacteroides* and *Ruminococcus*.

Some lactococci can increase the production of trefoil factor peptides in mice, and these substances are able to increase the viscosity of the mucin (Cummings and Macfarlane, 1991). Administration of dietary prebiotics appears to thicken the mucus layer and increase its secretion by goblet cells. The precise mechanism is not well understood, but there is an association with the presence of bacteria and a global increase in the synthesis and secretion of mucin.

5.2.3   Phytic Acid and Mineral Bioavailability

Phytic acid is found mostly in legumes, and possesses apparent anti-carcinogenic properties, providing it reaches the colon without degradation. Phytate is a molecule with six charged phosphate groups, which allows it to bind mineral cations such as zinc, iron and calcium, making them unavailable for absorption by the body. Iron deficiency is associated with anemia, calcium deficiency is associated with osteoporosis, and zinc is needed for skeletal growth and maturation. Research shows that there is a strong inverse relationship between the amount of phytic acid in the diet and iron absorption. Only very small amounts of phytic acid (0.7%) in the diet are needed to halve iron absorption. However, prebiotics are known to have stimulatory effects on iron absorption in the large bowel, by increasing the soluble fraction of iron found in caecal cells, and enhancing absorption by as much as 23% ($P < 0.001$).

The ratio of phytic acid to zinc in the diet is associated with progressive inhibition of metal ion absorption, and in human studies, phytate is strongly associated with decreased zinc uptake in healthy young people, elderly subjects, and in rats. Addition of FOS to the diet restores zinc absorption by enhancing zinc bioavailability ($P < 0.05$).

Studies in the rat caecum have shown that bacterial metabolism results in phytate hydrolysis, and that prebiotic consumption is associated with enhanced breakdown of the molecule (60%), and significantly reduced excretion in feces. This is also linked to greater cation absorption, which is associated with increased bacterial metabolic activities. Organic acids produced during fermentation, particularly SCFA, form soluble ligands with cations to prevent the formation of insoluble mineral phytates, as well as inducing phytase enzymes. The direct effect of SCFA on pH has also been associated with decreased solubility of phytate-mineral complexes.

There have been fewer studies on whether fiber or prebiotics can affect copper bioavailability. Administration of FOS with high and low degrees of polymerization decreased the absorption of copper and other minerals in rats (Delzenne et al., 1995). However, in human studies, administration of FOS increased copper absorption by 45% ($P < 0.05$, Lopez et al., 2000) and 10 g/day short-chain FOS significantly enhanced its absorption in postmenopausal women (Ducros et al., 2005). This increased bioavailability of copper may also depend on the presence of phytic acid in the diet, since this has been shown to decrease total body retention of this metal ion in rats by 57%. However, in one human investigation, feeding high fiber or phytic acid to young men did not have a marked effect on calcium absorption (Turnlund et al., 1985), and phytic acid enhanced copper bioavailability in copper deficient rats (Lee et al., 1988).

5.2.4 Release of Bone-Modulating Factors

FOS increase the bone-preserving effects of phyto-oestrogens in non-ovariectomized mice and rats. Prebiotics also stimulate the growth and metabolic activities of some bifidobacteria and lactobacilli, which can increase synthesis of the luminal bacterial enzyme β-glycosidase, which hydrolyses the glycosidic bond of isoflavone conjugates. This allows flavonoids to be more rapidly absorbed in their free aglycone form, rather than as an intact glycoside. Improvement of isoflavone bioavailability as an isoform with a higher bone-preserving potential may be an important mechanism that improves bone preservation.

Polyamine production by bacteria in the gut modulates changes in bone structure in rats, in a dose-dependent manner. In ovariectomized rats fed a prebiotic diet, studies using microcomputed tomography to assess bone composition analysis showed that higher levels of polyamines resulted in greater trabecular numbers and areas.

5.3   Modulation of the Gut Microbiota

Evidence from human feeding trials, animal models and *in vitro* modeling systems has shown that prebiotics affect the composition of the gut microbiota, leading to an increase in health-promoting organisms such as bifidobacteria and lactobacilli. These bacteria are generally regarded as safe because they mainly ferment carbohydrates, are not pathogenic and are non-toxigenic, while they have a role in colonization resistance and frequently manifest immunomodulatory properties in the host. Some species are also able to ferment prebiotics to SCFA such as acetate and butyrate, which are important sources of energy for the host. While bifidobacteria do not produce butyrate, they have been shown to stimulate butyrate producing bacterial species such as eubacteria in the gut (Belenguer et al., 2006). SCFA also play a role in regulating growth and cellular differentiation, colonic epithelial cell transport processes, and hepatic control of lipid and carbohydrate metabolism. One advantage that prebiotics have over probiotics is that the target bacteria are already present in the host; however, it should be noted that if the organisms required to promote health are not already present in the gut, due to disease, for example, the prebiotic might manifest no useful effects. Studies with prebiotics have shown that in certain cases, they are able to reduce the numbers of some groups of bacteria in the gut, such as clostridia, bacteroides, enterococci and enterobacteria, some members of which may have a detrimental role in host health. Some of these organisms, particularly the clostridia, are directly toxigenic and are able to breakdown proteins, and ferment their component amino acids, resulting in the production of toxic metabolites such as indoles, phenols, ammonia, thiols, H_2S and amines which may be involved in colorectal cancer (see later).

However, to date the selectivity of most prebiotics has not really been proven since the majority of studies have only looked at changes in a few large groups of bacteria, at group or genus level, which does not really show what is happening in a global sense. The sugar composition, and degree of polymerization of the prebiotic, together with the availability of other carbohydrates, all affect the way in which bifidobacteria (and other saccharolytic species) are able to grow on these substances, since individual species and strains have specific substrate preferences. Consequently, their utilization can vary markedly in different species. Increases in other groups of bacteria such as eubacteria and roseburia have been found in studies with inulin, while species belonging to other genera such as clostridia and bacteroides have been reported to be able to ferment FOS and GOS, respectively. The increasing use of molecular methods of microbial community analysis, which are able to detect difficult-to-culture species in the gut microbiota, will further highlight the specificities of individual prebiotics. Although much work has been done, there is still a real need for detailed qualitative and quantitative assessments of the gut microbiota to be made, so that the effects that changing microbiota composition has on the nutrition and immunity of the host can be fully understood.

In addition, it may be useful to look at the effects of prebiotics not only on fecal populations, but on mucosal microbiotas in the gut. Recently, a daily combination of 7.5 g inulin and 7.5 g FOS was shown to increase levels of mucosal bifidobacteria and eubacteria (Langlands et al., 2004). This may be useful in the treatment of gastrointestinal diseases such as ulcerative colitis (UC), where decreased levels of bifidobacteria and a general dysbiosis in mucosal bacterial populations have been shown to exist (Macfarlane et al., 2004).

5.4   Immune System

There is increasing interest in modulation of the immune system using prebiotics, which may be particularly useful in inflammatory conditions, or in children and the elderly. Evidence so far suggests that prebiotics can have significant effects on the immune system (⊙ *Table 5.1*). It is however, unknown if these are direct or indirect effects resulting from stimulation by immuno-modulatory bacteria, or production of SCFA, which are known to have immunomodulatory properties, and can bind to SCFA G protein coupled receptors on immune cells within gut-associated lymphoid tissues (Brown et al., 2003).

Addition of FOS and lactulose to the diet has been shown to increase mucosal immunoglobulin production, mesenteric lymph nodes, Peyer's patches and altered cytokine formation in the spleen and intestinal mucosa (Schley and Field, 2002). Investigations on the effects of prebiotics on the immune system

⬛ Table 5.1

Modulation of immune function by prebiotics

Beneficial effects on gut associated lymphoid tissue (GALT) and mucosal immune system (MALT)
Increase in mucosal immunoglobulin, altered cytokine and lymphocyte expression, increase in secretory IgA
Indirect effects
Stimulation of immunomodulatory bacteria (bifidobacteria, lactobacilli)
Increased production of SCFA and other fermentation products
Butyrate suppresses cytokine induced and constitutive expression of NF_KB in HT29 cell lines
Propionate is anti-inflammatory to colon cancer cells
Acetate increases peripheral blood antibody production and natural killer cell activity in cancer patients
Pyruvate is anti-inflammatory and decreases NF_KB expression
Through specific G-protein coupled SCFA receptors (GPR41 and GPR43) on immune cells
Direct effects
Little data showing direct effects of prebiotics on immune function

require careful assessments of the choice of markers, which will vary, and be dependent on the condition under study.

Animal studies have indicated that mice fed with FOS and inulin have an improved response to salmonella vaccine. At 1 week post-immunization, splenic cell cultures were shown to have increased production of cytokines such as INF-γ, IL-12 and TNF-α as well as salmonella-specific blood IgG, and fecal IgA. Overall, the implication is that the FOS/inulin mix stimulated host mucosal immunity to produce a greater response to the salmonella vaccine.

5.5   Lipid Metabolism

Experimental investigations in animals, and a few human studies, have shown interesting cholesterol and/or triglyceride lowering effects, and have raised the question of possible lipidaemia and cardiovascular benefits associated with prebiotic consumption (Jackson et al., 1999; Vigne et al., 1987). Human studies have been small in scope, few and far between, and have focused on the relationship between the intake of prebiotics and serum lipid levels. The results have been inconsistent, and any mechanisms of action unclear.

Rats supplemented with oligofructose have significantly lower serum phospholipid and triacylglycerol levels, and in particular, significantly reduced serum very low density lipoprotein (VLDL), which is produced in the liver by assembling lipids and apoproteins. This effect was caused by a reduction in hepatic synthesis of triacylglycerol, because hepatocytes from these animals have a 40% lower capacity to synthesize this molecule (Fiordaliso et al., 1995; Kok et al., 1996). This is due to a 50% reduction in the activities of key hepatic lipogenesis enzymes; FAS, malic enzyme, ATP citrate lyase, acetyl-coA carboxylase, and glucose-6-phosphate 1-dehydrogenase, as well as a reduction in fatty acid synthase mRNA. The proposed mechanism of action, therefore, is that prebiotics, in some way, modify gene expression of lipogenic enzymes.

In rats fed a high fat, oligofructose-supplemented diet, postprandial triglyceridaemia was halved, and there were no raised free cholesterol levels in the plasma, which would usually be expected after a high fat meal (Kok et al., 1998). These findings implicate an extrahepatic effect of lipid metabolism, which as yet is not fully understood. However, it is postulated that there is a link with prebiotic effects on insulin, which potentiates the gene expression effects. Another possible mechanism is the production of propionate during fermentation, which has been shown to inhibit fatty acid synthesis *in vivo*. When acetate enters the hepatocyte, it is activated by enzymes and then enters the lipogenesis pathways, but propionate competes with the protein that promotes acetate entry to the hepatocytes. Consequently a proposed mechanism of action is the prebiotic's ability to alter the acetate to propionate ratio in the cell.

In obese animals fed a fructan-supplemented diet, no effect was seen on postprandial triglyceridaemia, but there was a reduction in hepatic steatosis. Reductions in fat mass and body weight were also observed, and were presumed to be due to a reduced availability of fatty acids from adipose tissue (Daubioul et al., 2000). Fat digestibility is significantly decreased in dogs receiving an oligosaccharide diet, which may be another mechanism of reducing lipid levels, and the general effects seen in other studies.

A direct effect of alterations in the composition of the microbiota may also be partly responsible for prebiotic manifestations in lipid metabolism. The evidence for this rests with probiotic studies in which lactobacilli and bifidobacteria given in dairy products are known to have a cholesterol-lowering action, which may be due to the production of propionate from lactate, or by enhancement of bile acid deconjugation. However, this notion has to be interpreted with a degree of caution, because in some trials, probiotics have had no demonstrable effects on serum triacylglycerol or cholesterol, while using the same bacteria in synbiotic studies have demonstrated cholesterol-lowering effects.

5.6　　Mineral Absorption

One of the most significant health effects of prebiotics on mammalian physiology is their abilities to improve calcium, magnesium, iron and zinc absorption, and the attendant enhancement of bone mineralization and can be seen in ❯ *Figure 5.1*, several mechanisms have been proposed for prebiotic action in mineral absorption. Although human studies have been limited and small in scale, this could be beneficial in preventing osteoporosis, a common and often painful disease, as well as in avoiding diet-related anemia and enhancing micronutrient absorption to avoid states of malnutrition.

Intestinal calcium absorption has been observed to increase by up to 20% in animal and human studies, and *in vivo* measurements also show increased calcium retention (Coudray et al., 1997; Griffin et al., 2002). In humans, calcium is mostly absorbed in the small intestine, and prebiotic feeding studies ileostomists have failed to demonstrate increased calcium absorption, suggesting that prebiotics were affecting these processes in the large intestine. Several investigations have confirmed findings suggesting that some calcium is absorbed from the colon, and prebiotic metabolism is thought to increase large intestinal calcium uptake.

There are a number of mechanisms whereby this could occur. Fermentation of prebiotics, as discussed earlier, acts to lower intraluminal pH in the large

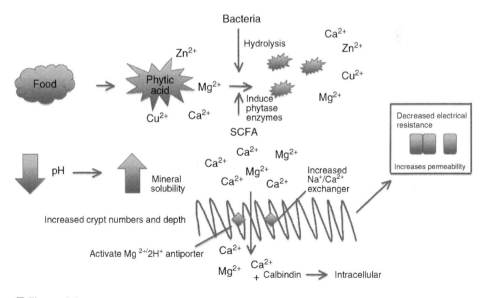

□ Figure 5.1
Prebiotics mechanisms in mineral absorption.

bowel, thereby enhancing calcium solubility and bioavailability for absorption (Dupuis et al., 1978). Calcium moves across the epithelium via calcium channels, and by ATPase extrusion. Increased calcium solubility increases the gradient across the epithelium and promotes passive uptake, and most studies have shown that net calcium absorption only occurs when a downhill concentration gradient prevails. However, detailed rat experiments have looked at lowering pH, with and without the use of SCFA, and it was found that the calcium flux across the epithelium was only increased in the SCFA experiment, implicating other potential mechanisms of divalent ion transport (Raschka and Deniel, 2005). These authors used paracellular markers, and recorded transepithelial electrical resistance. It was found that the addition of SCFA to rat mucosae increased the permeability of the marker, and decreased electrical resistance. They concluded that some prebiotics interact with the tight epithelial cell junctions, and thereby increase the paracellular permeability of minerals.

Other potential mechanisms that have been proposed for enhanced divalent metal ion uptake have been based on *in vitro* studies on calcium absorption from the ovine rumen. They suggest direct uptake of calcium, with colonic uptake of SCFA, and a direct fermentation effect on gene expression of proteins that are linked to sequestration and mucosal ion binding. This has been confirmed by experiments measuring mRNA levels of mucosal target genes, which found increased transcription of $Na+/Ca^{2+}$ exchanger and calbindin, which would allow increased intracellular binding of calcium, and the basolateral membrane extrusion rate.

Studies looking at the effects of different chain length, and types of branching of inulin-type fructans have found no significant differences in calcium or magnesium absorption, with the exception of the combination of oligofructose and HP-inulin, which appeared to work synergistically to significantly increase the absorption of calcium (Coudray et al., 2003). A mixed diet of inulin and FOS in ovariectomized rats showed a significantly suppressed bone resorption rate relative to the formation rate. However, the mechanism of suppressed resorption is not clear, and requires further investigation.

Magnesium absorption has been specifically linked to the lactate pool in the gut, and low pH, but not the presence of SCFA. Lactic acid is more acidic than SCFA, implying that the mechanism is the act of lowering the pH directly absorption. Despite this, one of the postulated mechanisms of enhanced magnesium absorption with prebiotics is a direct contribution by SCFA via a cation exchange mechanism, where SCFA stimulate the flux of magnesium ions by activating the $Mg^{2+}/2H^{+}$ antiport. This has been demonstrated in *in vitro* studies only, using rat hindgut segments. Butyrate was shown to be the most effective SCFA in stimulating the magnesium flux in the mucosal to serosal direction (Kashimura et al., 1996). Absorption of iron from the gut has also been correlated with the presence of SCFA, and to increased weight of the caecal wall, but the precise mechanisms of enhanced iron absorption need to be clarified (Asvarujanon et al., 2005).

Assuming these mechanisms of increasing calcium absorption are operative in humans, prebiotic consumption may be an effective adjuvant to oral calcium for osteopenia and osteoporosis, or they could be administered early to enhance peak one mass prior to the onset of demineralization, which has been observed in rats (Zafar et al., 2004).

5.7   Infants

Oligosaccharides are prebiotic factors in human milk, and increased levels of bifidobacteria in breast-fed babies compared to bottle-fed infants is thought to be due to their ability to utilize these substances. The putative ability of GOS to resemble glycoconjugate receptors on cell surface receptors may also offer protection from pathogenic microorganisms. Prebiotics have been used in infant formulas in Japan over the last 2 decades, and in Europe for the last 5 years. A large number of trials have been carried out in infants, with the majority aimed at determining the ability of the prebiotic to increase levels of fecal bifidobacteria. After repeated trials in babies, the Scientific Committee on Food for the European Commission published a statement to the effect that the addition of 0.8 g/dl of a mixture of 10% short chain FOS and 90% long chain infant formulas was safe to add to infant formula.

5.7.1   Atopic Disease

At birth, babies have acquired high levels of Th2 cytokines from the mother. The Th1/Th2 balance is redressed as the sterile gut becomes colonized after birth. Allergic disease in infants is based on IgE-mediated food allergies, and a Th2-biased response and lower numbers of bifidobacteria are found in allergic infant feces (He et al., 2001; Kalliomaki et al., 2001). Based on this, a logical mechanism of action of prebiotics in allergic disease is their bifidogenic and immune-mediated effects. This idea has been supported by studies administering raffinose and alginate-based oligosaccharides to allergic infants, which induce a reduced Th2 response (Nagura et al., 2002; Yoshida et al., 2004), presumably by stimulating the Th1-response, and rebalancing the immune response to increase anti-inflammatory cytokines such as IL-10 and TGF-β (Kelly et al., 2005).

Prebiotic feeding studies in allergic infants have demonstrated significant reductions in the incidence of atopic dermatitis, and this was associated with increased numbers of fecal bifidobacteria, although the stools had been frozen prior to analysis, which may have affected bacterial recoveries (Bonten et al., 1997; Moro et al., 2006).

Part of the mucosa-associated immune system (MALT) of the gut includes the large amounts of sIgA, which protect against pathogenic bacteria and viruses adhering to, or invading the intestinal mucosa. IgA secretions are partly degraded by intestinal bacteria, but the antibody coats the organisms, preventing the host having immune reactions against the commensal species i.e., immune tolerance. Formula-fed babies are more prone to allergic disease than those that have been breast-fed, and they are also known to have lower concentrations of sIgA. These levels can be increased by bifidobacteria, lactobacilli and FOS in animals (Moreau and Gaboriau-Routhiau, 2000; Nakamura et al., 2004). A feeding study by Scholtens et al. (2008) involved 215 healthy infants in the first 26 weeks of life, where the infants were randomized to a combination of GOS and FOS or placebo, and used ELISA to measure fecal sIgA, found significantly higher levels in the prebiotic group, at 719 mg/g, compared to 263 mg/g in the placebo group ($P < 0.001$), and higher levels of bifidobacteria ($P < 0.04$, Scholtens et al., 2008). There were also lower numbers of clostridia in the prebiotic group ($p = 0.006$), which was interpreted as evidence of a positive effect on mucosal immunity.

Possible mechanisms of action are for organisms supported by the prebiotic to immunomodulate the host pathway, as described earlier, and INF-γ is known to stimulate the expression of the secretory component of IgA by epithelial cells. However, in prebiotic feeding studies in the mouse, no correlation was found between caecal sIgA concentrations, and changes seen in interferon production, arguing against a potential role for this cytokine.

5.7.2   Necrotising Enterocolitis

In premature infants, bifidobacterial colonization is delayed in favor of high levels of enterobacteria and clostridia, and some premature illnesses, for example necrotising enterocolitis (NEC) are associated with these organisms e.g., klebsiella, and *E. coli*. Breast-feeding has always been thought to protect against NEC, and in rat models, bifidobacterial colonization reduces the risk of NEC by modulating the inflammatory cascade. In the quail model, use of pre-biotics and probiotics reduces caecal lesions in NEC, but there have been no human studies using prebiotics alone. The animal mechanisms have only been studied in so far as to note changes in the bacterial populations, specifically an increase in lactobacilli and bifidobacteria. Presumably, many of the local actions of prebiotics would also be helpful in prevention of this disease, such as the antimicrobial effects, and increased secretory function.

5.7.3 Infection Prevention

Studies in children aimed at prevention of infection have also had mixed results. The addition of 1.1 g of oligofructose daily to cereal in a DBRCT of 123 infants (4–24 months) was associated with reduced episodes of fever and medical visits. The control group had more sick days, and a higher intake of antibiotics. The treatment group had fewer episodes of emesis, regurgitation and perceived discomfort, but there were no changes in growth, constipation and flatulence. However, in a study of 140 children aged 1–2 years, who were receiving antibiotics for bronchitis, the subjects were randomized to a prebiotic or control formula for 3 weeks. Significant increases in fecal bifidobacteria were observed in the prebi-otic group, but there were no differences in gastrointestinal side-effects associated with antibiotic use in either group. In another double-blind randomized control trial (DBRCT) involving 134 infants fed prebiotics (8 g/l GOS/FOS) for the first 6 months of life, principally looking at allergic disease, the subjects were followed up until they were 2 years of age. Growth was assessed together with infectious episodes as a secondary endpoint (Arslanoglu et al., 2008). Not only was the incidence of atopic disease reduced in the prebiotic group ($P < 0.05$), but there were fewer episodes of physician diagnosed infections ($P < 0.01$), fewer episodes of fever ($P < 0.00001$), and fewer antibiotic prescriptions ($P < 0.05$). However, the study did not demonstrate mechanisms of action for any of the effects seen, in what was a relatively well-powered investigation.

The main problem with most prebiotic studies in children is that they have largely relied on demonstrating increased levels of bifidobacteria and lactobacilli as evidence of effectiveness. In reality, the bifidogenic effect is well documented. What are now needed are studies linking clinical benefits to more precise mechanisms of action, be they local, systemic, immune or microbiological.

5.8　Gastrointestinal Effects

5.8.1　Irritable Bowel Syndrome

Irritable bowel syndrome (IBS) has been linked to intestinal bacteria in a number of different trials. Culture-based techniques have shown lower levels of bifidobacteria and lactobacilli in IBS patients, but this has not been confirmed in DNA studies. IBS is a difficult disease to investigate, because the symptoms are highly subjective and difficult to quantify, making open label studies inadequate and of limited usefulness.

The main products of prebiotic metabolism in the large bowel are bacterial cell mass, SCFA, and the gases carbon dioxide and hydrogen. Production of gaseous by-products can be a significant disincentive to prebiotic usage, and unwanted gas-related symptoms have been widely reported in prebiotic feeding studies. There have been mixed results in breath hydrogen studies, some demonstrating no change after a 10 g challenge with FOS, and others showing dose-related increases in breath hydrogen, borborygmi and mild flatulence, with intakes ranging from 5 to 20 g/day. Perhaps unsurprisingly, inulin has been reported to have much the same effects as FOS, and doses of 14 g can cause significant increases in flatulence, together with colicky abdominal pain and bloating, which was considered to be unacceptable to 12% of the volunteers taking part in these trials. Undoubtedly there is widespread variation amongst subjects in response to the potential adverse effects of prebiotic fermentation, but there is no question that the production of gas as a major byproduct of fermentation can cause abdominal discomfort, belching, bloating and flatulence. These are amongst the principal symptoms of IBS.

Prebiotic studies in this area have reported symptomatic improvements in general health and nausea ($P = 0.042$), indigestion and flatulence ($P = 0.008$) and diarrhea ($P = 0.003$), but, as yet, there are no major investigations using prebiotics in IBS. It remains likely that there is a subset of patients, possibly those individuals with constipation-predominant IBS, who may benefit from prebiotics, but since many have visceral hypersensitivity, prebiotic fermentation and gas production may exacerbate their symptoms, and have adverse affects on health and wellbeing.

5.8.2　Constipation

Laxative effects have been documented with a number of prebiotics, although as a treatment for constipation, their benefits are limited and unclear. The mechanism of increasing bowel habit depends on the fermentation of prebiotics to produce SCFA, as well as an increase in bacterial cell mass, leading to stimulation of gastrointestinal peristalsis. The influence of prebiotics on constipation has been assessed in several trials. In two studies by the same group on elderly constipated volunteers, one showed a significant increase in stool weight from 32.4 to 69 g/day with 10 g daily consumption of oligofructose, in the other investigation, a 70% increase in stool weight was found with isomalto-oligosaccharides (Chen et al., 2000, 2001). Gibson et al. (1995) showed that 15 g of FOS could signifi-cantly increase stool output from 136 to 154 g/d in a small group of subjects (n = 8). However, this is contrasted by other studies that failed to show any increase in fecal weight when the volunteers were given galacto-oligosaccharides. With the exception of lactulose, the prebiotics studied so far in human trials have been shown to have little effect on managing constipation, and to have only mildly laxative properties. This variability in outcome may be due to the difficulty in measuring daily fecal outputs, while the methods used to measure constipation are mostly qualitative, using bowel habit diaries and patient questionnaires.

5.8.3   Infectious and Antibiotic-Associated Diarrhea

There have been a number of animal studies investigating the efficacy of a prebiotic diet in preventing colonization by pathogenic microorganisms, showing the inhibition of *Salmonella* Typhimurium survival in the gut lumen, and reduced pathogen densities in Peyer's patches. One postulated mechanism involves loss of adhesion sites for pathogenic bacteria, due to the increased number of bifido-bacteria in the gut, and by the prebiotics acting directly and blocking adhesion to epithelial cells by functioning as receptor analogues (◉ *Figure 5.2*). This has been translated into human studies looking at the prevention of traveler's diarrhea in a DBRCT, which showed a non-statistically significant reduction (P = 0.08) in the frequency of diarrhea in the prebiotic group (11.2%), compared to the placebos (19.5%) (Cummings et al., 2001). A large study of Peruvian children assessed the frequency and severity of diarrhea in a DBRCT using prebiotic supplementation, and found no benefit in the prebiotic group (Duggan et al., 2003). One reason for the failure to demonstrate any difference in this population is the large numbers of mothers who breast fed, because breast milk contains naturally contains high levels of prebiotics.

Clostridium difficile infections are currently of great interest, and preventative studies for antibiotic-associated diarrhea (AAD) have found that prebiotics

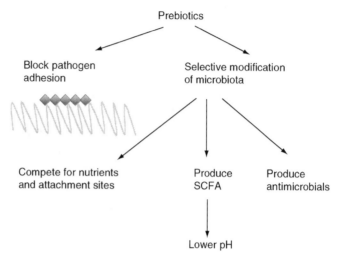

■ Figure 5.2
Mechanisms of prebiotic action against pathogens.

reduce episodes of AAD and *C. difficile* diarrhea. These investigations have been criticized for excluding the most at risk population groups, and have failed to identify mechanisms of action, but the results are interesting and appear to suggest an ability to prevent relapse, and possibly primary disease, and will inevitably lead to more trials (Lewis et al., 2005).

5.8.4   Inflammatory Bowel Disease

Inflammatory bowel diseases (IBD) such as ulcerative colitis and Crohn's disease are multifactorial disorders associated with reduced levels of bifidobacteria, and increased levels of putatively pathogenic micro-organisms such as *E. coli* and peptostreptococci. Immune tolerance in the bowel mucosa is usually strictly maintained, together with the balance of Th1 and Th2 response profiles. The inflammatory cascade is activated in IBD but then remains inappropriately activated. Therefore, potential mechanisms of prebiotic action in IBD include the bifidogenic effects already described, as well as immunomodulation of the Th1 and Th2 immune responses.

Prebiotic feeding in animal studies with induced colitis, using inulin, FOS or GOS have found improvements in levels of mucosal inflammation. This is associated with reduced inflammatory markers, including TNF-α, tissue myeloperoxidase activities (an index of neutrophil infiltration), leukotriene B4, IL-1β, thromboxane B2 and prostaglandin E2 ($P < 0.05$, Videla et al., 2001), reduced pH ($P < 0.001$), increased fecal lactate ($P = 0.02$) and butyrate concentrations ($p < 0.001$), and increased proliferation of lactobacilli and bifidobacteria. Murine studies have also demonstrated histological improvements using prebiotics, including reduced severity of crypt damage, that are associated with increased caecal butyrate levels, an attenuation in NFκB activation, serum IL-6 levels and mucosal STAT3 expression (Kanauchi et al., 2003). Treatment of colitis-induced Sprague-Dawley rats with dietary prebiotics have found accelerated colonic epithelial repair and increased caecal butyrate concentrations.

Human prebiotic feeding trials have been limited, and have focused on the outcome rather than the mechanism of action. There are only three reports, one using inulin, one using FOS, and the other employing a combination of the two. They have shown endoscopic and histological improvements in inflammation, which were associated with increased butyrate levels, reduced pH and significantly increased fecal bifidobacterial load ($P < 0.001$). In the mixed prebiotic study, there was an increase in the percentage of IL-10 positive dendritic cells ($P = 0.06$), and of dendritic cells expressing TLR2 and TLR4 ($P = 0.08$ and $P < 0.001$, respectively).

A prebiotic study in 20 patients with pouchitis, which is a clinical condition in patients who have had an ileal pouch-anal anastomosis after total colectomy, used 24 g of inulin per day, for 3 weeks. This study showed improvements in histological and endoscopic scores that were associated with increases in fecal butyrate, and reductions in bacteroides counts ($P < 0.05$) (Welters et al., 2002).

Synbiotic studies have also demonstrated immune changes in IBD, but it is unclear as to what can be attributed to the probiotic, the prebiotic, or to synergy between the two is unclear, and requires delineation. For example, the use of *Bifidobacterium longum* and Synergy I as a synbiotic in a pilot study of UC showed that mucosal TNF-α was significantly reduced, as were inducible human beta defensins 2, 3 and 4, which are specific epithelial markers of inflammation in epithelial cells (Furrie et al., 2005). IgA alterations are discussed in the infants and allergic disease section of this chapter, but there is no record of fecal or MALT sIgA being measured in any IBD studies.

In summary, the use of prebiotics in IBD is a relatively neglected area, in that most studies have tended to focus on the therapeutic benefits of probiotics. As to the mechanistic effects, there are well documented immunological and bifidogenic changes associated with prebiotic use in animal models, and to a lesser extent, in humans. It seems likely that some of the local actions of prebiotics improve the condition of IBD patients, but as yet, mucosal changes are just referred to as reduced damage, or histological improvements, rather than specific assays of crypt numbers and depth.

5.9   The Elderly

As people get older they often have a greater susceptibility to diseases and suffer from an increase in gastrointestinal infections, malnutrition, constipation and diarrhea. There is evidence for a decrease in immune function and a reduction in numbers and species diversity of beneficial bacteria such as bifidobacteria, along with an increase in potentially harmful organisms such as clostridia, enterococci and enterobacteria in older people (Mitsuoka, 1992; Woodmansey et al., 2004). There have been few studies on the use of prebiotics in the elderly, however, in one investigation involving feeding a synbiotic comprising 6 g of the prebiotic Synergy 1 given in combination with a capsule containing 10^{10} cells each of *Bif. bifidum* and *Bif. lactis* to nine healthy elderly volunteers twice per day, the numbers and diversity of fecal bifidobacterial populations increased significantly in the synbiotic group, compared to the placebos who were given maltodextrin (Bartosch et al., 2005).

5.10   Cancer

Colon cancer is the third most common cancer worldwide, and evidence suggests that along with genetic and lifestyle factors, it is associated with diets high in animal fats and proteins and reduced intakes of fruits and vegetables. Evidence indicates that intestinal bacteria are intimately involved in disease aetiology, and many species have been shown to be able to produce genotoxic and mutagenic products from food components. Thus, supplementing the diet with prebiotics to increase the availability of fermentable carbohydrate, reduce proteolysis and modulate bacterial species to decrease the production of mutagenic and toxic metabolites may protect against large bowel cancer. Stimulation of SCFA production by intestinal bacteria is a mechanism whereby prebiotics could play a role in reducing colorectal cancer (CRC). Butyrate has been shown to stimulate apoptosis, and may be protective in cancer prevention, while propionate has been shown to be anti-inflammatory to colon cancer cells. Prebiotics may also suppress the activities of microbial enzymes involved in genotoxicity and production of cancerogenic metabolites in the large gut. *In vivo* studies administering 4 g of FOS

(Neosugar) per day found that feeding the prebiotic decreased the activities of β-glucuronidase and glycocholic acid hydroxylase (Buddington et al., 1996), which are associated with carcinogen formation. In other fecal studies, 12.5 g of FOS per day had no effect on β-glucuronidase, nitroreductase or azoreductase, although the metabolic significance of fecal enzyme activity measurements is difficult to interpret. But in a larger study involving 53 healthy subjects in a 4 week randomized crossover study, the combination of lactulose and inulin significantly reduced fecal β-glucuronidase activity (De Preter et al., 2008).

Results from several investigations have demonstrated that prebiotics can reduce aberrant crypt foci, decrease cellular proliferation, and lower the incidence of tumors in animal models of colonic cancer. However, to date, there have been no studies on the use of prebiotics alone in human trials, which may be due to the lack of, until recently, suitable markers of disease. In one placebo-controlled study of patients at high risk of colon cancer, a synbiotic combination of 12 g of Synergy 1, and two probiotic bacteria, *Lactobacillus* GG and *Bifidobacterium* BB12, for 3 months. Biomarkers of colon cancer risk and DNA damage in mucosal tissue were reduced by 60% in patients fed the synbiotic, compared to the placebo group (Rafter et al., 2007). Since the distal colon and rectum are the main sites of cancer formation, the use of Synergy 1, which contains rapidly fermentable short-chain FOS and slower fermented long-chain inulin, which prolong the effects of the prebiotic along the length of the gastrointestinal tract, may explain some of the protective effects of the prebiotic in CRC.

5.11   Other Areas

5.11.1   Diabetes

Dietary interventions in diabetes are the first line treatment in type 2 diabetes, aimed at reducing body weight, and altering hyperglycaemia, hyperlipidaemia and insulin resistance, which are the principal mechanisms of pathophysiology in this disease. Because prebiotics are non-digestible low energy bulking ingredients, this makes any beneficial properties they have in diabetic control appealing to a patient group who are often overweight. The proposed mechanisms of action in diabetic disease involve SCFA, particularly acetate and propionate. High concentrations of free fatty acids in plasma lower the use of glucose in tissues, and induce insulin resistance. Acetate has been shown to lower free fatty acids in the plasma, while propionate, as a long-term dietary supplement in rats and humans, has been demonstrated to decrease blood glucose.

Feeding prebiotics to broiler chickens increases glucose absorption in the jejunal mucosa by 70%. In a study with healthy human subjects in which they ingested 20 g of FOS daily for 4 weeks, it was found that the prebiotic reduced hepatic basal glucose production by 6%, but failed to affect insulin-stimulated glucose metabolism or fasting plasma glucose. Studies in type 2 diabetics have variously reported that a FOS diet either failed to change fasting plasma glucose, or lowered it. In a crossover study involving 12 type 2 diabetics taking FOS for 4 weeks, Luo et al. (2000) looked at assessing mechanisms as well as actions, using the glucose tolerance test (GTT) to determine insulin resistance, which was unaffected by the FOS diet. This is contrary to animal studies, where in rats a 3-week 10% FOS diet prevented insulin resistance, and in baboons, a propionic acid supplemented diet caused a lower glycaemic response to the oral GTT. This study also failed to find any difference in basal hepatic glucose production.

From the small number of studies done so far, it is hard to determine whether the apparent differences in animal and human studies are species-related, or are due to the different doses of prebiotic used. What is clear, however, is that more research is needed to clarify the effects and mechanisms of prebiotic action in human type 2 diabetics. These studies will require subjects to take the prebiotic for longer periods of time, and where possible, in higher doses.

5.11.2   Rheumatoid Arthritis

Adjuvant-induced arthritis in Wistar rats and type II collagen-induced arthritis DBA/1J mice were fed galacto-oligosaccharides orally, which reduced erythema, joint swelling and histopathological findings. These changes were directly associated with a reduction in plasma nitrite/nitrate levels in the rats. In cell culture systems using peritoneal rat macrophages, the prebiotic increased IL-1 production. Therefore, by modulating the intestinal microbiota in animals with arthritis, an immunomodulating effect can be achieved in such a way as to alter inflammatory joint symptoms. In a study involving 16 rats, half of which were fed Synergy 1, the prebiotic significantly reduced inflammatory scores, pro-inflammatory cytokines and increased caecal bifidobacteria and lactobacilli (Hoentjen et al., 2005).

5.11.3   Obesity

Prebiotics are currently of particular interest to the food and drink industry because of their low calorific values. GOS are stable at high temperatures and have low metabolic value at 1.73 kcal/g, while FOS are similar at 1.5 kcal/g. The non-digestibility of prebiotics in the upper GI tract and their long and safe use by the food industry demonstrates that they have excellent potential to act as substitutes for sucrose, and to be used as sweeteners. Feeding studies (Cani et al., 2006) in ten healthy subjects fed 8 g of oligofructose per day showed that they promoted satiety, and reduced food intake ($P < 0.05$). There is also evidence from animal work implicating intestinal bacteria in the pathogenesis of obesity, and theoretically, selective modification of the microbiota may impact on this. In terms of clinical trials in the obese population, this has yet to be investigated, but these early results show promising potential for use in the so-called obesity epidemic that is occurring in Western countries.

5.11.4   General Wellbeing

The promotion of wellbeing is one of the main claims made by manufacturers for many functional foods. However, few studies have been undertaken in humans on the effects of prebiotics on wellbeing or enhancements in quality of life. This may be due to the fact that wellbeing is difficult to measure, largely due to the number of parameters that need to be taken into account to substantiate this, such as those related to energy, mood and cognitive function. In one recent study on 142 healthy volunteers, where patients filled out questionnaires on a range of factors related to wellbeing (Aspiroz, 2005), consumption of 10 g/day inulin was shown to have no effect compared to the placebo group, in mood, sleep quality or bowel function, however, increased wind, bloating and stomach cramps occurred in some subjects.

5.11.5   Other Developing Areas

The mechanism of hepatic encephalopathy is incompletely understood, but it has been attributed to ammonia levels in the plasma, and an altered aromatic amino acid, branched chain amino acid ratio. A longstanding basic treatment of hepatic encephalopathy has been regular and frequent high doses of lactulose, with patients expected to move their bowels at least two or three times per day. Unsurprisingly, the basic mechanisms of action are, as with constipation, an increased bacterial cell mass and SCFA production to cause peristalsis. The act of increased peristalsis itself has been presumed to remove deaminating bacteria, and reduce intestinal uptake of toxic bacterial metabolites such as ammonia. These suppositions are just that, and have as yet to be followed up by confirmatory research.

There has been some work investigating the possibility of preventing infections in critically ill patients using prebiotics. One trial looked at whether prebiotics affected gastrointestinal permeability, using urinary excretion of sucrose and the lactulose/mannitol ratio in burns patients (Olguin et al., 2005). This group failed to demonstrate any improvement in barrier function, but this method is more useful for assessing upper gut permeability rather than lower that in the large bowel, which is where prebiotics would be expected to have their main effects.

5.12 Summary

- Prebiotics have a long history of safe commercial use, and are consumed on a daily basis by most adults. With the potential for prebiotic use to promote health being so wide, it is inevitable that research in this area will be exponential over the next decade.
- The bifidogenic effects of prebiotics are well documented, and their immunomodulatory properties are becoming better understood. However, what is now clear is that simply documenting a bifidogenic effect is no longer an acceptable way to explain any health benefits that may accrue from prebiotic consumption.
- Well-planned human prebiotic trials are needed in numerous areas, either to clarify conflicting results from previous studies, or to assess the basic health effects of these substances.
- Researchers conducting investigations need to take into account the importance of demonstrating the mechanisms involved, rather than simply reporting phenomenological observations.
- It is becoming increasingly evident that the mechanisms of prebiotic action are more complex than first thought, and involve genetic and systemic effects that we are only just beginning to understand, even with the sophisticated investigative techniques that are now available.
- The reasons for changes in cytokine formation in IBD studies, changes in bacterial cell mass in laxation, or the metabolic activities of prebiotics in lipid and diabetic studies need to be determined.

- In light of animal studies showing the potential for harm by prebiotic consumption by increasing salmonella translocation due to increased intestinal permeability (Bovee-Oudenhoven et al., 2003), it is important that mechanisms of action are understood as completely as possible to enable their safe and appropriate selection for human trials.
- A study looking at the gene expression found that FOS significantly induced the expression of 177 mitochondria-related colonic genes in rats (Rodenburg et al., 2008), emphasizing how much more we have to discover about prebiotic actions in health.
- More clarity is needed concerning the mechanisms of prebiotic impact on health, and it is essential that researchers and industry back up any health claims, with factual scientific evidence of how prebiotics work.

List of Abbreviations

AAD	antibiotic-associated diarrhea
CRC	colorectal cancer
DBRCT	double-blind randomized controlled trial
ELISA	enzyme linked immuno absorbent assay
FOS	fructo-oligosaccharide
GOS	galacto-oligosaccharide
GTT	glucose tolerance test
IBD	inflammatory bowel disease
IBS	irritable bowel syndrome
IL-10	interleukin 10
INF-γ	interferon gamma
MALT	mucosa-associated immune system
NDO	non-digestable oligosaccharides
NEC	necrotising enterocolitis
SCFA	short chain fatty acid
TLR-2	toll receptor 2
TNF-α	tumor necrosis factor alpha
UC	ulcerative colitis
VLDL	very low density lipoprotein

References

Arslanoglu S, Moro G, Schmitt J, Tandoi L, Rizzardi S, Boehm G (2008) Early dietary intervention with a mixture of prebiotic oligosaccharides reduces the incidence of allergic manifestations and infections during the first two years of life. J Nutr 138(6):1091–1095

Asvarujanon P, Ishizuka S, Hara H (2005) Promotive effects of non-digestible disaccharides on rat mineral absorption depend on the type of saccharide. Nutrition 21: 1025–1035

Azpiroz F (2005) Intestinal perception: mechanisms and assessment. Br J Nutr 93:S7–S12

Bartosch S, Woodmansey EJ, Paterson JC, McMurdo ME, Macfarlane GT (2005) Microbiological effects of consuming asynbiotic containing *Bifidobacterium bifidum, Bifidobacterium lactis*, and oligofructose in elderly persons, determined by real-time polymerase chain reaction and counting of viable bacteria. Clin Infect Dis 40:28–37

Belenguer A, Duncan SH, Calder AG, Holtrop G, Louis P, Loblet GE, Flint HJ (2006) Two routes of metabolic cross feeding between *Bifidobacterium adolescentis* and butyrate-producing anaerobes from the gut. Appl Environ Microbiol 72: 3593–3599

Blottiere HM, Buecher B, Galmcihe JP, Cherbut C (2003) Molecular analysis of the effect of short-chain fatty acids on intestinal cell proliferation. Proc Nutr Soc 62:101–106

Bonten MJM, Nathan C, Weisein RA (1997) Recovery of nosocomial faecal flora from frozen stool specimens and rectal swabs. Comparison of preservatives for epidemiological studies. Diagn Microb Infect Dis 27:103–106

Bovee-Oudenhoven IM, ten Bruggencate SJ, Lettink-Wissink ML, van der Meer R (2003) Dietary fructo-oligosaccharides and lactulose inhibit intestinal colonisation but stimulate translocation of salmonella in rats. Gut 52:1572–1578

Brown AJ, Goldsworthy SM, Barnes AA, Eilert MM, Tcheang I, Daniels D, Muir AI, Wigglesworth MJ, Kinghorn I, Fraser NJ, Pike NB, Strum JC, Steplewiski KM, Murdock PR, Holder JC, Marshall FH, Szekeres PG, Wilson S, Ignar DM, Foord SM, Wise A, Dowell SJ (2003) The orphan G protein-coupled receptors GPR41 and GPR43 are activated by propionate and other short chain carboxylic acids. J Biol Chem 278: 11312–11319

Buddington RK, Williams CH, Chen S-C, Witherley SA (1996) Dietary supplementation of neosugar alters the fecal flora and decreases activities of some reductive enzymes in human subjects. Am J Clin Nutr 63:709–716

Cani PD, Joly E, Horsman Y, Delzenne NM (2006) Oligofructose promotes satiety in healthy humans: a pilot study. Eur J Clin Nutr 60:567–572

Chen H-L, Lu Y-H, Lin J-J, Ko L-Y (2000) Effects of fructooligosaccharide on bowel function and indicators of nutritional status in constipated elderly men. Nutr Res 20:1725–1733

Chen H-L, Lu Y-H, Lin J-J, Ko L-Y (2001) Effects of isomalto-oligosaccharides on bowel functions and indicators of nutritional status in constipated elderly men. J Am Coll Nutr 20:44–49

Coudray C, Bellanger J, Castiglia-Delavaud C, Remsesy C, Vermorel M, Rayssignuier Y (1997) Effect of soluble or partly soluble dietary fibres supplementation on absorption and balance of calcium, magnesium, iron and zinc in healthy young men. Eur J Clin Nutr 51:375–380

Coudray C, Tressol JC, Gueux E, Rayssiguier Y (2003) Effects of inulin-type fructans of different chain length and type of branching on intestinal absorption and balance of calcium and magnesium in rats. Eur J Nutr 42(2):91–98

Cummings JH, Christie S, Cole TJ (2001) A study of fructo-oligosaccharides in the prevention of travellers' diarrhoea. Aliment Pharmacol Ther 15:1139–1145

Cummings JH, Macfarlane GT (1991) The control and consequences of bacterial fermentation in the human colon. J Appl Bacteriol 70:443–459

Daubioul CA, Taper HS, de Wispelaere L, Delzenne NM (2000) Dietary oligofructose lessens hepatic steatosis, but does not prevent hypertriglyceridemia in obese Zucker rats. J Nutr 130:1314–1319

Delzenne N, Aertssens J, Verplaetse H, Roccaro M, Roberfroid M (1995) Effect of fermentable fructo-oligosaccharides on mineral, nitrogen and energy digestive balance in the rat. Life Sci 57:1579–1587

De Preter V, Raemen H, Cloetens L, Houben E, Rutgeerts P, Verbeke K (2008) Effect of dietary intervention with different pre- and probiotics on intestinal bacterial enzyme activities. Eur J Clin Nutr 62: 225–231

Ducros V, Arnaud J, Tahiri M, Coudray C, Bornet F, Bouteloup-Demange C, Brouns F, Rayssiguier Y, Roussel AM (2005) Influence of short-chain fructo-oligosaccharides (sc-FOS) on absorption of Cu, Zn, and Se in healthy postmenopausal women. J Am Coll Nutr 24:30–37

Duggan C, Penny ME, Hibberd P, Gil A, Huapaya A, Cooper A, Coletta F, Emenhiser C, Kleinman RE (2003) Oligofructose-supplemented infant cereal: 2 randomised, blinded, community-based trials in Peruvian infants. Am J Clin Nutr 77:937–942

Dupuis Y, Digaud A, Fournier P (1978) The relations between intestinal alkaline phosphatase and carbohydrates with regard to calcium absorption. Arch Int Physiol Biochem 86:543–556

Fiordaliso M, Kok N, Desager JP, Goethals F, Deboyser D, Roberfroid M, Delzenne N (1995) Dietary oligofructose lowers triglycerides, phospholipids and cholesterol in serum and very low density lipoproteins of rats. Lipids 30:163–167

Furrie E, Macfarlane S, Kennedy A, Cummings JH, Walsh SV, O'Neil DA, Macfarlane GT (2005) Synbiotic therapy (*Bifidobacterium longum*/Synergy 1) initiates resolution of inflammation in patients with active ulcerative colitis: a randomised controlled pilot trial. Gut 54:242–249

Gibson GR, Beatty ER, Wang X, Cummings JH (1995) Selective stimulation of bifidobacteria in the human colony by oligofructose and inulin. Gastroenterology 108: 975–982

Gibson GR, Roberfroid M (1995) Dietary modulation of the human colonic microbiota: introducing the concept of prebiotics. J Nutr 125:1401–1412

Gibson GR, Probert HM, Van Loo J, Rastall RA, Roberfroid M (2004) Dietary modulation of the human colonic microbiota: updating the concept of prebiotics. Nutr Res Rev 17:259–275

Griffin IJ, Davilla PM, Abrams SA (2002) Nondigestible oligosaccharides and calcium absorption in girls with adequate calcium intake. Br J Nutr 87(S2):S187–S191

He F, Ouwehand AC, Isolauri E, Hashimoto H, Benno Y, Salminen S (2001) Comparison of mucosal adhesion and species identification of bifidobacteria isolated from healthy and allergic infants. FEMS Immunol Med Microbiol 30:43–47

Hoentjen F, Welling GW, Harmsen HJ, Zhang X, Snart J, Tannock GW, Lien K, Churchill TA, Lupicki M, Dieleman LA (2005) Reduction of colitis by prebiotics in HLA-B27 transgenic rats is associated with microflora changes and immunomodulation. Inflamm Bowel Dis 11:977–985

Ichikawa H, Sakata T (1997) Effect of L-lactic acid, short-chain fatty acids and pH in cecal infusate on morphometric and cell kinetic parameters of the rat caecum. Dig Dis Sci 42:1598–1610

Jackson KG, Taylor GRL, Clohessy AM, Williams CM (1999) The effect of the daily intake of inulin on fasting lipid, insulin and glucose concentrations in middle-aged men and women. Br J Nutr 82:23–30

Kalliomaki M, Kirjavainen P, Eerola E, Kero P, Salminen S, Isolauri E (2001) Distinct patterns of neonatal gut microflora in infants in whom atopy was and was not developing. J Allerg Clin Immunol 107:129–134

Kanauchi O, Serizawa I, Araki Y, Suzuki A, Andoh A, Fujiyama Y, Mitsuyama K, Takaki K, Toyonaga A, Sata M, Bamba T (2003) Germinated barley foodstuff, a prebiotic product, ameliorates inflammation of colitis through modulation of the enteric environment. J Gastroenterol 38(2):200–201

Kashimura J, Kimura M, Itokawa Y (1996) The effects of isomaltulose, isomalt and isomaltulose-based oligomers on mineral absorption and retention. Biol Trace Elem Res 54:239–250

Kelly D, Conway S, Aminov R (2005) Commensal gut bacteria: mechanisms of immune modulation. Trends Immunol 26:326–333

Kok N, Roberfroid M, Delzenne N (1996) Involvement of lipogenesis in the lower VLDL secretion induced by oligofructose in rats. Br J Nutr 76:881–890

Kok N, Taper H, Delzenne N (1998) Oligofructose modulates lipid metabolism alterations induced by a fat rich diet in rats. J Appl Toxicol 18:47–53

Langlands SJ, Hopkins MJ, Coleman N, Cummings JH (2004) Prebiotic carbohydrates modify the mucosa associated flora of the large bowel. Gut 53:1610–1616

Lee DY, Scroeder J 3rd, Gordon DT (1988) Enhancement of Cu bioavailability in the rat by phytic acid. J Nutr 118:712–717

Lewis S, Burmeister S, Brazier J (2005) Effect of the prebiotic oligofructose on relapse of *Clostridium difficile*-associated diarrhoea: a randomized, controlled study. Clin Gastroenterol Hepatol 3:442–448

Lopez HW, Coudray C, Levrat-Verny MA, Feiller-Coudray C, Demigné C, Rémésy C (2000) Fructooligosaccharides enhance mineral apparent absorption and counteract the deleterious effects of phytic acid on mineral homeostasis in rats. J Nutr Biochem 11:500–508

Luo J, Van Ypersalle M, Rizkalla SW, Rossi F, Bornet FRJ, Slama G (2000) Chronic consumption of short-chain fructooligosaccharides does not affect basal hepatic glucose production or insulin resistance in type 2 diabetics. J Nutr 130:1572–1577

Macfarlane GT, Steed H, Macfarlane S (2008) Bacterial metabolism and health-related effects of galacto-oligosaccharides and other prebiotics. J Appl Microbiol 104:305–344

Macfarlane S, Furrie E, Cummings JH, Macfarlane GT (2004) Chemotaxonomic analysis of bacterial populations colonising the rectal mucosa in patients with ulcerative colitis. Clin Infect Dis 38:1690–1699

Mineo H, Amano M, Minaminida K, Chiji H, Shigematsu N, Tomita F, Hara H (2006) Two-week feeding of difructose anhydride III enhances calcium absorptive activity with epithelial cell proliferation in isolated rat cecal mucosa. Nutrition 22:312–320

Mitsuoka T (1992) Intestinal flora and aging. Nutr Rev 50:438–446

Moreau MC, Gaboriau-Routhiau V (2000) Influence of resident intestinal microflora on the development and functions of the intestinal-associated lymphoid tissue. In: Fuller R, Perdigon G (eds) Probiotics. Kluwer Academic Publishers, Doordrecht, pp 69–114

Moro GE, Arslanoglu S, Stahl B, Jelinek U, Wahn U, Boehm G (2006) A mixture of prebiotic oligosaccharides reduces the incidence of atopic dermatitis during the first six months of age. Arch Dis Child 91:814–819

Nagura T, Hachimura S, Hashiguchi M, Ueda Y, Kanno T, Kikuchi H, Sayama K, Kaminogawa S (2002) Suppressive effect of dietary raffinose on T-helper 2 cell-mediated activity. Br J Nutr 88:421–427

Nakamura Y, Nosaka M, Suzuki S, Nagafuchi T, Takahashi T, Yajima N, Takenouchi-Ohkubo N, Iwase T, Moro I (2004) Dietary fructooligosaccharides up-regulate immunoglobulin A response and polymeric immunoglobulin receptor expression in intestines of infant mice. Clin Exp Immunol 137:52–58

Olguin F, Araya M, Hirsch S, Brunser O, Ayala V, Rivera R, Gotteland M (2005) Prebiotic ingestion does not improve gastrointestinal barrier function in burns patients. Burns 31:484–488

Poldbeltsev DA, Nikitiuk DB, Pozdniakov AL (2006) Influence of prebiotics on morphological structure of the mucous membrane of intestinum crassum of rats. Vopr Pitan 75:26–29

Rafter J, Bennett M, Caderni G, Clune Y, Hughes R, Karlsson PC, Klinder A, O'Riordan M, O'Sullivan GC, Pool-Zobel B, Rechkemmer G, Roller M, Rowland I, Salvadori M, Thijs H, Van Loo J, Watzl B, Colins JK (2007) Dietary synbiotics reduce cancer risk factors in polypectomized and colon cancer patients. Am J Clin Nutr 85:488–496

Raschka L, Deniel H (2005) Mechanisms underlying the effects of inulin-type fructans on calcium absorption in the large intestine of rats. Bone 37:728–735

Rodenburg W, Keijer J, Kramer E, Vink C, van der Meer R, Bovee-Oudenhoven IM (2008) Impaired barrier function by dietary fructo-oligosaccharides (FOS) in rats is accompanied by increased colonic mitochondrial gene expression. BMC Genomics 9:144

Schley PD, Field CJ (2002) The immune-enhancing effects of dietary fibres and prebiotics. Br J Nutr 87:S221–S230

Scholtens PA, Alliet P, Raes M, Alles MS, Kroes H, Boehm G, Knippels LM, Knol J, Vandenplas Y (2008) Fecal secretory immunoglobulin A is increased in healthy infants who receive a formula with short-chain galacto-oligosaccharides and long-chain fructo-oligosaccharides. J Nutr 138(6):1141–1147

Turnlund JR, King JC, Gong B, Keyes Wr, Michel MC (1985) A stable isotope study of copper absoption in young men: effect of phytate and alpha-cellulose. Am J Clin Nutr 42:18–23

Videla S, Vilaseca J, Antolin M, Garcia-Lafuente A, Guarner F, Crespo E, Casalots J, Salas A, Malagelada JR (2001) Dietary inulin improves distal colitis induced by dextran sodium sulphate in the rat. Am J Gastroenterol 96:1486–1493

Vigne JL, Lairon D, Borel P, Portugal H, Pauli AM, Hauton JC, Lafont H (1987) Effect of pectin, wheat bran and cellulose on serum lipids and lipoproteins in rats fed on a low- or high-fat diet. Br J Nutr 58:405–413

Welters CF, Heineman E, Thunnissen FB, van den Bogaard AE, Soeters PB, Baeten CG (2002) Effect of dietary inulin supplementation on inflammation of pouch mucosa in patients with an ileal pouch-anal anastomosis. Dis Colon Rectum 45:621–627

Woodmansey EJ, McMurdo ME, Macfarlane GT, Macfarlane S (2004) Comparison of compositions and metabolic activities of fecal microbiotas in young adults and in antibiotic-treated and non antibiotic-treated elderly subjects. Appl Environ Microbiol 70:6113–6122

Yoshida T, Hirano A, Wada H, Takahashi K, Hattori M (2004) Alginic acid oligosaccharide suppresses Th2 development and IgE production by inducing IL-12 production. Int Arch Allergy Immunol 133:239–247

Zafar TA, Weaver CM, Zhao Y, Martin BR, Wastney ME (2004) Nondigestible oligosaccharides increase calcium absorption and suppress bone resorption in ovariectomized rats. J Nutr 134:399–402

6 Fructan Prebiotics Derived from Inulin

Douwina Bosscher

6.1 Introduction

Inulin, as well as the shorter form oligofructose, is a nondigestible carbohydrate (fructan) that has been part of the daily food of mankind for centuries. Inulin-type fructans naturally occur in many edible plants as storage carbohydrates. They are present in leek, onion, garlic, wheat, chicory, artichoke, and banana. It is estimated that an average North American consumes about 1–4 g/day of inulin or oligofructose. In Western Europe, the average intake varies between 3 and 10 g/day. Occasionally, people can have higher intakes, e.g., after consuming a bowl of French onion soup, salsify dish, etc., and intakes can then exceed easily 10 g. This illustrates that via the normal diet some, and at certain times, all populations consume relatively high quantities of inulin-type fructans. It also follows that wheat, onion, and banana, and to a lesser extend garlic are the most important sources of inulin-type fructans in the diet. Although inulin-type fructans are nutritive substances and part of our daily diet, these compounds are currently not taken up in food composition tables.

On an in industry scale, inulin-type fructans are obtained from the roots of the chicory plant. After the extraction process, inulin and oligofructose are processed and purified for use as functional food ingredients in a variety of foods. Foods that can contain inulin and oligofructose are bakery products and breakfast cereals, watery systems such as drinks, dairy products and table spreads, and even tablets. Inulin is often used in the manufacture of low-fat dairy products, such as milk drinks, fresh cheeses, yoghurts, creams, dips, and dairy desserts. Other low-fat products in which inulin can be applied are table spreads, butter-like products, dairy spreads, cream cheeses, and processed cheeses. Inulin allows the replacement of significant amounts of fat (up to 100%) and the stabilization of the emulsion while improving the foods' organoleptic characteristics, upgrading both taste and mouthfeel, in a wide range of applications.

Oligofructose is also often formulated in dairy products through incorpora-tion into the added fruit preparation (e.g., fruit yoghurts). Other (low-calorie) dairy products in which oligofructose can be applied are frozen desserts and meal replacers. In such food products oligofructose is often used as sugar replacer on account of its sweet taste. Its incorporation into baked goods allows, besides sugar replacement, also better moisture retention. The use of oligofructose in recipes is quite straightforward and mostly only requires minor adaptations of the produc-tion process. Additionally, new food technical applications of inulin-type fructans are being studied for their potential as carriers for probiotic cultures improving their stability and viability in food products during food processing and storage, but also during gastrointestinal passage and residence in the host microbiota. Probiotic foods are often dairy-based products in which inulin-type fructans can be easily applied. The use of inulin-type fructans together with probiotic cultures is often referred to as "synbiotic," given the combination of probiotics and prebiotics to act in a dual beneficial way. The term "prebiotic" will be explained further in the text.

Inulin-type fructans are nowadays more and more used in foods, and more particularly in the manufacture "functional foods," because of their nutritional advantages. Inulin-type fructans have various physiological, metabolic, hor-monal, and immunological effects, being beneficial to the health and well-being of the host. The origins of these effects lie in their fermentation by the endoge-nous microbiota of the lower intestinal tract and the characterized "prebiotic" effect. They are not digested, nor absorbed, in the upper gastrointestinal tract. In the large intestine, they are selectively fermented by the endogenous micro-biota, increasing growth and/or activity of mainly bifidobacteria and lactobacilli, which are bacterial genera that are considered as biomarkers of a healthy colon microbiota. The prebiotic effect of inulin-type fructans in humans has been demonstrated in randomized, double-blind, and placebo-controlled studies in children, adults, and elderly and more recently in neonates. Given the well-described prebiotic effect of inulin-type fructans, these compounds are nowadays considered as the reference prebiotics and, therefore, frequently used in studies to compare the prebiotic effect of potential new compounds with.

There exists a large body of evidence, ranging from *in vitro* work to studies in animal models and humans, not only on the effects of inulin-type fructans on intestinal function, intestinal infection and inflammation, colonic cancers, immuno-modulation activity, and mineral absorption and accretion but also on the effects outside the intestine affecting sugar and lipid metabolism, oxidative stress, appetite, and subsequent energy intake. The systemic effects of inulin-type fructans have their origins in the intestinal tract, and new insights in the role of the colonic microbiota in overall health and disease (e.g., obesity) explain the potential of compounds modulating the microbiota to have extra-intestinal effects too. In certain fields the evidence on the benefits of prebiotics is well documented, whereas in others still some discrepancy exists. However, it is clear that the role of the colonic microbiota in health and disease, as well as the strategies of modulating it, is a most exciting area of research and that more knowledge can be expected to be generated in the coming years.

6.2   Natural Occurrence

Inulin-type fructans are naturally occurring oligosaccharides that represent the carbohydrate reserve in plants. After starch, fructans are the most abundant nonstructural polysaccharides found in nature. Plants containing inulin-type fructans primarily belong to the Liliales, e.g., leek, onion, garlic, and asparagus; or the Compositae, such as Jerusalem artichoke (*Helianthus tuberosus*), dahlia, and chicory (*Cichorium intybus*). The following text gives some examples of the inulin content and the distribution in chain length (degree of polymerization, DP) of the inulin chains in a selection of food items.

Samples were analyzed by HPLC by Van Loo et al. (1995). For more information about the analytical methodologies to measure the inulin content in foods, the reader is referred to Section 4. The inulin content of commercial onion types (*Allium cepa*) ranges from 1.1 to 7.1 g/100 g (fresh weight) with a range distribution of DP of the inulin chains of 1–12 (most occurring DP is 5). The Jerusalem artichoke has an inulin content ranging from 17 to 20.5 g/100 g (fresh weight) with 74% of the chains of DP 2–19, 20% of DP 19–40, and 6% of DP > 40. Chicory contains 15.2–20.5 g/100 g (fresh weight) inulin with a distribution of 55% of DP 2–19, 28% of DP 19–40, and 17% of DP > 40.

Most of the inulin produced on an industrial scale is nowadays derived from chicory roots. Of interest to the reader is that the chicory used for inulin production (*C. intybus*) is of the same type as the one used for the production of the coffee substitute. The roots look like small, oblong sugar beets. Their inulin content is high and fairly stable over time for a given region, with values ranging from 16.2, 17.0, 16.1, 14.7, and 14.5 g/100 g from August till November, respectively. The distribution in chain lengths, however, is more prone to variation and decreases during harvest from DP of 13.6–9.5 over the same period respectively.

6.3 Chemical Structure

Inulin is a polydisperse carbohydrate material consisting mainly, if not exclusively, of ß (2–1) fructosyl–fructose links. A starting glucose moiety can be present. Fructan is a more general name used for any compound in which one or more fructosyl–fructose links constitute the majority of linkages (e.g., covering both inulin and levan). Inulin-type fructans can be represented as both GF_n and F_m. In chicory inulin, n, the number of fructose units linked to a terminal glucose, can vary from 2 to 70 units. This means that inulin is a mixture of oligomers and polymers. The molecular structure of inulin compounds is shown in ◉ Figure 6.1.

Native inulin is a mixture of oligomers and polymers with a DP ranging from 3 to 70 (average 10). "Native" refers to the inulin as it is extracted from the fresh roots. The lengths of the inulin chains can be reduced by means of an *endo-inulinase* to a DP between 2 and 8 (average DP = 4). The resulting product, called "oligofructose," is a mixture of GFn (G = glucose, F = fructose, and n = number of fructose monomers) and Fm fragments, which are respectively, the sucrose endings and the fragmented polymer tail. By physical separation of the longer chains, a long-chain inulin product can be manufactured (DP ranges between

◻ Figure 6.1
Chemical structure of inulin compounds.

12 and 65; average DP = 25). Oligofructose-enriched inulin (a patented mixture which is commercially named Synergy1) is made of a mixture of oligofructose and long-chain inulin and comprises a preparation with a selected range of chain lengths.

6.4   Quantitative Analysis

Several methodologies for the determination of inulin-type fructans have been established. These include HP-anion exchange chromatography with pulsed amperometric detection (HPAEC-PAD), high-temperature gas chromatography (GC), or liquid chromatography (LC). In all these analytical methods, however, other oligo- and polysaccharides present in the sample interfere with the inulin peaks, and therefore allow only partial quantification. Therefore, techniques that hydrolyze the inulin chains prior to analysis of their constituent monosaccharides by GC, LC, HPAEC-PAD, or thin-layer chromatography are most preferred and in common use nowadays.

Inulin-type fructans can be quantitatively determined by the AOAC method no. 997.08 (Hoebregs, 1997). This method is a reliable enzymatic and chromatographic method for the quantification of inulin and oligofructose, but is very labor intensive and requires expensive chromatographic equipments. An outline of the method is given in ◉ Figure 6.2. If inulin is the only compound present in the sample, the method consists only of steps 1 and 3. The inulin is extracted from a substrate at 85°C for 10 min; part of the extract is kept apart for determination of free fructose, glucose, and sucrose by any reliable chromatographic method available (HPLC, HGC, or HPAEC-PAD) and the other part is submitted to an enzymatic hydrolysis. After the hydrolysis step, the resulting fructose and glucose are determined again by chromatography. By subtracting the initial glucose, fructose, and sucrose content from the final ones, the following formula can be applied: $Inu = k(G_{inu} + F_{inu})$, where k (<1) depends on the DP of the inulin analyzed and corrects for the water gain after hydrolysis. G_{inu} and F_{inu} are, respectively, the glucose and fructose strictly originating from inulin. If a complex sample needs to be analyzed, as is often the case dealing with food products, an amyloglucosidase treatment must be included before step 3 and an extra sugar analysis performed to avoid overestimation of the glucose originating from starch or maltodextrins present.

A faster and easier method is an enzymatic and spectrophotometric analysis (AOAC Method 999.03). This method can be used for a correct quantitative

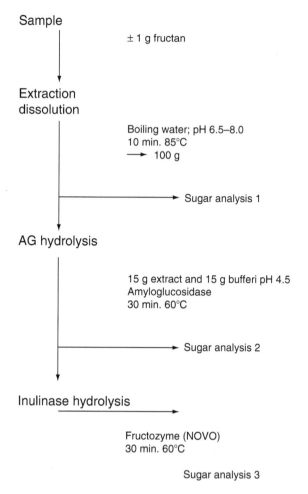

■ Figure 6.2
Flow diagram of enzymatic fructan method.

determination of the inulin content of foods. In case of oligofructose, however, the technique causes a considerable underestimation of the result (about 20%), and therefore cannot be generally applied. Therefore, a new enzymatic and spectrophotometric method was developed that is simple and accurate and quantifies both inulin and oligofructose from any matrix using only standard laboratory equipment. The principle is an enzymatic hydrolysis of both inulin and oligofructose by a fructanase into glucose and fructose. Sugars are determined in the initial extract and subsequent hydrolysates by a spectrophotometer. Subsequent calculation of the contents of inulin and oligofructose in the matrix are based "only" on fructose measurements. Extensive validation of the method has yielded high recovery (96.0 ± 5.3%) and good accuracy and precision data (5.9%) (Steegmans et al., 2004).

6.5 Physical and Chemical Properties

Chicory inulins are available as white and odorless powders with different particle distribution and density. The taste is neutral or slightly sweet, without any off-flavor or aftertaste. The shorter chains have about 10% sweetness compared to sugar, whereas long-chain inulin has no sweetness. It behaves like a bulk ingredient, and contributes to body and mouthfeel, provides a better-sustained flavor with reduced aftertaste (e.g., in combination with high-potency sweeteners), and improves stability. Inulin is moderately soluble in water (maximum 10% solubility at room temperature), which allows its incorporation into watery systems (other fibers often precipitate). In very acidic conditions, the $\beta(2-1)$ bonds between the fructose units in inulin can be (partially) hydrolyzed and fructose formed. At high concentration (>25% in water for native inulin and >15% for long-chain inulin) it has gelling properties and forms a particle gel network after shearing. When thoroughly mixed with water, or another aqueous liquid, a white, creamy structure results, which can easily be incorporated into foods to replace fat by up to 100%. Inulin also improves the stability of foams and emulsions, such as aerated dairy desserts, ice creams, table spreads, and sauces. It can, therefore, replace other stabilizers in different food products (Franck, 2002). Oligofructose is much more soluble than inulin (up to 85% solubility in water at room temperature). It is fairly sweet (35% compared to sucrose) and has a sweetening profile closely approaching that of sugar with a clean taste (no lingering effect). Oligofructose combines with intense sweeteners (such as aspartame and acesulfam K), providing mixtures with rounder mouthfeel and improved flavor with reduced aftertaste. Oligofructose has technological properties close to sucrose and glucose syrups, and, therefore, together with its sweetness profile, is frequently used as a sugar alternative, e.g., in low glycaemic or diabetic foods (Franck, 2002).

Dairy products are frequently used as food vehicles for probiotics. Frequently used probiotics belong to the genera of bifidobacteria and lactobacilli, which most often are able to ferment inulin-type fructans. From this perspective, many food manufacturers have studied the functionality of selected synbiotic combinations in food matrices with respect to probiotic viability not only in the food product itself but also their survivability in the gastrointestinal tract.

In the PROTECH (Nutritional Enhancement of Probiotics and Prebiotics: Technology Aspects on Microbial Viability, Stability, Functionality and Prebiotic Function; QLK1-CT-2000-30042) project, the strategies to address and overcome such specific scientific and technological hurdles that impact on the performance of functional foods based on probiotic–prebiotic interactions were the subject of investigation (Ananta et al., 2004). Increased growth of probiotic bacteria, enhancement in their activity, and good retention of their viability during food storage under different conditions (duration, cooling, pH conditions, etc.) and during passage in the gastrointestinal tract in the presence of inulin or oligofructose have been demonstrated for some strains. In addition, progress in the development of probiotic delivery systems by the use of prebiotic encapsulation materials has also offered new insights into the ways to maximize probiotic viability and survivability. A good matching of the probiotic with an appropriate prebiotic is a prerequisite to obtain maximum beneficial effect of the synbiotic combination with respect to its nutritional properties as well as technical applications and, therefore, optimal utilization in the respective food matrix.

6.6   Nutritional Properties

6.6.1   Caloric Value

The digestive enzymes in mammals are not able to hydrolyze the inulin polymer (or its oligofructose oligomers), and therefore the compound passes unaltered through the mouth, the stomach, and the small intestine. Studies in ileostomized volunteers have demonstrated that orally ingested inulin enters the colon almost quantitatively (>90%), where is it subsequently completely metabolized by the endogenous colonic microbiota.

In the colon, inulin-type fructans are completely converted by the microbiota into bacterial biomass, organic acids such as lactic acid and short-chain fatty acids (SCFA: acetic, propionic, and butyric acid), and gases (CO_2, H_2, CH_4). SCFA and lactate contribute to the host's energy metabolism. SCFA and lactate are partly used by the bacteria themselves and partly taken up by the host. Still, SCFA and lactate are less effective energy substrates than sugars. These factors together explain the low caloric value of inulin-type fructans (compared to its constituent monosaccharide moieties).

On the basis of ^{14}C studies in humans, a caloric value of 1.5 kcal/g was calculated for short-chain fructo-oligosaccharides. Experimental *in vitro* (fermentation) and *in vivo* data (rat experiments) allowed Roberfroid et al. (1993) to calculate the caloric value of inulin and oligofructose to be 1.4 kcal/g, according to basic biochemical principles. Other scientific observations have even suggested lower caloric values. Currently, a caloric value between 1 and 1.5 kcal/g inulin-type fructans is being used for food labeling practices.

6.6.2 Acceptability

Because of their indigestibility, nondigestible oligosaccharides that pass into the colon can induce osmotic effects. These are mainly induced by smaller molecules and lead to an increased presence of water in the colon. This is probably the reason why, e.g., lactulose has a higher laxative potential than inulin. Other determinants of acceptability are the production of gases that results from bacterial fermentation in the colon. Slowly fermenting compounds have been shown to be easier to tolerate than more fast fermenting analogs. This can explain why inulin is easier to tolerate than polyols and short-chain oligofructose. Flatulence is a well-known and often accepted side effect of higher intake of dietary fibers in general.

6.6.3 Intestinal Function, Metabolism, and Microbiota

6.6.3.1 Intestinal Function

Inulin-type fructans, through their presence and subsequent fermentation in the large bowel, influence the colonic metabolism in its lumen and the integrity and functioning of the epithelial cell lining. Randomized, double-blind, and placebo-controlled studies in humans have observed a significant increase in weighted stool output (also called the "fecal bulking effect") upon inulin intake with an (approximate) increase in stool weight of 1.5–2.0 g per gram of inulin intake (referred to as "bulking index") (Den Hond et al., 2000; Gibson et al., 1995a). This level of stool bulking is more or less comparable to that of other soluble dietary fibers that are well-known bulking agents, such as pectins and gums. The increase in stool bulking upon inulin and oligofructose consumption was found in randomized, double-blind, and placebo-controlled studies in subjects with low stool frequency patters or in constipated patients to result in a significant increase in the number of stools per week and to exert a laxative effect reducing functional constipation (Den Hond et al., 2000; Gibson et al., 1995a; Kleessen et al., 1997). In a recent randomized, double-blind, and placebo-controlled multicenter study in patients with minor functional disorders, its was found that oligofructose supplementation significantly decreased the intensity of digestive disorders (Paineau et al., 2008). In the CROWNALIFE ("Crown of Life" Project on Functional Foods, Gut Microflora and Healthy Ageing; QLK1-2000-00067) project, it was demonstrated that the administration of a synbiotic supplement containing a probiotic *Bifidobacterium animalis* and an oligofructose-enriched inulin preparation to elderly subjects significantly increased stool frequency and reported to improve the well-being and the quality of life (Zunft et al., 2004).

6.6.3.2 Prebiotic Effect

The term "prebiotic" was defined by Gibson and Roberfroid (1995b) as "a nondigestible food ingredient that selectively stimulates growth and/or activity of one or a limited number of bacteria in the colon, thereby improving host health." As research progressed, three criteria were accepted that a food ingredient should fulfill before it could be classified as prebiotic: first, it should be nondigestible and resistant to gastric acidity, hydrolysis by intestinal (brush border/pancreatic) digestive enzymes, and gastrointestinal absorption; second, it should be fermentable; and third, it should in a selective way stimulate growth and/or metabolic activity of intestinal bacteria that are associated with health and well-being. Inulin-type fructans do fulfill all the above criteria and are generally accepted prebiotics (Gibson et al., 2004). Indeed, the prebiotic properties of inulin-type fructans are well documented among various age groups and in people living in different regions around the world.

The intestinal microbiota can be considered as a metabolically adaptable and rapidly renewable organ of the body. The induction of a prebiotic effect with inulin-type fructans can, therefore, be rather rapidly achieved within a few days of administration. A prospective, randomized, and double-blind two-center study on the prebiotic effect of an oligofructose-enriched inulin-supplemented infant formula found a significant increase in the levels of bifidobacteria already after two weeks of supplementation and remained over the full 4 weeks of the intervention, whereas numbers of lactobacilli, bacteroides, and clostridia remained stable. Interestingly, the inclusion of a nonrandomized breast-fed group showed that bifidobacteria levels in the colon of neonates supplemented with the inulin milk reached those levels found in breast-fed infants. It has been previously demonstrated that feeding breast milk creates an environment favoring the development of a simple flora dominated by bifidobacteria to which various health benefits have been ascribed. Formula-fed infants, on the other hand, have a more complex pattern in which bacteroides, clostridia, and streptococci are prevalent with lower levels of bifidobacteria. These observations have led to the current practice of including prebiotics into infant formula milks to more closely resemble the microbiota and intestinal functioning of the breast-fed infant. And, even more, as our general knowledge about the multiple health benefits of bifidobacteria progressed and our belief of the importance of a well-balanced flora for overall health and well-being became confirmed in well-conduced research, high levels of bifidobacteria in the colon are considered favorably at all ages during life.

The administration of oligofructose to infants at older age (after their weaning period) has also been shown to increase the numbers of bifidobacteria (up to 9.5 log of colony-forming units per gram of feces). The increase in bifidobacteria was higher in infants with lower initial colonization levels. This has been demonstrated also in previous studies. Interestingly, the numbers of staphylococci and clostridia in the feces and the number of infants colonized with staphylococci appread to be lower in the oligofructose group. However, any observed difference disappeared after stopping the intervention, which indicates that the supplementation needs to be continued as part of the diet (Waligora-Dupriet et al., 2007).

During childhood, the bacterial populations residing in the microbiota increase in numbers and complexity, a process that has been initiated during weaning as solid foods are introduced in the infant's diet. At that time, numbers of enterobacteria and enterococci have increased sharply with higher levels of colonization by bacteroides, clostridia, and streptococci, resembling more and more the adult-like flora. In adulthood, however, day-to-day variations in microbial populations are less pronounced and the microbiota becomes more stable and constant over time. In healthy adult volunteers, the prebiotic properties of inulin and oligofructose have been demonstrated in numerous well-controlled, randomized human intervention studies using varying doses (ranging from 5 to 40 g/day) of short- and long-chain inulins over various periods of supplementation and by measuring microbial changes in the luminal as well as the mucosa-associated microbiota (Gibson et al., 1995a; Langlands et al., 2004; Rao, 2001; Tuohy, 2001).

In elderly people, the administration of a synbiotic, containing *Bifidobacterium bifidum* and *Bifidobacterium lactis* together with oligofructose-enriched inulin, in a randomized, double-blind, controlled manner induced significant higher numbers of bifidobacteria (Bartosh et al., 2005). Similar results were also obtained in other studies and accompanied by beneficial effects on intestinal functioning and well-being (Kleessen et al., 1997; Zunft et al., 2004). These changes in microbiota composition toward higher levels of bifidobacteria are of particular interest, as with aging structural changes occur in the intestinal tract and the numbers of putative pathogenic bacteria (e.g., clostridia and enterobacteria) have been found to increase.

Interestingly, the magnitude of the prebiotic effect of inulin-type fructans (increase in log colony-forming units of bifidobacteria per gram colonic content) is dependent on the initial bifidobacteria level of an individual rather than the intake dose.

In vitro studies with three-stage fermentor vessels that mimick the different parts of the colon (ascendents, transversum, and descencents) have demonstrated that depending on the chain length of the inulin-type fructan a prebiotic effect can be induced at different sites along the colon. Oligofructose is more rapidly fermented, and therefore has an impact on the composition of the intestinal flora in the proximal part (vessel 1). The longer inulin chains, being fermented more slowly, have their effect in the mid colon and more distal parts (vessels 2 and 3). In humans, combining short (oligofructose) and long inulin chains (long-chain inulin) has been demonstrated to increase levels of bifidobacteria in the mucosa-associated flora of biopsies taken from proximal and distal colon mucosa (Langlands et al., 2004). Given the so widely studied prebiotic effects of inulin-type fructans, these compounds are nowadays accepted as the gold standard of a prebiotic food ingredient.

6.6.3.3 Barrier Function

In some conditions such as old age, the use of antibiotics, or in case of (critical) illnesses (acute or chronic such as inflammatory bowel diseases and cancer), the intestinal barrier functions less and gastrointestinal dysfunction can occur. As a result, increased bacterial translocation may happen, leading to systemic illness.

The interdigestive intestinal motility (e.g., migrating myoelectric/motor complex) is one physiological mechanism that prevents bacterial overgrowth and translocation in the gut. As there appears to be a relationship between the intestinal motility and the composition of the intestinal microbiota, the effect of synbiotics on the motility response was tested in rats by implanting electrodes and recording duodenojejunal electromyography. Administration of *Lactobacillus rhamnosus* GG, *Bifidobacterium lactis* Bb12, and oligofructose-enriched inulin to elderly rats regularized the occurrence of intestinal contractions of high amplitude, which are more effective in propelling the residual food, debris, secretions, and bacterial cells (Lesniewska et al., 2006). Other animal experiments to test the potential of modulating the microbiota to efficiently discriminate and eliminate pathogenic organisms showed decreased translocation of bacteria (total aerobic, anaerobic, and the Enterobacteriaceae) to the mesenteric lymph nodes and liver, after oral administration of probiotics (*Bifidobacterium infantis*) and/or prebiotics (oligofructose-enriched inulin) in DSS-colitis-induced rats (Osman et al., 2006). These data are indicative of an improved epithelial barrier function and in agreement with earlier studies. In mice infected (intraperitoneally) with virulent strains of systemic pathogens (*Listeria monocytogenes* and *Salmonella typhimurium*), mortality rates were much lower upon inulin feeding (Buddington et al., 2002). Other studies in rats, also infected with *Salmonella*, showed lower pathogen colonization in the intestines with oligofructose; however, the authors observed an increase in translocation rate to the spleen and liver. These observations can most likely be ascribed to the low calcium diet used in this model, which by itself damaged barrier function, as the authors demonstrated that increasing the calcium level of the diets was accompanied by a decrease in the rate of translocation (Ten Bruggencate et al., 2004).

In humans, no effect on barrier function of (high dose) oligofructose was found in healthy volunteers. Barrier function was measured by the levels of intestinal epitheliolysis and the excretion of O-linked oligosaccharides in stools. The latter refers to the production of glycoproteins which build up the mucus gel layer that covers the intestinal epithelium. The authors, however, did observe a lower level of cytotoxicity of the fecal water with oligofructose (Scholtens et al., 2006). More extensive research has been performed on the cytotoxicity of the fecal water as a biomarker for the risk of colonic cancer in the SYNCAN (Synbiotics and Cancer Prevention; QLK-1999-00346) project, which will be described in Section 10.6.5. In the SYNCAN project, the effect of oligofructose-enriched inulin (as synbiotic) on the epithelial barrier function was studied. The trans-epithelial resistance (*ex vivo*) of cell lines subjected to the fecal water from polypectomized volunteers supplemented with oligofructose-enriched inulin was measured as an indicator of barrier functioning. The fecal water is the fecal fraction in most intimate contact with the colonic epithelium and mediates its functioning. A common observed effect of tumor promoters is the reduction in barrier function of the epithelium inducing lower protection of the mucosa to carcinogenic substances. Interestingly, the synbiotic intervention increased the barrier function of the epithelium, which was shown by the significantly increased percentage in trans-epithelial resistance of the Caco-2 cell monolayer when subjected to the fecal water of polyp patients receiving the synbiotic (Rafter et al., 2007).

6.6.3.4   Butyrogenic Effect

The link between the consumption of inulin-type fructans, their bifidogenic effect, and the increase in butyrate production in the colon was, until recently, unclear. Inulin-type fructans are prebiotics and selectively fermented by bifido-bacteria (and to a minor extend also lactobacilli). This colonic fermentation process generates organic acids to which several beneficial effects linked to the ingestion of inulin-type fructans are (to some extent) attributed. Organic acids that are produced by bifidobacteria are primarily lactate and acetate, and these organisms have not been reported to produce butyrate. Nonetheless, studies (both *in vitro* and *in vivo*) have demonstrated that the colonic fermentation of inulin-type fructans increases the production of butyrate, which is the so-called butyrogenic effect.

Eubacteria (*E. hallii*-like strains) are butyrate-producing bacteria and can account for 4% of the bacterial flora in feces. Recently, some previously unknown butyrate-producing colonic bacteria have been identified. These belong to the clostridia cluster XIVa, which is one of the most abundant bacterial groups in human feces. Contributing species such as *Anaerostipes caccae* and *Roseburia intestinalis* have been shown to be efficient lactate and/or acetate converters, and bacteria related to these species have been reported to compose up to 3% of the colonic microbiota. Cross-feeding between these microorganisms and inulin degraders (e.g., bifidobacteria) might play a key role in the gut ecosystem with important consequences for human health.

Fermentation studies (*in vitro*) using simple and complex bacterial cultures or fecal slurries offer a valuable tool to study individual bacterial metabolism and interspecies interactions. Kinetic analyses of co-cultures with *Bifidobacterium* spp. and butyrate-producing colonic bacteria in the presence of oligofructose revealed distinct types of cross-feeding reactions that were strain dependent. In such studies, butyrate-producing bacteria (e.g., *A. caccae* and *R. intestinalis*) were unable to degrade oligofructose, whereas in the presence of bifidobacteria and/or fermentation metabolites (acetate) or breakdown products, degradation did occur with corresponding butyrate production (Falony et al., 2006). Studies with stable isotopes, which enable the following of carbon flows, showed that indeed *Bifidobacterium* spp. in the presence of oligofructose produce lactate (and/or breakdown products), which in turn are converted into butyrate in the presence of butyrate-producing bacteria (e.g., *R. intestinalis* and *E. halli*) (Belenguer et al., 2006). Fecal batch cultures found that the addition of oligo-fructose significantly increased butyrate production. About 80% of the newly synthesized butyrate derived from oligofructose fermentation originated from the interconversion of extracellular acetate and lactate. Also, Duncan et al. (2004) found that the contribution of external acetate to butyrate formation from oligofructose fermentation ranged from 82% (fecal slurry batch culture) to 87% (continuous cultures). The increased flux of extracellular acetate to butyrate upon oligofructose fermentation in mixed (fecal) slurries is in agreement with butyryl CoA:acetyl CoA transferase being the dominant butyrate-producing pathway. It appears that this pathway is selectively activated upon oligofructose fermentation, with concomitant butyrate production. These cross-feeding mechanisms could play an important role in the colonic ecosystem and contribute to the combined bifidogenic and butyrogenic effect observed after addition of inulin-type fructans to the diet.

6.6.4   Intestinal Infection and Inflammation

6.6.4.1   Intestinal Infection

Several mechanisms have been postulated by which inulin-type fructans and their interaction with the resident bacterial flora improve resistance of the intestinal tract against (exogenous) pathogen invasion (called "resistance to pathogen colonization") (Bosscher et al., 2006a). By restricting the availability of substrates within the lumen and adhesion sites on epithelial cells, endogenous bifidobacteria and lactobacilli may inhibit pathogen survival and adherence. Additionally, upon inulin fermentation, lactic acid bacteria produce organic acids (SCFA and lactate), thereby creating an environment unfavorable of pathogen growth. Moreover, butyric acid has been shown to support barrier function of epithelia.

Exploratory *in vitro* work with fecal slurries in the early 1990s indicated that inulin and oligofructose are completely fermented by the colonic microbiota and selectively stimulate bifidobacteria and lactobacilli growth and activity at the expense of pathogenic bacteria (e.g., clostridia). Protective effects of bifidobacteria have been demonstrated in (gnotobiotic) quails against the development of necrotizing enterocolitis (NEC)-like lesions when inoculated with a pathogenic flora (containing *Clostridium butyricum* and *C. perfingens*) from premature newborns. Lesions occurred rapidly after establishment of the NEC flora (e.g., thickening of the caecal wall with gas cysts, hemorrhagic ulcerations, necrotic areas), whereas they were fewer in the presence of *Bifidobacterium infantis* and *B. longum* (Butel et al., 1998). Supplementing the quails' diet with oligofructose induced an increase in the level of bifidobacteria which prevented overgrowth of bacteria implicated in NEC (e.g., *E. coli*, *Clostridium perfringens.*, *C. difficile*, and *C. ramosum*) and reduced NEC-like lesions caused by polymicrobial infection (Butel et al., 2001). Other experiments in mice showed that supplementation with inulin-type fructans reduced intestinal yeast densities after oral challenge of mice with *Candida albicans*, resulting in an enhanced survival rate (Buddington et al., 2002). A combination of oligofructose and *Lactobacillus paracasei* also was shown to suppress pathogens (*Clostridium*, enterococci, and enterobacteria) in weaning pigs (Bomba et al., 2002). Furthermore, in pigs with cholera toxin-induced secretory diarrhea, oligofructose suppressed the presence of pathogens and increased the number of lactobacilli (Oli et al., 1998).

Clinical studies in humans have also shown that inulin-type fructans can protect against pathogen colonization and infection. Critically ill patients have a gut microbial ecology that is in dysbalance and is characterized by high numbers of potential pathogens. Such patients, at risk for developing sepsis (at intensive care units) when receiving oligofructose (as a synbiotic), had fewer pathogens in their nasogastric aspirates. Treatment with antibiotics, on the other hand, also changes the gut microbiota and disrupts normal ecological balance, which often leads to antibiotic-associated diarrhea. In the study of Orrhage et al. (2000), antibiotic treatment of patients induced a marked decrease in the anaerobic microbiota, mainly with a loss of bifidobacteria and an overgrowth in enterococci. Oligofructose administration (as synbiotic) in those patients restored their numbers of lactobacilli and bifidobacteria. Also, in patients with *C. difficile*-associated diarrhea, which frequently occurs after antibiotic therapy, oligofructose suppressed colonization with *C. difficile* and increased bifidobacteria levels. These changes were accompanied by a lower rate of relapse of diarrhea and reduced the length of hospital stay (Lewis et al., 2005).

6.6.4.2   Intestinal Inflammation

Chronic inflammatory bowel diseases such as ulcerative colitis, Crohn's disease, and pouchitis are thought to have their etiology to some extent linked to the composition of the colonic microbial community and its activities. Although members of the gut microbiota normally do not induce disease, in genetically susceptible hosts chronic intestinal inflammation can develop in response to commensal bacteria. It appears that bacteria are crucial in the pathogenesis of colitis, but not all bacteria are equal in their capacity to induce chronic intestinal inflammation. An altered immune response toward normal commensal organisms is thought to drive the inflammatory process towards a state of chronic inflammation.

The effect of inulin-type fructans in modulating the disease process has been repeatedly demonstrated in experimental models in which inflammation was induced by chemical agents such as dextran sodium sulphate (DSS) (Videla et al., 2001) or 2,4,6-trinitrobenzenesulfonic acid (TNBS) (Cherbut et al., 2003). In each of these, administration of inulin-type fructans (alone or as synbiotic) to the diets of animals reduced the inflammatory process (e.g., MPO, IF-γ, PGE2) and improved clinical and histological markers with a reduction in corresponding lesions. The HLA-B27 transgenic (TG) rat is a well-characterized model of chronic intestinal inflammation. The model spontaneously develops colitis. Oral administration of oligofructose-enriched inulin to HLA-B27 TG rats decreased gross cecal and inflammatory histological scores in the caecum and colon and altered mucosal cytokine profiles (decreased IL-1β and increased TGF-β levels). Cytokine responses of mesenteric lymph node (MLN) cells were also studied *in vitro* by their response to cecal bacterial lysates (CBL). Stimulation of MLN cells by CBL from oligofructose-enriched inulin-treated TG rats induced a lower IF-γ response (Hoentjen et al., 2005).

Ulcerative colitis (UC) is a relapsing inflammatory disease of the colon whose etiology has not been clarified yet. In patients suffering from ulcerative colitis, it has been found that bifidobacteria populations are about 30-fold lower compared to those in healthy individuals. This led to the hypothesis that restoring bifidobacteria populations in these patients by the use of pre- or synbiotics may influence the disease process. Supplementation of the diet of patients with UC with oligofructose-enriched inulin together with a probiotic (*Bifidobacterium longum*) for 1 month resulted in a 42-fold increase in bifidobacteria numbers in mucosal biopsies. A clinical intervention study in ulcerative patients was performed and patients were supplemened with the same synbiotic as indicated above. The level of inflammation was assessed both by traditional methods (endoscopy and biopsy examination) and new sensitive methods (changes in gene expression of antimicrobial peptides called human beta defensins, hBD) and cytokine profiles. hBD are uniquely expressed by epithelial cells. hBD2 and 3 are upregulated in UC and their production is positively correlated with the severity of the active disease, making them excellent targets for assessing inflammatory responses in UC epithelia after therapy. Administration of the synbiotic improved the full clinical appearance of chronic inflammation, as evidenced by a reduction in sigmoidoscopy scores, reduction in acute inflammatory activity (cytokines that drive the inflammatory process, e.g., TNF-α and IL1-α), and regeneration of the epithelial tissue. Expression levels of inducible hBD2 and hBD3 in synbiotic patients were lower in the post-treatment period (Furrie et al., 2005). In another placebo-controlled clinical trial in patients with UC, the effects of oligofructose-enriched inulin were measured using noninvasive inflammatory markers. Fecal calprotectin is a quantitative marker of intestinal inflammation (presence of leukocytes) and has been shown to successfully predict relapse of the disease. Patients receiving oligofructose-enriched inulin had lower levels of calprotectin in their feces, thereby improving the patients' response to therapy by mitigating intestinal inflammation (Casellas et al., 2007). A reduction of the inflammation and associated factors was observed also in patients with an ileal pouch-anal anastomosis after therapy with inulin-type fructans (Welters et al., 2002). Moreover, in patients with active ileo-colonic Crohn's disease, dietary intervention with a combination of inulin and oligofructose has been shown to lead towards an improvement of the disease activity (reduction in Harvey Bradshaw Index) as well as enhanced lamina propria denritic cell IL-10 production and TLR2 and TLR4 expression. Strikingly different changes in mucosa microbiota following inulin supplementation were observed between patients who entered remission and those that did not. Patients who entered remission had an increase in mucosal levels of bifidobacteria (Lindsay et al., 2006).

6.6.5   Colonic Cancer

Diet has a strong influence on the etiology of colorectal cancers, and appropriate changes in dietary habits are therefore expected to have a major impact on its prevalence. Evidence pinpoints the role of the colonic microbiota in this process. Intestinal bacterial metabolism can generate substances with genotoxic, carcinogenic, and tumor-promoting potential, and human feces have been shown to be genotoxic and cytotoxic to the colonic cells. Some bacteria carry specific enzymes that generate carcinogenic substances from food compounds (e.g., clostridia and bacteroides). Studies *in vitro*, animal chemoprevention models, and humans at risk have shown the importance of a balanced gut microbiota in terms of reducing colorectal cancer risk.

Studies in rat models of colonic cancer have demonstrated that inulin and oligofructose reduce the number of aberrant crypt foci (ACF). The increased metabolic activity of bifidobacteria and lactobacilli due to the selective fermentation of inulin and oligofructose is thought to be the basis of their anti-carcinogenic properties. ACF are pre-neoplastic lesions found in the etiology of most colon cancers. ACF show different phenotypic characteristics over normal crypts, e.g., presence of one or more large crypts that appear as a single focus, have thickened epithelia, and appear elevated compared with normal crypts when viewed under a microscope. ACF contain numerous crypts that multiply with time after treatment with a carcinogen. These ACF can develop into polyps and possibly into colon cancer. Injection with azoxymethane (AOM) can induce ACF in rats as early as 2 weeks after the carcinogen injection and mimics the adenoma–carcinoma sequence in humans. Formation of foci of aberrant crypts in the rat colon can be used as a biomarker to evaluate short-term effects of inulin-type fructans and investigate their efficacy as chemopreventive agents. Administration of weanling rats with different types of inulin-type fructans induced a reduction in the number of ACF in the proximal, distal, and total colon. Such reductions in the distal parts of the colon (and the whole colon) were most pronounced when rats were fed oligofructose-enriched inulin, which resulted in the lowest numbers of colonic ACF (Verghese et al., 2005). Long-term studies with probiotics, prebiotics, and synbiotics in rats with AOM-induced colon cancer showed a reduction in the number of colon carcinomas when supplemented with oligofructose-enriched inulin either alone or given as a synbiotic (with *Lactobacillus rhamnosus* GG and *Bifidobacterium lactis* Bb12) (Roller et al., 2004). Treatment with the carcinogen AOM suppressed the rats' natural killer (NK-) cytotoxicity in the Peyer's patches (PP). NK cells are involved in both the recognition and subsequent elimination of tumor cells. Suppression of this NK-cell activity may subsequently contribute to tumor growth. Interestingly, the changes in tumor formation upon the intervention coincided with a stimulation of immune functions within the gut-associated lymphoid tissue (GALT) and that of PP which are the primary lymphoid tissues responsive to oral intake of prebiotics or synbiotics. The supplementation with oligofructose-enriched inulin (alone or as a synbiotic) prevented such carcinogen-induced NK-cell suppression in PP. After 33 weeks of treatment, immunological investigation of the rat's PP revealed significantly higher NK cell-like activity after the intake of the pre- or synbiotic. Other immunological markers in PP cells that differed upon both interventions were the stimulation in IL-10 production. This increase in IL-10 cytokine production in PP was also found in a previous study by the same authors after short-term exposure of AOM rats to prebiotics, probiotics, and synbiotics (Roller et al., 2004).

The challenges we still face are in the understanding of the mechanism through which prebiotics affect pathways of tumor initiation. For this, the cytotoxicity of the fecal water was measured after fermentation of inulin-type fructans. Rats were supplemented with prebiotics, probiotics, and synbiotics for a given time and aqueous extracts of feces were taken and tested for genotoxicity in colonic cell lines (HT29). Human colon cells were used as target because they are considered to be surrogates of the human colonic epithelium *in vivo*. DNA damage in the human cells was determined by the "comet assay." Damaged DNA is visualized as a "comet" as consequence of the migration of the damaged DNA within an electrical field. Overall, the prebiotic diet yielded fecal water that was less genotoxic, indicating that inulin-based diets reduce exposure to genotoxins in the gut, and therefore could lead to the prevention of tumorogenesis (Klinder et al., 2004). More *in vitro* evidence on the potential of inulin fermentation products in preventing the early stages of cancer onset was in the observations on gene expression profiles in nontransformed human colonocytes. Chirurgically obtained human colonocytes were incubated with oligofructose-enriched inulin fermentation supernatants. This resulted in a 2–3-fold increase of SCFA and induced tropic effects. The supernatants modulated the expression of several glutathione S-transferases (GST) isoforms: GSTM2 (2 fold) and GSTM5 (2.2 fold). GST are phase II enzymes of biotransformation that detoxify many carcinogens and, therefore, could protect the cells from carcinogenicity (Sauer et al., 2007).

Other substances in the diet have also been shown to extert antioxidative and anticarcinogenic functions. Biologically active compounds such as flavonoids, phenolic acids, and phyto-estrogens occur in plants mainly as glycosylated compounds. However, before these compounds can be absorbed by the human body, they need to be deconjugated by the intestinal microbes. The glycosidase activities of bifidobacteria have been demonstrated in some cases to be able to liberate the aglycones from their glycosidic linkages and allow these compounds to be absorbed in the body. Inulin-type fructans therefore could, through their stimulation of bifidobacteria, indirectly improve the absorption and bioavailability of such plant bioactive compounds and/or potentiate or intensify the activity of bioactive products. One example is soy-derived isoflavones (a subclass of the more ubiquitous flavonoids) that contain genistein and daidzein. Inulin-type fructans have been shown to increase plasma and urinary concentrations of soy-derived genistein and daidzein and their aglycone forms. Another colonic metabolite, which is derived from daidzein, is equol, which has been shown to have enhanced antioxidant activity compared with daidzein. In rats it has been shown that feeding oligofructose-enriched inulin in combination with soy potentiated the chemopreventive effect against AOM-induced aberrant crypt foci development. Concomitently, the combination diet enhanced the activity of antioxidative and detoxifying enzymes, indicating a potential mechanism of the observed reduction in AOM-induced ACF.

In 2001, the SYNCAN project started. SYNCAN was a collaborative European Network involving eight different partners from seven countries. The core feature of the project involved a phase-II anticancer study, randomized, double-blinded, and placebo-controlled, in 80 patients with a history of colon cancer or polyps and supplemented with oligofructose-enriched inulin and *Bifidobacterium lactis* Bb12 and *Lactobacillus rhamnosus* GG for 12 weeks. The synbiotic preparation increased the level of bifidobacteria and lactobacilli. This was accompanied by a decrease in the numbers of pathogens (coliforms and *Clostridium perfringens*). The altered composition of the colonic bacterial ecosystem beneficially affected the metabolic activity in this organ. This was obvious from the decreased DNA damage in the colonic mucosa (measured by the comet assay) and the tendency to lower the level of colorectal proliferation (surrogate biomarker for colon cancer risk) in polyp patients (no measures were taken in cancer patients). Other effects were the decreased cytotoxicity of the fecal water. The fecal water of synbiotic-fed polyp patients also showed a lower level of cell necrosis as demonstrated by the lower cytotoxic potential in HCT116 cell types. This indicates that the synbiotic effectively prevented cell death of the colonic epithelium (Rafter et al., 2007).

6.6.6 Modulation of Immune Function

In the above sections, the potential of inulin-type fructans to modulate immune parameters were described. These studies involved different disease conditions, e.g., chronic inflammation and cancer, and suitable animal models thereof. To study the effect of foods and food ingredients on immunity in healthy conditions, however, vaccination strategies (challenge models) can be applied, or studies in populations at risk (e.g., environments with high burden of infections, malnourished children) can be performed and clinical outcomes (e.g., incidence of illness) investigated.

Inulin and oligofructose have been demonstrated in mice to improve immune responses to oral vaccination with a suboptimal dose of live attenuated *Salmonella* Typhimurium vaccine, thereby contributing to an enhanced oral vaccine efficacy. Specific anti-salmonella antibody responses increased post immunization and were much higher in the inulin and oligofructose mixture as indicated by the high specific blood salmonella immunoglobulin G and fecal immunoglobulin A responses. These changes were concomitant with an increase in the survival rate upon later challenge with a virulent salmonella strain. The vaccination of mice led to a 40% protection, and the rate of protection improved to 73% on the inulin mix feeding regime (Benyacoub et al., 2008).

Infants vaccinated with measles vaccine and fed on weaning foods enriched with a synbiotic, containing inulin and oligofructose, have been shown to improve their vaccination-induced immune response. Post vaccination, specific IgG-antibody levels were higher with the synbiotic, indicating an enhanced immune response to vaccination (Firmansyah et al., 2001). Further investigations in elderly persons confirmed that adequate nutritional supplements can modulate immune responses. Aging is associated with alterations in the immune responsiveness that increase susceptibility to infections, reduce the response to immunization, and increase the incidence of autoimmune diseases. The preservation of an adequate NK cell function is considered part of successful aging. Healthy elderly volunteers received a nutritional supplement containing proteins, vitamin E, vitamin B12, folic acid, *Lactobacillus paracasei*, as well as a mixture of oligofructose and inulin for 4 months prior to vaccination against influenza and pneumococcus. After 4 months, the nutritional supplement increased the NK cell activity. NK cell activity is one of the first-line defense mechanisms against viral infections. Cells with NK activity are a special subset of T cells that respond to glycolipid antigens, are capable of promoting T helper 1 response, and have a role in regulating autoimmunity and rejecting tumor cells. At 4 months, all persons were vaccinated. After vaccination, the production of IL-2 by peripheral blood mononuclear cells (PBMC) remained in the supplemented individuals and did not lower as seen in those not receiving the supplement. One of the functions of IL-2 is to stimulate NK cell activity and to elicit a T-helper 1 immune response. Most persons had a positive antibody response to the vaccine, but no additional effect of the supplement was observed. Interestingly, during the 1-year follow-up, the volunteers that took the supplement reported fewer infections (Bunout et al., 2004).

Infants too are very vulnerable to infections and gastrointestinal tract disturbances. The likelihood of these occurring is ever increasing when attending daycare centers, where infectious diseases are frequent and easily disseminate from one child to another. Saavedra and Tschernia (2002) examined the long-term (6 months) effect of daily intake of oligofructose (supplemented to cereal) in infants (4–24 months of age) attending daycare centers. The consumption of the prebiotic cereal was associated with a decrease in the severity of diarrhea disease. General gut status was improved with decreased bowel movement discomfort, vomiting, and regurgitation. Furthermore, consumption of the prebiotic cereal resulted in adequate growth and led to a reduction in the number of febrile events and cold symptoms, antibiotic prescription (associated with respiratory illness), and daycare absenteeism. Such observations were later confirmed in a smaller scale study in infants with nearly similar age. Oligofructose intake improved the intestinal bacterial colonization, as characterized by higher levels of bifidobacteria, especially in those infants with lower baseline levels, and a decrease in clostridia. Daily oligofructose administration was well tolerated. Additionally, number of febrile events and gastrointestinal illness symptoms, such as flatulence, diarrhea, and vomiting, were less often observed (Waligora-Dupriet et al., 2007). Fisberg et al. (2000) evaluated the incidence and duration of sickness in 626 mild to moderately malnourished children (1–6 years of age) who received a nutritional supplement with a synbiotic (oligofructose together with *Lactobacillus acidophilus* and *Bifidobacterium infantis*). In a subgroup of children (aged 3–5 years), the number of sick days were fewer with symbiotic administration as were days of constipation.

6.6.7 Absorption and Accretion of Minerals

6.6.7.1 Influence on Calcium Metabolism

Calcium is an integral part of the skeleton, and increasing the amount of calcium absorbed from the diet is an important strategy to improve bone metabolism at all ages. Over the past 10 years, much research has been performed on the effects of inulin and oligofructose on calcium absorption as well as bone metabolism and its mineralization. This extensive amount of research done in animals and humans has repeatedly shown that inulin-type fructans increase calcium absorption from the diet, modulate makers of bone metabolism, and improve calcium sequestration in the bone (Bosscher et al., 2006b).

Two key animal models have been used to determine the impact of prebiotics on calcium absorption and bone mineral density: (1) young growing animals that represent the adolescent phase in humans, and (2) adult ovariectomized rats that represent the postmenopausal phase in humans. The "apparent calcium absorption" is then measured by subtracting the fecal calcium content from the actual calcium intakes of the animals. First experiments were performed in growing rats fed on inulin-type fructans in the diet, which found increased apparent calcium absorption (Delzenne et al., 1995). Later on, similar results were seen with oligofructose-enriched inulin in ovariectomized rats. Moreover, kinetic data on bone metabolism showed improved bone balance and suppression of bone turnover (Zafar et al., 2004). It is clear that inulin-type fructans improve calcium absorption; but does this translate to advances in bone mineralization and strength? This was evaluated in the two animal models described above. Roberfroid et al. (2002) showed that feeding growing rats with inulin resulted in an increase in whole-body bone mineral content (BMC) and whole-body bone mineral density (BMD). In this rat model, long-term inulin administration appeared to impact mainly on the trabecular bone network. In adult ovariectomized rats, oligofructose increased BMC in the femur and lumbar vertebra and prevented the ovariectomy-induced loss of bone in the trabecular structure (Scholz-Ahrens et al., 2002).

Following this, a number of intervention studies in adolescents were carried out. The best method for determining calcium absorption is believed to be the use of stable isotopes, as this excludes the confounding variable of endogenous calcium excretion. A measure of "true calcium absorption" is then provided, which cannot be obtained by the balance method. If isotopes are administered as part of a metabolic study, one can also see the individual components of calcium metabolism: i.e., absorption, excretion, endogenous secretion, bone formation rates, and bone resorption rates. Only studies in which stable isotopes and supplementation with oligofructose-enriched inulin were used will be described here. In adolescent girls (mean age 11.8 years), supplementation with oligofructose-enriched inulin increased true calcium absorption, whereas urinary excretion of calcium remained unaffected. As the response to treatment differed between subjects, further study was initiated to evaluate the subject's characteristics (Griffin et al., 2003). The new study had a protocol similar to that carried out before in order to merge the two sample sizes. An extra 25 adolescent girls were recruited, bringing the overall sample to 54 (mean age 12.4 years) across two centers in Texas and Nebraska. As with the previous study, there was a significant increase in calcium absorption with the oligofructose-enriched inulin. A range of characteristics was studied: e.g., age, weight, height, Tanner stage, and ethnicity. However, the most consistent determinant of a beneficial effect of the oligofructose-enriched inulin on calcium absorption was the fractional calcium absorption of the girls at baseline. Thus, girls with lower baseline levels of calcium absorption responded with the greatest increase when supplemented with oligofructose-enriched inulin. This finding is important, as these adolescents are most likely to benefit from the higher amount of calcium absorbed from the diet. Since the all subjects received a similar calcium load from the test meals, the variation cannot be explained by the presence of enhancers or inhibitors of calcium absorption in the diet. Rather, it is likely that the subjects' genotypes governed both their baseline calcium absorption and their response to the oligo-fructose-enriched inulin consumption. Interestingly, the subjects' polymorphisms of the Fok1 vitamin D receptor gene were determined, and an interaction with Fok1 genotype was present such that ff genotype subjects had the least initial response to the inulin supplementation. This shows that effects of dietary factors on calcium absorption may be modulated by genetic factors including ethnicity and specific vitamin D receptor gene polymorphisms (Abrams et al., 2005).

In a subsequent long-term (1 year) intervention study in 100 pubertal girls and boys in early puberty (9–12 years of age), the impact of oligofructose-enriched inulin on bone accretion was studied. The participating subjects were selected for Tanner stage 2 or 3. This increased the likelihood that calcium accretion was at its highest during the study when it would have the capacity to determine bone health at a later age. Against a background of normal calcium intakes (900–1000 mg/d), subjects received 8 g/day of oligofructose-enriched inulin added to milk or a calcium-fortified orange juice. As ◉ *Figure 6.3* shows, 2 months into the intervention, subjects receiving oligofructose-enriched inulin had higher true calcium absorption. The enhanced calcium absorption was maintained throughout the intervention so that at 1 year, the differences were maintained. Correspondingly, subjects receiving oligofructose-enriched inulin had greater change in whole-body BMC at 1 year (see ◉ *Figure 6.4*). Assuming that the fraction of calcium in BMC is 32%, these values correspond to a daily skeletal calcium accretion of 218 mg/d for the oligofructose-enriched inulin group (and 189 mg/d for the controls). This can be estimated as an additional net accretion of 30 mg/d. The change in whole body BMD was also greater in subjects receiving oligofructose-enriched inulin. This corresponded to an increment in whole-body BMD after 1 year of 15 $\mu g/cm^2$/year (see ◉ *Figure 6.5*). The interpretation of this study is that oligofructose-enriched inulin increased calcium absorption during pubertal growth and enhanced bone mineralization, leading to a greater bone mass accretion during adolescence (Abrams et al., 2005).

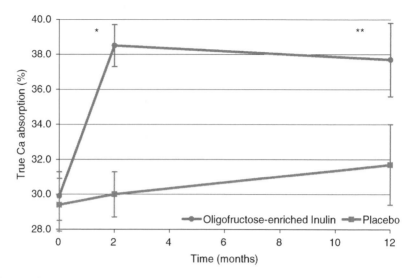

◻ Figure 6.3

True calcium absorption (%) of the subjects (mean ± SD) after supplementation of the diets with oligofructose-enriched inulin or placebo (maltodextrin) for 8 weeks and 1 year (asterisks represent significant difference compared to the placebo * $P < 0.001$, ** $P = 0.04$).

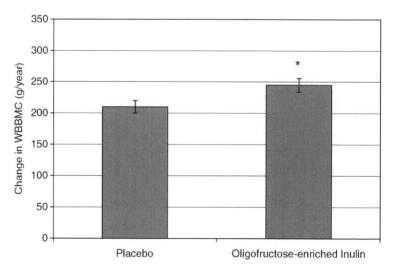

◻ Figure 6.4

Change in whole-body bone mineral content (WBBMC) (g/year) of the subjects (mean ± SD) after supplementation of the diets with either oligofructose-enriched inulin or placebo (maltodextrin) for 1 year (asterisks represent significant difference compared to the placebo $P = 0.03$).

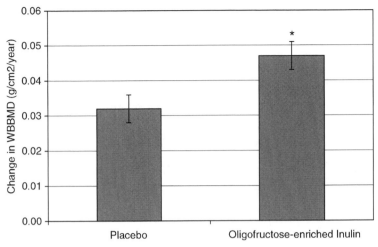

Figure 6.5
Change in whole-body bone mineral density (WBBMD) (g/cm^2/year) of the subjects (mean ± SD) after supplementation of the diets with either oligofrucose-enriched inulin or placebo (maltodextrin) for 1 year (asterisks represent significant difference compared to the placebo $P = 0.01$).

Other studies on the impact of oligofructose-enriched inulin were performed in elderly women. In women, the menopause is a time when estrogen deficiency leads to accelerated bone resorption and negative bone balance. A subsequent intervention trial with oligofructose-enriched inulin in postmenopausal women (mean 72 years of age) was carried out studying markers of bone turnover. The supplementation improved calcium and magnesium absorption, even when vitamin D status was adequate and calcium intake was good. Markers of bone turnover showed a short-term decrease (urinary deoxypyridinoline crosslinks) in bone resorption and a clear increase in bone formation (osteocalcin), and this was most pronounced in those women with lower initial spine BMD (Holloway et al., 2007).

Another interesting feature of inulin-type fructans that may affect bone spearing in elderly women is their effects on the intestinal metabolism of iso-flavones. Isoflavones are found in soybeans and behave as estrogen mimics and, therefore, are implicated for use in the prevention of gonadal-induced osteopenia at the time of menopause. Isoflavones mainly occur in plants as glycosides, which have to be hydrolyzed before they can be absorbed. Glucosidases of intestinal bacteria, such as lactobacilli and bifidobacteria, are estimated to be involved in this process. In ovariectomized rats, fructo-oligosaccharide supplementation to a soy-based diet was shown to increase plasma levels of genistein, daidzein, and equol and improved the protective effect of isoflavones against bone loss secondary to castration (Mathey et al., 2004). In postmenopausal women, inulin has indeed been demonstrated to enhance blood levels of isoflavones (daidzein and genistein) after soy intake. However, whether this results in improved bone protection (less bone resorption) in this age group is unclear and needs further study.

The underlying mechanisms of the effect of inulin-type fructans on calcium absorption seem to be diverse and affect the different stages of calcium absorption. Calcium moves across the intestinal epithelium in two different ways. The trans-cellular route comprises apical calcium entry via calcium channels, calcium binding and sequestration through the cytosol, mostly by calbindin, and basolateral calcium extrusion mostly by an ATPase. In the para-cellular route, calcium moves through tight junctions where claudins are thought to form channel-like structures. The major fraction of calcium is absorbed in the small intestine; however, part of the dietary calcium is also absorbed in the large intestine. The fermentation of inulin-type fructans and the corresponding acid formation are thought to play a role essentially on the para-cellular route in the colon. This has been shown to lead to a reduced luminal pH with a marked increase in the luminal pools for total, soluble, and ionized calcium.

By the use of Ussing chambers (*in vitro*) of the rats' large intestine, the effect of oligofructose-enriched inulin and its fermentation products was studied on trans-epithelial calcium transport. The presence of SCFA increased trans-epithelial calcium transport independently of the luminal pH. It might be hypothesized that increased protonation of SCFA anions in a more or less acidic environment allows their permeation into the epithelium by non-ionic diffusion. Once in the cell, dissociation might occur, and due to the higher pH release their protons that can then be exchanged with luminal calcium. Alternatively, SCFA anions could also be absorbed in exchange with HCO_3^-, produced by carbonic anhydrase (CA) from H_2CO_3, and with apical secretion of the proton that can be exchanged for calcium. Alternatively, it was hypothesized that inulin-type fructans and/or their fermentation products could also directly interact with the tight junctions. All will ultimately result in an increase in net calcium uptake. The observed changes in the mRNA level of genes linked to trans- and para-cellular calcium transport support the above hypotheses. The upregulation of the CA transcript levels in the proximal colon supports the coupling theory of SCFA movement and calcium flux. CA could provide more intracellular HCO_3^- anions for exchange with SCFA anions, while the remaining H^+ could be extruded by $Ca^{2+}/2H^+$ and Na/H^+ exchangers. Interestingly, and indicative of an involvement of the para-cellular routs as well was the high expression level of claudin, is a tight junction protein that acts as a calcium channel (Raschka and Daniel, 2005). An upregulation of the calbindin-D9k protein and its associated transcription factors VDR and cdx-2 in the rats' colorectal segment upon oligo-fructose feeding has been demonstrated by others (Fukushima et al., 2005).

Apart from such molecular alterations, dietary administration of inulin-type fructans has also been found in animal models to lead to a net enlargement in total tissue surface (but not thickness) of the intestinal wall. The possible cause for such tissue enlargement might be the trophic actions of the butyrate formed during the fermentation process.

6.6.7.2   Influence on Iron Metabolism

While the effects of inulin-type fructans on calcium absorption and metabolism are well documented, recent studies also pinpoint to the effects on iron absorption, particularly in a state of iron deficiency. Young weaning piglets are considered a suitable model for studying human iron nutrition because of their similarities in the anatomy of the gastrointestinal tract and digestive physiology. Addition of oligofructose-enriched inulin to the diet of iron-deficient and anemic pigs has been found to dose-dependently increase soluble levels of iron in the colon as well as improve blood hemoglobin levels and hemoglobin repletion efficiency. Sulfide levels were found to be lower. High levels of sulfide occur mostly as a result of protein fermentation and can bind iron in the intestinal tract, rendering it unavailable for absorption. Attenuated levels of sulfide in the colon might be suggestive of a potential mechanism (Yasuda et al., 2006).

6.6.8   Body Weight, Appetite, Energy Intake, and Metabolism

6.6.8.1   Body Weight Modulation

Obesity is one of the greatest public-health challenges of the 21st century. The condition is caused by a persistent imbalance between (high) energy intake and (low) energy expenditure and the development of fat mass as storage of the excess in energy. Changes in lifestyles are seen as the major cause, as diets high in nutrients and complex carbohydrates gave way to energy-dense and high-fat diets.

The long-term effects of oligofructose-enriched inulin on the maintenance of body weight were observed in (primarily nonobese) 100 pubertal adolescents (9.0–13.0 years of age). Those findings were part of a bigger study which also included measurements of calcium absorption and bone mineralization. The bone-related findings are discussed in Section 10.6.7.1. Adolescents that received oligofructose-enriched inulin (8 g/day) over a period of 12 months had lower increments in body mass index (BMI), body weight, and fat mass after one year. In other words, those adolescents that received oligofructose-enriched inulin benefited in the maintenance of an appropriate BMI during their pubertal growth. They had an increase in BMI of about 0.7 kg/m^2 during the supplementation year, which is consistent with expected increases during puberty. Subjects not receiving the prebiotic (placebo) had a higher increase (of 1.2 kg/m^2). Interestingly, the effect of oligofructose-enriched inulin on BMI was in a nonlinear fashion modified by dietary intake of calcium, such that the benefit of oligofructose-enriched inulin was greater when low calcium intakes were avoided. Follow-up data were available for 89 subjects. Interestingly, BMI values for the prebiotic supplemented subjects remained lower after stopping the supplement for one year, indicating that the effect was persistent after discontinuation of the supplementation. The lower increase in BMI in the prebiotic group implies an overall regulatory effect on energy intake associated with the oligofructose-enriched inulin supplementation (Abrams et al., 2007). The overall greater increase in BMI during the year in the nonsupplemented group is likely to be nonideal and is related to the overall trend toward increased BMI during adolescence currently.

6.6.8.2 Appetite Regulation and Energy Intake

Our body is equipped with powerful and complex interacting pathways (often mediated by hormones) that can initiate and terminate food intake and reflect body adiposity and body energy balance by integrating signals to the hypothalamus and brain stems. Several of such hormones have their origins in the gut, and this is often referred to as the "brain-to-gut axis."

Studies starting in the late 1990s using various animal models have consistently shown that the fermentation of inulin-type fructans in the colon can modulate the expression of hormones involved in appetite and having their origins in the gut (e.g., proglucagon mRNA). This in turn can modulate their levels in the blood, affecting appetite and food intake. This has been demonstrated in normal rats when fed on a diet supplemented with inulin-type fructans. The levels of GLP-1 (7–36) amide in plasma were higher, and proglucagon mRNA in the proximal colonic mucosa showed a marked upregulation in case oligofructose or oligofructose-enriched inulin was added. The expression of proglucagon takes place in the large intestine where it is cleaved into peptides (besides glucagon) such as GLP-1. GLP-1 is a biologically active peptide secreted into the portal circulation with anorexigenic properties. The opposite was true for the plasma levels of the hormone ghrelin, which remained significantly lower in the oligo-fructose and oligofructose-enriched inulin fed animals (Cani et al., 2004). Energy intake was consequently lower. This led to a decrease in (epidydimal) fat mass after several weeks of supplementation in the oligofructose and oligofructose-enriched supplemented animals. In a subsequent study undertaken in rats put on a diet high in fat (i.e., to mimic the Western diet), oligofructose lowered energy intake and reduced high-fat induced hyperphagia. Consequently, body weight gain on the high-fat diet was lower and the total weight of the adipose tissue was twofold lower than in the animals' not receiving oligofructose. In addition, the plasma triglyceride levels were reduced, and hepatic triglyceride levels were about 30% lower in those receiving oligofructose than those that did not receive it. The levels of proglucagon (mRNA) and GLP-1 in cecum and colonic tissues and GLP-1 in the portal vein were higher with the oligofructose diet (Cani et al., 2005a). Also in other models of obesity, such as genetically obese (fa/fa Zucker) rats, the same effects of oligofructose on body weight (wt) and steatosis were observed (Daubioul et al., 2002). This was found to be linked to a decrease in hepatic de novo fatty acid synthesis, which is thought to be due to the inhibition of the fatty acid synthase activity (Kok et al., 1998). When other dietary fibers, e.g., nonfermentable fibers, were added, those effects were not observed. No protec-tion against the development of steatosis was observed when cellulose was added to the diet. Also, obese rats fed on cellulose had significantly higher food intake when compared to those receiving oligofructose (Daubioul et al., 2002).

A pilot study in healthy men and women (aged 21–39 years) with normal BMI values showed that oligofructose (2 times 8 g/day) given at beatkast and dinner affected satiety scores and energy intakes. Appetite ratings of hunger, satiety, and fullness after each meal were recorded by visual analog scales. Visual analog scales are standard 100-mm lines anchored at each end by a phrase denoting the most extreme appetite sensations. The visual analog scales are a well-accepted tool to assess satiety in its intensity and duration. Oligofructose-supplemented volunteers experienced higher levels of satiety, reduced hunger, and prospective food consumption (see ❷ *Figure 6.6*). Subjects were also invited to a day of

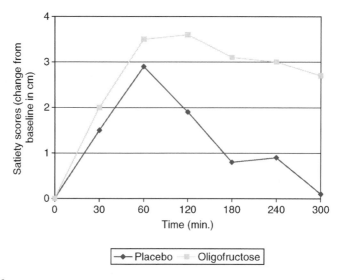

Satiety scores after dinner (measured by Visual Analogue Scale) in healthy subjects after 2 weeks of supplementation with oligofructose (8 g twice daily) or placebo. The results are presented as change from baseline scores and are means ± SEM, n = 10 subject for each intervention phase.

free-choice buffets during which their food and drink intakes were monitored. The energy intake at breakfast and lunch was lower with oligofructose, which resulted in a significant reduction in the total energy intake during the day (see ◗ *Figure 6.7*) (Cani et al., 2006a). Also in other human studies the effects of oligofructose, either alone or in combination with other dietary fibers (pea fiber), on satiety and food intake have been documented. Addition of oligofructose to enteral formula also containing pea fiber was shown to result into higher subjective measures Visual Analogue Scale, (VAS) of fullness and satiety feelings (Whelan et al., 2006).

The satiating power of inulin-type fructans together with their technological benefits to replace fat, while remaining the palatability of foods, could be a dual way to manufacture foods with high "satiety power" while at the same time having a low energy density. In fact, strategies focused on lowering fat content (and energy density) of individual foods have failed in some cases, as poor palatability and the lack of satiating power of such low-fat versions led to the (over)compensation for the reduced energy intake during the rest of the day. Replacing half of the fat by inulin in foods (e.g., reduced-fat patty instead

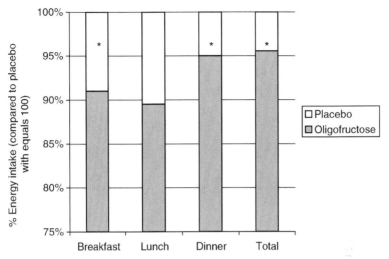

◘ Figure 6.7
The percentage energy intake relative to placebo at breakfast, lunch, dinner, and over the day in healthy subjects after 2 weeks of oligofructose supplementation (8 g twice daily). Values are means ± SEM, *n* = 10 subjects (asterisk Indicates a significant difference from the placebo treatment $P < 0.05$ **, $P < 0.01$).

of full-fat sausage patty), while otherwise remaining similar in protein and carbohydrate content, reduced the fat content by 36% and energy content by 15% and showed good sensorial acceptance of the breakfast patties among healthy men. Although satiety scores of the men participating in the study did not differ after consumption of the low-fat inulin version compared with the full-fat version, the authors were suggestive of an effect of inulin on post-meal satiety given that the inulin breakfast provided less energy than the full-fat breakfast. Moreover, energy intakes during the whole day were lower with the inulin breakfasts, indicating that full compensation for the lower energy content did not occur (Archer et al., 2004).

Nowadays, more evidence has become available about the role of fermentable dietary fibers in enhancing satiety and the relationship with high blood levels of glucagon-like peptide 1 (7–36) amide (GLP-1) with a corresponding upregulation of proglucagon expression in the (proximal) colonic mucosa (the latter coming from animal studies). GLP-1 is a peptide produced by intestinal endocrine L cells through a specific post-translational processing of the proglucagon gene. Feeding rats with oligofructose has been shown to almost double the numbers of GLP-1 positive L cells in the proximal colon, which was associated with the differentiation of cell precursors (stem cells) into mature L cells (Cani et al., 2007a). GLP-1 is considered a key peptide in the control of glucose tolerance, insulin secretion by pancreatic β-cells, and the transmission of satiety signals to the brain. The current available evidence indicates that the effect of oligofructose is largely dependent on GLP-1 and that the GLP-1 peptide appears to constitute the link between the fermentation of inulin-type fructans in the colon and their modulation of lipid and glucose metabolism, as well as appetite regulation and food intake.

6.6.8.3   Adipose Tissue and Inflammation

Obesity and metabolic disorders (insulin resistance, hyperlipaemia) are tightly linked to a chronic low-grade state of inflammation (elevated levels of circulating inflammatory markers such as IL-6, and C-reactive protein). The adipose tissue is an endocrine organ actively releasing a number of immune active factors (TNF-α, IFN-γ, IL1-β, IL-6, IL-8, IL-10, acute-phase proteins, etc.). Inflammatory factors are known to be involved in insulin resistance, favoring hyperinsulinemia and excessive hepatic and adipose tissue lipid storage. It appears that an altered gut microbiota in the obese state could contribute towards low-grade inflammation resulting in the development of metabolic diseases associated with the condition (e.g., diabetes, cardiovascular disease, etc.). However, the factors triggering such metabolic alterations remain to be determined.

In the obese, lower levels of Bacteroidetes and higher levels of the phylum Firmicutes in the colonic microbiota as compared to lean counterparts are found (Ley et al., 2006). These observations have been associated with increased gut fermentation and calorific bioavailability to the host. Moreover, feeding high-fat diets have been demonstrated to alter dramatically the microbiota composition in mice with reduction in the quantities of dominant Gram-positive groups, e.g., *Bifidobacterium* spp. and *E. rectale, C. coccoides* groups, and the murine Gram-negative group, *Bacteroides MB.*

Recent studies in animal models have shown that such changes within the microbial ecology or functional activities of the gut microbiota can induce a metabolic shift toward a proinflammatory phenotype, whole-body, liver and adipose tissue weight gain, and impaired glucose metabolism. Factors of microbial origins (e.g., bacterial lipopolysaccharides) are hypothesized to be basis of such effects. In mice, high-fat feeding led to (low level of) metabolic endotoxemia, low inflammatory tone, increasing macrophage infiltration in adipose tissue, and dysregulating lipid and glucose metabolism. Multiple correlation analyses showed that the level of endotoxaemia was negatively correlated with *Bifidobacterium* spp., but no relationship was seen between any other bacterial groups. On the other hand, restoration of the levels of bifidobacteria in the intestine of mice upon oligofructose supplementation lowered endotoxaemia and the level of microbial toxins and improved mucosal barrier function. Interestingly, the lower body weight and visceral adipose tissue mass in the oligofructose group (compared with the non-supplemented high-fat fed mice) showed a positive correlation with the endotoxin plasma levels and a negative one with the levels of bifidobacteria. Moreover, levels of mRNA of IL-1, TNF-α, and plasminogen activator inhibitor type-1 (Pai-1, or Serpine-1) in adipose tissue were increased in high-fat fed mice, whereas the levels were blunted with oligofructose feeding. In addition, a normalization of IL-1α and IL-6 cytokines was observed upon oligofructose feeding. These data indicate that a lower fat mass and body weight "only" are not a prerequisite for a lower inflammatory tone and that this effect is accompanied by prebiotic changes in the microbiota. Plasma cytokines were positively correlated with plasma endotoxin levels and negatively with bifidobacteria levels (Cani et al., 2007b). In diabetic mice, feeding oligofructose reduced hepatic levels of phosphorylate IKK-β and NF-κB, suggestive of a reduction in the hepatic inflammatory status which might relate to an improvement of the insulin sensitivity (Cani et al., 2005b).

6.6.8.4   Glucose and Lipid Metabolism

High glucose and blood cholesterol levels and hyperlipaemia are major risk factors for development of the metabole syndrome, associated diabetes, and cardiovascular diseases, and are most often linked to wrong food choices (e.g., high-fat and low-fiber diets). In the above sections, the beneficial effects of inulin-type fructans on lipid and glucose metabolism have already been indicated briefly; however, more evidence is currently available, especially from animal models.

The effects of oligofructose on satiety and energy intake have been clearly related to the levels of GLP-1. However, as GLP-1 is also a key peptide in the control of glucose tolerance and insulin secretion, effects of oligofructose supplementation on glucose metabolism can be estimated. Indeed, further research in diabetic rats (treated iv with streptozotocin to induce postprandial hyperglycemia) showed that oligofructose supplementation reduced high blood glucose (due to the diabetic condition) and even normalized glycaemia to normal levels. Alternatively, low plasma insulin levels (due to the diabetic condition) were increased and also normalized to normal (nondiabetic) levels. Moreover, oligofructose supplementation elevated pancreatic insulin levels and beta cell mass (as these were drastically reduced because of the diabetic condition), indicating improved beta-cell mass and function. Consistent with earlier reports, these physiological changes were linked with higher levels of GLP-1 in the portal blood of the rats. Oral glucose tolerance tests (OGTT) in these diabetic animals revealed a lower area under the curve (AUC) in rats fed with oligofructose. The plasma insulin response to glucose during the OGTT, on the other hand, was markedly increased. Interestingly, the AUC for insulin was similar for diabetic rats fed the oligofructose as with normal (nondiabetic) rats (Cani et al., 2005b). The effects of oligofructose are largely dependent on the action of GLP-1. Extensive investigations in diabetic rats indeed showed that the disruption of the GLP-1 receptor function, either by infusing Ex-9 (GLP-1 R antagonist) or by using GLP-1 receptor−/− mice, completely prevented the majority of the beneficial metabolic effects observed following oligofructose supplementation. In the presence of the antagonist, the effect of oligofructose on body weight, glucose tolerance, fasting blood glucose, glucose-stimulated insulin secretion, and insulin-sensitive hepatic glucose production disappeared. Also, GLP-1R −/− mice appeared to be totally insensitive to the systemic effects of oligofructose (Cani et al., 2006b).

Various animal models (hamsters, mice, and rats) have consistently shown the lipid and cholesterol lowering actions of inulin-type fructans. In rats fed on a high-fat diet, oligofructose suppressed the postprandial increase in triglyceride levels and hepatic triacylglycerol load (originating from the diet) (Kok et al., 1998). In genetically obese (fa/fa Zucker) rats, a similar pattern was observed, with lower body weight, fat mass, and steatosis development when inulin-type fructans were part of the diet. Detailed biochemical studies in isolated hepatocytes demonstrated that inulin-type fructans reduce the activities of key hepatic enzymes related to lipogenesis (e.g., de novo synthesis of fatty acid synthesis). Further research revealed that an altered gene expression was the cause of the downregulation, which might have been in response to hormonal changes induced by inulin-type fructans (Kok et al., 1998). More recent evidence comes from studies in ApoE-deficient mice, a genetically modified mice model that spontaneously develops atherosclerosis. Feeding mice with inulin reduced plasma levels of cholesterol, triacylglycerols, hepatic cholesterol, and triacylglycerols. Histo-morphometry of the aortic sinus showed less plaque formation in mice receiving long-chain inulin or oligofructose-enriched inulin, with a mean reduction in the lesion area of 35% and 25%, respectively, when compared to mice not receiving the supplements. It is most likely that the inhibition of atherosclerotic plaque formation was related to the changes observed in lipid and cholesterol metabolism upon inulin intake (Rault-Nania et al., 2006).

Also, in humans inulin-type fructans have been shown to affect lipid metabolism, although the data are less consistent compared to the results from animal studies. This might be due to differences in methodological setup, type of subjects, as well as duration of the intervention, dose, and type (short or long chain) of the inulin-type fructan used in the study. It has been reported that the consumption of inulin-type fructans reduced serum triglycerides and in some cases also cholesterol (mostly LDL fraction) in healthy volunteers who are (slightly) hyperlipidemic. Lipid parameters in healthy (normolipidemic) young adults, on the contrary, appear to be unaffected. Also, studies indicate that such effects, if any, take some time to become established, urging for a longer term intervention (Brighenti et al., 1999; Davidson and Maki, 1999; Jackson et al., 1999). Although much uncertainty still exists about the mechanisms that are responsible for those effects, it is already clear that those include various interdependent biochemical pathways that may take place in the liver, pancreas, intestine, and peripheral tissues (e.g., adipose tissue). Research in this field has evolved with primary focus on the endocrine activity in the gut, which has been described in earlier sections.

6.7   Outlook and Perspectives

The number of publications on prebiotics and inulin-type fructans has increased dramatically once the concept was established in 1995. Especially, the number of publications demonstrating the nutritional effects of inulin-type fructans is overwhelming. This is expected to continue since the importance of a well-balanced colonic microbiota as being key in the modulation of human immunity, metabolism, and endocrine activities is being more and more recognized among the general scientific population. As new insights are being elucidated about the composition of the microbiota and its species diversity, the role the microbiota in the origins of disease, and the mechanisms of action, interest will continue to rise. Together with this, it is of paramount importance to develop strategies to modulate this microbiota in a way to reduce the risk of developing disease through dietary means. Dietary strategies require a combination of technological means and well-established nutritional evidence as well as legal responsibility to benefit health and well-being of the population. Functional foods with prebiotics and probiotics offer great value in this regard and, although many health benefits have been demonstrated today, more can be expected in the near future.

List of Abbreviations

ACF	aberrant crypt foci
AOM	azoxymethane
AUC	area under the curve
BMC	bone mineral content
BMD	bone mineral density
BMI	body mass index
CA	carbonic anhydrase
CBL	cecal bacterial lysates
DP	degree of polymerization
GALT	gut-associated lymphoid tissue
GC	gas chromatography
GST	glutathione S-transferases
hBD	human beta defensins
HPAEC-PAD	HP-anion exchange chromatography with pulsed amperometric detection
LC	liquid chromatography
MLN	mesenteric lymph node
NEC	necrotizing enterocolitis
NK	natural killer
OGTT	oral glucose tolerance tests
PP	Peyer's patches
PBMC	peripheral blood mononuclear cells
SCFA	short-chain fatty acids
TG	transgenic
UC	ulcerative colitis
WBBMC	whole-body bone mineral content
WBBMD	whole-body bone mineral density

References

Abrams SA, Griffin IJ, Hawthorne KM, Ellis KJ (2007) Effect of prebiotic supplementation and calcium intake on body mass index. J Pediatr 151:293–298

Abrams SA, Griffin IJ, Hawthorne KM, Liang L, Gunn SK, Darlington G, Ellis KJ (2005) A combination of prebiotic short- and long-chain inulin-type fructans enhances calcium absorption and bone mineralization in young adolescents. Am J Clin Nutr 82:471–476

Ananta E, Birkeland S-E, Corcoran B, Fitzgerald G, Hinz S, Klijn A, Mättö J, mercenier A, Nilsson U, Nyman M, O'Sullivan E, Parche S, Rautonen N, Ross RP, Saarela M, Stanton C, Stahl U, Suomalainen T, Vincken J-P, Virkajärvi I, Voragen F, Wesenfeld J, Wouters R, Knorr D (2004) Processing effects on the nutritional advancement of probiotics and prebiotics. Microb Ecol Health Dis 16:113–124

Archer BJ, Johnson SK, Devereux HM, Baxter AL (2004) Effect of fat replacement by inulin or lupin-kernel on sausage patty acceptability, post-meal perceptions of satiety and food intake in men. Br J Nutr 91:591–599

Bartosch S, Woodmansey EJ, Paterson JCM, McMurdo ET, Macfarlane GT (2005) Microbiological effects of consuming a synbiotic containing Bifidobacterium bifidum, Bifidobacterium lactis, and oligofructose in elderly persons, determined by real-time polymerase chain reaction and counting of viable bacteria. CID 40:28–37

Belenguer A, Duncan SH, Calder AG, Holtrop G, Louis P, Lobley GE, Flint HJ (2006) Two routes of metabolic cross-feeding between Bifidobacterium adolescentis and butyrate-producing anaerobes from the human gut. Appl Environ Microbiol 72:3593–3599

Benyacoub J, Rochat F, Saudan K-Y, Rochat I, Antille N, Cherbut C, von der Weid T, Schiffrin EJ, Blum S (2008) Feeding a diet containing a fructooligosaccharide mix can enhance Salmonella vaccine efficacy in mice. J Nutr 138:123–129

Bosscher D, Van Loo J, Franck A (2006a) Inulin and oligofructose as prebiotics in the prevention of intestinal infections and diseases. Nutr Res Rev 19:216–226

Bosscher D, Van Loo J, Franck A (2006b) Inulin and oligofructose as functional ingredients to improve mineralization. Int Dairy J 16:1092–1097

Brighenti F, Casiraghi MC, Canzi E, Ferrari A (1999) Effect of consumption of a ready-to-eat breakfast cereal containing inulin on the intestinal milieu and blood lipids in healthy male volunteers. Eur J Clin Nutr 53:726–733

Buddington KK, Donahoo JB, Buddington RK (2002) Dietary oligofructose and inulin protect mice from enteric and systemic pathogens and tumor inducers. J Nutr 132:472–477

Bunout D, Barrera G, Hirsch S, Gattas V, Pia de la Maza M, Haschke F, Steenhout P, Klassen P, Hager C, Avendano M, Petermann M, Munoz C (2004) Effects of a nutritional supplement on the immune response and cytokine production in free-living Chilean elderly. J Parenter Enteral Nutr 28:348–354

Butel M, Catala I, Waligora-Dupriet A, Taper H, Tessedre A, Durao J, Szylit O (2001) Protective effect of dietary oligofructose against cecitis induced by clostridia in gnotobiotic quails. Microb Ecol Health Dis 13:166–172

Butel MJ, Roland N, Hibert A, Popot F, Favre A, Tessedre AC, Bensaada M, Rimbault A, Szylit O (1998) Clostridial pathogenicity in experimental necrotising enterocolitis in gnotobiotic quails and protective role of bifidobacteria. J Med Microbiol 47(5):391–399

Cani P, Daubioul C, Reusens B, Remacle C, Catillon G, Delzenne N (2005b) Involvement of endogeneous glucagon-like peptide-1(7–36) amide on glycemia-lowering effect of oligofructose in streptozotocin-treated rats. J Endocrin 185:457–465

Cani P, Dewever C, Delzenne M (2004) Inulin-type fructans modulate gastrointestinal peptides involved in appetite regulation (glucagon-like peptide-1 and ghrelin) in rats. Br J Nutr 92: 521–526

Cani PD, Hoste S, Guiot Y, Delzenne NM (2007a) Dietary non-digestible carbohydrates promote L-cell differentiation in the proximal colon of rats. Br J Nutr 98:32–37

Cani P, Joly E, Hormans Y, Delzenne N (2006a) Oligofructose promotes satiety in healthy human: a pilot study. Eur J Clin Nutr 60:567–572

Cani P, Knauf C, Iglesias M, Drucker D, Delzenne N, Burcelin R (2006b) Improvement of glucose tolerance and hepatic insulin secretion by oligofructose requires a functional glucagon-like peptide 1 receptor. Diabetes 55:1484–1490

Cani PD, Neyrinck AM, Fava F, Knauf C, Burcelin RG, Tuohy KM, Gibson GR, Delzenne NM (2007b) Selective increases of bifidobacteria in gut microflora improve high-fat-induced diabetes in mice through a mechanism associated with endotoxaemia. Diabetologica 50:2374–2383

Cani P, Neyrinck A, Maton N, Delzenne N (2005a) Oligofructose promotes satiety in rats fed a high-fat diet: involvement of glucagon-like peptide-1. Obes Res 13: 1000–1007

Casellas F, Borruel N, Torrejón A, Varela E, Antolin M, Guarner F, Malagelada J-R (2007). Oral oligofructose-enriched inulin supplementation in acute colitis is well tolerated and associated with lower faecal calprotectin. Aliment Pharmacol Ther 25: 1061–1067

Cherbut C, Michel C, Lecannu G (2003) The prebiotic characteristics of fructooligosaccharides are necessary for reduction of TNBS-induced colitis in rats. J Nutr 133: 21–27

Daubioul C, Rousseau N, Demeure R, Gallez B, Taper H, Declerck B, Delzenne N (2002) Dietary fructans, but not cellulose, decrease triglyceride accumulation in the liver of obese Zucker fa/fa rats. J Nutr 132:967–973

Davidson MH, Maki KC (1999) Effects of dietary inulin on serum lipids. J Nutr 129: 1474S–1477S

Delzenne N, Aertssens J, Verplaetse H, Roccaro M, Roberfroid M (1995) Effect of fermentable fructo-oligosaccharides on mineral, nitrogen and energy digestive balance in the rat. Life Sci 57:1579–1587

Den Hond E, Geypens B, Ghoos Y (2000) Effect of high performance chicory inulin on constipation. Nutr Res 20:731–736

Duncan SH, Holtrop G, Lobley GE, Calder AG, Stewart CS, Flint HJ (2004) Contribution of acetate to butyrate formation by human faecal bacteria. Br J Nutr 91:915–923

Falony G, Vlachou A, Verbrugge K, De Vuyst L (2006) Cross-feeding between Bifidobacterium longum BB536 and acetate-converting, butyrate-producing colon bacteria during growth on oligofructose. Appl Environ Microbiol 72:7835–7841

Firmansyah A, Pramita G, Fassler C, Haschke F, Link-Amster H (2001) Improved humoral immune response to measles vaccine in infants receiving infant cereal with fructooligosaccharides. J Pediatr Gastroenterol Nutr 31:A521

Fisberg M, Maulen I, Vasquez E, Garcia J, Comer GM, Alarcon PA (2000) Effect of oral supplementation with and without synbiotics on catch-up growth in preschool children. J Pediatr Gastroenterol Nutr 31:A521

Franck A (2002) Technological functionality of inulin and oligofructose. Br J Nutr 87: S287–S291

Fukushima A, Ohta A, Sakai K, Sakuma K (2005) Expression of calbindin-D9k, VDR and Cdx-2 Messenger RNA in the process by which fructo-oligosaccharides increase calcium absorption in rats. J Nutr Sci Vitaminol 51:426–432

Furrie E, Macfarlane S, Kennedy A, Cummings JH, Walsh SV, O'Neil DA, Macfarlane GT (2005) Synbiotic therapy (Bifidobacterium longum/Synergy1) initiates resolution of inflammation in patients with active ulcerative colitis: a randomized controlled pilot trial. Gut 54:242–249

Gibson GR, Beatty ER, Cummings J (1995a) Selective fermentation of bifidobacteria in the human colon by oligofructose and inulin. Gastroenterology 108:975–982

Gibson GR, Probert HM, Van Loo J, Rastall RA Roberfroid MB (2004) Dietary modulation of the human colonic microbiota:

updating the concept of prebiotics. Nutr Res Rev 17:259–275

Gibson GR, Roberfroid MB (1995b) Dietary modulation of the human colonic microbiota - Introducing the concept of prebiotics. J Nutr 125:1401–1412

Griffin IJ, Hicks PMD, Heaney RP, Abrams SA (2003) Enriched chicory inulin increases calcium absorption mainly in girls with lower calcium absorption. Nutr Res 23: 901–909

Hoebregs H (1997) Fructans in foods and food products, ion-exchange chromatographic method: collaborative study. J AOAC 5: 80–102

Hoentjen F, Welling GW, Harmsen HJM, Zhang X, Snart J, Tannock GW, Lien K, Churchill TA, Lupicki M, Dieleman LA (2005) Reduction of colitis by prebiotics in HLA-B27 transgenic rats is associated with microflora changes and immunomodulation. Inflamm Bowel Dis 11:977–985

Holloway L, Moynihan S, Abrams SA, Kent K, Hsu AR, Friedlander AL (2007) Effects of oligofructose-enriched inulin on intestinal absorption of calcium and magnesium and bone turnover markers in postmenopausal women. Br J Nutr 97:365–372

Jackson KG, Taylor GR, Clohessy AM, Williams CM (1999) The effect of the daily intake of inulin on fasting lipid, insulin and glucose concentrations in middle-aged men and women. Br J Nutr 82:23–30

Kleessen B, Sykura B, Zunft H-J, Blaut M (1997) Effects of inulin and lactose on feval microbiota, microbial activity, and bowel habit in elderly constipated persons. Am J Clin Nutr 65:1397–1402

Klinder A, Förster A, Caderni G, Femia AP, Pool-Zobel NL (2004) Fecal water genotoxicity is predictice of tumor-preventive activities by inulin-like oligofructoses, probiotics (Lactobacillus rhamnosus and Bifidobacterium lactis), and their synbiotic combination. Nutr Cancer 49:144–155

Kok NN, Morgan LM, Williams CM, Roberfroid MB, Thissen JP, Delzenne NM (1998) Insulin, glucagon-like peptide

1, glucose-dependent insulinotropic polypeptide and insulin-like growth factor I as putative mediators of the hypolipidemic effect of oligofructose in rats. J Nutr 128:1099–1103

Langlands SJ, Hopkins MJ, Coleman N, Cummings JH (2004) Prebiotic carbohydrates modify the mucosa associated microflora of the human large bowel. Gut 53:1610–1616

Lesniewska V, Rowland I, Laerke HN, Grant G, Naughton PJ (2006) Relationship between dietary-induced changes in intestinal commensal microflora and duodenojejunal myoelectric activity monitored by radiotelemetry in the rat in vivo. Exp Physiology 91:229–237

Lewis S, Burmeister S, Brazier J (2005) Effect of the prebiotic oligofructose on relapse of Clostridium difficile-associated diarrhea: a randomized, controlled study. Clin Gastroenterol Hepatol 3:442–448

Ley et al. (2006) Human gut microbes associated with obesity. Nature 444:1022–1023

Lindsay JO, Whelan K, Stagg AJ, Gobin P, Al-Hassi HO, Raiment N, Kamm MA, Knight SC, Forbes A (2006) Clinical, microbiological and immunological effects of fructo-oligosaccharides in patients with Crohn's disease. Gut 55:348–355

Mathey J, Puel C, Kati-Coulibaly S, Bennetau-Pelissero C, Lebecque P, Horcajada MN, Coxam V (2004) Fructo-oligosaccharides maximize bone-spearing effects of soy isoflavone-enriched diet in ovariectomized rat. Calcif Tissue Int 75:169–179

Oli MW, Petschow BW, Buddington RK (1998) Evaluation of fructooligosaccharide supplementation of oral electrolyte solutions for treatment of diarrhea: Recovery of the intestinal bacteria. Dig Dis Sci 43:138–147

Orrhage K, Sjostedt S, Nord CE (2000) Effects of supplements with lactic acid bacteria. J Antimicrob Chemother 46:603–611

Osman N, Adawi D, Molin G, Ahrne S, Berggren A, Jeppsson B (2006) Bifidobacterium infantis strain with and without a combination of oligofructose-enriched inulin (OFI) attenuate inflammation in DSS-induced colitis in rats. BMC Gastroenterology 6:31 (online publication)

Paineau D, Payen F, Panserieu S, Coulombier G, Sobaszek A, Lartigau I, Brabet M, Galmiche J-P, Tripodi D, Sacher-Huvelin S, Chapalain V, Zourabichvilli O, Respondek F, Wagner A, Bornet FRJ (2008) The effects of regular consumption of short-chain fructo-oligosaccharides on digestive comport of subjects with minor functional bowel disorders. Br J Nutr 13:311–318

Rafter J, Bennett M, Caderni G, Clune Y, Hughes R, Karlsson PC, Klinder A, O' Riordan M, O'Sullivan GC, Pool-Zobel B, Rechkemmer G, Roller M, Rowland I, Salvadori M, Thijs H, Van Loo J, Watzl B, Collins JK (2007) Dietary synbiotics reduce cancer risk factors in polypectomized and colon cancer patients. Am J Clin Nutr 85:488–496

Rao A (2001) The prebiotic properties of oligofructose at low intake levels. Nutr Res 21:843–848

Raschka L, Daniel H (2005) Mechanisms underlying the effects of inulin-type fructans on calcium absorption in the large intestine of rats. Bone 37:728–735

Rault-Nania MH, Gueux E, Demougeot C, Demigne C, Rock E, Mazur A (2006) Inulin attenuates atherosclerosis in apolipoprotein E-deficient mice. Br J Nutr 96:840–844

Roberfroid MB, Cumps J, Devogelaer JP (2002) Dietary chicory inulin increases whole-body bone mineral density in growing male rats. J Nutr 132:3599–3602

Roberfroid M, Gibson G, Delzenne N (1993) Biochemistry of oligofructose, a non-digestible fructooligosaccharide: an approach to estimate its caloric value. Nutr Rev 51:137

Roller M, Femia AP, Caderni G, Rechkemmer G, Watzl B (2004) Intestinal immunity of rats with colon cancer is modulated by oligofructose-enriched inulin combined with Lactobacillus rhamnosus and Bifidobacterium lactis. Br J Nutr 92:931–938

Saavedra J, Tschernia A (2002) Human studies with probiotics and prebiotics: clinical implications. Br J Nutr 87:S241–S246

Sauer J, Richter KK, Pool-Zobel BL (2007) Products formed during fermentation of the prebiotic inulin with human gut flora enhances expression of biotransformation genes in human primary colon cells. Br J Nutr 97:928–937

Scholtens PAM, Alles MS, Willemsen LEM, van der Braak C, Bindels JG, Boehm G, Govers MJAP (2006) Dietary fructo-oligosaccharides in healthy adults do not negatively affect faecal cytotoxicity: a randomised, double-blind, placebo-controlled crossover trial. Br J Nutr 95:1143–1149

Scholz-Ahrens KE, Acil Y, Schrezenmeir J (2002) Effect of oligofructose or dietary calcium on repeated calcium and phosphorus balances, bone mineralization and trabecular structure in ovariectomized rats. Br J Nutr 88:365–377

Steegmans M, Iliaens S, Hoebregs H (2004) Enzymatic, spectrophotometric determination of glucose, fructose, sucrose and inulin/oligofructose in foods. J AOAC Int 87:1200–1207

Ten Bruggencate SJ, Bovee-Oudenhoven IM, Lettink-Wissink ML, Katan MB, Van der Meer R (2004) Dietary fructo-oligosaccharides and inulin decrease resistance of rats to Salmonella: protective role of calcium. Gut 53:530–535

Tuohy KM (2001) A human volunteer study on the prebiotic effects of HP-inulin-faecal bacteria enumerated using fluorescent in situ hybridisation (FISH). Anaerobe 7:113–118

Van Loo J, Coussement P, De Leenheer L, Hoebregs H, Smits G (1995) On the presence of inulin and oligofructose as natural ingredients in the Western diet. Crit Rev Food Sci Nutr 35:525–552

Verghese M, Walker LT, Shackelford L, Chawan CB (2005) Inhibitory effects of nondigestible carbohydrates of different chain lengths on azoxymethane-induced aberrant crypt foci in Fisher 344 rats. Nutr Res 25:859–868

Videla S, Vilaseca J, Antolin M, Garcia-Lafuente A, Guarner F, Crespo E, Casalots J, Salas A, Malagelada JR (2001) Dietary inulin improves distal colitis induced by dextran sodium sulfate in the rat. Am J Gastroenterol 96:1486–1493

Waligora-Dupriet A-J, Campeotto F, Nicolis I, Bonet A, Soulaines P, Dupont C, Butel M-J (2007) Effect of oligofructose supplementation on gut microflora and well-being in young children attending a day care centre. Int J Food Microbiol 113:108–113

Welters CF, Heineman E, Thunnissen FB, van den Bogaard AE, Soeters PB, Baeten CG (2002) Effect of dietary inulin supplementation on inflammation of pouch mucosa in patients with an ileal pouch-anal anastomosis. Dis Colon Rectum 45:621–627

Whelan K, Efthymiou L, Judd PA, Preedy VR, Taylor MA (2006) Appetite during consumption of enteral formula as a sole source of nutrition: the effect of supplementing pea-fibre and fructo-oligosaccharides. Br J Nutr 96:350–356

Yasuda K, Roneker K, Miller D, Welch R, Lei XG (2006) Supplemental dietary inulin affects the bioavailability of iron in corn and soybean meal to young pigs. J Nutr 136:3033–3038

Zafar TA, Weaver CM, Zhao Y, Martin BR, Wastney ME (2004) Nondigestible oligosaccharides increase calcium absorption and suppress bone resorption in ovariectomized rats. J Nutr 134:399–402

Zunft H-JF, Hanisch C, Mueller S, Koebnick C, Blaut M, Doré J (2004) Synbiotic containing *Bifidobacterium animalis* and inulin increases stool frequency in elderly healthy people. Asia Pac J Clin Nutr 13:112

7 Galacto-Oligosaccharide Prebiotics

George Tzortzis · Jelena Vulevic

7.1 Introduction

The wide recognition of bifidobacteria as health promoting bacteria (Boesten and de Vos, 2008) has attracted a lot of interest in identifying substances that can selectively promote their growth. Many studies using conventional culture and molecular techniques for bacterial identification have shown that breast-fed infants are characterized by an intestinal microbiota that is dominated by bifidobacteria (Benno et al., 1984), which is different from that of infants fed on cow's milk in that their microbiotas are characterized by lower counts of bifidobacteria, with greater numbers of more potentially harmful organisms such as clostridia and enterococci (Lunderquist et al., 1985). As a result of this difference in the microbiota composition, higher levels of ammonia, amines and phenols and other potentially harmful substances have also been found in infants fed cow's milk products (Lunderquist et al., 1985).

This has led, at the beginning of the last century, to the belief that there are molecules in the human milk that can promote the growth of this specific type of intestinal bacteria, leading to attempts to isolate and characterize those bifidogenic factors (Hamosh, 2001). Although bifidogenic nucleotides have been detected in human milk the predominance of bifidobacteria in breast-fed babies is thought to result from their abilities to utilize the oligosaccharides fraction of breast milk (Sela et al., 2008). Oligosaccharides are the third largest component of human milk and high levels are found in the colostrum where these substances constitute up to 24% of total colostrum carbohydrates (Bode, 2006). Total oligosaccharides in breast milk can reach concentrations as high as 8–12 g/l, which is 100 times greater than in cow's milk, and their concentrations steadily decrease to between 19% and 15% in the first 2 months after birth (Kunz et al., 1999). Milk contains a greater proportion of neutral compared to acidic oligosaccharides and the principal sugar components of oligosaccharides are sialic acid, N-acetylglucosamine, L-fucose, D-glucose and D-galactose, resulting in a complex mix of over 130 different oligosaccharides, due to the great variety of different sugar combinations that are possible (Bode, 2006). Amongst those oligosaccharides, human milk contains a large amount of galactose with the backbone structure based on lactose (galactose–glucose) plus a further external galactose residue that leads to the formation of three galactosyl-lactoses, $1{\rightarrow}3$, $1{\rightarrow}4$ and $1{\rightarrow}6$-galactosyl-lactose, with $1{\rightarrow}6$ galactosyl-lactose being found in amounts ranging between 2.0 and 3.9 mg/l. The total concentration of lactose-derived oligosaccharides in human milk has been estimated to approximately 1 g/l (Boehm et al., 2005).

The ability of those oligosaccharides to replicate the bifidogenic properties of breast milk (Knol et al., 2005) and to resemble glycoconjugate structures on cell surface receptors used by pathogens for adherence in the gut (Rudloff et al., 2002, Morrow et al., 2005) has attracted a lot of interest in further studying their physiological properties and production. As a result, galacto-oligosaccharides have attracted significant interest for their inclusion in various foods as health promoting ingredient, particularly in Japan and Europe.

7.2   Production Process

7.2.1   Transgalactosylation Reaction

Galacto-oligosaccharides are defined as a mixture of those substances produced from lactose, comprising of between two and eight saccharide units, with one of these units being a terminal glucose and the remaining saccharide units being galactose, and disaccharides comprising of two units of galactose.

They can be synthesized by classical chemical synthesis methods from simple sugars, but the preferred mode for their synthesis is by enzymatic catalysis from lactose using an appropriate β-galactosidase enzyme. The two types of enzyme that can be used in the preparation of GOS are the glycosyltransferases (EC 2.4) and the glycohydrolases (EC 3.2.1).

Galactosyltransferases catalyze the transfer of sugar moieties from activated donor molecules to specific acceptor molecules, forming glycosidic bonds. These enzymes are highly regio-and stereo-selective and can produce high yield of GOS, but the fact that they are not readily available and their requirement for sugar nucleotides, make their use for industrial GOS production prohibiting due to the cost involved.

Galactohydrolases are much more readily available than glycosyltransferases but are generally less stereo-selective. They transfer the galactosyl moiety of a substrate to hydroxyl acceptors and are able to catalyze hydrolysis or transgalactosidation depending on the relative concentration of hydroxyl acceptors from water or other carbohydrates respectively. Synthesis of GOS with β-galactosidases is a general characteristic of retaining galactohydrolases that during lactose hydrolysis use a double-displacement mechanism to form a covalent glycosyl-enzyme intermediate, which is subsequently hydrolyzed via oxocarbenium ion-like states. The active site of the enzyme contains a pair of carboxylic acids, serving as a proton donor and a nucleophilic base, with an average distance of ≈5.5 Å apart. Acid base catalysis is important for this enzyme class and is provided by a single carboxyl group at the active site, which is functioning as the acid catalyst for the first glycosylation step and as the base catalyst for the second deglycosylation step. During the glycosylation step one carboxyl acid, the acid catalyst, protonates the glycosidic oxygen, whereas the other carboxyl acid mediates the aglycon departure by acting as the nucleophile. During the deglycosylation step, the produced glycosyl-enzyme intermediate is hydrolyzed by a water molecule activated by the first carboxyl acid which now behaves as the base catalyst. However, in the presence of other carbohydrates in the reaction mixture, especially at elevated concentrations, they can act as acceptors for the glycosyl moiety resulting in the elongation of the carbohydrate acceptor to a higher degree of polymerization.

The enzymatic conversion of lactose into GOS by β-galactosidases is thus a kinetically controlled reaction during which the thermodynamically favored hydrolysis of the substrate generates D-galactose and D-glucose in competition to the transferase activity that generates a complex mixture of various galactose based di- and oligosaccharides of different structures.

In addition to this transferase activity, another mechanism of GOS synthesis that leads directly to formation of the disaccharide allolactose is by the direct internal transfer of galactose from the position 4, found in lactose, to the position 6 of the glucose moiety without the release of the glucose moiety from the active site. Quantitatively, allolactose is one of the major oligosaccharides formed by neutral pH β-galactosidases and although this mechanism has been demonstrated only for β-galactosidase enzymes from *E. coli*, it has been proposed also for other β-galactosidases with similarities to the LacZ enzyme.

During this enzymatic reaction, the amount and nature of the formed oligosaccharide mixture is affected by the ratio of hydrolytic and transferase activities of the enzyme. This ratio depends on the enzyme source, the concentration and nature of the substrate and the reaction conditions (pH, temperature and time) following the general principle that the transferase activity is favored at high lactose concentration, elevated reaction temperature and lower water activity. The source of the enzyme is directly influencing the type of glycoside bond formed between the galactose moieties of the produced GOS, and is also setting the range of pH and temperature conditions available for the synthesis reaction.

7.2.2 Microbial β-Galactosidase

The most extensively studied β-galactosidases for GOS synthesis are of microbial origin (β-galactohydrolase, EC 3.2.1.23). Enzymes from species belonging to *Kluyveromyces*, *Aspergillus*, *Bacillus*, *Streptococcus* and *Cryptococcus* have been used for the synthesis of GOS from lactose showing differing requirements for reaction conditions in terms of pH and temperature and differing product formation in terms of the glycoside bonds formed between the galactose moieties and the degree of polymerization (DP) of the synthesized oligosaccharides. Usually 55% of the initial lactose is converted into a mixture of products containing glucose and galactose due to the hydrolytic activity of the enzyme, un-reacted lactose, disaccharides of galactose and glucose with different β-glycoside bonds from lactose due to direct internal transfer, and trans-galactosylation products such as galactobiose, galactotriose, galactosyl lactose, tetra- to octasaccharides of similar rearrangement and/or side chain formations (Playne, 2002).

6′-galactosyl lactose is the main product when the yeast β-galactosidase from *Kluyveromyces* (*K. marxianus* subsp. *lactis*, *K. fragilis*) is used, whilst 3′- and 6′-galactosyl lactose are formed when the fungal lactase of *Aspergillus oryzae* is used. Enzymes from *Bacillus circulans* or *Cryptococcus laurentii* form mainly 4′-galactosyl lactose and enzymes from *Streprococcus thermophilus* 3′-galactosyl lactose. Another interesting approach for sourcing microbial β-galactosidase, has been explored in species of probiotic bacteria. The rationale behind the use of β-galactosidases from probiotic bacteria is that since the origin of the enzyme used in this type of manufacturing is important in the final GOS mixture composition and therefore functionality, the use of enzymes originating from probiotic bacteria as synthetic catalyst will produce oligosaccharide mixtures that will be more readily metabolized by the producing organism, resulting so in higher selectivity towards that organism. Following this approach, enzymes from various *Bifidobacterium bifidum* strains have been used to produce mixtures of linear 3′-galactosyl lactose as well as branched oligosaccharides.

The pH conditions of the reaction vary from very acidic to neutral with the fungal β-galactosidase from *A. oryzae* showing optimum GOS formation at pH as low as 3.5, whilst the enzymes from yeast or bacterial origin have more neutral pH optima between 6 and 7.5.

The origin of β-galactosidase is also influencing the temperature tolerability of the enzyme. Generally high temperatures are preferred in order to speed the reaction, to increase the solubility of the lactose substrate and prevent its crystallization and also to reduce the viscosity of the reaction mixture so that the transferase activity is favored over the hydrolytic activity. At the same time, the higher lactose concentration is reducing the water activity of the reaction solution which in turn is influencing the degree of polymerization of the formed GOS products, since lower water activity conditions are favoring the production of trisaccharides and higher water activity is required to synthesize GOS of greater length. Therefore, considerable efforts have been focused on sourcing thermostable β-galactosidases as a mean for the improvement of the reaction yields.

Another approach to improve the reaction yields through improving the transferase/hydrolytic ratio of β-galactosidases has been attempted through genetic engineering. Jørgensen et al. (2001) investigated the functional importance of the C-terminal part of BIF3 β-galactosidase by deletion mutagenesis and expression of truncated variants using *E. coli* cells as the host. Deletion of approximately 580 amino acid residues from the C-terminal part converted the enzyme from a normal hydrolytic β-galactosidase into a highly efficient transgalactosylating enzyme. Quantitative analysis showed that the truncated β-galactosidase utilized approximately 90% of the reacted lactose for production of GOS while hydrolysis constituted a 10% side reaction. This 9:1 ratio of transgalactosylation to hydrolysis was maintained at lactose concentrations ranging from 10% to 40% (w/w), suggesting that the truncated β-galactosidase is behaving as a true transgalactosylase even at low lactose concentrations.

7.2.3   Production Process

Various reactor designs and configurations have been reported for GOS synthesis, including the batch reactor, continuous stirred-tank reactor (CSTR), CSTR coupled with crossflow filtration, hollow fiber membrane reactor, fixed-bed and fluidized-bed reactor (Boon et al., 2000). The batch modes of operation prevail so far in the scientific literature on trans-galactosylation processes, mainly because of their ease of operation, but also the reduction in the risk of possible microbial contaminations that long-term continuous processes, especially under realistic reaction conditions of moderate temperatures, have or fouling that can occur in systems where membranes are used for the retention of the enzyme.

Continuous systems, however, have been proposed as a way for reducing the process cost by offering a more efficient usage of the enzymes that can be very expensive. In the batch process, the initially added to the reaction mixture enzyme is usually lost at the end of the reaction. In the continuous process re-usage of the enzyme can be achieved by immobilizing it on a carrier and thus limiting the loss of enzyme activity or by retaining the soluble enzyme in the reactor with the use of an ultra-filtration membrane.

Another very interesting aspect, regarding the mode of operation, is the effect on the composition of the produced mixture. In batch systems, the composition and concentration of possible galactosyl acceptors are changing constantly over the reaction time whilst in the steady state of a continuous system the concentrations of possible galactosyl acceptors stay constant over the entire reaction time. This leads to more defined GOS mixtures being expected in a continuous process system and has been suggested as the reason why a larger fraction of trisaccharides is formed in batch production compared to continuous ones. Once the conversion of lactose to GOS mixture has been completed, the efficient removal/inactivation of the β-galactosidase is an important factor in order to prevent product hydrolysis that takes place if the reaction is continued beyond the peak of oligosaccharide formation.

For the commercialization of GOS-based products the purification of the produced oligosaccharides from the reaction mixture is a significant and challenging step. On a larger scale, monosaccharides can be separated using chromatographic applications such as ion-exchange resins or activated carbon. In the case of ion exchange chromatography, cation-exchange resins are mainly used since they have the highest affinity for monosaccharides and therefore oligosaccharides are the first to elute from the column. Activated carbon has a higher affinity for oligosaccharides, compared to mono- and disaccharides, which makes their operation at industrial level more preferable, since regeneration can take place off-line without large substrate losses. The separation of lactose from the disaccharide fraction of the GOS products has been proven to be extremely difficult and usually results to large losses of GOS products. Lactose-free GOS mixtures are of great interest considering that 70% of the world population lack β-galactosidase in the small intestine and are therefore sensitive to lactose. An approach based on the selective enzymatic oxidation of lactose into lactobionic acid using a fungal cellobiose dehydrogenase has been described

(Splechtna et al., 2001). The produced lactobionic acid can then be easily separated from the non-ionic sugars of the mixture with the use of anion exchange chromatography, whilst the monosaccharides are subsequently removed in a single chromatographic step, yielding to purified GOS mixture with only minor losses of the main components of interest.

7.2.4   Commercially Available GOS

GOS have been used as food ingredients in Japan and Europe for at least 30 years and their application is currently expanding rapidly. At present, Japanese companies still dominate worldwide galacto-oligosaccharide production and development activity, although, European interest in GOS based products is increasing with several companies currently producing or planning to produce GOS mixtures. In contrast, GOS production in the USA at present remains negligible. The major companies manufacturing GOS are still located in Japan. Yakult Honsha (Tokyo, Japan), Nissin Sugar Manufacturing Company (Tokyo, Japan) and Snow Brand Milk Products (Tokyo, Japan) together with Friesland Foods Domo (ex Borculo Domo ingredients) in the Netherlands and Clasado Ltd in the UK are the main manufacturers. Most of the manufacturers produce several classes of products in terms of GOS purity in either syrup and/or powder format.

Yakult is producing three GOS product: Oligomate 55 in syrup form, Oligomate 55P in powder form and TOS-100 a purified version of 99% oligosaccharide content. Nissin is producing Cup-Oligo in syrup (Cup-Oligo H70) and powder format (Cup-Oligo P) and Snow Brand produces GOS that is incorporated into its infant milk formula P7L, without offering sales outside its organization.

In Europe, Friesland Food Domo is offering Vivinal GOS in a syrup format containing 57% oligosaccharides on dry matter and in a powder format containing 29% oligosaccharides on dry matter. Clasado Ltd is offering mainly a powder GOS product, Bimuno, with 52% galacto-oligosaccharide content on dry matter, as well as a syrup version of that product.

Besides the differences in the purity amongst the commercially offered products, there are differences also in the linkages of the oligosaccharide chain due to the different enzymes used in their production. The Oligomate range is produced with enzymes originating from *Aspergillus oryzae* offering mainly β 1–6 linkages, the Bimuno product is produced using enzymes from *Bifidobacterium bifidum* and contains mainly β 1–3 linkages whilst Cup-Oligo and Vivinal offer mainly β 1–4 linkages as a result of the activity of enzymes from *Bacillus circulans* for the latter and *Cryptococcus laurentii* for the former GOS product. Yakult is also considering dual enzymes systems combining the activity of enzymes from *A. oryzae* and *B. circulans* to produce GOS mixtures of β 1–4 and β 1–6 linkages.

Although different enzymes are used in the production of the various commercially available GOS products the overall process flow chart of them is very similar (⊙ *Figure 7.1*). 20–40% (w/v) lactose solution is incubated with the β-galactosidase enzyme in either batch or continuous reactor until the optimum of lactose conversion into oligosaccharides has been reached. The solution is then decolorized and demineralized before being further processed for removal of

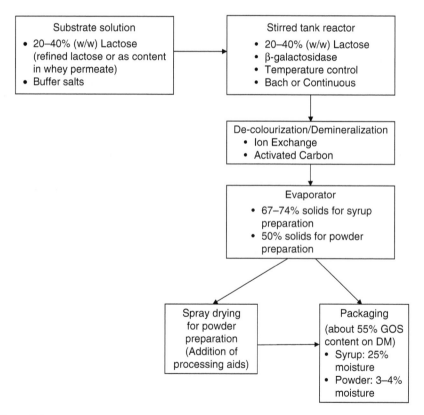

☐ Figure 7.1

Process flow schematic for the manufacturing of GOS from lactose.

monosaccharides, either by chromatographic separation or nanofiltration, to increase the oligosaccharide purity. The resultant solution is concentrated by evaporation usually to 67–75% total solids to produce the syrup format or spray dried to produce the powder format.

A wide range of granted patents or patent applications related to the conditions as well as problem solving during the production of GOS have taken place and are available on-line (www.uspto.gov, www.epo.org).

7.3   Safety-Toxicity of GOS

In terms of safety and toxicity, GOS has been evaluated in both acute and chronic toxicity tests in rats. Oral administration of GOS did not show any toxicity as a single dose of 20 g/kg of body weight and a daily dose of 1.5 g/kg of body weight for 6 months. Using the Ames' test and the Rec assay no mutagenicity was found and the only known adverse effect of GOS is transient diarrhea due to osmotic pressure, which occurs when excess oligosaccharides are consumed. In the case of GOS, the amount of oligosaccharides which does not induce osmotic diarrhea is estimated to be approximately 0.3–0.4 g/kg body weight, or about 20 g/human body. When GOS syrup was administered to rats by gavage at 2,500 or 5,000 mg/kg of body weight daily for 90 days, no significant adverse toxicological effects attributable to treatment were noted. Clinical signs were unremarkable, and there were no ocular findings in any of the animals. Statistical analysis of clinical pathologies, including blood biochemistries, hematology, urinalysis and coagulation did not reveal any significant effects (Anthony et al., 2006).

Galacto-oligosaccharides have a generally regarded as safe (GRAS) status in the USA, a non Novel Food status in the EU and are regarded as foods of specific health use (FOSHU) in Japan.

7.4   Physicochemical Properties

GOS provide several physicochemical and health benefits, which make their use as food ingredients particularly attractive. Since commercially available food grade GOS are mixtures their specific physicochemical and physiological properties will to some extend vary depending on the mixture of oligosaccharides of different degree of polymerization, the un-reacted lactose and the generated monosaccharides. Overall, galacto-oligosaccharides are colorless, water soluble with a viscosity similar to that of high fructose syrup leading to improved body and mouth-feel. They are stable to pH 2 at 37°C for several months making their application in non-refrigerated fruit juice matrices possible, whilst the presence of β type linkages makes them ingredients of increased resistance to high temperature in acidic medium. They remain unchanged after treatment at 160°C for 10 min at neutral pH and after treatment at 120°C for 10 min at pH 3 or 100°C for 10 min at pH 2, offering potential for a wide range of food applications. They are mildly sweet, typically 0.3–0.6 times the sweetness of sucrose and can be used in very sweet foods as bulking agent to enhance other food flavors. GOS are resistant to salivary degradation and are not utilized by the oral microbiota and can therefore be used as low cariogenic sugar substitutes in chewing gums and confectionary. They are not hydrolyzed by pancreatic enzymes and gastric juice passing the small intestine offering reduced glycemic index and a calorific value lower than 50% that of sucrose, which makes them suitable for low-calorie diet food and for consumption by individuals with diabetes. Galacto-oligosaccharides have a high moisture retaining capacity preventing excessive drying that can be useful in baked goods especially bread where GOS are not broken down during fermentation with yeast and the baking process, providing the bread with excellent taste and texture. Their low water activity can help in controlling microbial contamination and depending on the molecular weight of the oligosaccharide content, they can alter the freezing temperature of frozen foods and reduce the amount of coloring, due to Maillard reactions, in heat processed food as relatively fewer reducing moieties are available.

Analysis of galacto-oligosaccharides in different foods matrices is of high significance in terms of inclusion in various food products and a method, based on the principle of the quantification of FOS in food matrices, is available (AOAC 2001.02). This two stage method relies on the enzymatic treatment of a test solution with a β-galactosidase enzyme, followed by the quantitative determination of galactose by high-performance anion exchange chromatography with pulsed amperometric detection (HPAEC-PAD), in order to overcome the lack of sensitivity and selectivity that HPLC suffers. In the first stage, the free galactose and lactose are determined in the initial test solution and in the second stage, the total amount of galactose released from GOS and lactose is determined in the β-galactosidase treated solution. The GOS content is then calculated from the concentrations of lactose and galactose before and after the enzymatic treatment. The method has been tested and evaluated in dairy, fruit juice, pet candy, biscuit and infant formula matrices.

7.5   Functional Properties

Apart from the advantages provided to food manufacturers by the physicochemical properties of galacto-oligosaccharides, GOS offer a number of health promoting properties mainly due to their ability to modulate the balance of the colonic microbiota and promote the proliferation of intestinal bifidobacteria. Especially this bifidogenicity is making the inclusion of GOS on its own or as part of a mixture with FOS in infant milk formula, follow on formula and infant food the major application of GOS at the current time.

7.5.1   GOS Bifidogenicity

From the early steps of research into bifidobacteria, it has been recognized that those bacteria express higher activity of β-galactosidase than many other members of the colonic microbiota (Desjardins and Roy, 1990). This has been further supported by the isolation and characterization of various β-galactosidases from different *Bifidobacterium* species and subsequently confirmed at the genome level through the sequencing and annotation of numerous β-galactosidase encoding genes present in many *Bifidobacterium* strains of human origin. Within the available literature it is apparent that most bifidobacterial strains make use of more than one β-galactosidase isoenzyme for their growth. Tochikura et al. (1986) isolated and characterized two different β-galactosidases from *B. longum* biovar *longum* 401, whereas the genome sequence of *B. longum* biovar *longum* NCC2705 (Schell et al., 2002) revealed the existence of four different isoenzymes belonging to either GH 2 or GH 42 family. Furthermore, the cloning and sequencing of three genes from *B. bifidum* DSM20215 (Møller et al., 2001) and two genes from *B. longum* biovar *infantis* HL96 (Hung et al., 2001) confirm the ability of bifidobacteria to make extensive use of these enzymes for their growth. Amongst those bifidobacterial β-galactosidases many have been proven to be lactases, hydrolyzing mainly lactose into its simpler glucose and galactose moieties, but some of them, the ones belonging to the GH 42 family seem to preferentially hydrolyze galacto-oligosaccharides other than lactose, as it has been demonstrated with a β-galactosidase extracted from *B. adolescentis* DSM20083. This enzyme was shown to preferentially hydrolyze mainly β-D-(1→4) linked di- and galacto-oligosaccharides with reasonable Km values (2.2–6.4 mM) and be able to liberate galactose from β-D-(1→3), β-D-(1→6) and β-D-(1→1) galacto-biose, but not from lactose. The presence of β-galactosidases with different activity within the same cell of *Bifidobacterium* strains enables them to better utilize GOS mixtures by working in a synergistic fashion. Enzymes belonging to the GH 42 family liberate galactose moieties from the galacto-oligosaccharide molecule whereas enzymes belonging to the GHF 2 family act on lactose backbone of the molecule. The ability of bifidobacteria to utilize GOS molecules if further supported by the presence of appropriate membrane transport mechanisms which facilitate the internalization of the released carbohydrates into the cell. Although, not much research has been carried out, the presence of glucose and/or galactose transport systems has been identified in many species of bifidobacteria. In *B. bifidum* DSM20082, galactose crosses through the membrane by diffusion, whereas glucose is incorporated by a cation symport which is regulated by K^+ ions. Two glucose transport systems have been identified in *B. longum* biovar *longum* NCC2705, whereby one of them additionally participates in the incorporation of galactose even though it is repressed by the presence of lactose. Those transport systems are rapidly fed with simple carbohydrate moieties that exist in the environment of the cells through the action of extracellular enzymes that are able to degrade non-digestible galacto-oligosaccharides. *B. bifidum* DSM20215 has a putative extracellular β-galactosidase (GHF 2) which contains at their N-terminal part signal peptides that enable it to be extracellularly translocated by the cells, whereas the C-terminal part consists of domains that most probably mediate its attachment to the cell wall. Another simpler isoenzyme, in terms of protein domain structure (GHF 42), has been identified as being extracellular in *B. adolescentis* (van Laere et al., 2000). Extracellular location of the different β-galactosidase isoenzymes allows the cells to have better access and ability to degrade galacto-oligosaccharides into their simpler moieties (galactose and glucose) and subsequently internalize them via the described transport mechanisms.

However, many studies have demonstrated the preference of bifidobacteria towards di- or even oligosaccharides over their simpler moieties, suggesting that they have developed a mechanism for internalizing complex oligosaccharides into the cell and thus complement the extracellular degradation. Kim et al. (2005) and Parche et al. (2006) demonstrated the preference of *B. longum* biovar *longum* for lactose over glucose in growth culture experiments containing both carbohydrates. They showed that lactose was consumed first whereas assimilation of glucose was repressed until all lactose disappeared from the growth medium. This mechanism involves a transcriptional down-regulation of the glucose transport system in the presence of lactose that is most probably internalized by a lactose transferase. Gopal et al. (2001) and Amaretti et al. (2007) also demonstrated the preference of *B. animalis* biovar *lactis* and *B. adolescentis* for oligosaccharides over glucose, suggesting the induction and expression of permeases specific for tri- and tetrasaccharides. Lactose transport systems have been investigated in *B. bifidum* DSM20082 and proved to be most probably based on a proton symporter. Moreover, using proteome analysis of *B. longum* biovar *longum* NCC2705 it was revealed the involvement of 19 permeases for diverse carbohydrates uptake (Parche et al., 2007). Amongst them, three putative lactose transport systems were identified. Through the function of these transporter mechanisms bifidobacteria can internalize a variety of galactooligosaccharide molecules and subsequently degrade them in the cell, avoiding thus the competition of nutrients and cross feeding of other bacteria that can occur when glucose and galactose moieties are liberated in the extracellular cell environment.

Although little data are available, the expression of β-galactosidases, as well as, galactoside transporters seems to occur in a controlled fashion in the *Bifidobacterium* cell. Growth of *B. adolescentis* DSM20083 on galactooligosaccharides results in the induction of β-galactosidase activity. When glucose, galactose or lactose were present, the expression of a lactose hydrolyzing enzyme was induced, whereas in the presence of galacto-oligosaccharides *B. adolescentis* DSM20083 expressed a galacto-oligosaccharide hydrolyzing isoenzyme (van Laere et al., 2000). In *B. bifidum* DSM20082, the presence of lactose slightly stimulated the β-galactosidase activity (1.5 times) that was accompanied by higher lactose incorporation into the cell, suggesting higher expression of a lactose permease. Induction of β-galactosidase activity by the substrate has also been reported for *B. longum* biovar *longum* CCRC15708 which contains four putative lactose operons in its genome (Schell et al., 2002). Two of those operons consist of a β-galactosidase gene, an LacI-type transcriptional regulator and ABC-type oligosaccharide transporters (BL0258-BL0260 and BL1167-BL1169). The third operon consists of a putative lactose permease (BL0976) in the opposite direction to a β-galactosidase gene (BL0978; *lacZ*), whereas the fourth, which is probably not functional, consists of a *lacI*-type repressor gene (BL1774) and a cryptic β-galactosidase-like gene (BL1775) (Parche et al., 2006; Schell et al., 2002). Microarray data analysis has indicated that those genes are either constitutively expressed or induced by the presence of lactose but none of them is significantly repressed by glucose (Parche et al., 2006). However, contrary to the above observations which indicate constitutive or inducible expression of β-galactosidases, the gene annotation of *B. longum* biovar *longum* NCC2705 has shown that this strain predominantly uses repressors for the negative transcriptional regulation of β-galactosidase gene expression, something that is in contrast to most prokaryotes, but has been justified a way to allow a quicker and higher stringent response to environmental changes.

7.5.2   Non-digestibility of GOS

To exert an effect and to come into contact with bacteria growing in the colon, any prebiotic, by its definition, must escape digestive processes in the stomach and small intestine. The non-digestibility of prebiotics can be demonstrated *in vitro* by subjecting them to treatment with pancreatic and other gastrointestinal digestive enzymes. In the case of GOS, several *in vitro* experiments have shown its non-digestibility and stability to hydrolysis by such enzymes. A European consortium studied the effects of various non-digestible oligosaccharides and concluded that more than 90% of GOS arrives into the colon (van Loo et al., 1999).

The non-digestibility of prebiotics *in vivo* for human subjects can be demonstrated with ileostomized volunteers, because the digestion in these individuals is limited to the small intestine and the remainder of the food bolus can be collected from the pouch. However, very few prebiotics have been tested in this way and most of *in vivo* data is available by means of the hydrogen (H_2) breath tests. In general, studies report increased breath H_2 excretion following ingestion of GOS (Tanaka et al., 1983). These studies, therefore, give an indication that GOS is fermented by the colonic bacteria. However, the measurements of breath H_2 excretion do not provide information on the amount of GOS that actually escapes digestion and arrives into the colon intact. In the study by Bouhnik et al. (1997), reduced breath H_2 excretion was observed after daily dose of 10 g of GOS was administered for 21 days to eight volunteers. Nevertheless, the bifidobacterial numbers were increased, demonstrating that GOS was fermented by bacteria in the colon.

The calorific value of prebiotics, such as GOS, has been estimated to be 1–2 kcal/g (Cummings et al., 1997). According to the Japanese standard methods, calorific value of GOS has been calculated to be 1.73 kcal/g.

7.5.3   Prebiotic Properties of GOS

Along with inulin, FOS and lactulose, GOS is one of the prebiotics that has been the most thoroughly investigated and its prebiotic effect has been proven.

At present, this prebiotic effect is defined as the selective stimulation of the growth of bifidobacteria in complex microbial communities that exist in the large intestine. Bifidobacteria and lactobacilli are recognized as health-promoting organisms and are widely used as probiotics. Although, their specific health promoting effects are yet to be fully explained, no adverse effects related to their consumption have been reported.

7.5.3.1 *In Vitro* Effects

Methods for the assessment of prebiotic fermentation *in vitro*, range from simple static batch cultures to the multiple-stage continuous cultures inoculated with either single/mixed bacterial strains or fecal homogenates. Such fermentation studies describe the prebiotic capabilities through switches in the composition of purportedly beneficial and detrimental microbiota of fecal homogenates and the fermentation of prebiotics in pure cultures. Summary of *in vitro* data for the prebiotic effect of GOS is shown in ⊙ *Table 7.1*.

Early studies used pure cultures and showed that GOS was selectively metabolized by all the *Bifidobacterium* strains tested compared to lactulose and FOS whose specificity was less remarkable (Ohtsuka et al., 1989). However, Hopkins et al. (1998) tested abilities of various bifidobacterial isolates to utilize various prebiotics and showed that substrate utilization was highly variable with considerable interspecies and interstrain differences. This finding was confirmed in a study by Gopal et al. (2001), where various strains of lactic acid bacteria were tested for their abilities to utilize GOS, and in addition to bifidobacterial species, some lactobacilli and pediococci strains were also able to utilize these substrates. Authors showed a perfect correlation between the ability of strains to utilize GOS and the presence of β-galactosidase, as well as the ability of *B. lactis* to preferentially utilize tri- and tetrasaccharides whereas *L. rhamnosus* preferred di- and monosaccharides when grown with GOS as growth substrate. Similar finding was observed in another study where *B. adolescentis* was able to degrade tri- and tetrasaccharides, while *B. longum* bv. *infantis* and *L. acidophilus* could only utilize those with a DP < 3 (van Laere et al., 2000). The ability of *B. adolescentis* to utilize GOS more efficiently was attributed to the presence of a novel β-galactosidase with activity towards GOS but not lactose. Indeed, strains belonging to bacteroides and clostridia have also been shown to utilize GOS in pure cultures (Tanaka et al., 1983).

While pure culture studies can offer an insight into various mechanisms involved in the utilization or growth processes, they can not be used to show a

7

◘ Table 7.1

Effect of GOS on the colonic microbiota *in vitro* (Cont'd p. 223)

Method	Substrate	Effect	Reference
Test tubes, pure cultures (various strains)	GOS, glucose, inulin, lactose, lactulose	GOS selectively utilized by all *Bifidobacterium* strains tested and to a lesser extent by *L. acidophilus* but not by other bacteria, compared with lactulose and inulin whose specificity was less remarkable	Ohtsuka et al. (1989)
Static, temperature controlled batch culture (pure *Bifidobacterium* strains)	FOS, GOS, inulin, pyrodextrin, SOS, XOS	Substrate utilization highly variable between species	Hopkins et al. (1998)
Test tubes, pure cultures (various strains of lactic acid bacteria and bifidobacteria)	GOS	Other than bifidobacteria, some lactobacilli and pediococci strains able to utilize GOS. Perfect correlation observed between the ability of strains to utilize GOS and presence of β-galactosidase. *B. lactis* utilizes tri- and tetrasaccharides, and *L. rhamnosus* prefers mono- and disaccharides	Gopal et al. (2001)
Stirred, pH/temperature controlled, anaerobic batch cultures, fecal homogenates	FOS, GOS, IMO, inulin, lactulose, SOS, XOS	All prebiotics increased bifidobacteria. XOS and lactulose produced the highest population of bifidobacteria and FOS resulted in the largest increase in lactobacilli. GOS resulted in the largest decrease in clostridia	Rycroft et al. (2001)
Stirred, pH/temperature controlled, anaerobic batch cultures, fecal homogenates	FOS, fructose, GOS	No differences in the number of bifidobacteria between substrates. *B. adolescentis*, *B. longum* and *B. angulatum* comprised the dominant bifidobacterial species	Sharp et al. (2001)

◻ Table 7.1

Method	Substrate	Effect	Reference
MPE (measurement of changes in dominant bacteria, substrate assimilation and SCFA), using batch cultures and fecal homogenates	FOS, V-GOS, IMO, SOS, ChOS, Psyllium, Benefiber, Sunfiber, FOS:V-GOS (50:50), FOS: Benefiber (90:10), V-GOS:Benefiber (90:10)	V-GOS in combination with FOS and alone resulted in highest MPE, suggesting the best prebiotic effect of the tested substrates	Vulevic et al. (2004)
Three-stage gut model, pH/temperature controlled to represent proximal, transverse and distal colons	B-GOS	In vessels corresponding to proximal and transverse colons, strong bifidogenic effect was observed. There was no effect on bifidobacterial population in the third vessel, due to complete utilization of substrate in previous vessels	Tzortzis et al. (2005)

prebiotic effect. As mentioned above, under these conditions GOS, as indeed most oligosaccharides, can support the growth of various bacteria and thus the selectivity can not be demonstrated. Batch and continuous culture studies using fecal homogenates offer a better *in vitro* assessment of a prebiotic potential.

Rycroft et al. (2001) used various prebiotics as substrates, in pH/temperature controlled anaerobic batch cultures inoculated with fecal homogenates, and showed that all prebiotics increased the numbers of bifidobacteria but GOS resulted in the largest decrease of clostridial population. In another study, fecal homogenates were grown in batch cultures in the presence of FOS, fructose or GOS and no differences were observed in the number of bifidobacteria between the treatments (Sharp et al., 2001). However, quantitative analysis, termed a measure of the prebiotic effect (MPE), which takes into an account a number of dominant bacterial groups, fermentation end products such as short-chain fatty acids (SCFA) and substrate assimilation was developed by Vulevic et al. (2004). Here it was shown that out of 11 substrates tested, V-GOS on its own or in combination with FOS (V-GOS:FOS, 50:50) resulted in the highest MPE suggesting the best prebiotic effect of the tested substrates. Recently, B-GOS mixture (1%, w/v) was shown to have a strong bifidogenic effect in a three-stage gut model, in vessels corresponding to the proximal and transverse colons but not in the third vessel (Tzortzis et al., 2005). The lack of the effect in the last gut model vessel, which represents distal colon, was attributed to low molecular weight and complete assimilation of B-GOS by the bacteria in previous vessels.

The results obtained from *in vitro* fermentation with fecal homogenates are important in prebiotic investigations, particularly when choosing or evaluating substrates to be used *in vivo*. However, *in vitro* models do not provide a full picture of the complex ecosystem that exists within the large intestine.

7.5.3.2 *In Vivo* Human Studies

⊙ *Table 7.2* summarizes *in vivo* human studies investigating the prebiotic effect of GOS. Several studies showed that administration of various GOS mixtures to healthy humans result in significant increases in the population numbers of bifidobacteria following ingestion (Bouhnik et al., 1997; Gopal et al., 2003; Ito et al., 1993; Tanaka et al., 1983). This increase in bifidobacteria was sometimes accompanied by increased lactobacilli numbers (Gopal et al., 2003; Tanaka et al., 1983) and decreased bacteroides numbers (Tanaka et al., 1983) without any significant effect on the other bacterial groups assessed. Initially higher doses (10 g/day) of GOS were used (Bouhnik et al., 1997; Tanaka et al., 1983), however Ito et al. (1993) showed that lower dose of GOS (2.5 g/day) was sufficient to observe a bifidogenic effect when the initial number of the bifidobacteria population is low. This was confirmed in a study by Gopal et al. (2003), where it was shown that 4 weeks of supplementation with 2.4 g of GOS per day resulted in increased numbers of both bifidobacteria and lactobacilli. However, Bouhnik et al. (2004) studied the effect of 1 week intake of C-GOS (Cup-oligo) at concentrations of 0, 2.5, 5.0, 7.5 and 10.0 g/day and observed no dose-dependent effect, although the low initial numbers of bifidobacteria were associated with overall better prebiotic effect compared to the other oligosaccharides tested. In addition, this study also showed that the effect of GOS on bifidobacterial population level was higher than that observed with other prebiotics (i.e., inulin, FOS, lactulose) used in the study (Bouhnik et al., 2004). In a recent study, the administration of B-GOS mixture (3.6 g/day) was recently shown to result in a better bifidogenic effect than V-GOS (4.9 g/day) after 1 week of intake by healthy humans (Depeint et al., 2008). Moreover in the same study, a different composition of bifidobacteria was suggested after the two GOS treatments which was indicated through the isolation of different fecal *Bifidobacterium* species The difference in the magnitude and quality of the effect on bifidobacteria between the two GOS mixture was attributed to the difference in structures of the oligosaccharide present in the mixtures due to the different enzymes used for their production. Similar effect on bifidobacterial population of B-GOS was

◻ Table 7.2

Bacteriological changes after GOS administration in human volunteer studies *in vivo*

Dose per day	Number of subjects	Treatment period	Effect	Method	Reference
3 or 10 g Oligomate	5	3 weeks	Increased bifidobacteria and lactobacilli. Decreased bacteroides	Agar plate counts	Tanaka et al. (1983)
2.5 g Oligomate	12	3 weeks	Increased bifidobacteria	Agar plate counts	Ito et al. (1993)
8.1 g V-GOS	10	2 weeks	No effect on endogenous bifidobacterial population	Qualitative DGGE analysis and sequencing	Satakori et al. (2001)
2.4 g BPOligo	10	4 weeks	Increased bifidobacteria and lactobacilli. No effect on clostridia and enterobacteria	Agar plate counts	Gopal et al. (2003)
2.5, 5.0, 7.5, 10.0 g Cup-oligo, FOS, IMOS, inulin, lactulose, SOS	8	1 week	10 g of GOS resulted in a best effect on bifidobacterial numbers. No dose-dependent effect was observed, but lower initial bifidobacterial numbers associated with better prebiotic effect	Agar plate counts	Bouhnik et al. (2004)
4.9 g V-GOS	29	1 week	Increased bifidobacteria with both but B-GOS significantly higher than V-GOS. Dose-dependent effect observed in bifidobacterial numbers with B-GOS	FISH	Depeint et al. (2008)
1.9, 3.6 g B-GOS	30				

observed even at a lower dose of 1.9 g/day in this study, and the effect was found to be dose dependent in the type of microbiota with the initial numbers of bifidobacteria being within normal range. It is clear that not all GOS mixtures necessarily result in the same effect on the microbiota, since differences in their degree of polymerization as well as structure composition can become significant when it comes to their assimilation by bifidobacteria in the complex colonic ecosystem.

Additionally, attempts have been made to perform the qualitative analysis (using DGGE and subsequent DNA sequencing) of bifidobacterial population following human intake of GOS. Satokari et al. (2001) showed that 2 weeks of supplementation with a relatively high dose (8.1 g/day) of V-GOS did not affect the qualitative composition of the endogenous bifidobacteria.

7.5.4   Metabolism in the Colon

The major end-products of carbohydrate fermentation are SCFAs, of which acetate, propionate and butyrate are quantitatively the most important in the human colon. All SCFAs are rapidly absorbed from the large intestine and stimulate salt and water absorption: principally, the gut epithelium, liver and muscle metabolize them, with virtually none appearing in the urine and only small amounts appearing in the feces. The three major SCFAs are trophic when infused into the colon, and these trophic properties have important physiological implications in addition to maintaining the mucosal defense barrier against invading organisms. The amount and type of SCFA produced in the colon will depend on the type of substrate as well as the composition of the microbiota. Because SCFA are rapidly absorbed from the gut, measurements in feces do not provide useful information about the fermentative abilities of different substrates. This has been demonstrated in human trials where the administration of GOS resulted in increased bifidobacteria but with no effect on fecal SCFA concentrations (Ito et al., 1993). However, *in vitro* studies, using fecal homogenates, can be useful models for studying the fermentation profiles. Bouhnik et al. (1997) showed that the addition of 10 g of GOS to batch cultures inoculated with human fecal homogenates resulted in increased acetic and lactic acid productions which were not observed in control fermenters. The authors suggested the change of microbiota composition (i.e., increase in the population numbers of bifidobacteria) was responsible for this. Similar findings were reported in other studies using batch cultures. For example, at concentrations of 10 g/l, FOS and GOS were shown to increase acetate and butyrate formation, with transient accumulation of lactic and succinic acids (Hopkins and Macfarlane, 2003). Studies with rats inoculated with human fecal microbiota show clear demonstration of decreased pH and increased SCFA and other organic acids in the caecal contents following consumption of GOS (Kikuchi et al., 1996). In these models, the major increases are observed in the production of lactic and succinic acids suggesting that these may be responsible for the lowering of the pH. In a study with piglets

(Tzortzis et al., 2005), inclusion of 4% (w/w) B-GOS in the diet resulted to the significant decrease of pH in the proximal colonic content, which was attributed to the significant increase of acetic acid and total SCFA concentration. This effect though could not be seen in the distal colonic content as well.

7.5.5 Physiological Effects of GOS

7.5.5.1 Stool Improvement

GOS has been shown to beneficially affect fecal nature and improve some parameters related to constipation. Deguchi et al. (1997) studied the effect of administering 2.5 and 5 g of GOS (Oligomate) daily for 1 week to 75 women who had a tendency to be constipated. Higher dose of GOS was found to improve the defecation frequency significantly. Similar findings were reported by Korpela and colleagues who performed series of studies in both healthy adults and constipated elderly (Sairanen et al., 2007; Teuri et al., 1998). This research group showed that consumption of yoghurt containing high doses (9–15 g) of V-GOS daily for 2–3 weeks, can increase the defecation frequency, however in younger adult subjects gastrointestinal symptoms, such as flatulence, also increased (Teuri et al., 1998).

The frequency of defecation does not provide information on the improvement of stool weight, however, and studies that have examined the effect of GOS on mean daily stool weight did not report any significant changes regardless of the dose (Bouhnik et al., 1997). Furthermore, no differences between of inulin, FOS and V-GOS at an intake of 15 g on mean daily stool weight were also reported (van Dokkum et al., 1999). It is worth noting here that both studies examining the effect of GOS on mean daily stool weight have used healthy adults and not constipated subjects who should provide a better group to study the laxative effect of any prebiotic. The possibility that GOS may have an improving effect in the treatment of constipation may, therefore, not be excluded and further trials are necessary to fully answer this question.

7.5.5.2 Mineral Absorption

Calcium and magnesium can be absorbed from both small and large intestines and their adequate supply and bioavailability are important factors determining the healthy bone structure. Deficiency of these minerals can lead to problems, such as osteoporosis, later in life and especially in postmenopausal women. It has been noted that these minerals when present in the large intestine exist in forms that are poorly absorbable and that acidification caused by bacterial fermentation increases their solubility and thus absorption. Since prebiotics, including GOS, have an effect on reducing the pH and supporting the growth of lactic acid-producing bacteria, it has been shown that this types of substrate have an enhancing effect on the metabolism of calcium and magnesium as well.

Current works involving GOS are summarized in ❯ *Table 7.3*, and as shown the majority of evidence comes from animal models. Chonan and colleagues conducted series of experiments in rats and showed that calcium absorption was stimulated by ingestion of GOS when normal but not low dietary concentrations were used. This calcium absorption in turn increased bone mass and calcium content (Chonan and Watanuki, 1996; Chonan et al., 1995, 1996). Chonan and colleagues have also investigated and showed increased magnesium absorption

◻ Table 7.3

Effect of GOS supplementation on mineral metabolism and bone mineralization

Dose per day	Study	Aim/method	Result	Reference
0.5, 5.0 g per 100 g	Rats	Ca absorption with normal and high dietary levels of GOS	Ca absorption stimulated when normal but not low dietary concentrations were used	Chonan and Watanaki (1996)
5 g per 100 g	Rats	Effect of GOS in overiectomized rats on Ca absorption and bone weight	Ca absorption stimulated and bone mass and Ca content increased	Chonan et al. (1995)
5 g per 100 g	Rats	Effect in Mg deficient rats on Mg absorption, concentration of Mg in bones and accumulation of Ca in the kidney and heart	Mg absorption and bone concentration increased and Ca accumulation in the kidney and heart reduced	Chonan et al. (1996)
15 g V-GOS, FOS, inulin	Humans (n = 12) 3 weeks	Effect on Ca and Fe absorption in healthy man using 24 h urine collections for Ca isotopes measurements	No effect on Ca and Fe absorption	van den Heuvel et al. (1998)
20 g V-GOS	Humans (n = 12) 9 days	Effect on Ca absorption in menopausal women using urine collections for administered Ca isotopes	Increased Ca absorption	van den Heuvel et al. (2000)

and bone concentration in rats fed GOS which in turn suppressed the calcification of the heart and kidney under conditions of high phosphorus and calcium
dietary conditions (Chonan et al., 1996).

Although the rat models show clear beneficial effects of GOS on mineral
absorption, human studies are scarce and with inconsistent results. In one study,
male volunteers were fed 15 g/day of FOS, inulin and V-GOS for 3 weeks and no
effect on iron and calcium absorption were observed with any of the tested
prebiotics (van den Heuvel et al., 1998). In another study, postmenopausal
women received 15 g/day of V-GOS for 9 days and calcium absorption was
significantly increased (van den Heuvel et al., 2000). Therefore, current results
offer promise, especially in relation to osteoporosis, however more human trials
are needed to offer definitive answers. It is worth noting that FOS has been used
more widely in human studies with good results.

7.5.5.3 Lipid Metabolism

Several animal studies have shown that administration of prebiotics, namely
inulin and FOS or fermented milk products is effective in lowering blood
cholesterol levels. However, *in vivo* results are variable, with some studies reporting lowering effects and others no effect on total serum cholesterol levels. Thus
far, GOS has been used in two trials where serum cholesterol levels were investigated and in both no significant effect could be seen. In one trial, the effects of
administration of 15 g/day of V-GOS, FOS and inulin were compared in healthy
humans but no significant changes in serum lipids or glucose absorption were
observed (van Dokkum et al., 1999). Recently, it was shown that 5.5 g of a GOS
mixture administration to healthy elderly had no effect upon total serum cholesterol and HDL cholesterol levels (Vulevic et al., 2008). However, most studies
reporting a decrease in the levels of serum cholesterol and/or increase in the levels
of HDL cholesterol, following either pro- or prebiotic administration, have used
subjects with initial elevated serum cholesterol levels. To common knowledge,
GOS has not been used in this context and it is, therefore, not excluded that it
may have an effect in these subjects.

7.5.5.4 Carcinogenesis

It has long been suggested that the human gut microbiota plays an important role
in the metabolism and toxicity of both dietary and endogenous substrates.

In many cases the products of bacterial metabolism in the large intestine are associated with detrimental effects on the host and could lead to initiation and/or promotion of colon carcinogenesis. In particular, bacterial enzymes such as β-glucosidase, β-glucuronidase and nitroreductase have been associated with production of carcinogens. Since prebiotics are known to support the growth of probiotic bacteria and lower the pH in the colon it is to be expected that they would have an effect on the production of genotoxic enzymes. However, there are very few studies with GOS that have looked at these effects and most data originates from *in vitro* and animal models (❯ *Table 7.4*).

An *in vitro* study using a three-stage continuous gut model system showed that 5% (w/v) V-GOS was only weakly bifidogenic, but it strongly supported the growth of lactobacilli in the vessel corresponding to the proximal colon (McBain and Macfarlane, 2001), without any significant effects being observed in the other two vessels corresponding to transverse and distal colons. Fermentation of V-GOS decreased the activities of β-glucosidase and β-glucuronidase in all three vessels, however nitroreductase and azoreductase activities were increased. At the same time, this study showed that inulin exerted similar effects on the beneficial microbiota, nitroreductase and azoreductase activities, but inulin increased enterobacteria and *C. perfringens* and had no significant effect on β-glucosidase and β-glucuronidase (McBain and Macfarlane, 2001).

Studies in human-microbiota associated rats have shown that GOS (5%, w/w) administration for 4 weeks, increases caecal bifidobacteria and lactobacilli and decreases enterobacteria and the pH (Rowland and Tanaka, 1993). This change in the microbiota was followed by significant decreases in the activities of β-glucuronidase and nitroreductase, but not of β-glucosidase activity that showed to increase during GOS fermentation. Kikuchi et al. (1996) showed increased levels of β-galactosidase, lactic acid and caecal pH in human-microbiota associated rats fed GOS (10%, w/w) accompanied by a decrease in the levels of ammonia and β-glucuronidase activity.

The above results from *in vitro* and animal models suggest that GOS may have reducing effect upon the production and activities of some genotoxic enzymes. However, to date, there is only one human *in vivo* study that has looked at the effect of feeding healthy humans with 15 g/day of inulin, FOS or V-GOS on β-glucuronidase production (van Dokkum et al., 1999). Indeed, inulin and V-GOS were found to have a reducing effect upon β-glucuronidase concentration, whereas FOS had no effect in this study.

The effect of V-GOS was also assessed on the development of aberrant crypt foci in rats and it was found that at increased intake levels of V-GOS (10%, w/w),

◘ Table 7.4

Studies on the effect of GOS on colon carcinogenesis

Dose	Study	Aim/method	Result	Reference
5% (w/v) V-GOS, inulin	*In vitro*	Effect on microbiota and activities of genotoxic enzymes in three-stage gut model	V-GOS and inulin weakly bifidogenic but increased lactobacilli. Inulin also increased enterococci and *C. perfringens*. V-GOS decreased β-glucosidase and β-glucoronidase, inulin had no effect. Both substrates increased azoreductase and nitroreductase	McBain and Macfarlane (2001)
5% (w/w) Oligomate	Rats	Effect of feeding human-microbiota colonized rats for 4 weeks on microbiota and genotoxic enzymes	Increased bifidobacteria, lactobacilli and β-glucosidase. Decreased enterobacteria, cecal pH, β-glucoronidase and nitroreductase	Rowland and Tanaka (1993)
10% (w/w) Oligomate	Rats	Effect of feeding human-microbiota colonized rats for 4 weeks on fermentation metabolites and genotoxic enzymes	Increased β-galactosidase and lactic acid. Decreased cecal pH, β-glucoronidase and ammonia	Kikuchi et al. (1996)
5, 10% (w/w) V-GOS	Rats	Fischer rats injected twice with AOM to induce colorectal tumors and fed low or high concentration V-GOS. At 4 weeks, 18 rats in each group were killed and scored, half of the remaining group swapped to the other diet. Six weeks later 9 rats in each group were killed, the rest were killed after 10 months	At first count, number of aberrant crypt foci reduced with higher compared to low V-GOS. After diet change total number of aberrant crypt foci increased, but less with animals swapped onto high V-GOS diet	Wijnands et al. (2001)
15 g per day V-GOS, FOS, inulin	Humans (n = 12) 3 weeks	Originally aimed at assessing the effect of prebiotic administration to healthy young men on serum lipids	Decreased β-glucoronidase with inulin and V-GOS but not FOS	van Dokkum et al. (1999)

it had a reducing effect, thus suggesting its potential as a protective agent against the development of colorectal tumors (Wijnands et al., 2001). However, human studies are still lacking in this respect and indeed in respect of identifying the specific anti-cancer effects of GOS and prebiotics in general, in order to draw any definite conclusions at the present time. It is worth mentioning that one of the possible and very important effects of GOS, in respect of anti-cancer as well as anti-inflammatory potential, may be its immunomodulatory properties.

7.5.5.5 Immunomodulation

The immune system protects the body against foreign agents and invasion by pathogens. It can be divided into the innate or non-specific and adaptive or acquired (specific) immune system. The innate immune system acts as first line of defence and it comprises physical barriers such as skin, and phagocytic, dendritic, inflammatory and natural killer (NK) cells as well as soluble mediators such as complement proteins and cytokines. The adaptive immune system becomes active in response to a challenge to the innate immune system. This system is more antigen-specific and it consists of two major cell types, T- and B-lymphocytes. T-lymphocytes develop into functionally different cell types with specific cytokine patterns, whereas B-lymphocytes are a part of the memory of the immune system and they produce only one type of antibody matching a specific type of antigen. T-lymphocytes can further be divided into those who mediate immunity to intracellular pathogens (Th1 cells) and those responsible for extracellular pathogens (Th2 cells).

The largest component of the immune system is situated in the gut. It is called the gut-associated lymphoid tissue (GALT), and it contains about 60% of all lymphocytes in the body. It also contains large amounts of secretory immunoglobulin A (sIgA), which plays a key role in the defence of the gut against adherence and invasion of pathogenic bacteria and viruses. GALT is constantly in contact with the microbiota and their metabolic by-products, thus dietary substrates reaching the large intestine that are able to influence the microbiota should affect the GALT.

7.5.5.6 GOS and the Immune System

The idea that prebiotics could help the intestinal defence system originated from the observations that newborn babies, who have an underdeveloped intestinal host defence system, lack an appropriate capacity to defend themselves against intestinal infections. Furthermore, infants consuming their mother's milk were found to have a greatly reduced risk of diarrhea diseases, and a lower risk of respiratory and other infections. Human milk contains various protective components and active ingredients, including non-digestible oligosaccharides, which represent the third largest component of human milk and have been identified as the main factors involved in the development of an appropriate colonization process in infants, which in turn stimulates the maturation of intestinal host defenses.

Although it is known that human milk oligosaccharides can exert a prebiotic effect (Sela et al., 2008), research into the immunomodulatory actions of prebiotics is very recent, with most data originating from animal models and in relation to FOS and its prebiotic effect (i.e., bifidogenicity and SCFA production). However, there are few studies and lines of evidence that either suggest or demonstrate the effect of GOS on the immune system. This effect could either be direct in the form of interactions with immune, mucosal or epithelial cells and/ or indirect through the species or even strain selective modulation of the microbiota and their metabolic products.

As outlined earlier, several studies have shown that GOS increases the population numbers of bifidobacteria, lactobacilli and subsequent SCFA production what are known to have immunomodulatory effect. Butyric acid is known to suppress lymphocyte proliferation, inhibit cytokine production of Th1-lymphocytes and upregulate IL-10 production, to suppress the expression of the transcription factor NF-κB and upregulate Toll-like receptors (TLR) expression, as well as to protect against colon cancer as it inhibits DNA synthesis and induces cell differentiation (Hoyles and Vulevic, 2008). Pharmacological doses of acetic acid when administered intravenously to healthy individuals and cancer patients increase NK cell activity and peripheral blood antibody production (Ishizaka et al., 1993).

Specific probiotic species belonging to either bifidobacteria or lactobacilli, when administered orally, are known to increase the secretion of sIgA in the small intestine and the feces, and to stimulate Peyer's patches (PP) B lymphocyte IgA production (Hoyles and Vulevic, 2008). In one recent study 57 infants, split into three groups, were fed either standard formula alone, standard formula containing a probiotic (*B. animalis* – Bb12) or formula containing a prebiotic mixture (V-GOS:inulin; 90:10) for 8 months (Bakker-Zierikzee et al., 2006). Measurements of fecal sIgA were made at regular intervals during the course of the study, and it was shown that babies fed the probiotic formula had variable levels of sIgA, whereas babies receiving the prebiotic mixture demonstrated a trend towards higher levels of sIgA compared to the controls. Apart from formula fed infants, another group that could potentially benefit from the immunomodulatory effects of GOS is the elderly who are known to generally have reduced levels of beneficial bacteria and impaired immune system. A very recent study investigated the effect of feeding 5.5 g of GOS (B-GOS) to 44 healthy elderly volunteers on the microbiota composition and immune function (NK cells, phagocytosis and cytokines) (Vulevic et al., 2008). This study showed that B-GOS administration led to a significant decrease in the numbers of the less beneficial bacteria (i.e., bacteroides, *C. perfringens*, *Desulfovibrio* spp., *E. coli*) and a significant increase in the numbers of the beneficial bacteria, especially bifidobacteria. The study also found significant positive effects upon the immune response, evidenced by an improvement in NK cell activity and phagocytosis, increased secretion of the anti-inflammatory cytokine, IL-10, and decreased secretion of pro-inflammatory (IL-6, IL-1β and TNF-α) cytokines by stimulated PBMC. Additionally, a clear positive correlation between the number of bifidobacteria and both NK cell activity and phagocytosis was demonstrated. This was the first time that the immunomodulatory effect of GOS has been demonstrated in humans and it was shown that dietary intervention, such as B-GOS, can be an effective and attractive option for the enhancement of both gastrointestinal tract function and immune system (innate and adaptive) function.

7.5.5.7 Inflammatory Bowel Disease (IBD)

IBD principally includes Crohn's disease (CD) and ulcerative colitis (UC). Although, the cause of IBD is not yet known, it is generally accepted that a combination of factors such as genetic susceptibility, priming by enteric pathogens and immune-mediated tissue injury, result in its pathogenesis. Reduced numbers of beneficial bacteria accompanied by increases in the numbers of other less beneficial bacteria, such as *E. coli*, have been reported in both feces and mucosa microbiota of IBD patients. In patients with IBD Th1 and Th2 pattern of cytokine formation seem to be modified or increased in comparison to healthy individuals. For example, IBD patients seem to have increased production of tumor necrosis factor α (TNF-α) which triggers inflammation via the transcription factor NF-κB. There are very few studies with prebiotics in general, that have looked at their effect in human IBD. The use of V-GOS, however, as a therapy for immunomodulation in IBD was tested in one animal study.

The trinitrobenzene sulphonic acid (TNBS)-induced colitis rat model was used to test the effect of feeding rats 4 g/kg of body mass per day either whey or V-GOS, 10 days before the induction of colitis, or dexamethasone at colitis induction, as a control (Holma et al., 2002). Fecal bifidobacteria numbers, myeloperoxidase activity and macroscopic damage were assessed 72 h after the induction of colitis, and it was found that the bifidobacterial numbers increased with the V-GOS administration, but inflammation was not reduced. Although, this study did not show any effect from the V-GOS administration on IBD, this one animal model is not enough to draw any conclusions and more studies are needed to fully determine the potential of GOS in preventing or treating IBD.

7.5.6 Allergy

Improved hygiene and reduced exposure of infants to microorganisms are one of the suggested reasons for the observed increases in the incidence rates of allergic diseases in developed countries during recent decades. Studies indicate that there may be a link between the colonic microbiota and allergy, since reduced numbers of bifidobacteria have been found in the feces of allergic infants (Kirjavainen et al., 2002). These infants have IgE-mediated food allergies, and a Th2-biased immune response. Studies have found that FOS can reduce Th2 response in children, however GOS has not been used in this context and only indirect evidence for the effect of GOS currently exists.

Recently, 1,223 pregnant women carrying children at high risk of allergy, were divided into two groups, and fed for 2–4 weeks, before delivery, either a placebo or a mixture of 4 probiotics (*L. rhamnosus* GG, *L. rhamnosus* LC705, *B. breve*, *Propionbacterium freudenreichii* subsp. *shermanii* JS) (Kukkonen et al., 2007). Their infants received the same probiotic mix combined with 0.9 g V-GOS or a placebo for 6 months after birth, and at 2 years of age their IgE levels and the incidence of allergic diseases were evaluated. It was found that the synbiotic preparation had no effect on the incidence rates of allergy, however, incidences of eczema and atopic eczema were reduced, whilst a tendency towards the reduction of other atopic (IgE-associated) diseases was noted (Kukkonen et al., 2007).

In another study, the effect of a prebiotic mixture (V-GOS: inulin; 90:10) was assessed on the development of atopic dermatitis (AD), which is usually the first manifestation of allergy development during early infancy. Infants (259) at risk of atopy were fed a prebiotic mixture, or a placebo, from birth and for 6 months and they were examined for clinical evidence of AD (Moro et al., 2006). Additionally, a smaller group of infants (98) was selected to assess the effect of the treatments on the fecal microbiots. The results showed a significant decrease in the incidence of AD in the V-GOS:Inulin group with 10 infants developing AD compared to the 24 infant that developed AD in the placebo group. In terms of the effect on the fecal microbiota, a significantly higher, compared to placebo, number of bifidobacteria was observed in the prebiotic group without any significant effect on lactobacilli.

7.5.7   Anti-pathogenic Activity of GOS

Studies have suggested that prebiotics could directly be involved in protecting the gut from infection and inflammation through the inhibition of the attachment and/or invasion of pathogenic bacteria or their toxins to the colonic epithelium. This attachment is necessary before pathogens can colonize and cause disease, and it is mediated by glycoconjugates on glycoproteins and lipids present on the microvillus membrane. Certain prebiotic oligosaccharides contain structures, similar to those found on the microvillus membrane, that interfere with the bacterial receptors by binding to them and thus preventing bacterial attachment to the same sugar moiety of the microvillus glycoconjugates. For example, α-linked GOS, present in human milk, are known to have anti-adhesive properties and be capable of toxin neutralization (Newburg et al., 2005, Morrow et al., 2005). B-GOS contains an oligosaccharide in alpha anomeric configuration, and it was shown to significantly decrease the attachment of enteropathogenic *E. coli* (EPEC) and *Salmonella enterica* serovar Typhimurium to HT-29 epithelial cell line (Tzortzis et al., 2005). The same GOS mixture was further studied in an oral challenge experiment, during which BalbC mice were fed either a placebo or B-GOS prior to *Salmonella enterica* serovar Typhimurium infection (Searle et al., 2009). It was shown that the animals fed the GOS mixture did not develop clinical symptoms of salmonellosis, even though the pathogen could be recovered in the feces. Furthermore the histopathology and structure of the epithelium were completely protected and translocation of the pathogen to other organs was limited compared to the placebo. In another study, GOS (Oligomate) was shown to inhibit the adhesion of EPEC to Hep-2 and Caco-2 epithelial cell lines more effectively than inulin, FOS, lactulose or raffinose (Shoaf et al., 2006). However, the anti-adhesive properties may be a result of GOS binding to pathogens and not a direct modulation of host immune system.

7.5.8   Neonates and Infants

The significance of a bifidobacteria predominant microbiota in healthy breast fed infant is well accepted. Infants become progressively colonized over a series of weeks and months by different bacteria, and by 2 years of age, the establishment of a more adult microbiota composition begins. Until then, however, bifidobacteria predominate in breast-fed infants, while formula fed infants have a more diverse microbiota composition. As mentioned previously, human milk oligosaccharides are thought to be responsible for this pattern of colonization in breast-fed infants (Hamosh, 2001; Sela et al., 2008), and therefore attempts have been made to develop infant formulas containing GOS mixture that will promote a microbiota pattern similar to that of breast feeding. Thus far, the most studied formulation is the combination of GOS and inulin at a ratio 9:1.

Preterm infants of about 31 weeks' gestational age and about 1 week old, were studied to compare standard formula with formula containing 1% (w/v) of the GOS-inulin mixture whilst a separate group fed fortified human milk was studied in parallel. Within 1 month of feeding the number of fecal bifidobacteria and lactobacilli in the GOS-inulin formula group increased to levels similar to the breast-fed group. In addition to this, the difference in the composition of the fecal flora between the standard formula and the GOS containing formula group was highly significant, with the latter being closer to the breast fed one. Moreover, stool consistency and stool frequency were also found to be similar between the GOS fed group and breast-fed infants (Boehm et al., 2002).

A number of similar studies, using the same prebiotic mixture, were performed in term infants and even toddlers showing similar results on the fecal microbiota (Veereman-Wauters, 2005). Interestingly, FOS was also tested in term infants at concentrations of 200, 400 and 600 mg per bottle for 2 weeks and it was found to exert no effect (Veereman-Wauters, 2005). It is generally accepted by the Scientific Committee on Food of the European Commission that the addition of V-GOS:inulin prebiotic mixture at a concentration of 0.8 g/dl to infant formula is considered safe, and this prebiotic mixture has been widely used in the last few years in Europe.

Although, a vast number of studies suggest the ability of a prebiotic mixture containing inulin and V-GOS to increase fecal bifidobacterial populations in infants, relatively few studies have looked at other possible effects, such as disease prevention. Atopy, as explained earlier, is an area that has been studied in some details and an area that offers a promise for use of GOS-based prebiotic mixtures. However, further studies are required in order to identify the mechanism of action of these substrates, if GOS-based prebiotic mixtures are to be considered and used in a treatment strategy.

7.6   Under-researched and Possible Beneficial Properties of GOS

The full potential of health related and health promoting benefits of GOS is yet to be fully ascertained. As the interactions of the colonic microbiota, especially the beneficial effects of *Bifidobacterium* spp and *Lactobacillus* spp with the host are elucidated, the increased bifidogenicity of GOS is becoming a significant property for its application as a health promoting food ingredient. Consequently there are many potential areas of research where GOS has not been used and where, indeed, the prebiotics in general have not been studied.

Irritable bowel syndrome (IBS) is a common chronic functional gastrointestinal disorder that exhibits a broad spectrum of severity, ranging from mild symptoms to severe and intractable symptoms. IBS is characterized by recurrent abdominal pain and discomfort associated with alterations in bowel habit. The etiology and physiology of IBS are not fully understood, but it is most likely multifactorial. Alterations in gastrointestinal motility, visceral perception, and psychosocial factors contribute to overall symptom expression. Currently there is no single therapeutic modality of proven benefit to all IBS patients and treatment is based on the physician's understanding of the individual patient's symptom pattern and the associated psychosocial factors. Mixed probiotic combinations, mainly using specific *Bifidobacterium* species, have been used successfully in the treatment of IBS symptoms, however prebiotics have not. The possibility that GOS, through its increased bifidogenicity, could help ameliorate some of the symptoms associated with IBS as has been shown by Silk and colleagues (2009) could not be excluded and this should definitely create one area for the future research.

A common side effect of the use of antibiotics is antibiotic-associated diarrhea (AAD), which presents a particular problem in hospitalized, and especially vulnerable elderly, patients. Diarrhea associated with *Clostridium difficile* is a leading cause of hospital outbreaks of diarrhea and it considerably increases mortality and healthcare costs for inpatients. The condition occurs when patients are treated with antibiotics, for an underlying infection, which result in the disruption of the barrier of normally protective colonic microbiota. Probiotics have been successfully used to reduce the incidence of AAD. Lewis et al. (2005)

investigated the effect of 12 g of FOS on the incidence and relapse of *C. difficile* in elderly patients. Although differences between the prebiotic and placebo group were not observed on the incidence rate, there was a significant reduction in the rate of relapse in the prebiotic group accompanied by a significant increase in bifidobacteria numbers. It is clear that more prebiotic trials and involving of GOS are needed to fully explain the potential in treating or preventing AAD, as well as other GI pathogen related conditions such as travellers' diarrhea and these also offer potential areas for the future research.

Arthritis is associated with a broad spectrum of clinical and experimental intestinal inflammation, ranging from infections with intestinal pathogens, overgrowth and changes to the normal commensal colonic microbiota, injection of purified bacterial cell wall components and dietary manipulation to chronic idiopathic IBD. The common feature of all those conditions is increased exposure of the lamina propria and systemic circulation to colonic microbiota and their products, either through increased proliferation or mucosal permeability, pathogenic invasion, or immune modulation. Experimental data assessing the potential of prebiotics in arthritis is still lacking and limited to animal models. However, α-GOS has been used in adjuvant-induced arthritis Wistar rats, and type II collagen-induced arthritis in DBA/1 J mice. The dose-dependent beneficial effect was observed in erythema, swelling of limbs and on histological findings in the hind paw joints (Abe et al., 2004). This study also showed reduced levels of nitrite and nitrate in blood, although the production of IL-1 by macrophages was increased. The results indicate the potential of using GOS to immunomodulate the inflammation seen in arthritis through either direct effect or via the modulation of the colonic microbiota. However, more research and involving human subjects is needed to clarify the potential.

7.7   Summary

- Commercially available GOS products are mixtures of galactose based oligosaccharides of varying DP and linkage configuration with glucose, galactose and lactose.
- The oligosaccharide composition varies amongst GOS mixtures depending on the origin of the β-galactosidase enzyme used as well as the mode of production. Developments in the production of GOS aim to delivering purer and more efficient mixtures.
- GOS consist of oligosaccharides that are pH and heat stable offering a wide range of food application potential.

- GOS are safe well tolerated ingredients up to levels of 20 g intake per day, have a GRAS status in the USA and FOSHU status in Japan and could be included in the dietary fiber content of foods.
- GOS mixtures are well established prebiotic ingredients with increased selectivity towards *Bifidobacterium* species. Depending on their oligosaccharide composition, GOS products vary in terms of quantitative and qualitative bifidogenic properties.
- Besides through their bifidogenic effect, GOS contribute to the health of the host through their direct interaction, leading to increased immunomodulatory and anti-pathogenic properties.
- Infant and elderly nutrition offer the highest opportunity for GOS applications based on their functional properties (bifidogencicty, protection from pathogens, regulation of the immune function) and the host's requirements.

List of Abbreviations

AD	atopic dermatitis
CSTR	continuous stirred-tank reactor
DP	degree of polymerization
EPEC	enteropathogenic *Escherichia coli*
FOS	fructooligosaccharides
GALT	gut associated lymphoid tissue
GOS	galacto-oligosaccharides
NF κB	nuclear factor κB
NK cell	natural killer cells
SCFA	short chain fatty acids
TLR	toll like receptors
TNF-α	tumor necrosis factor α

References

Abe C, Fujita K, Kikuchi E, Hirano S, Kuboki H, Yamashita A, Hashimoto H, Mori S, OkadaM (2004) Effects of alpha-linked galactooligosaccharide on adjuvant-induced arthritis in Wistar rats and type II collagen-induced arthritis in DBA/1J mice. Int J Tissue React 26:65–73

Amaretti A, Bernardi T, Tamburini E, Zanoni S, Lomma M, Matteuzzi D, Rossi M (2007) Kinetics and metabolism of Bifidobacterium adolescentis MB 239 growing on glucose, galactose, lactose, and galactooligosaccharides. Appl Environ Microbiol 73:3637–3644

Anthony JC, Merriman TN, Heimbach JT (2006) 90-day oral (gavage) study in rats with galactooligosaccharides syrup. Food Chem Toxicol 819–826

Bakker-Zierikzee AM, van Tol EAF, Kroes H, Alles MS, Kok FJ, Bindels JG (2006) Faecal SIgA secretion in infants fed on pre- or probiotic infant formula. Pediatr Allergy Immunol 17:134–140

Benno Y, Sawada K, Mitsuoka T (1984) The intestinal microflora of infants: composition of fecal flora in breast-fed and bottle-fed infants. Microbiol Immunol 28:975–986

Bode L (2006) Recent advances on structure, metabolism, and function of human milk oligosaccharides. J Nutr 136: 2127–2130

Boehm G, Lidestri M, Casseta P, Jelinek J, Negretti F, Stahl B, Marini A (2002) Supplementation of a bovine milk formula with an oligosaccharide mixture increases counts of faecal bifidobacteria in preterm infants. Arch Dis Child Fetal Neonatal Ed 86:F178–F181

Boehm G, Stahl B, Jelinek J, Knol J, Miniello V, Moro GE (2005) Prebiotic carbohydrates in human milk and formulas. Acta Pediatr 94:18–21

Boesten RJ, de Vos WM (2008) Interactomics in the human intestine: Lactobacilli and Bifidobacteria make a difference. J Clin Gastroenterol 42(Pt 2):S163–S167

Boon MA, van't Riet K, Janssen AEM (2000) Enzymatic synthesis of oligosaccharides: Product removal during a kinetically controlled reaction. Biotechnol Bioeng 70:411–420

Bouhnik Y, Flourie B, D'Agay-Abensour L, Pochart P, Gramet G, Durand M, Rambaud JC (1997) Administration of transgalacto-oligosaccharides increases fecal bifidobacteria and modifies colonic fermentation metabolism in healthy humans. J Nutr 127:444–448

Bouhnik Y, Raskine L, Simoneau G, Vicaut E, Neut C, Flourie B, Brouns F, Bornet FR (2004) The capacity of nondigestible carbohydrates to stimulate fecal bifidobacteria in healthy humans: a double-blind, randomized, placebo-controlled, parallel-group, dose-response relation study. Am J Clin Nutr 80:1658–1664

Chonan O, Watanuki M (1996) The effect of 6'-galactooligosaccharides on bone mineralization of rats adapted to different levels of dietary calcium. Int J Vitam Nutr Res 66:244–249

Chonan O, Matsumoto K, Watanuki M (1995) Effect of galactooligosaccharides on calcium absorption and preventing bone loss in ovariectomized rats. Biosci Biotech Biochem 59:236–239

Chonan O, Takahashi R, Yasui H, Watanuki M (1996) Effects of beta 1–4 linked galactooligosaccharides on use of magnesium and calcification of the kidney and heart in rats fed excess dietary phosphorous and calcium. Biosci Biotech Biochem 60: 1735–1737

Cummings JH, Roberfroid MB, Anderson H, Barth C, Ferro-Luzzi A, Ghoos Y, Gibney M, Hermonsen K, James WPT, Korver O, Lairon D, Pascal G, Voragen AGS (1997) PASSCLAIM –Gut health and immunity. Eur J Clin Nutr 51:417–423

Deguchi Y, Matsumoto K, Ito T, Watanuki M (1997) Effects of b1-4 galacto-oligosaccharides administration on defecation of healthy volunteers with constipation tendency. Jpn J Nutr 55:13–22

Depeint F, Tzortzis G, Vulevic J, I'Anson K, Gibson GR (2008) Prebiotic evaluation of a novel galactooligosaccharide mixture produced by the enzymatic activity of Bifidobacterium bifidum NCIMB 41171, in healthy humans: a randomized, double-blind, crossover, placebo-controlled intervention study. Am J Clin Nutr 87:785–791

Desjardins ML, Roy D (1990) Uncoupling of growth and acids production in Bifidobacterium ssp. J Dairy Sci 73:299–307

van Dokkum W, Wezendonk B, Srikumar TS, van den Heuvel EGHM (1999) Effect of nondigestible oligosaccharides on large-bowel functions, blood lipid concentrations and glucose absorption in young healthy male subjects. Eur J Clin Nutr 53:1–7

Gopal KP, Sullivan PA, Smart JB (2001) Utilisation of galacto-oligosaccharides as selective substrates for growth by lactic acid bacteria including *Bifidobacterium lactis* DR10 and *Lactobacillus rhamnosus* DR20 Int Dairy J 11:19–25

Gopal KP, Prosad J, Gill HS (2003) Effects of the consumption of *Bifidobacterium lactis* HN019 (DR10TM) and galacto-oligosaccharides on the microflora of the gastrointestinal tract in human subjects. Nutr Res 23:1313–1328

van den Heuvel EGHM, Schaafsma G, Muys T, van Dokkum W (1998) Nondigestible oligosaccharides do not interfere with calcium and nonheme-iron absorption in young, healthy men. Am J Clin Nutr 67:445–451

van den Heuvel EGHM, Schoterman MHC, Muijs T (2000) Transgalactooligosaccharides Stimulate Calcium Absorption in Postmenopausal Women J Nutr 130:2938–2942

Hamosh M (2001) Bioactive factors in human milk. Pediatr Clin North Am 48:69–86

Holma R, Juvonen P, Asmawi MZ, Vapaatalo H, Korpela R (2002) Galacto-oligosaccharides stimulate the growth of bifidobacteria but fail to attenuate inflammation in experimental colitis in rats. Scand J Gastroenterol 37:1042–1047

Hopkins MJ, Macfarlane GT (2003) Nondigestible oligosaccharides enhance bacterial colonization resistance against Clostridium difficile in vitro. Appl Environ Microbiol 69:1920–1927

Hopkins MJ, Cummings JH, Macfarlane GT (1998) Inter-species differences in maximum specific growth rates and cell yields of bifidobacteria cultured on oligosaccharides and other simple carbohydrate sources. J Appl Microbiol 85:381–386

Hoyles L, Vulevic J (2008) GI Microbiota and Regulation of the Immune System. In: Huffnagle GB, Noverr MC (ed) Landes Biosciences, Texas, pp. 79–92

Hung MN, Xia Z, Hu NT, Lee BH (2001) Molecular and biochemical analysis of two beta-galactosidases from Bifidobacterium infantis HL96. Appl Environ Microbiol 67:4256–4263

Ishizaka S, Kikuchi E, Tsujii T (1993) Effects of acetate on human immune system. Immunopharmacol Immunotoxicol 15:151–162

Ito M, Deguchi Y, Matsumoto K, Kimura M, Onodera N, Yajima T (1993) Influence of galactooligosaccharides on the human fecal microflora. J Nutr Sci Vitaminol 39:635–640

Jørgensen F, Hansen OC, Stougaard P (2001) High-efficiency synthesis of oligosaccharides with a truncated beta-galactosidase from Bifidobacterium bifidum. Appl Microbiol Biotechnol 57:647–652

Kikuchi H, Andrieux C, Riottot M, Bensaada M, Popot F, Beaumatin P, Szylit O (1996) Effect of two levels of transgalactosylated oligosaccharide intake in rats associated with human faecal microflora on bacterial glycolytic activity, end-products of fermentation and bacterial steroid transformation. J Appl Bacteriol 80:439–446

Kim TB, Song SH, Kang SC, Oh DK (2005) Quantitative comparison of lactose and glucose utilization in *Bifidobacterium longum* cultures. Biotechnol Prog 19:672–675

Kirjavainen PV, Arvola T, Salminen SJ, Isolaurie E (2002) Aberrant composition of gut microbiota of allergic infants: a target of bifidobacterial therapy at weaning? Gut 51:51–55

Knol J, Scholtens B, Kafka C, Steenbakkers J, Gro S, Helm K, Klarczyk M, Schopfer H, Bockler HM, Wells J (2005) Colon Microflora in Infants Fed Formula with Galacto- and Fructo-Oligosaccharides: More Like Breast-Fed Infants. J Pediatr Gastroenterol Nutr 40:36–42

Kukkonen K, Savilahti E, Haahtela T, Juntunen Backman K, Korpela R, Poussa T, Turre T, Kuitunen M (2006) Probiotics and prebiotic galacto-oligosaccharides in the prevention of allergic diseases: A randomized, double-blind, placebo-controlled trial J Allergy Clin Immunol 119:192–198

Kunz C, Rudloff S, SchadW, Braun D (1999) Lactose-derived oligosaccharides in the milk of elephants: comparison with human milk. Br J Nutr 82(5):391–399

van Laere KM, Abee T, Schols HA, Beldman G, Voragen AG (2000) Characterization of a novel beta-galactosidase from Bifidobacterium adolescentis DSM 20083 active towards transgalactooligosaccharides. Appl Environ Microbiol 66:1379–1384

Lewis S, Burmeister S, Brazier J (2005) Effect of the prebiotic oligofructose on relapse of Clostridium difficile-associated diarrhea: a randomized, controlled study. Clin Gastroenterol Hepatol 3:442–448

van Loo J, Cummings J, Delzenne N, Englyst H, Franck A, Hopkins M, Kok N (1999) Functional food properties of non-digestible oligosaccharides: a consensus report from the ENDO project (DGXII AIRII-CT94-1095). Br J Nutr 81:121–132

Lunderquist B, Nord C, Winberg J (1985) The composition of the faecal microflora in breast-fed and bottle-fed infants from birth to eight weeks. Acta Paediatr Scand 74:45–51

McBain AJ, Macfarlane GT (2001) Modulation of genotoxic enzyme activities by non-digestible oligosaccharide metabolism in in-vitro human gut bacterial ecosystems. J Med Microbiol 50:833–842

Møller PL, Jørgensen F, Hansen OC, Madsen SM, Stougaard P (2001) Intra- and extracellular beta-galactosidases from Bifidobacterium bifidum and B. infantis: molecular cloning, heterologous expression, and comparative characterization. Appl Environ Microbiol 67:2276–2283

Moro GE, Arslanoglu S, Stahl B, Jelinek U, Wahn U, Boehm G (2006) A mixture of prebiotic oligosaccharides reduces the incidence of atopic dermatitis during the first six months of age Arch Dis Child 91:814–819

Morrow AL, Ruiz-Palacios GM, Jiang X, Newburg DS (2005) Human-milk glycans that inhibit pathogen binding protect breast-feeding infants against infectious diarrhea. J Nutr 135:1304–1307

Newburg DS, Ruiz-Palacios GM, Morrow AL (2005) Human milk glycans protect infants against enteric pathogens. Annu Rev Nutr 25:37–58

Ohtsuka K, Benno Y, Endo A, Ueda H, Ozawa O, Uchida T, Mitsouka T (1989) Effects of 49 galactosyllactose intake on human fecal microflora.Bifidus 2:143–149

Parche S, Amon J, Jankovic I, Rezzonico E, Beleut M, Barutçu H, Schendel I, Eddy MP, Burkovski A, Arigoni F, Titgemeyer F (2007) Sugar transport systems of Bifidobacterium longum NCC2705. J Mol Microbiol Biotechnol 12:9–19

Parche S, Beleut M, Rezzonico E, Jacobs D, Arigoni F, Titgemeyer F, Jankovic I (2006) Lactose-over-glucose preference in Bifidobacterium longum NCC2705: glcP, encoding a glucose transporter, is subject to lactose repression. J Bacteriol 188:1260–1265

Playne MJ (2002) The health benefits of probiotics. In Encyclopaedia of dairy sciences, Academic Press London pp. 1151–1158

Rowland IR, Tanaka R (1993) The effects of transgalactosylated oligosaccharides on gut flora metabolism in rats associated with a human faecal microflora. J Appl Bacteriol 74:667–674

Rudloff S, Stefan C, Pohlentz G (2002) Detection of ligands for selectins in the oligosaccharide fraction of human milk. Eur J Nutr 41:85–92

Rycroft CE, Jones MR, Gibson GR, Rastall RA (2001) A comparative in vitro evaluation of the fermentation properties of prebiotic oligosaccharides. J Appl Microbiol 91:878–887

Sairanen U, Piirainen L, Nevala R, Korpela R (2007) Yoghurt containing galacto-oligosaccharides, prunes and linseed reduces the severity of mild constipation in elderly subjects. Eur J Clin Nutr 61:1423–1428

Satokari RM, Vaughn EE, Akkermans ADL, Saarela M, de VOS WM (2001) Bifidobacterial diversity in human feces detected by genus-specific PCR and denaturing gradient gel electrophoresis. Appl Environ Microbiol 67:504–13

Schell MA, Karmirantzou M, Snel B, Vilanova D, Berger B, Pessi G, Zwahlen MC, Desiere F, Bork P, Delley M, Pridmore RD, Arigoni F (2002) The genome sequence of Bifidobacterium longum reflects its adaptation to the human gastrointestinal tract. Proc Natl Acad Sci USA 99:14422–14427

Searle LEJ, Best A, Nunez A, Salguero FJ, Johnson L, Weyer U, Dugdale AH, Cooley WA, Carter B, Jones G, Tzortzis G, Woodward MJ, La Ragione RM (2009) A mixture containing galactooligosaccharide, produced by the enzymatic activity of Bifidobacterium bifidum, reduces Salmonella enterica serovar Typhimurium infection in mice. J Med Microbiol 58:37–48

Sela DA, Chapman J, Adeuya A, Kim JH, Chen F, Whitehead TR, Lapidus A, Rokhsar DS, Lebrilla CB, German JB, Price NP, Richardson PM, Mills DA (2008) The genome sequence of Bifidobacterium longum subsp. infantis reveals adaptations for milk utilization within the infant microbiome. Proc Natl Acad Sci USA 105(48): 18964–18969

Sharp R, Fishbain S, Macfarlane GT (2001) Effect of short-chain carbohydrates on human intestinal bifidobacteria and Escherichia coli in vitro. J Med Microbiol 50:152–160

Shoaf K, Mulvey GL, Armstrong GD, Hutkins RW (2006) Prebiotic galactooligosaccharides reduce adherence of enteropathogenic Escherichia coli to tissue culture cells. Infect Immun 74:6920–6928

Silk DB, Davis A, Vulevic J, Tzortzis G, Gibson GR (2009) Clinical trial: the effects of a trans-galactooligosaccharide prebiotic on faecal microbiota and symptoms in irritable bowel syndrome. Aliment Pharmacol Ther 29 (5):508–518

Splechtna B, Petzelbauer I, Baminger U, Haltrich D, Kulbe DK, Nidetzky B (2001) Production of a lactose-free galacto-oligosaccharide mixture by using selective enzymatic oxidation of lactose into lactobionic acid Enz Micro Technol 29:434–440

Tanaka R, Takayama H, Morotomi M, Kuroshima T, Ueyama S, Matsumoto K, Kuroda A, Mutai M (1983) Effects of administration of TOS and Bifidbacterium breve 4006 on the human flora.Bifidobacteria Microflora 2:17–24

Teuri U, Korpela R, Saxelin M, Montonen L, Salminen S (1998) Increased fecal frequency and gastrointestinal symptoms following ingestion of galacto-oligosaccharide-containing yogurt. J Nutr Sci Vitaminol 44:465–471

Tochikura R, Sakai K, Fujiyoshi T, Tachiki T, Kumagai H (1986) p-Nitrophenyl glycoside-hydrolyzing activities in Bifidobacteria and characterization of β-D-galactosidase of Bifidobacterium longum 401 Agric Biol Chem 50:2279–2286

Tzortzis G, Goulas AK, Gee JM, Gibson GR (2005) A novel galactooligosaccharide mixture increases the bifidobacterial population numbers in a continuous in vitro fermentation system and in the proximal colonic contents of pigs in vivo. J Nutr 135:1726–1731

Veereman-Wauters G (2005) Application of prebiotics in infant foods. Br J Nutr 93: S57–S60

Vulevic J, Rastall RA, Gibson GR (2004) Developing a quantitative approach for determining the in vitro prebiotic potential of dietary oligosaccharides. FEMS Microbiol Lett 236:153–159

Vulevic J, Drakoularakou A, Yaqoob P, Tzortzis G, Gibson GR (2008) Modulation of the fecal microflora profile and immune function by a novel trans-galactooligosaccharide mixture (B-GOS) in healthy elderly volunteers. Am J Clin Nutr 88 (5): 1438–1446

Wijnands MVW, Schoterman HC, Bruijntjes JP, Hollanders VMH, Woutersen RA (2001) Effect of dietary galacto-oligosaccharides on azoxymethane-induced aberrant crypt foci and colorectal cancer in Fischer 344 rats Carcinogenesis 22:127–132

8 Prebiotic Potential of Xylo-Oligosaccharides

H. Mäkeläinen · M. Juntunen · O. Hasselwander

8.1 Introduction

Xylo-oligosaccharides (XOS) are chains of xylose molecules linked with β1–4 bonds (◉ *Figure 8.1*) with degree of polymerization ranging from 2 to 10.

XOS are naturally present in fruits, vegetables, bamboo, honey and milk, and can be produced at industrial scale by enzymatic hydrolysis from xylan, which is the major component of plant hemicelluloses and therefore readily available in nature (Alonso et al., 2003; Vázquez et al., 2000).

XOS are non-digestible carbohydrates and have been suggested to exert prebiotic activity. They were hence first used as food ingredient for gastrointestinal health in the 1990s in Japan. This chapter provides an overview of XOS with a particular focus on the prebiotic potential.

8.2 Manufacture of XOS

XOS can be produced commercially by hydrolysis of xylan, the most abundant hemicellulosic polymer. Possible lignocellulosic raw materials for XOS production include corn cobs, hardwoods, straws, bagasses, hulls, malt cakes and bran. Xylan can either be hydrolyzed enzymatically, by chemical methods (hydrothermal treatments) or a combination of both. The resulting crude XOS solutions require a sequence of purification steps to yield high purity XOS containing at least 70–95% XOS (Moure et al., 2006; Vázquez et al., 2000). Depending on the source of raw material, XOS may be branched and contain arabinose units or carry acetyl or uronic acid residues (Vázquez et al., 2000).

XOS have been used as food ingredient predominantly in Asia, particularly Japan and Suntory Limited (Japan) was the first commercial-scale producer

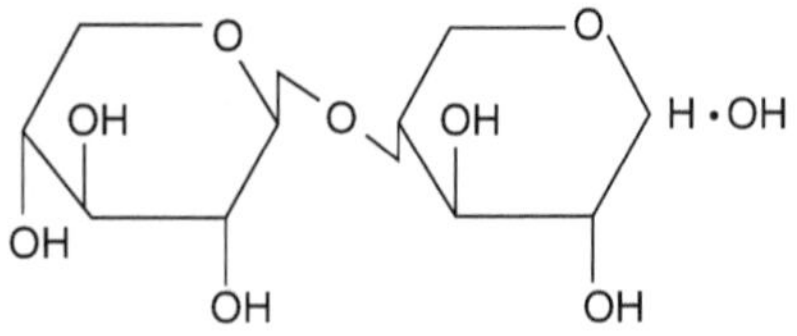

□ Figure 8.1
Disaccharide Xylobiose (β-D-xylopyranosyl (1→4)- D-xylopyranose).

applying enzymatic hydrolysis of xylan. The production volume of XOS was estimated at 650 tons annually in 2004 and the XOS price was 2,500 Yen/kg (Taniguchi, 2004). More recently, XOS were also offered by the Chinese producer Shandong Longlive Bio-technology co. In 2003, XOS represented only a small proportion (less than 3%) of the total Asian oligosaccharide market (Nakakuki, 2003).

The commercial XOS products are available in syrup or powder form and are predominantly composed of the disaccharide xylobiose and the trisaccharide xylotriose with small amounts of higher oligosaccharides also present.

8.3 XOS as Prebiotics

8.3.1 Resistance to Digestion

XOS are relatively stable in acidic conditions due to structural properties. This may endow protection from decomposition when passing through the stomach (Imaizumi et al., 1991). The degradation of XOS (xylobiose) in the gastrointestinal tract has been studied *in vitro* with an artificial model of digestive enzymes (α-amylase, pancreatin, gastric juice and intestinal brush border enzymes) and no hydrolysis of xylobiose was observed (Koga and Fujikawa, 1993; Okazaki et al., 1991).

The fate of xylobiose was also studied in humans after oral administration. Xylobiose was not excreted into feces or urine during 24-h following the ingestion, thus, supporting the fact that xylobiose is degraded *in vivo* not by the action of digestive enzymes, but by the gastrointestinal microbiota (Okazaki et al., 1991).

8.3.2 Fermentation by the Gastrointestinal Microbiota and Selective Stimulation of Growth and/or Activity of Intestinal Bacteria Associated with Health and Well-Being

8.3.2.1 Pure Culture Studies

Pure culture fermentation studies with single microbes and substrates can be used to identify bacterial strains that are able to degrade oligosaccharides. Since these fermentations do not resemble the competitive environment of the colon, the results can be used to gain knowledge of the fermentative capacity of individual strains within the intestinal microbial population, but not to study the effects that oligosaccharides have on the whole microbiota.

XOS are reported to be preferentially fermented by a relatively limited number of intestinal microbes *in vitro*. Several pure culture studies have indicated that XOS are well utilized by *Bifidobacterium* species, namely some strains of *B. bifidum, B. catenulatum, B. longum, B. animalis* and *B. adolescentis* (Crittenden et al., 2002; Jaskari et al., 1998; Moura et al., 2007; Palframan et al., 2003; Yamada et al., 1993). Furthermore, utilization of XOS seems to be strain-dependent, since not all studies have shown enhancement of for example all *B. longum* and *B. adolescentis* strains (Hopkins et al., 1998). Lactobacilli are not able to utilize XOS as a sole carbon source (Jaskari et al., 1998; Kontula et al., 1998), with the exception of *Lactobacillus brevis*, which growth was enhanced moderately by XOS (Crittenden et al., 2002; Moura et al., 2007). Some other intestinal microbes are also able to utilize XOS, but not to the same extent as bifidobacteria. Crittenden and co-workers studied the fermentation of a wide group of numerically dominant saccharolytic intestinal bacterial species and demonstrated that besides many *Bifidobacterium* strains only some *Bacteroides* isolates were efficiently fermenting XOS. *Escherichia coli*, enterococci, *Clostridium difficile* and *Clostridium perfringens* were not able to ferment XOS (Crittenden et al., 2002). Jaskari et al. found that XOS was metabolized by bifidobacteria, but also moderately by *Bacteroides thetaiotaomicron, Bacteroides vulgatus* and *Clostridium difficile*. However, these strains mainly utilized the monosaccharide xylose fraction of the XOS mixture, which in humans does not reach the colon (Jaskari et al., 1998).

Unpublished research (carried out by M. Juntunen at Danisco's Health and Nutrition Center in Kantvik, Finland) suggests a selective utilization of XOS by *Bifidobacterium lactis* strains (❍ *Figure 8.2*). Other tested microbes showed poor growth on commercial (XOS Longlive 95P, Shandong Longlive Bio-technology

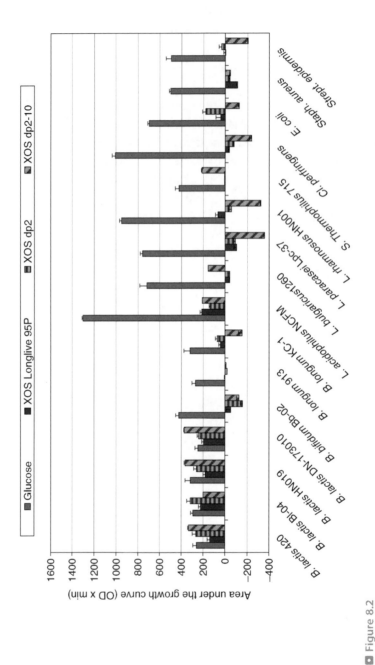

Figure 8.2

The growth of a selection of bifidobacteria, lactobacilli and other microbial strains on XOS. The growth was measured as the change in absorbance (600 nm) of liquid samples and represented the area under the growth curve (OD x minutes) obtained during the 24-h growth experiment (Jaskari et al., 1998). The area of the control medium without added carbohydrates was subtracted from results.

co., China) and hardwood-derived XOS (XOS dp2 and XOS dp2–10) when compared to growth obtained on glucose during 24-h incubation.

Interestingly, pure culture studies have also shown that for a number of *Bifidobacterium* strains the bacterial growth was higher on oligosaccharides in comparison to their monosaccharide constituents (Hopkins et al., 1998), thus, indicating that there may be a specific membrane transport mechanism for XOS but not xylose. Some bifidobacteria might import XOS before hydrolyzing it (Crittenden et al., 2002), which could offer a competitive advantage against cross-feeding by other microbes in the intestine.

8.3.2.2   Batch Culture Studies

Effects of oligosaccharide fermentation on predominant intestinal microbial groups can be monitored in batch culture fermentations and continuous or semi-continuous color simulations using colonic microbiota from fecal samples. The effects of different prebiotic oligosaccharides (FOS, inulin, lactulose, galacto-oligosaccharides (GOS), isomalto-oligosaccharides (IMO), soybean oligosaccharides (SOS), and XOS) on microbiota were compared in batch fermentations (Rycroft et al., 2001). Fermentation properties of different oligosaccharides varied and resulted in different responses in the composition and activity of microbiota. Fermentation of all oligosaccharide compounds increased the numbers of bifidobacteria and most decreased clostridia, but XOS and lactulose produced the highest increase in bifidobacteria, whereas FOS were the most effective substrate for lactobacilli. A similar significant increase in numbers of bifidobacteria, and also lactobacilli, was seen as a result of XOS fermentation in a semi-continuous simulator system inoculated with adult feces (Zampa et al., 2004). Microbial fermentation of XOS moderately increased the production of gases, and the concentrations of lactate, acetate, and propionate in both studies (Rycroft et al., 2001; Zampa et al., 2004). Zampa and co-workers also reported beneficial effects of XOS not only derived from increased populations of bifidobacteria and lactobacilli, but also from reduced concentrations of secondary bile acids, which exert negative actions on the colon and present a dose dependent toxic potential related to their co-mutagenic and tumor-promoting properties (Moure et al., 2006).

8.3.2.3   Animal Studies

The bifidogenic effects of XOS have also been observed in animal studies (◉ *Table 8.1*). Studies in rats have demonstrated that XOS significantly stimulate

▣ Table 8.1
Animal studies on effects of XOS

Animal studies	Major findings	Reference
Diabetic rats (n = 8), dietary sucrose and corn starch replaced with XOS for 5 weeks	Improvement in diabetic symptoms such as elevated serum glucose, cholesterol and triglycerides	Imaizumi et al. (1991)
Rats (n = 50) were fed with fiber-free diet, or diet containing 7.5% of oat fiber, gum arabic, FOS or XOS for 17 days	Oligosaccharides and gum arabic all increased the cecal wall and contents weight and decreased cecal pH with concomitant increase in SCFA. Excretion of nitrogen in feces was increased, thus, blood urea and renal nitrogen excretion decreased	Younes et al. (1995)
Rats (n = 44) and mice (n = 52) were assigned to control, FOS, XOS or gum arabic group for 14 days	XOS did not affect microbiota of animals, and only moderate effects on cecal cell proliferation were found. FOS increased significantly bifidobacteria in rats	Howard et al. (1995)
Rats (n = 10 in each group) were fed with control, cellulose, oligofructose, FOS or XOS containing diet for 14 days	Fecal pH was lowest and bifidobacteria numbers highest in the XOS group. Consumption of XOS increased most the weight of colon and cecum	Campbell et al. (1997)
Rats (n = 40) were fed with basal, XOS or FOS diet for 35 days, and treated with 1,2-dimethylhydrazine (DMH) to induce colon carcinogenesis	Oligosaccharides decreased cecal pH, increased cecal weight and bifidobacteria population, but XOS had a stronger effect than FOS. Both oligosaccharides reduced the formation of precancerous lesion	Hsu et al. (2004)
Mice (n = 16 in each group), diet supplemented (1%) with nine different oligosaccharides for a 6- month study period	XOS was most efficient prebiotic in increasing lactobacilli and bifidobacteria counts, and in reducing sulphite-reducing clostridia	Santos et al. (2007)

the growth of cecal and fecal bifidobacteria. Furthermore, XOS induced an even larger increase in bifidobacteria numbers than an equivalent dose of FOS, which had a greater effect on the *Lactobacillus* population (Campbell et al., 1997; Hsu et al., 2004). A more recent feeding-trial with mice (Santos et al., 2006) compared the long-term effects of various prebiotics on the microbial populations. In this study, 1% of FOS, inulin, lactulose, XOS, SOS, IMO, or transgalacto-oligosaccharides (TOS) were administered to the basal diet and the effects on small and large intestinal microbiota were determined after 6 months of intervention. From all the prebiotics tested, XOS was the most efficient substrate to increase bifidobacteria, lactobacilli and total anaerobic microbial numbers in the colon. Of all the oligosaccharides, XOS supplementation also resulted in the strongest reduction of sulphite-reducing clostridial strains, thus affecting the colonic microbiota in an overall favorable manner.

8.3.2.4 Human Intervention Studies

Reports from few human interventions, mostly conducted in Japan, regarding colonic fermentation of XOS have been published to date (◉ *Table 8.2*). Okazaki and co-workers fed 5 g of XOS daily for 3 weeks to healthy men, and found a significant increase in fecal bifidobacteria numbers. The effect of XOS

◻ Table 8.2

Human intervention studies conducted on XOS

Human intervention study	Major findings	Reference
Healthy men (n = 9), 5 g/day XOS for 3 weeks	Significant increase in *Bifidobacterium* and *Megasphaera*, other microbes unaffected. Increased fecal acetic acid and decreased pH	Okazaki et al. (1990a)
Healthy men (n = 5), 1 and 2 g/day XOS for 3 weeks each	Significant increase in *Bifidobacterium*, other microbes unaffected, Consumption of 2 g was more effective, but even dose of 1 g of XOS increased bifidobacteria	Okazaki et al. (1990b)
Healthy men (n = 10), 2, 5 and 10 g/day XOS	Occurrence of diarrhea decreased with concomitant increase in *Bifidobacterium* with daily doses of 2 and 5 g	Kobayshi et al. (1991)
Healthy men (n = 9), 5 g/day XOS for 3 weeks	Putrefaction products in feces (*p*-cresol, indole, skatol) decreased	Fujikawa et al. (1991)
Constipated women (n = 40), 0.4 g/day XOS for 4 weeks	Defecation frequency and stool quantity increased, self reported quality of life improved	Iino et al. (1997)
Constipated pregnant women (n = 29), 4.2 g/day XOS for 4 weeks	XOS effective in reducing severe constipation and normalizing stool consistency. Clinical symptoms scores improved	Tateyama et al. (2005)
Healthy elderly (n = 9 in control group; n = 13 in XOS group), 4 g/day XOS for 3 weeks	Significant increase in Bifidobacteria and fecal moisture content, decreased fecal pH	Chung et al. (2007)

administration on the growth of several other microorganisms was also studied, and only *Megasphera* numbers were significantly affected (increased) in addition to bifidobacteria, indicating a very selective proliferation effect (Okazaki et al., 1990a). Fecal pH reduced and acetic acid concentrations increased during this intervention. In another trial by the same researchers, *Bifidobacterium* numbers were increased in the feces of subjects consuming only 1 and 2 g of XOS per day, although the higher daily dose resulted in a more significant increase (Okazaki et al., 1990b). Even as low a daily dose as 0.4 g of XOS was shown to increase the numbers of bifidobacteria in human fecal samples, without effects on other microbial groups (Iino et al., 1997). Furthermore, decreased concentrations of putrefactive products such as *p*-cresol, indole and skatole were measured concomitantly with an increased ratio of fecal bifidobacteria (Fujikava et al., 1991). In a more recent human intervention study in elderly subjects (Chung et al., 2007), a daily XOS dose of 4 g increased significantly the *Bifidobacterium* numbers from 10^6 to 10^8 cfu/g of wet feces during a 3 week study period. *Clostridium perfringens* levels remained unchanged and the fecal pH was decreased and fecal moisture content increased during the study period. No difference in adverse gastrointestinal symptoms (flatulence, discomfort, stool consistency, defecation frequency) between the control and study group was recorded, suggesting that XOS was well tolerated by the elderly.

The limitations of the human interventions studies carried out to date are relatively small number of subjects, the (in most cases) uncontrolled study-design and the use of plating method in enumeration of microbes in feces, thus, larger controlled human intervention studies are needed to confirm the prebiotic status of XOS.

8.3.2.5   Effect of Substitution and Origin of XOS

The effect of differently substituted XOS on fermentability was first recognized by Van Laere et al., who included a linear XOS and arabinoxylo-oligosaccharides (AXOS) in a fermentation study of complex plant cell wall derived oligosaccharides (Van Laere et al., 2000). It was shown that the linear XOS were fermented by more of the tested intestinal microbial strains tested compared to the branched AXOS. The number and nature of the substitutes in XOS molecule appears to affect the fermentation speed and the metabolites produced. This was in particular demonstrated in a batch fermentation study (Kabel et al., 2002), where non-substituted XOS (nXOS) and arabino-XOS (AXOS) were reported to be fermented more quickly than structurally more complex acetylated XOS (AcXOS) and XOS containing a 4-O-methylglucuronic acid group (GlcA(me)XOS). In this study, the fermentation of the least substituted molecules XOS and AXOS resulted in quick production of acetate and lactate, whereas the fermentation of AcXOS and GlcA (me)XOS increased the production propionate and butyrate. These results highlight the importance of detailed elucidation of the structural features of non-digestible oligosaccharides in relation to their fermentation properties.

8.3.3   Effects on Health

The formation of preneoplastic lesions (aberrant crypt foci, ACF) in the distal colon is used as a biomarker for colon carcinogenesis. In a study with rats (Hsu et al., 2004), the ACF formation was induced with 1,2-dimethylhydrazine (DMH, a tumor promoter) and the rats were fed with basal diet or diet containing XOS or FOS. Dietary supplementation with both oligosaccharides inhibited the development of precancerous colonic lesions and simultaneously lowered the cecal pH level through increased concentrations of short chain fatty acids (SCFA) and increased the *Bifidobacterium* population and cecal weight. The decreased ACF and increased cecal weight could be due to the normalization of epithelial cell proliferation via increased concentrations of SCFA from oligosaccharide fermentation by *Bifidobacterium*. In this study, XOS was more efficient in increasing the bifidobacteria numbers and relative colonic and cecal wall weight than FOS, leading the investigators to conclude that XOS could be more effective in increasing and controlling the epithelial cell proliferation through a trophic effect of produced SCFA. These findings are supported by other animal trials (Campbell et al., 1997; Howard et al., 1995; Younes et al., 1995), although Howard and co-workers found only moderate effects of XOS on cell density and cecal crypt depths in mice, and no effects on microbiota (Howard et al., 1995). Campbell et al. showed increased cecal and colonic weight in rats together with decreased pH, increased SCFA (especially lactate and acetate) and bifidobacteria as a result of XOS and FOS supplementation. In the same study, XOS supplementation resulted in more significant changes in measured parameters than FOS supplementation. Younes and co-workers found an increase in total cecal weight (wall and content weight) after rats consumed XOS and FOS. The two oligosaccharides reduced the pH in cecum significantly more than oat fiber or control diet, and increased cecal SCFA concentrations. The ratios of SCFA differed considerably between oligosaccharides with XOS producing more acetate and FOS butyrate. Also, fecal nitrogen excretion increased and urinary nitrogen excretion decreased as a results of dietary oligosaccharide supplementations (Younes et al., 1995). Results from these animal trials indicate that fermentation of XOS could play a role in maintaining normal mucosal differentiation and, thus, the integrity of the colonic mucosa.

A study on diabetic rats found that when simple carbohydrates in the diet were replaced with XOS (10 g per day), the increased serum cholesterol and triglyceride levels seen in diabetes were reduced and liver triglycerides levels to a comparable level seen in healthy rats (Imaizumi et al., 1991). The researchers concluded that XOS could be applicable to foods as a sweetener replacing sucrose, which would benefit diabetic patients. However, these findings reported in animal trials have not been confirmed in humans yet.

In some human interventions, effects of XOS on intestinal microbiota and gastrointestinal function have been studied at the same time. Intervention in constipated pregnant women showed marked improvements in the defecation frequency and stool consistency during and after dietary supplementation with 4.2 g of XOS for 4 weeks. Before XOS administration the subjects recorded, in average, to defecate only once per week. During and after the intervention, defection frequency increased significantly to 6–7 times per week with concomitant improvement in subject's self-reported symptoms (Tateyama et al., 2005). Similar findings with adult women have been reported previously (Iino et al., 1997). Defecation frequency and abdominal symptoms improved simultaneously with increased bifidobacteria numbers, and persisted 2–4 weeks after XOS ingestion was completed. Kobayashi et al. found that administration of 2 g of XOS per day decreased the frequency of diarrhea in men (Kobayashi et al., 1991).

8.4   Safety and Regulatory Status

XOS were tested for mutagenicity, acute and subchronic toxicity. XOS were found to be non-mutagenic and showed no acute toxicity. Safety was also confirmed in a 90-day subchronic toxicity study in rats. These studies are mentioned in a Suntory product brochure and were conducted with Suntory's Xylo-oligo70 product (Biotec Suntory – Xylo-oligosaccharide brochure).

Limited data on digestive tolerability are available; however, volume of gas produced by human fecal bacteria during 24-h fermentation *in vitro* was similar for FOS and XOS (Rycroft et al., 2001) and no adverse effects have been reported in the human interventions carried out in healthy subjects (Okazaki et al., 1990a,b) including pregnant women (Tateyama et al., 2005) and elderly (Chung et al., 2007).

In Japan, XOS are approved as ingredients for Foods for Specified Health Uses (FOSHU), specifically for foods to modify gastrointestinal conditions at a recommended daily dose of 1–3 g (Japanese Ministry of Health, Labour and Welfare, www.mhlw.go.jp/english/topics/foodsafety/fhc/02.html). Use of XOS as food ingredient outside Japan and other Asian countries where XOS-containing products are currently marketed may require specific regulatory approval.

8.5   Market Information and Application

XOS were first used as food ingredient in the 1990s in Japan and since 1997, 32 product launches (including new formulations and varieties) have been recorded by Mintel's global new products database (Mintel, 2008). Use of XOS in dietary supplements accounted for 38% of all launches, use in dairy for 25%, use in sugar and gum confectionery for 16%, use in non-alcoholic beverages for 13% and use in baby food and soup for the remaining 8%. XOS is predominantly used in Asia (Japan, China, South Korea, Vietnam, Taiwan) with Japan accounting for more than half of the new product launches since 1997.

According to Suntory's product brochure (Biotec Suntory – Xylo-oligosaccharide brochure), XOS are acid- and heat-resistant. XOS remain intact after heating for 1 h at 100°C within a pH range from 2.5 to 8. Heat stability has also been confirmed at 120°C. Sweetness is claimed to be approximately 40% of sugar after comparable refinement and viscosity should allow easy use in various food applications (Biotec Suntory – Xylo-oligosaccharide brochure).

8.6   Summary

- XOS are resistant to digestion and are fermented by gastrointestinal microbiota.
- Fermentation of XOS by microbiota increases the concentrations of SCFA in colon, especially acetate, propionate and lactic acid. Gas volumes produced are moderate.
- Results from *in vitro* and *in vivo* studies are promising and suggest that XOS may selectively stimulate growth and/or activity of intestinal bacteria associated with health and well-being, particularly bifidobacteria.
- Additional controlled human intervention studies using molecular techniques to determine changes in fecal microbiota are needed to confirm the prebiotic potential of XOS and the efficacious dose, which has been suggested to be as low as 1 g/day.
- XOS can be considered an emerging prebiotic, as the scientific evidence is still not sufficient to classify XOS as an established prebiotic compound (Gibson et al., 2004).

List of Abbreviations

ACF	aberrant crypt foci
AcXOS	acetylated xylo-oligosaccharides
AXOS	arabinoxylo-oligosaccharides
DMH	1,2-dimethylhydrazine
dp	degree of polymerization
FOS	fructo-oligosaccharides
GlcA(me)XOS	xylo-oligosaccharides containing a 4-O-methylglucuronic acid group
GOS	galacto-oligosaccharides
IMO	isomalto-oligosaccharides
nXOS	non-substituted xylo-oligosaccharides
OD	optical density
SCFA	short chain fatty acids
SOS	soybean oligosaccharides
TOS	transgalacto-oligosaccharides
XOS	xylo-oligosaccharides

References

Alonso JL, Domínguez H, Garrote G, Parajó JC, Vázquez MJ (2003) Xylo-oligosaccharides: properties and production technologies. Electron J Environ Agric Food Chem 2:230–232

Campbell JM, Fahey GC, Jr, Wolf BW (1997) Selected indigestible oligosaccharides affect large bowel mass, cecal and fecal short-chain fatty acids, pH and microflora in rats. J Nutr 127:130–136

Chung YC, Hsu CK, Ko CY, Chan YC (2007) Dietary intake of xylooligosaccharides improves the intestinal microbiota, fecal moisture, and pH value in the elderly. Nutr Res 27:756–761

Crittenden R, Karppinen S, Ojanen S, Tenkanen M, Fagerström R, Mättö J, Saarela M, Mattila-Sandholm T, Poutanen K (2002) In vitro fermentation of cereal dietary fibre carbohydrates by probiotic and intestinal bacteria. J Sci Food Agric 82:781–789

Fujikava S, Okazaki M, Matsumoto N (1991) Effect of xylooligosaccharide on growth of intestinal bacteria and putrefaction products. J Jpn Soc Nutr Food Sci 44:37–40 (in Japanese, abstract in English)

Gibson G, Probert H, Van Loo J, Roberfroid MB, Rastall RA (2004) Dietary modulation of thehuman colonic microbiota: updating the concept of prebiotics. Nutr Res Rev 17:259–275

Hopkins M Cummings JH, Macfarlane GT (1998) Inter-species differences in maximum spesific growth rates and cell yields of bifidobacteria cultured on oligosaccharides and other simple carbohydrate sources. J Appl Microbiol 85:381–386

Howard M, Gordon D, Garleb KA, Kerley MS (1995) Dietary fructooligosaccharide, xylooligosaccharide and gum arabic have variable effects on cecal and colonic microbiota and epithelial cell proliferation in mice and rats. J Nutr 125: 2604–2609

Hsu C, Liao J, Chung Y, Hsieh CP, Chan YC (2004) Xylooligosaccharides and fructooligosaccharides affect the intestinal microbiota and precancerous colonic lesion development in rats. J Nutr 134: 1523–1528

Iino T, Nishijima Y, Sawada S, Sasaki H, Harada H, Suwa Y, Kiso Y (1997) Improvement of constipation by a small amount of xylooligosaccharides ingestion in adult women. J Jpn Assoc Dietary Fiber Res 1:19–24 (in Japanese, abstract in English)

Imaizumi K, Nakatsu Y, Sato M, Sedarnawati Y, Sugano M (1991) Effects of Xylooligosaccharides on Blood Glucose, serum and liver lipids and cecum short-chain fatty acids in diabetic rats. Agric Biol Chem 55:199–205

Jaskari J, Kontula P, Siitonen A, Jousimies-Somer H, Mattila-Sandholm T, Poutanen K (1998) Oat beta-glucan and xylan hydrolysates as selective substrates for Bifidobacterium and Lactobacillus strains. Appl Microbiol Biotechnol 49: 175–181

Kabel M, Kortenoeven L, Schols HA, Voragen AG (2002) In vitro fermentability of differently substituted xylo-oligosaccharides. J Agric Food Chem 50:6205–6210

Kobayashi T, Okazaki M, Fujikawa S, Koga K (1991) Effect of Xylooligosaccharides on Feces of Men. J Jpn Soc Biosci Biotech Agrochem 65:1651–1653 (in Japanese, abstract in English)

Koga K, Fujikawa S (1993) Xylo-oligosaccharides In: Nakakuki T (ed) Oligosaccharides: Production, Properties and Applications, Japanese Technology Reviews. Gordon and Breach Science Publishers, Yverdon, pp. 130–143

Kontula P, von Wright A, Mattila-Sandholm, T (1998) Oat bran beta-gluco- and xylo-oligosaccharides as fermentative substrates for lactic acid bacteria. Int J Food Microbiol 45:163–169

Mintel, Global new products database (gnpd), www.gnpd.com. Accessed on 10th of July 2008

Moura P, Barata R, Carvalheiro F, Gírio F, Loureiro-Dias M, Paula Esteves M. (2007) In vitro fermentation of xylooligosaccharides from corn cobs autohydrolysis by Bifidobacterium and Lactobacillus strains. LWT 40:963–972

Moure A, Gullón P, Domínguez H, Parajó JC (2006) Advances in the manufacture, purification and applications of xylooligosaccharides as food additives and nutraceuticals. Process Biochem 41: 1913–1923

Nakakuki T (2003) Development of Functional Oligosaccharides in Japan. Trends Glycosci Glycotechnol 15:57–64

Okazaki M, Fujikava S, Matsumoto N (1990a) Effect of xylooligosaccharide on the growth of Bifidobacteria. Bifidobacteria Microflora 9:77–86

Okazaki M, Fujikava S, Matsumoto N. (1990b) Effects of Xylooligosaccharides on growth of bifidobacteria. J Jpn Soc Nutr Food Sci 43:395–401 (in Japanese, abstract in English)

Okazaki M, Koda H, Izumi R, Fujikava S, Matsumoto N. (1991) In vitro digestibility and in vivo utilization of xylobiose. J Jpn Soc Nutr Food Sci 44:41–44 (in Japanese, abstract in English)

Palframan R, Gibson GR, Rastall RA (2003) Carbohydrate preferences of befidobacterium species isolated from the human gut. Curr Issues Intest Microbiol 4:71–75

Rycroft C, Jones M, Gibson GR, Rastall RA (2001) A comparative in vitro evaluation of the fermentation properties of prebiotic oligosaccharides. J Appl Microbiol 91:878–887

Santos A, San Mauro M, Diaz DM (2006) Prebiotics and their long-term influence on the microbial populations of the mouse bowel. Food Microbiol 23:498–503

Taniguchi H. (2004) Carbohydrate research and industry in Japan and the Japanese society of applied glycoscience. Starch 56:1–5

Tateyama I, Hashi K, Johno I, Iino T, Hirai, K, Suwa Y, Kiso Y (2005) Effects of xylooligosaccharide intake on severe constipation in pregnant women. J Nutr Sci Vitaminol 51:445–448

Van Laere K, Hartemink R, Bosveld M, Schols HA, Voragen AG (2000) Fermentation of plant cell wall derived polysaccharides and their corresponding oligosaccharides by intestinal bacteria. J Agric Food Chem 48:1644–1652

Vázquez M, Alonso J, Dominguez H, Parajó JC (2000) Xylo-oligosaccharides: manufacture and applications. Trends Food Sci Technol 11:387–393

Yamada H, Itoh K, Morishita Y, Taniguchi H (1993) Structure and properties of oligosaccharides from wheat bran. Cereal Foods World 38:490–492

Younes H, Garleb K, Behr S, Remesy C, Demigne C (1995) Fermentable fibers or oligosaccharides reduce urinary nitrogen excretion by increasing urea disposal in the rat cecum. J Nutr 125:1010–1016

Zampa A, Silvi S, Fabiani R, Morozzi G, Orpianesi C, Cresci A(2004) Effects of different digestible carbohydrates on bile acid metabolism and SCFA production by human gut micro-flora grown in an in vitro semi-continuous culture. Anaerobe 10:19–26

9 Resistant Starch and Starch-Derived Oligosaccharides as Prebiotics

A. Adam-Perrot · L. Gutton · L. Sanders · S. Bouvier · C. Combe · R. Van Den Abbeele · S. Potter · A. W. C. Einerhand

Dietary fiber has long been recommended as part of a healthy diet based on the observations made by Burkitt and Trowell (1975). Since then, epidemiological evidence has consistently shown that populations consuming higher levels of foods containing fiber have decreased risk of a variety of chronic health disorders such as cardiovascular disease, type II diabetes, and certain cancers. Average fiber intake in the United States is approximately 13 g/day for women and 18 g/day for men (National Academy of Sciences, 2006). The FDA recommends a minimum of 20–35 g/day for a healthy adult depending on calorific intake. In many EU countries including France, Germany and the UK (see ◉ *Figure 9.1*), fiber intakes are much lower than authorities recommend for men and women (Buttriss and Stokes, 2008; Gray, 2006). Thus, there is a need to increase fiber consumption and many newly isolated or developed fibers can easily be added to beverages and processed foods. The reasons for such low compliance is somewhat complex, however the most basic rationale for not consuming fiber-rich foods is perceived bad taste and mouthfeel and the availability of conventional food items containing fiber.

Dietary fibers confer a wide range of health effects, from alleviation of constipation to reduction of cholesterol (Buttriss and Stokes, 2008). The physiological effects of dietary fibers in humans depend on the physico-chemical properties of fiber (viscosity, fermentability, bulking properties) and on the human gastro-intestinal (GI) tract (gut microbiotia, GI transit time). A specific subset of dietary fibers, so called prebiotics, convey health benefits by selectively stimulating the growth and/or activity of one or a limited number of bacteria like bifidobacteria and lactobacilli in the colon (Gibson and Roberfroid, 1995).

There is considerable industry and public interest in the capacity of foods and food components to promote health and lower risk of non-infectious

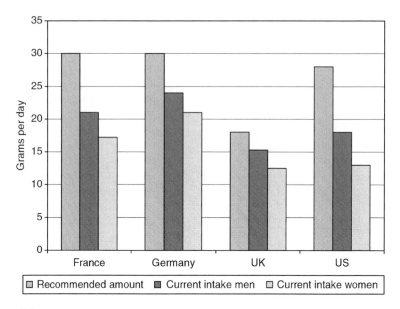

■ Figure 9.1
Daily intake of fibers.

diseases related to diet and lifestyle. Industry is thus challenged to develop fibers that can overcome the problems with stability under manufacturing conditions, functionality in various food systems, and taste in order to give the consumer more options when it comes to getting fiber in their diet. Newly derived starch fibers meet these requirements. In this chapter, prebiotic properties of newly derived fibers from maize and other starches will be reviewed with a specific focus on PROMITOR™ fibers, which were designed for optimal taste and texture and have prebiotic properties. The PROMITOR™ line includes a Soluble Gluco Fiber (PROMITOR™ Soluble Corn Fiber in US) and an insoluble Resistant Starch that are classified as food ingredients.

9.1   Introduction

The human gut microbiota constitutes a dynamic and ecologically diverse environment. The large intestine is by far the most heavily colonized region of the digestive tract, with up to 10^{12} bacteria for every gram of gut contents containing more than 400 different species of bacteria. The number of bacteria in the colon outnumber (10-fold) the number of human cells making it a very powerful target for nutritional interventions. Through the process of fermentation, colonic bacteria are able to produce a wide range of compounds that have both positive and negative effects on the gut physiology as well as other systemic influences. For instance, certain microbial populations present in the gut provide an efficient barrier to invading pathogens (Macfarlane, 2008). Competition for nutrients and ecological niches, production of antimicrobial compounds, lowering of intestinal pH through production of short chain fatty acids and stimulation of the immune system play a role in limiting the ability of pathogens to colonize the gut and potentially cause disease. Many of these microbiota-associated activities have a direct impact on host health. While prebiotics are selectively interacting with the intestinal microbiota, they are being fermented by the bacteria into many different metabolites. As the composition of the microbiota is modified, the types of fermentation metabolites into which prebiotic substrates are converted are also modified. Some of these metabolites are utilized by the cells lining the intestine, while others are absorbed into the blood of the host and pass the blood barrier to enter the systemic body space, where they interact with many physiological processes in all vital organs and peripheral tissues of the host (Lenoir-Wijnkoop et al., 2007).

9.2   Resistant Starches

Resistant starches are defined as the sum of starch and products of starch degradation not absorbed in the small intestine of healthy individuals (Asp, 1997). There are four main groups of resistant starches: RS1, RS2, RS3 and RS4. RS1 is physically inaccessible starch (i.e., starch in whole grains), RS2 is granular starch i.e., starch in green bananas), RS3 is retrograded starch (i.e., starch in cooked and cooled potatoes) and RS4 is a chemically-modified starch (i.e., an esterified starch). PROMITOR™ Resistant Starch is classified as a type 3 resistant starch.

9.2.1   Introduction to a Type 3 Resistant Starch

9.2.1.1   Regulatory Status

Resistant starches occur naturally in many foods and thus have been safely consumed across the globe for years. PROMITOR™ Resistant Starch is a food ingredient in the US and EU and can be labeled as "Resistant Starch,"

"Starch," "Maize Starch" or "Corn starch." It is a non genetically-modified source of dietary fiber which enables, where relevant regulatory conditions are met; the use of fiber nutrient content claims (contains fiber, source of fiber, high fiber).

The European definition of dietary fiber published in Directive 2008/100/EC recognizes as dietary fiber, carbohydrate polymers with three or more monomeric units, which are neither digested nor absorbed in the human small intestine and belong to the following categories:

- edible carbohydrate polymers naturally occurring in the food as consumed
- edible carbohydrate polymers which have been obtained from food raw material by physical, enzymatic or chemical means and which have a beneficial physiological effect demonstrated by generally accepted scientific evidence
- edible synthetic carbohydrate polymers which have a beneficial physiological effect demonstrated by generally accepted scientific evidence

RS meets the European definition and can be incorporated into food products to meet the Regulation (EC) 1924/2006 that requires at least 3 g of fiber per 100 g or at least 1.5 g per 100 kcal for the nutrition claim "source of fiber"; and at least 6 g of fiber per 100 g or at least 3 g per 100 kcal for the nutrition claim "high in fiber." Levels necessary for nutrient content claims in the US are 2.5 g fiber/serving for a "good source" claim and 5.0 g fiber/ serving for an "excellent source" claim.

9.2.1.2   Dietary Fiber Content

PROMITOR™ Resistant Starch has an analysis of approximately 60–75% total dietary fiber per AOAC method 991.43. This AOAC method works well for fibers that are insoluble or nearly insoluble.

9.2.1.3   Calorific Value

RS has a calorific value of 1.70 kcal/g, as calculated using standard practices of subtracting the percent fiber content as analyzed by AOAC 991.43 from total carbohydrates. This is the calorific declaration used on specification sheets in the US, as specified by US regulations for insoluble fibers. EU calorific value will be 2 kcal/g as per Commission Directive 2008/100/EC.

The true metabolizable energy (TME) content was also determined in an *in vivo* avian model at the University of Illinois (Knapp et al., 2008). The model is a more precise model for calorific determination than *in vitro* models and collection of urine and feces is easier and often more precise than in humans. This method, using bomb calorimetry, determined the calorific value of RS to be in the range of 1.7–2.0 kcal. This suggests that either a small portion of the RS is digested, or that fermentation products generated in the colon adds a small fraction of calories. Fermentation is a more likely hypothesis based on human glycemic response data. The glycemic response for RS is approximately 10% that of a maltodextrin control (Kendall et al., 2008). Since maltodextrin and RS are both glucose polymers, a glycemic response of 10% suggests that the remaining 90% of RS remains undigested and enters the colon. Thus, it is likely the majority of the calories from RS may actually come from fermentation metabolites.

9.2.1.4   Digestive Tolerance

Type 3 RS was shown to be well tolerated up to doses as high as 45 g/d (Bouhnik et al., 2004). Stewart et al. (2008, 2009) conducted a study in which subjects (n = 20, 10 men, 10 women) consumed 12 g fiber/d for 14 consecutive days. Subjects were asked to report gastrointestinal symptoms (cramping, bloating, stomach noise, flatulence) on a 10 point scale (1 = minimal symptoms, 10 = severe symptoms). RS was shown to be well tolerated (Stewart et al., 2009). In a dose-response study, tolerance was assessed in 22 healthy volunteers consuming three different doses of RS (5 g, 15 g, 25 g) in acute conditions after an overnight fast (Kendall et al., 2009). Subjects were asked to report gastrointestinal symptoms (belching, bloating, flatulence, nausea and diarrhoea) on a 100 mm VAS scale at 0, 15, 30, 45, 60, 90 and 120 min after consumption. RS was shown to be well tolerated in these conditions.

9.2.1.5   Applications

RS acts as an insoluble fiber that can be added to baked products, cereal products or snacks, providing them with higher fiber levels as well as other health benefits. It is one of the most stable resistant starches on the market (unpublished data) and so can be used in extruded or sheared and baked products with less loss of fiber during processing. As a result the final product has a high fiber content while offering a range of health benefits. With only 1.70 kcal/g, it can also be used in low calorie products reducing both calories and carbohydrates when replacing flour or other cereal-based ingredients.

Applications for RS include puffed or sheeted snacks, chips, extruded breakfast cereals, pasta, muffins, cookies and biscuits, crackers, frozen dough, breads.

It can be used as a partial replacement for flour in bakery products that exhibit characteristics comparable to those achieved using conventional wheat flour (e.g., cookie spread, golden brown color, pleasant aroma, surface cracking). Thanks to its low water holding property, it also does not affect height and spread management of biscuits, cookies or other baked goods.

RS enhances crispiness of cookies and crackers as well as the surface of baked sheeted crackers and extruded products. Furthermore, the induced reduction of water activity and moisture content, enhance sensory characteristics as well as the shelf life of goods. Notably, it tends to decrease bulk density, improving expansion in extruded cereals and snacks. In fried snacks, fat uptake may be reduced by up to 25% when RS is used, helping to meet "high/rich in fiber" claims.

Moreover, with a thermal stability as high as 150°C, it will retain more fiber content and structure than other resistant starches, which start to break down below 120°C.

9.2.2   Prebiotic Properties of Various Resistant Starch Products (RS2 and RS3)

9.2.2.1   Effect on Microbiota Modulation

The ability of RS to favor growth of bifidobacteria and lactobacilli within the gut flora has been assessed *in vitro*, in animal models and in humans.

In a study Le Blay et al. (2003) showed the prebiotic properties of RS2. Eighteen rats were fed a low-fiber diet (Basal) or the same diet containing raw potato starch (RS2) (9%) or short-chain FOS (9%) for 14 days. Changes in wet-content weights, bacterial populations and metabolites were investigated in the caecum, proximal, distal colon and feces. Both substrates exerted a prebiotic effect compared with the Basal diet. All bacteriological analysis were performed within 2 h after sampling. Samples were diluted and dilution were applied on plates using both unselective and selective media. After incubation, single colonies were counted. FOS increased lactic acid-producing bacteria throughout the caecocolon and in feces, whereas the effect of RS2 was limited to the caecum and proximal colon. As compared with RS2, FOS doubled the pool of caecal fermentation products, while the situation was just the opposite distally. This difference was mainly because of the anatomical distribution of lactate, which accumulated in the caecum with FOS and in the distal colon with RS2. Feces reflected these impacts only partly, showing the prebiotic effect of FOS and the metabolite increase induced by RS2. In conclusion, this study demonstrates that FOS and RS2 exert complementary effects and combined ingestion could be beneficial by providing health-promoting effects throughout the colon.

Brown et al. (1997) also observed prebiotic effects in 12 young male pigs fed with high amylose cornstarch diet (RS2). Starch provided 50% of total daily energy either as a low amylose cornstarch or as a high amylose (amylomaize) cornstarch. Fecal output, fecal concentrations and excretion of total SCFA (notably propionate and butyrate), fecal culture-based bifidobacteria counts (expressed per gram of wet feces) and their daily fecal excretion were higher when pigs were fed the high amylose cornstarch.

Similar prebiotic effects have been reported for retrograded RS (RS3) in several animal models. Dongowski et al. (2005) and Jacobasch et al. (2006) have shown in rats fed with RS3 that the growth of bifidobacteria and the production of SCFA were increased throughout the digestive tract, favoring thus a decrease of the pH in the caecum, colon and feces.

Using another model of human flora-inoculated gnotobiotic rats (HFA), colonized with microbiotas from UK or Italian donors, Silvi et al. (1999) looked at the prebiotic effects of a RS3. Consumption of this RS3 (15 g/100 g diet) resulted in significant changes in both the UK and Italian flora-associated rats: numbers of lactobacilli and bifidobacteria were increased 10–100-fold, and there was a concomitant decrease in enterobacteria when compared with sucrose-fed rats (control). The induced changes in caecal microbiota of both HFA rat groups were reflected in changes in bacterial enzyme activities and caecal ammonia concentration. This RS3 markedly increased the proportion of n-butyric acid in both rat groups, lowered caecal ammonia concentration, caecal pH and beta-glucuronidase activity.

The prebiotic properties of RS3 have also been demonstrated in humans (Bouhnik et al., 2004). First, this study determined the bifidogenic properties of a RS3 at 10 g/d; Second, the dose-response relationship of the bifidogenic effects of this RS3 at doses ranging from 2.5 to 10 g/d in comparison with a placebo were assessed. Faecal samples were diluted and dilutions were applied on plate using different selective media. RS3 was shown to be bifidogenic at a dose of 10 g/d. However, bacteria counts increased at doses of 5 and 7.5 g/d, but decreased at doses of 2.5 and 10 g/d. Thus, no firm conclusions could be drawn from this study on bifidogenic properties of RS3 in humans, however the authors observed that only a low baseline bifidobacteria count was significantly associated with a bifidogenic response to treatment. This observation was recently corroborated by Roberfroid who highlighted that the baseline level of bifidobacteria together with the time of exposure to a prebiotic are more determinant factors than the amount of prebiotics consumed to assess the potential prebiotic properties of a fiber (Roberfroid, 2007). The lack of bifidogenic effect of RS3 in this study at 2.5 and 10 g/d in phase 2 might be primarily linked to an elevated baseline bifidobacteria count in the groups of volunteers.

In this respect, a recent *in vitro* study also showed that the RS3 crystalline polymorphism can impact the RS fermentability by human gut microbiota as well as the short chain fatty acids production. Human fecal pH-controlled batch cultures showed that RS induces an ecological shift in the colonic microbiota, with polymorph B being much more efficient in inducing *Bifidobacterium* spp. growth than polymorphism A. Interestingly, polymorph B also induced higher butyrate production to levels of 0.79 mM (Lesmes et al., 2008). Type-A crystalline polymorph is found in typical cereal starch granules while the type-B polymorph is found in potato and high amylose cereal starch granules. The A polymorph has a much lower water content in the crystal lattice compared to the B polymorph.

PROMITOR™ Resistant Starch has shown to be a potential prebiotic in *in vitro* studies at TNO (Maathuis et al., 2008). The aim of the study was to investigate the effect of newly developed maize-based fibers on the activity and composition of the microbiota in the colon. The tested fibers were glucose-based and have variable structures including two resistant starch preparations. The fibers were pre-digested, mono- and di-saccharides were removed, and the residual polymer was used to assess the production of microbial metabolites and changes in composition of the microbiota using a dynamic, validated, *in vitro* model of the large intestine (TIM-2). Microbial metabolite analysis showed an increase in health-promoting metabolites (short-chain fatty acids) and a reduction in toxic metabolites from protein fermentation compared to the poorly-fermentable control (cellulose). The lactate production was also relatively low, indicating that it is slowly fermented. This may contribute to its excellent tolerance and extend its health benefits throughout the entire large intestine. Using microarray technology to compare multiple species and groups of colonic microbiota, RS was found to promote the growth of beneficial bacteria such as bifidobacteria and lactobacilli (⊙ *Figure 9.2*). Further studies are underway to determine if these *in vitro* effects are also seen *in vivo*.

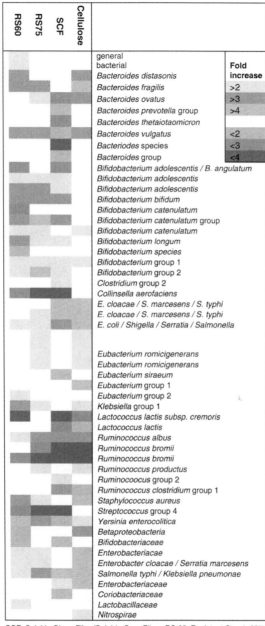

SGF: Soluble Gluco Fibre/Soluble Corn Fibre; RS 60: Resistant Starch 60%;
RS 75: Resistant starch 75%

◘ Figure 9.2

Use of the microarray technology to evaluate the effects of PROMITOR™ fibers on multiple species and groups of colonic microbito.

Based on the above-mentioned studies, it can be reasonably concluded that RS2 and RS3 have prebiotic activity. The mechanisms involved in promotion of bifidobacteria/lactobacilli growth may be different than those observed for the other prebiotics (e.g., FOS or inulin). A study with FOS and high amylose starch (RS2) in pigs fed a diet based on human foods showed that both raised fecal bifidobacteria numbers by approximately equal amounts when fed separately. When fed together there was an increase that exceeded the individual increases, suggesting that they operate by different mechanisms (Brown et al., 1997). If FOS acts as a metabolic substrate for bifidobacteria and lactobacilli, RS2 seems to function differently. Indeed, *in vitro* studies showed on one hand that pure bifidobacteria strains have limited capacity to use RS2 as a substrate (Topping et al., 2003); on the other hand, they also showed physical adhesion of several bifidobacteria species to RS2 (Topping et al., 2003), suggesting thus that the prebiotic properties of RS2 may be linked to its ability to confer physical protection on the bifidobacteria/lactobacilli throughout the upper digestive tract. The same reasoning may apply to RS3, as Brouns et al. (2007) have shown that breast-fed babies' microbiota, mainly composed of bifidobacteria was unable to use RS3 as a substrate, though several animal and human studies have shown prebiotic effects of RS3.

Nevertheless, it seems that a limited number of bifidobacteria strains has the ability to use RS2 as a substrate as shown in the study of Crittenden et al. (2001). In this study, 40 *Bifidobacterium* strains were examined to complement RS in a synbiotic yogurt. Only *B. lactis* Lafti B94 possessed all of the required characteristics. This isolate was the only one able to hydrolyze Hi-maize (RS2), to survive well in conditions simulating passage through the gastrointestinal tract and to possess technological properties suitable for yogurt manufacture. *Bifidobacterium lactis* Lafti B94 survived without substantial loss of viability in synbiotic yogurt containing Hi-maize during storage at 4°C for 6 weeks. In this study, RS2 was seen as a good complementary prebiotic ingredient for new synbiotic functional food products.

9.2.2.2   Health Benefits Associated with Prebiotic Properties of Resistant Starch

Several health benefits have been linked to prebiotics, notably to FOS, GOS and inulin. One of these benefits is enhancement of the body's natural immune defenses (Schley and Field, 2002; Vos et al., 2007). This effect is primarily localized to immune defenses in the gut, such as the gut-associated lymphoid tissue, however some studies have shown systemic effects. The precise mechanisms by which prebiotics exert their immune effects are unclear; whether it is through changes in the microbial population, through fermentation metabolites (i.e., SCFA) or through direct interaction of the prebiotics with mucosal membrane receptors. There have been very few investigations into the immune-modulating effects of resistant starch. Preliminary data from a recent study including PROMITOR™ Resistant Starch demonstrates a potential immune-modulating response in an animal model of inflammatory bowel disease. Animals supplemented with Resistant Starch had fewer macroscopic lesions in the gut and a reduction in the size of mesenteric lymph nodes (unpublished data) as compared to animals without Resistant Starch in the diet.

Another benefit seen with fermentable and prebiotic dietary fibers is enhancement of mineral absorption and/or increase in bone mineralization and bone density. There are numerous hypothesized mechanisms by which fermentable fibers may alter mineral absorption and impact bone density, a few of which include increased solubility of minerals, increased gastrointestinal surface area for absorption (by means of SCFA production), and alteration of the microbial population (Scholz-Ahrens et al., 2007). Many of the studies surrounding mineral absorption and bone formation have included prebiotics, such as inulin and FOS. However, RS has also been reported to enhance the ileal absorption of a number of minerals in rats and humans. Lopez et al. (2001) and Younes et al. (2001) reported an increased absorption of calcium, magnesium, zinc, iron, and copper in rats fed RS2-rich diets. Similar preliminary results have been seen in an animal study investigating PROMITOR™ Resistant Starch where it was shown to increase femur calcium concentration after 12 weeks of supplementation (Martin et al., 2009).

9.2.2.3   Gut Health Biomarkers

It has been shown that RS has health-promoting actions on the colonic microbiota beyond the prebiotic effect. For instance, studies in children with cholera-induced diarrhoea having consumed RS (high-amylose starch) plus the rehydration therapy have shown major reduction in fluid loss and a halving of time to recovery (Ramakrishna et al., 2000, 2008). This study has been replicated in babies with other forms of infectious diarrhoea where it was shown that both RS (as green bananas) and non-starch polysaccharides (NSP) facilitated recovery and improved intestinal permeability (Rabbani et al., 2001). Accelerated recovery from infectious diarrhoea has also been shown in animals. A specific micro-organism, *Brachyspira hyodysenteriae*, causes substantial economic losses in the commercial pig breeding industry through morbidity and mortality in the weaning period. The effect is expressed as diarrhoeal disease on the introduction of solid food. Feeding with cooked rice, an established source of RS (Marsono et al., 1993), lowers the incidence and severity of disease with a consequent reduction in mortality (Hampson et al., 2000). Part of the benefit seems to be due to increased fluid absorption through greater SCFA production, as these acids stimulate the uptake of water and cations (Na^+, K^+, and Ca^{2+}), particularly in the proximal colon. Several studies in humans (⊙ *Table 9.1*) as well as in animals (⊙ *Table 9.2*) have shown the ability of RS to enhance production of SCFA in the ceacum and throughout the colon. This outcome is an obvious mechanism for reversing diarrhoea-induced fluid loss. SCFA also appear to modulate the muscular activity of the large bowel and to promote the flow of blood through the viscera; both these actions could assist in lowering the severity of diarrhoea. In addition to these well-documented effects, it is possible that RS could limit the viability of the cholera organism in the gut. It may be hypothe-sized that the bacteria could adhere to the starch granules, very much in the same way as bifidobacteria, and thus be removed from the site of action (Topping et al., 2003). The production of SCFA (Cummings et al., 1996; Heijnen et al., 1998; Jenkins et al., 1998; Muir et al., 2004), the reduction of colonic pH (Birkett et al., 1996; Heijnen et al., 1998; Phillips et al., 1995) together with a beneficial change in microbiota metabolism pattern induced by RS intake (⊙ *Tables 9.1* and ⊙ *9.2*) are further means of biocontrol for pathogens and potential pathogens.

Diet and its interaction with the gut microbiota, and reduced protection from the microbiota with age are likely contributory factors to colorectal cancer (CRC). Genotoxic or carcinogenic metabolites produced or activated by the gut microbiota provide a diversity of environmental insults, which play a role in the initial stages of cancer (Tuohy et al., 2005). There is considerable interest in using microbiota-modulating tools such as prebiotics (Burns and Rowland, 2000) or RS (Cassidy et al., 1994) to protect against colonic tumor development or to maintain a good colonic physiology.

To date, several human studies have examined the effect of RS on human colonic function (⊙ *Table 9.1*). These intervention studies have evaluated differ-ent types of RS, in combinations or alone, at different amounts. Some studies have mimicked human diets in that they have used a range of food-based sources, whereas others have used single manufactured forms.

◻ Table 9.1

Human studies investigating the effects of RS intake on gut health biomarkers *(Cont'd p. 272)*

Sample size and study length	Intervention	Parameters measured / outcome	Authors
14 healthy subjects - 4 weeks	- 45g Hylon VII (62% RS) to a standard diet - standard diet low in RS	Cytotoxicity of fecal water on colon cell line: – Cell proliferation in rectal biopsies: – Fecal output: ++ Fecal SCFA: ++ Fecal bile acids: ++ Fecal secondary bile acids: –	Van Munster et al, 1994
11 healthy subjects in a randomized controlled cross-over study - 3 weeks	- High RS diet (raw banana flour + unprocessed wheat seeds + high-amylose maize cornbread, 26 to 50g/d) - Low RS diet (cooked banana flour + processed wheat seeds + low-amylose maize cornbread, 3 to 8g/d)	Butyrate: ++ Faecal pH: – Faecal excretion of NSP: ++ Faecal ammonia: – Fecal output: ++	Phillips et al, 1995
11 subjects in a randomized controlled cross-over study - 3 weeks	- High RS diet (≈40g/d, RS1, RS2, RS3 mix) - Low RS diet (5g/d)	Faecal nitrogen excretion: ++ Faecal ammonia: – Faecal phenols: – Faecal pH: –	Birkett et al, 1996
12 healthy subjects in a controlled cross-over study - 15d periods	- 17-30g RS2 diets - potato and banana starch - 17-30g RS3 diets - retrograded maize and wheat starch - RS-free diet	Faecal output: ++ Faecal SCFA: ++ Faecal excretion of NSP: ++ Fecal nitrogen excretion: ++ Transit time: –	Cummings et al, 1996

◘ Table 9.1

Sample size and study length	Intervention	Parameters measured / outcome	Authors
23 hypertriglyceridemic subjects in a randomized controlled cross-over study - 4 weeks	- High amylose maize starch diet (RS:17-25g/d) - Oat bran: low-RS diet	Faecal pH: – Faecal SCFA: ++, butyrate Secondary faecal bile acids excretion: –	Heijnen et al, 1998
12 subjects in a controlled, cross-over study - 4 weeks	- Controlled basal diet enriched in RS (amylomaize starch, 55g/d RS) - Controlled basal RS-free diet	Faecal output: ++ Faecal SCFA: no effect β-glucuronidase: – Faecal bile acids excretion: – Faecal secondary bile acid: – Faecal concentration of neutral sterols: –	Hylla et al, 1998
24 healthy subjects in a controlled, randomized, cross-over study - 2 weeks	- RS2 (≈20g/d) - RS3 (≈25g/d) - Low fibre diet	Faecal output : ++ Butyrate: ++	Jenkins et al, 1998
57 babies (5-12 months) with persistent diarrhoea	- Rice-based diet containing green banana (250g/L diet) - Rice-based diet	Intestinal permeability: ++ Diarrhoea: –	Rabbani et al, 2001
23 patients with recently removed colonic adenomas in a randomized, controlled, parallel study - 4 weeks	- 45g amylomaize (28g/d RS2) in capsules - 45g maltodextrin in capsules	Cell proliferation: no effect Faecal output: no effect Faecal pH: no effect Faecal SCFA: no effect Faecal bile acids excretion: – Faecal secondary bile acids: –	Grubben et al, 2001

Subjects/Study	Intervention	Results	Reference
12 healthy volunteers in a controlled, cross-over study - 4 weeks	- High RS diet: Hylon VII in foods (RS: ≈ 55g/d) - Low RS (cornstarch, RS:2.8-3.7g/d)	Cell proliferation: no effect DNA adducts: ++	Wacker et al, 2002
20 volunteers with a family history of colorectal cancer in a randomized, cross-over study - 3 weeks	- Basal diet + RS (22g/d) + wheat bran (12g/d) - Basal diet + Wheat bran (12g/d) - Control diet	Feacal output: ++ Transit rate: ++ Faecal pH: −, only for RS Faecal SCFA: ++, only for RS Feacal phenols and ammonia: ++	Muir et al, 2004
8 healthy volunteers per fibre tested	- Phase 1: RS3 (10g/d) - Phase 2: RS3 (2.5, 5, 7.5 and 10g/d)	Prebiotic effects: ++ in phase 1 Prebiotic effects: ++ at 5 and 7.5g/d only in phase 2	Bouhnik et al, 2006
12 healthy volunteers with 3-14 bowel movements/ week, in a cross-over, randomized, double-blind, controlled study	- 25g/d Promitor™ Resistant Starch - Low fiber diet (<2g/d)	Fecal output: ++ Tolerance: good	Maki et al, 2009 (submitted)

++: increased or improved, −: decreased

● Table 9.2

Animal studies investigating the effects of RS intake on luminal and epithelial colorectal cancer biomarkers (Cont'd p. 276)

Animal model	Intervention	Parameters measured	Author
Wistar rats (azoxymethane)	Dietary carbohydrate replaced with sucrose, cornstarch or RS (RS$_2$, 67g/100g)	ACF: −	Thorup et al, 1995
Sprague Dawley rats (Dimethylhydrazine)	3 or 10g/100g cellulose RS$_3$ (hydrolyzed high amylose maize starch)	Faecal output: ++ SCFA and butyrate production: ++ Tumor incidence: no effect Tumor size and multiplicity: no effect	Sakamoto et al, 1996
Sprague Dawley rats (dimethylhydrazine)	- Low RS, low fibre diet - 14.4g/100g diet RS (RS$_2$)	ACF: ++ Cell proliferation: ++ Faecal output: ++ Tumor size and multiplicity: ++	Young et al, 1996
C57BL/6 min mice	- RS-free diet (2% cellulose, no RS) - RS$_3$ (high amylose cornstarch, 18g/100g)	Gut-associated lymphoid tissue: no effect Tumor incidence: no effect	Pierre et al, 1997
Cannulated piglets	- RS-free diet - RS$_2$ (uncooked high amylose cornstarch) diet, 17% - RS$_3$ (retrograded high amylose cornstarch) diet, 17%	Fecal nitrogen excretion: −, for RS$_3$ only Urine nitrogen excretion: −, for RS$_3$ only	Heijnen et al, 1997

Sprague Dawley rats (dimethylhydrazine)	- Retrograded high-amylose cornstarch (RS_3), 25% - RS_3, 25% + vitamin A diet Control diet	ACF: – Faecal output: ++ Faecal pH: – Butyrate: ++	Cassand et al, 1997
Sprague Dawley rats (dimethylhydrazine)	- RS-free diet (2% cellulose) - 25g/100g RS_3 (retrograded amylose maize starch)	ACF: no effect Caecal pH: – Faecal weight and output: ++ β-glucuronidase: –	Mazière et al, 1998
Human-flora associated Fisher rats	- Retrograded amylose (15g/100g diet) - RS free diet	Ammonia: – β-glucuronidase: – Caecal SCFA: ++ Cell proliferation: ++ Prebiotic effect: ++	Silvi et al, 1999
Min mice	RS free diet 1:1 potato RS and high-amylose maize diet	Tumor incidence: ++	Williamson et al, 1999
BDIX rats	- Low-fiber control diet - RS_3 (30%)	ACF: – Caecal butyrate: ++ Mucosal cell proliferation: no effect	Perrin et al, 2001
Wistar rats (IQ)	- RS free diet - Potato starch and high amylose maize starch (35%)	Excretion of the food carcinogen, Faecal output: ++ IQ: ++ Transit : ++	Ferguson et al, 2003

□ Table 9.2

Animal model	Intervention	Parameters measured	Author
Sprague Dawley rats	15g/100g casein with or without ≈50% Hi-Maize or 25g/100g casein with or without ≈50% Hi-Maize	Caecal SCFA: ++ DNA damage: − p-cresol: − Thinning of mucosal layer: −	Toden et al, 2005
Sprague Dawley rats (azoxymethane)	Low RS diet + 1% lyophilized Lactobacillus acidophilus and Bifidobacterium lactis - 10% RS diet (Hi-Maize) +1% lyophilized probiotics	Acute apoptotic response: ++ Cecal pH: − Cell proliferation: ++ Crypt column height: ++ Fecal and caecal SCFA: ++ Prebiotic effect : ++	Le Leu et al, 2005
Rats	- RS-free diet - Retrograded potato debranched maltodextrin (RSA(RS3), 5.7g/100g - RSB(RS3), 5.7g/100g)	Bile acids fecal excretion: ++ Caecal pH: − Caecal SCFA: ++ Fecal output: ++, for RSA	Dongowski et al, 2005
Sprague Dawley rats (DMH)	- RS-free diet - Heat-treated Novelose 330 (RS3, 10%)	Apoptosis: ++ Epithelial proliferation: − Tumor incidence: − Mucin properties: ++ for acidic mucin	Bauer-Marinovic et al, 2006
Wistar rats	- Heat-treated Novelose 330 - Normal Novelose 330	Caecal and colonic pH: − Crypt length: ++ Formation of secondary bile acids: − Fecal bile acids: ++ SCFA, butyrate: ++ Prebiotic effect: ++	Jacobash et al, 2006

++: increased or improved, −: decreased

In most of the studies, RS intake (17 g/d up to 50 g/d) was shown to increase fecal output, to increase fecal SCFA and to decrease fecal pH (Birkett et al., 1996; Cummings et al., 1996; Heijnen et al., 1998; Hylla et al., 1998; Jenkins et al., 1998; Maki et al., 2009; Muir et al., 2004; Phillips et al., 1995; van Munster et al., 1994).

The effect of RS consumption on fecal output is dose-dependent and increasing doses are associated with increasing recoverable starch in stools (Phillips et al., 1995). An increased fecal output is generally seen as being beneficial to gut health as it helps dilute the carcinogenic compounds and reduce thus their time contact with the epithelium.

The reduction of the fecal pH reflects an acidification of the colonic lumen due to the production of SCFA, notably butyrate linked to the fermentation of RS by the microbiota. The fact that RS favours butyrate production has been largely demonstrated *in vitro* (Brouns et al., 2007; Fassler et al., 2006, 2007) and in animal studies (Cassand et al., 1997; Dongowski et al., 2005; Jacobasch et al., 2006; Le Leu et al., 2005; Perrin et al., 2001; Sakamoto et al., 1996; Silvi et al., 1999; Toden et al., 2005). Butyrate plays a critical role in mucosal physiology and metabolism, by providing 50% of the daily energy requirements of the colonic mucosa, while also being implicated in cellular differentiation and proliferation. It has been seen *in vitro* that butyrate could be involved in cancer prevention through hyperacetylation of histone proteins in the cell nucleus (Tran et al., 1998) and selectively induce apoptosis in colon cancer cells but not in healthy colonocytes (Ruemmele et al., 2003).

The fermentation of RS induces changes in the microbiota metabolism pattern reducing either the numbers and/or activities of putrefactive bacteria in the colon, leading to a reduction in enzyme activities involved in carcinogenic pathways (e.g., β-glucuronidase) as well as toxic metabolites such as amines, indoles, p-cresol (Birkett et al., 1996; Heijnen and Beynen, 1997; Hylla et al., 1998). The use of protein fermentation products as a source of nitrogen for the microbiota to grow favours a reduction of fecal ammonia and phenols (Birkett et al., 1996; Muir et al., 2004). As protein fermentation products seem likely to be carcinogenic, this might be an additional mechanism by which RS might be protective. A low colonic pH is also associated with a decreased conversion rate of primary to secondary bile acids, which are thought together with other toxic metabolites like ammonia, phenol, amines, etc, to impact fecal water cytoxicity. In this respect, some human studies have shown that RS intake, due to its fermentation in the colon, can help reduce the production of secondary bile acids (Grubben et al., 2001; Hylla et al., 1998; van Munster et al., 1994). Seeing the fact that RS intake helps in reducing the amount of several toxic compounds in the colon lumen via several mechanisms, a reduction of the fecal water cytoxicity is expected and this has been observed in a human study (van Munster et al., 1994) as well as in *in vitro* studies (Fassler et al., 2007).

One study in humans did consistently not show any effect (Grubben et al., 2001). That study was conducted in subjects with recently removed colonic adenomas, and the RS was administered in capsule form. They also failed to observe any changes in fecal fermentation, which suggests that the capsules did not effectively release the RS.

Generally speaking, similar effects as in humans have been observed in animals for luminal biomarkers of gut health. That is to say that in all animal studies, RS intake has favored butyrate production, increased fecal output, decreased pH, enzyme activity, ammonia, p-cresol and other toxic metabolites (Dongowski et al., 2005; Ferguson et al., 2003; Jacobasch et al., 2006; Maziere et al., 1998; Perrin et al., 2001; Sakamoto et al., 1996; Silvi et al., 1999; Toden et al., 2005; Young et al., 1996).

In rats, challenged with 1,2-dimethylhydrazine or azoxymethane or again in genetically modified mice models (min mice), the effects of RS intake on colorectal neoplasia and on epithelial markers of colorectal cancer have also been investigated (◉ *Table 9.2*). However, depending on the type of animal model, on the nature and the amount of carcinogenic compounds used to induce colorectal cancer in rats, on the feeding time period, the effects of RS intake on aberrant crypt foci (ACF), cell proliferation, DNA damages, tumor incidence are highly variable. It is therefore very difficult to draw any conclusions on RS intake on markers of colorectal-cancers in animal studies. However, whether these particular animal models are relevant to investigate nutritional benefits of RS on gut health is a question mark as they induce drastic effects that do not reflect what occurs in physiological conditions and may wipe out any potential bioactive-associated preventive effects.

Toden et al. (2005) in this respect used a more appropriate rat model, which consisted in feeding rats with a western type diet enriched in proteins. They demonstrated that RS fermentation in the colon is beneficial to health in the sense that it helps counteract the deleterious effect induced by a high protein intake. The high protein diet resulted in a twofold increase in damage to colonocyte DNA compared to a low-protein diet. This was associated with thinning of the colonic mucosa barrier and increased level of p-cresol. The addition of

RS (high amylose starch) to the diet increased SCFA and attenuated DNA damages, suggesting protection against genotoxic agent and lesser genotoxocity of the fecal water.

These observations have been corroborated by a recent study that investigated the effects of *in vitro* fermentation products of *in-vitro*-digested or *in-vivo*-digested RS2 and RS3 on Caco-2 cells (Fassler et al., 2007). Compared to control, the cytotoxicity, anti-genotoxicity against hydrogen peroxide (comet assay) and the effect on barrier function measured by trans-epithelial electrical resistance of fermented samples of RS were determined. Batch fermentation of RS resulted in an anti-genotoxic activity ranging from 9–30% decrease in DNA damage for all the samples. Additionally, *in vitro* batch fermentation of RS caused an improvement in integrity across the intestinal barrier by approximately 22% for all the samples.

9.3 Other Starch-Derived Fibers with Potential Prebiotic Effects

Many new starch-derived prebiotic candidates are now available (e.g., Nutriose®, Fibersol-2® and PROMITOR™ Soluble Gluco Fiber).

Made from starch, Nutriose® can be described as a resistant dextrin. During the process of dextrinisation, the starch undergoes a degree of hydrolysis followed by repolymerization. It is this repolymerization step that makes starch become indigestible, due to many α 1,6, α 1,2 and α 1,3 linkages. According to the AOAC method 2001–2003, Nutriose® contains 85% fiber (Lefranc-Millot, 2008). The calorific value of Nutriose® has been reported to be 1.7 kcal/g (Lefranc-Millot, 2008) and is claimed to be consistent with clinical determination in healthy young men (Vermorel et al., 2004) and to be in agreement with the consensual calorific value of soluble dietary fibers (Livesey, 1992). Only 15% is enzymatically digested in the small intestine, while 75% reaches the colon where it is slowly fermented and 10% is excreted in fecal matters (van den Heuvel et al., 2005).

Fibersol-2® is a spray-dried powder produced by a pyrolysis and a controlled enzymatic hydrolysis of cornstarch. It was estimated that most of Fibersol-2 escapes digestion in the upper gastrointestinal tract and that 90% reaches the colon, where half of this fraction is fermented by the microbiota (Flickinger et al., 2000).

9.3.1 Introduction to PROMITOR™ Soluble Gluco Fiber (SGF)

9.3.1.1 Regulatory Status

PROMITOR™ Soluble Gluco Fiber is a regular food ingredient in the EU and can be labeled as "Soluble Gluco Fiber," "Glucose Syrup" or "Dried Glucose Syrup." The name "Soluble Gluco Fiber" is consistent with EU Directive 2000/13/EU as amended on labeling, presentation and advertising of foodstuffs as it indicates both the precise nature of the food in that it is a glucose type of food and distinguishes the fiber content of the glucose syrup. SGF is a non-genetically modified source of dietary fiber, which enables the use of fiber nutrient content claims (contains fiber, source of fiber, high fiber) where relevant regulatory conditions are met.

In the US, PROMITOR™ Soluble Corn Fiber (SGF in EU) is GRAS and can be labeled as "Soluble Corn Fiber," "Corn Syrup," or "Corn Syrup Solids."

9.3.1.2 Dietary Fiber Content

SGF is obtained from a partially hydrolyzed starch-made glucose syrup, using an existing production process that yields approximately 70–85% fiber with exceptional color, clarity and flavor (**◎** *Figure 9.3*).

SGF has a typical analysis of approximately 72% total dietary fiber per AOAC method 2001.03. Highly water soluble fibers such as SGF and some resistant maltodextrins contain digestion-resistant material that is not precipitated by the addition of ethanol as prescribed in the 991.43 method.

In three different human trials SGF has demonstrated a consistent glycemic response, approximately 30% that of rapidly digestible carbohydrates (i.e., glucose, maltodextrin) (Kendall et al., 2009). This correlates well to the amount of digestible carbohydrate (~70% fiber, ~30% digestible carbohydrate) based on the AOAC 2001.03 method.

9.3.1.3 Calorific Value

SGF, has a calorific content of 2 kcal/g (Fastinger et al., 2007). True metabolizable energy content, was determined in an *in vivo* avian model which utilizes bomb calorimetry of the food prior to consumption by the animal and

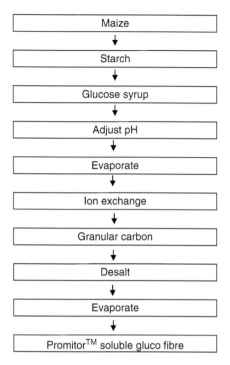

◻ Figure 9.3
Production process of PROMITOR™ Soluble Gluco Fiber.

subsequent bomb calorimetry of the collected waste (urine and feces) after consumption. Additionally, glycemic response studies in humans support this calorific value (Kendall et al., 2007, 2008). Using the calculation (4 kcal/g × 30% digestible), SGF would have 1.2 kcal/g. However, energy yield from fermentation cannot be estimated by the blood glucose response and likely yields a small amount of additional kcals (<1 kcal). *In vitro* analysis has also shown it to be approximately 25% digestible, 50% fermented and 25% unchanged. Using a fermentation value of 2 kcal/g, as suggested by (Oku and Nakamura, 2002), the following calculations yield a calorific value for Soluble Gluco Fiber of 2 kcal/g.

$$
\begin{aligned}
\text{Soluble Gluco Fiber kcal/g} &= (4\ \text{kcal/g}) \times (25\%\ \text{digestible}) \\
&\quad + (2\ \text{kcal/g}) \times (50\%\ \text{fermentable}) \\
&\quad + (0\ \text{kcal/g}) \times (25\%\ \text{unchanged}) \\
&= 2\ \text{kcal/g}
\end{aligned}
$$

9.3.1.4   Digestive Tolerance

The gastrointestinal tolerance has been assessed in human trials and was shown to be well tolerated up to doses as high as 25 g/d. Stewart et al conducted a study in which subjects (n = 20, 10 men, 10 women) consumed 12 g fiber/d SGF for 14 consecutive days (Stewart et al., 2008, 2009). Subjects were asked to report gastrointestinal symptoms (cramping, bloating, stomach noise, flatulence) on a 10 point scale (1 = minimal symptoms, 10 = severe symptoms). SGF was shown to be well tolerated (Kendall et al., 2009; Sanders et al., 2008). In an acute dose-response study (Kendall et al., 2009), tolerance was assessed in 22 healthy volunteers consuming three different doses (5 g, 15 g, 25 g) in acute conditions after an overnight fast. Subjects were asked to report gastrointestinal symptoms (belching, bloating, flatulence, nausea and diarrhoea) on a 100 mm VAS scale at 0, 15, 30, 45, 60, 90 and 120 min after consumption. It was also shown to be well tolerated in these conditions up to 25 g.

9.3.1.5   Applications

SGF is an easy-to-formulate fiber which functions like glucose syrup in most systems. It can be used as a partial or total replacement for regular glucose syrup, other sweeteners, or low calorie bulking agents such as polyols, whilst maintaining texture and mouthfeel.

With 2 kcal/g, it also reduces calories when replacing sugar/glucose syrups (4 kcal/g). It has little to no sweetness and so can be used in combination with high intensity sweeteners to balance sweetness to the level of the standard product (in accordance with the EU legislation on sweeteners use in foodstuffs – Directive EU 94/35). As a result of its high solubility and stability in acidic conditions it is suitable for high solids low pH food matrices such as fruit filling. Its low impact on flavor and viscosity – similar to glucose syrup (❷ *Figure 9.4*) – allows to reduce sugar and calories significantly, while adding fiber which does not affect the food product's organoleptic quality.

Main applications include cereals bars and breakfast cereals, cookies and biscuits, snacks, beverages, yogurts, ice creams, desserts, fruit fillings, sauces, confections and processed meats.

Beverages
SGF is a 100% water soluble, fiber source which does not cause any sedimentation or dramatically increase viscosity as some hydrocolloids do. These negative effects

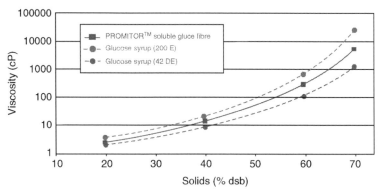

■ Figure 9.4
Viscosity of PROMITOR™ Soluble Gluco Fiber at 20°C.

can occur with other ingredients before or after heat treatment, dependent on the force and the type of shear applied. It also does not create cloudiness or turbidity. As a result, replacing sugar, either partially or totally, has no significant impact on the total "dry matter" or viscosity of the beverage, maintaining the original body and mouthfeel and avoiding the creation of a long or slimy structure.

As the energy is 2 kcal/g, it can help to deliver a significant sugar reduction. The sweetness is about 20% of the sweetness of sucrose, so adding a high intensity sweetener will create the desired taste. As a practical and simple example, replacing 4.5 g of sucrose per 100 mL in flavored and sweetened water with SGF will reduce energy load from 17.2 kcal/100 ml to less than 10 kcal/100 mL whilst adding 3 g fiber/100 mL. It can also be added to an existing beverage formulation and will then enhance its mouthfeel whilst simultaneously improving taste.

SGF is a maize starch derivative with a very bland flavor and does not contain any compounds responsible for off flavors sometimes caused during extraction and other processing stages. Its impact on the flavor of energy reduced drinks is not normally noticeable, with a smooth and non-grainy or powdery perception.

During beverage processing, pasteurization or flash pasteurization is often used to obtain microbiological stability. The combination of a heat treatment together with a low pH (as most beverages are) can damage some ingredients leading to breakdown before and during storage. SGF, however is stable throughout all normal processes in the beverage industry, which means there is no need to "overdose" it in formulations.

Bakery

Cereals bars are mainly composed of glucose syrups which work as a binder to maintain the cohesive structure of the particles of cereal and fruit. It is therefore particularly effective to replace glucose syrups with SGF, which can work as the sole binder thanks to its cohesive properties. Any reduction in sweetness can be compensated by ingredients such as fructose, although in many cases the dried part of the bar provides enough sweetness already. In addition SGF permits to achieve a fiber content of up to 35 g/100 g, corresponding to approximately 10 g/serving.

In moist matrices such as muffins, it decreases the water activity of the system and helps to maintain the stability of the product. In some cases a humectant (e.g., glycerol, sorbitol) can help to maintain the same water activity, ensuring that the product is preserved.

In biscuit-like and soft dough products, the low sweetness and the textural role of sugar, mean that the percentage of SGF incorporated does not generally exceed 50% of the total sugar.

Combining SGF and RS is a simple way to increase fiber content to reach the fiber amount necessary to make a "high fiber" claim, without affecting sensory characteristics of biscuits or muffin-like cakes whose regular flour based equivalents in the market have less than 1.5 g fiber/100 g.

Dairy

Because SGF behaves in a similar way to glucose syrup, it can effectively replace sugar while adding fiber in dairy products. At the same time it can contribute to a richer texture and a similar mouthfeel in low fat or non fat dairy products compared to full fat references.

Dairy processing typically includes heat treatment and homogenization steps which are synonymous to high shear, high temperature conditions. The stability in these harsh conditions makes it ideal for such dairy products.

Completely soluble at acidic and neutral pH, SGF adds texture and fibers with a smooth non sandy or powdery mouthfeel. For instance, it will compensate for the lack of body of a low fat and/or low sugar dairy dessert mousse, while increasing fiber and enhancing sensory characteristics.

It is also suitable for formulating fermented products due to its survivability during fermentation and can be added at the beginning of the process with no loss of fibers. (● *Figure 9.5*).

In fruit preparations, it can replace the sweetener (either partly or totally) without impacting texture, while maintaining the pleasant mouthfeel stemming from its glucose syrup-like viscosity.

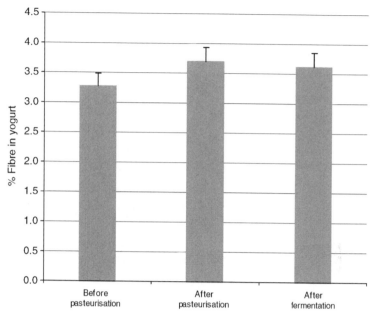

🔲 Figure 9.5

Stability of PROMITOR™ Soluble Gluco Fiber under Yogurt processing.

As SGF demonstrates high stability at low pH (pH < 4), it is particularly suitable for fruit preparations – where typical pH is around 3.8 to ensure good microbial stability during shelf-life – offering shelf-life stability with no loss of fiber content. Any sweetness loss is usually compensated by a high intensity sweetener such as sucralose.

In desserts and ice creams, SGF will add texture while replacing sweeteners (sugar or glucose syrup) and/or some fat, and will improve the nutritional profile of the end product. For instance, it is possible to achieve 30% fat and 60% sugar reductions in ice creams. In such formulations, it will help to keep a creamy taste and a mouthfeel similar to the full fat and sugar alternative.

9.3.2   Growing and Preliminary Evidence on Prebiotic Properties of New Starch-Derived Fibers

Fifteen grams per day Nutriose® intake over 2 weeks has been reported to decrease fecal pH and to reduce significantly *Clostridium perfringens* in humans (Lefranc-Millot, 2008).

Fibersol-2® has been shown to increase both bowel regularity, fecal volume and to favor growth of bifidobacteria in 20 healthy volunteers with a fecal frequency lower than three times a week (Satouchi et al., 1993). In this study, it has been observed that consumption of 3.75 g/d of Fibersol-2® for 5 days increased weekly fecal frequency from 2.6 times to 4.0 times and doubled fecal output. Bifidobacteria counts were also significantly increased. The prebiotic properties of Fibersol-2® have also been demonstrated in dogs (Flickinger et al., 2000). Though, a recent study, during which 38 healthy volunteers consumed 7.5 and 15 g Fibersol-2® over 3 weeks showed that resistant maltodextrin supplementation altered ($p < 0.05$) bacterial populations from baseline to treatment and increased butyrate production, but failed showing a significant effect ($p = 0.12$) on fecal *Bifidobacterium* populations during the treatment period (Fastinger et al., 2008).

SGF has been shown to be a potential prebiotic in *in vitro* studies at TNO (Maathuis et al., 2008). As described previously, the aim of the study was to investigate the effect of five newly developed maize-based fibers on the activity and composition of the microbiota in the colon. The fibers were pre-digested, mono- and di-saccharides were removed, and the residual polymer was used to assess the production of microbial metabolites and changes in composition of the microbiota using a dynamic, validated, *in vitro* model of the large intestine (TIM-2). Microbial metabolites analysis showed an increase in health-promoting metabolites (short-chain fatty acids) and a reduction in toxic metabolites from protein fermentation compared to the poorly-fermentable control (cellulose). Using microarray technology to compare multiple species and groups of colonic microbiota, it was found to promote the growth of beneficial bacteria such as bifidobacteria (⊙ *Figure 9.2*).

Human studies are ongoing to confirm the prebiotic properties of SGF.

9.4 Conclusion

The prebiotic effects of RS are promising. Existing data suggest that RS would be able to positively impact immune response, modulate inflammation, improve mineral absorption and help maintain a good colic function. Preliminary data on PROMITOR™ Resistant Starch suggest similar properties as those that have been reported so far for RS2 and RS3.

New soluble non-viscous fibers like PROMITOR™ Soluble Gluco Fiber are easy to integrate into new or existing formulations without compromising flavor or texture. Very well tolerated and clean tasting, the new generation of fibers can help develop new health-plus versions of products in a wide range of food categories. In particular their low calorific value (0–2 kcal/g), makes them preferable alternatives to high calorific ingredient (sugar 4 kcal/g, fat 9 kcal/g). The fact that they can compensate for a lack of body and texture in many low calorie – sugar or fat reduced – products means that they are frequently used in the dairy and bakery sectors in particular. They have also shown high process and acid stability allowing manufacturers to formulate and guarantee fiber content (and other associated benefits) throughout the entire shelf life of products.

References

Asp NG (1997) Resistant starch – an update on its physiological effects. Adv Exp Med Biol 427:201–210

Bauer-Marinovic M, Florian S, Müller-Schmehl K, Glatt H, Jacobasch G (2006) Dietary resistant starch type 3 prevents tumour induction by 1,2-dimethylhydrazine and alters proliferation, apoptosis and dedifferentiation in rat colon. Carcinogenesis 27:1849–1859

Birkett A, Muir J, Phillips J, Jones G, O'Dea K (1996) Resistant starch lowers fecal concentrations of ammonia and phenols in humans. Am J Clin Nutr 63:766–772

Bouhnik Y, Raskine L, Simoneau G, Vicaut E, Neut C, Flourie B, Brouns F, Bornet FR (2004) The capacity of nondigestible carbohydrates to stimulate fecal bifidobacteria in healthy humans: a double-blind, randomized, placebo-controlled, parallel-group, dose-response relation study. Am J Clin Nutr 80:1658–1664

Brouns F, Arrigoni E, Langkilde AM, Verkooijen I, Fassler C, Andersson H, Kettlitz B, van Nieuwenhoven M, Philipsson H, Amado R (2007) Physiological and metabolic properties of a digestion-resistant maltodextrin, classified as type 3 retrograded resistant starch. J Agric Food Chem 55:1574–1581

Brown I, Warhurst M, Arcot J, Playne M, Illman RJ, Topping DL (1997) Fecal numbers of bifidobacteria are higher in pigs fed Bifidobacterium longum with a high amylose cornstarch than with a low amylose cornstarch. J Nutr 127:1822–1827

Burkitt DP, Trowell HC (1975) Refined carbohydrate foods and disease: some implications of dietary fibre. Academic Press, London, p. 369

Burns AJ, Rowland IR (2000) Anticarcinogenicity of probiotics and prebiotics. Curr Issues Intest Microbiol 1:13–24

Buttriss JL, Stokes CS (2008) Dietary fibre and health: an overview. Nutr Bull 33:186–200

Cassand P, Maziere S, Champ M, Meflah K, Bornet F, Narbonne JF (1997) Effects of resistant starch- and vitamin A-supplemented diets on the promotion of precursor lesions of colon cancer in rats. Nutr Cancer 27:53–59

Cassidy A, Bingham SA, Cummings JH (1994) Starch intake and colorectal cancer risk: an international comparison. Br J Cancer 69:937–942

Crittenden RG, Morris LF, Harvey ML, Tran LT, Mitchell HL, Playne MJ (2001) Selection of a Bifidobacterium strain to complement resistant starch in a synbiotic yoghurt. J Appl Microbiol 90:268–278

Cummings JH, Beatty ER, Kingman SM, Bingham SA, Englyst HN (1996)

Digestion and physiological properties of resistant starch in the human large bowel. Br J Nutr 75:733–747

Dongowski G, Jacobasch G, Schmiedl D (2005) Structural stability and prebiotic properties of resistant starch type 3 increase bile acid turnover and lower secondary bile acid formation. J Agric Food Chem 53:9257–9267

Fassler C, Arrigoni E, Venema K, Brouns F, Amado R (2006) In vitro fermentability of differently digested resistant starch preparations. Mol Nutr Food Res 50:1220–1228

Fassler C, Gill CI, Arrigoni E, Rowland I, Amado R (2007) Fermentation of resistant starches: influence of in vitro models on colon carcinogenesis. Nutr Cancer 58:85–92

Fastinger N, Knapp B, Guevara M, Parsons C, Swanson K, Fahey G (2007) Glycemic response and metabolizable energy content of novel maize-based soluble fibers F4–809, F4–810 and F4–810LS using canine and avian models. FASEB J 21:A744

Fastinger ND, Karr-Lilienthal LK, Spears JK, Swanson KS, Zinn KE, Nava GM, Ohkuma K, Kanahori S, Gordon DT, Fahey GC Jr (2008) A novel resistant maltodextrin alters gastrointestinal tolerance factors, fecal characteristics, and fecal microbiota in healthy adult humans. J Am Coll Nutr 27:356–366

Ferguson LR, Zhu S, Kestell P (2003) Contrasting effects of non-starch polysaccharide and resistant starch-based diets on the disposition and excretion of the food carcinogen, 2-amino-3-methylimidazo[4,5-f]quinoline (IQ), in a rat model. Food Chem Toxicol 41:785–792

Flickinger EA, Wolf BW, Garleb KA, Chow J, Leyer GJ, Johns PW, Fahey GC Jr (2000) Glucose-based oligosaccharides exhibit different in vitro fermentation patterns and affect in vivo apparent nutrient digestibility and microbial populations in dogs. J Nutr 130:1267–1273

Gibson GR, Roberfroid MB (1995) Dietary modulation of the human colonic microbiota: introducing the concept of prebiotics. J Nutr 125:1401–1412

Gray J (2006) Dietary fibre: definition, analysis, physiology & health. ILSI Europe Consise monograph series, pp. 1–44

Grubben MJ, van den Braak CC, Essenberg M, Olthof M, Tangerman A, Katan MB, Nagengast FM (2001) Effect of resistant starch on potential biomarkers for colonic cancer risk in patients with colonic adenomas: a controlled trial. Dig Dis Sci 46:750–756

Hampson DJ, Robertson ID, La T, Oxberry SL, Pethick DW (2000) Influences of diet and vaccination on colonisation of pigs by the intestinal spirochaete Brachyspira (Serpulina) pilosicoli. Vet Microbiol 73:75–84

Heijnen ML, Beynen AC (1997) Consumption of retrograded (RS3) but not uncooked (RS2) resistant starch shifts nitrogen excretion from urine to feces in cannulated piglets. J Nutr 127:1828–1832

Heijnen ML, van Amelsvoort JM, Deurenberg P, Beynen AC (1998) Limited effect of consumption of uncooked (RS2) or retrograded (RS3) resistant starch on putative risk factors for colon cancer in healthy men. Am J Clin Nutr 67:322–331

Hylla S, Gostner A, Dusel G, Anger H, Bartram HP, Christl SU, Kasper H, Scheppach W (1998) Effects of resistant starch on the colon in healthy volunteers: possible implications for cancer prevention. Am J Clin Nutr 67:136–142

Institute of Medicine, Food and Nutrition Board (2006) Dietary Reference Intakes for Energy, Carbohydrates, Fiber, Fat, Fatty acids, Cholesterol, Protein and Amino acids. National Academy of Sciences

Jacobasch G, Dongowski G, Schmiedl D, Muller-Schmehl K (2006) Hydrothermal treatment of Novelose 330 results in high yield of resistant starch type 3 with beneficial prebiotic properties and decreased secondary bile acid formation in rats. Br J Nutr 95:1063–1074

Jacobasch G, Dongowski G, Schmiedl D, Muller-Schmehl K (2006) Hydrothermal treatment of Novelose 330 results in high yield of resistant starch type 3 with beneficial prebiotic properties and decreased secondary bile acid formation in rats. Br J Nutr 95:1063–1074

Jenkins DJ, Vuksan V, Kendall CW, Wursch P, Jeffcoat R, Waring S, Mehling CC, Vidgen E, Augustin LS, Wong E (1998) Physiological effects of resistant starches on fecal bulk, short chain fatty acids, blood lipids and glycemic index. J Am Coll Nutr 17:609–616

Kendall C, Esfahani A, Hoffman A, Evans A, Sanders LJ,AR, Vidgen E, Potter S (2008) Effect of novel maize-based dietary fibers on postprandial glycemia and insulinemia. J Am Coll Nutr 27:711–718

Kendall C, Esfahani A, Sanders L, Potter S, Jenkins D (2009) Resistant starch reduces postprandial glycemic and insulinemic response and increases satiety in humans. FASEB J, in press

Kendall C, Josse A, Potter S, Hoffman A, Jenkins D (2007) Effect of novel maize-based dietary fibers on postprandial glycemia. FASEB J 21:A177

Knapp BK, Parsons CM, Swanson KS, Fahey GC Jr (2008) Physiological responses to novel carbohydrates as assessed using canine and avian models. J Agric Food Chem 56:7999–8006

Le Blay GM, Michel CD, Blottiere HM, Cherbut CJ (2003) Raw potato starch and short-chain fructo-oligosaccharides affect the composition and metabolic activity of rat intestinal microbiota differently depending on the caecocolonic segment involved. J Appl Microbiol 94:312–320

Le Leu RK, Brown IL, Hu Y, Bird AR, Jackson M, Esterman A, Young GP (2005) A synbiotic combination of resistant starch and Bifidobacterium lactis facilitates apoptotic deletion of carcinogen-damaged cells in rat colon. J Nutr 135:996–1001

Lefranc-Millot C (2008) NUTRIOSE® 06: a useful soluble dietary fibre for added nutritional value. Nutr Bull 33:234–239

Lenoir-Wijnkoop I, Sanders ME, Cabana MD, Caglar E, Corthier G, Rayes N, Sherman PM, Timmerman HM, Vaneechoutte M, Van Loo J, Wolvers DA (2007) Probiotic and prebiotic influence beyond the intestinal tract. Nutr Rev 65:469–489

Lesmes U, Beards EJ, Gibson GR, Tuohy KM, Shimoni E (2008) Effects of resistant starch type III polymorphs on human colon microbiota and short chain fatty acids in human gut models. J Agric Food Chem 56:5415–5421

Livesey G (1992) The energy values of dietary fibre and sugar alcohols for man. Nutr Res Rev 5:61–84

Lopez HW, Levrat-Verny MA, Coudray C, Besson C, Krespine V, Messager A, Demigne C, Remesy C (2001) Class 2 resistant starches lower plasma and liver lipids and improve mineral retention in rats. J Nutr 131:1283–1289

Maathuis A, Hoffman A, Evans A, Sanders L, Venema K (2008) Digestibility and prebiotic potential of nondigestible carbohydrate fractions from novel maize-based fibers in a dynamic in vitro model of the human intestine. FASEB J 22:1089–1087

Macfarlane S (2008) Microbial biofilm communities in the gastrointestinal tract. Journal of clinical gastroenterology 42(Suppl. 3) Pt 1:S142–S143

Maki K, Sanders L, Reeves M, Kaden V, Cartwright Y (2009) Effects of resistant starch vs wheat bran on laxation in healthy adults. FASEB J, in press

Marsono Y, Illman RJ, Clarke JM, Trimble RP, Topping DL (1993) Plasma lipids and large bowel volatile fatty acids in pigs fed on white rice, brown rice and rice bran. Br J Nutr 70:503–513

Martin B, Lachcik P, Story J, Weaver C (2009) Calcium absorption, retention and bone density are enhanced by different fibers in male Sprague Dawley rats. FASEB J, in press

Maziere S, Meflah K, Tavan E, Champ M, Narbonne JF, Cassand P (1998) Effect of resistant starch and/or fat-soluble vitamins A and E on the initiation stage of aberrant crypts in rat colon. Nutr Cancer 31:168–177

Muir JG, Yeow EG, Keogh J, Pizzey C, Bird AR, Sharpe K, O'Dea K, Macrae FA (2004) Combining wheat bran with resistant starch has more beneficial effects on fecal indexes than does wheat bran alone. Am J Clin Nutr 79:1020–1028

Oku T, Nakamura S (2002) Digestion, absorption, fermentation, and metabolism of functional sugar substitutes and their available energy. Pure Appl Chem 74:1253–1261

Perrin P, Pierre F, Patry Y, Champ M, Berreur M, Pradal G, Bornet F, Meflah K, Menanteau J (2001) Only fibres promoting a stable butyrate producing colonic ecosystem decrease the rate of aberrant crypt foci in rats. Gut 48:53–61

Phillips J, Muir JG, Birkett A, Lu ZX, Jones GP, O'Dea K, Young GP (1995) Effect of resistant starch on fecal bulk and fermentation-dependent events in humans. Am J Clin Nutr 62:121–130

Rabbani GH, Teka T, Zaman B, Majid N, Khatun M, Fuchs GJ (2001) Clinical studies in persistent diarrhoea: dietary management with green banana or pectin in Bangladeshi children. Gastroenterology 121:554–560

Ramakrishna BS, Subramanian V, Mohan V, Sebastian BK, Young GP, Farthing MJ, Binder HJ (2008) A randomized controlled trial of glucose versus amylase resistant starch hypo-osmolar oral rehydration solution for adult acute dehydrating diarrhoea. PLoS ONE 3:e1587

Ramakrishna BS, Venkataraman S, Srinivasan P, Dash P, Young GP, Binder HJ (2000) Amylase-resistant starch plus oral rehydration solution for cholera. N Engl J Med 342:308–313

Roberfroid M (2007) Prebiotics: the concept revisited. J Nutr 137:830S–837S

Ruemmele FM, Schwartz S, Seidman EG, Dionne S, Levy E, Lentze MJ (2003) Butyrate induced Caco-2 cell apoptosis is mediated via the mitochondrial pathway. Gut 52:94–100

Sakamoto J, Nakaji S, Sugawara K, Iwane S, Munakata A (1996) Comparison of resistant starch with cellulose diet on 1,2-dimethylhydrazine-induced colonic carcinogenesis in rats. Gastroenterology 110:116–120

Sanders L, Kendall C, Maki K, Stewart M, Slavin J, Potter S (2008) A novel maize-based dietary fiber is well tolerated in humans. FASEB J 22:lb761

Satouchi M, Wakabayashi S, Ohkuma K, Fuyuwara K, Matsouka A (1993) Effects of indigestible dextrin on bowelmovements. Jpn J Nutr 51:31–37

Schley PD, Field CJ (2002) The immune-enhancing effects of dietary fibres and prebiotics. Br J Nutr 87(Suppl. 2): S221–S230

Scholz-Ahrens KE, Ade P, Marten B, Weber P, Timm W, Acil Y, Gluer CC, Schrezenmeir J (2007) Prebiotics, probiotics, and synbiotics affect mineral absorption, bone mineral content, and bone structure. J Nutr 137:838S–846S

Silvi S, Rumney CJ, Cresci A, Rowland IR (1999) Resistant starch modifies gut microflora and microbial metabolism in human flora-associated rats inoculated with faeces from Italian and UK donors. J Appl Microbiol 86:521–530

Stewart M, Nikhanj S, Timm D, Thomas W, Slavin J (2009) Four different fibers from maize and tapioca are well tolerated in a placebo-controlled study in humans. FASEB J, in press

Toden S, Bird AR, Topping DL, Conlon MA (2005) Resistant starch attenuates colonic DNA damage induced by higher dietary protein in rats. Nutr Cancer 51:45–51

Topping DL, Fukushima M, Bird AR (2003) Resistant starch as a prebiotic and synbiotic: state of the art. Proc Nutr Soc 62:171–176

Tran CP, Familari M, Parker LM, Whitehead RH, Giraud AS (1998) Short-chain fatty acids inhibit intestinal trefoil factor gene expression in colon cancer cells. Am J Physiol 275:G85–G94

Tuohy KM, Rouzaud GC, Bruck WM, Gibson GR (2005) Modulation of the human gut microflora towards improved health using prebiotics – assessment of efficacy. Curr Pharm Des 11:75–90

van den Heuvel EG, Wils D, Pasman WJ, Saniez MH, Kardinaal AF (2005) Dietary supplementation of different doses of NUTRIOSE FB, a fermentable dextrin, alters the activity of faecal enzymes in healthy men. Eur J Nutr 44:445–451

van Munster IP, Tangerman A, Nagengast FM (1994) Effect of resistant starch on colonic fermentation, bile acid metabolism, and mucosal proliferation. Dig Dis Sci 39:834–842

Vermorel M, Coudray C, Wils D, Sinaud S, Tressol JC, Montaurier C, Vernet J, Brandolini M, Bouteloup-Demange C, Rayssiguier Y (2004) Energy value of a low-digestible carbohydrate, NUTRIOSE FB, and its impact on magnesium, calcium and zinc apparent absorption and retention in healthy young men. Eur J Nutr 43:344–352

Vos AP, M'Rabet L, Stahl B, Boehm G, Garssen J (2007) Immune-modulatory effects and potential working mechanisms of orally applied nondigestible carbohydrates. Crit Rev Immunol 27:97–140

Wang X, Brown IL, Khaled D, Mahoney MC, Evans AJ, Conway PL (2002) Manipulation of colonic bacteria and volatile fatty acid production by dietary high amylose maize (amylomaize) starch granules. J Appl Microbiol 93:390–397

Williamson SL, Kartheuser A, Coaker J, Kooshkghazi MD, Fodde R, Burn J, Mathers JC (1999) Intestinal tumorigenesis in the Apc1638N mouse treated with aspirin and resistant starch for up to 5 months. Carcinogenesis 20:805–810

Younes H, Coudray C, Bellanger J, Demigne C, Rayssiguier Y, Remesy C (2001) Effects of two fermentable carbohydrates (inulin and resistant starch) and their combination on calcium and magnesium balance in rats. Br J Nutr 86:479–485

Young GP, McIntyre A, Albert V, Folino M, Muir JG, Gibson PR (1996) Wheat bran suppresses potato starch–potentiated colorectal tumorigenesis at the aberrant crypt stage in a rat model. Gastroenterology 110:508–514

10 Oligosaccharides Derived from Sucrose

Pierre F. Monsan · Francois Ouarné

10.1   Introduction

Sucrose is a non-reducing disaccharide, consisting of an α-D-glucopyranosyl residue and a β-D-fructofuranosyl residue linked covalently by their respective anomeric carbons (α-D-glucopyranosyl-1,2-β-D-fructofuranoside). It is not just a simple disaccharide, among others: in fact, the energy of its glycosidic bond is higher than that of a usual glycosidic bond. It is equal to 27.6 kJ/mol, which is similar to the energy of a nucleotide-sugar bond as in UDP-glucose or ADP-glucose. This means that sucrose is a protected and activated form of D-glucose (as well as of D-fructose), which plays a key role in the metabolism of plants, for a wide variety of synthesis reactions.

Sucrose is essentially produced by extraction from cane (75% of sugar production is form cane which contains 20% sucrose by weight) and beet (25% production; 15% sucrose by weight). The total production is higher than 120 million metric tons per year. The crystallization-purification process yields a highly pure compound (>99.9%), at a very reasonable price.

Such characteristics make sucrose a very interesting renewable raw material for synthesis reactions, and particularly for the synthesis of prebiotic oligosaccharides, using either fructosyl transferases or glucosyl transferases.

10.2   Fructo-Oligosaccharides or Fructans

Like starch and sucrose, fructans are naturally present in many plants as reserve carbohydrates. Fructans may be involved in the protection of plants from water-related stresses: drought, salt and cold stress.

They can also be produced by a wide range of microorganisms, bacteria, yeast or fungi.

In this chapter, we give an overview of fructan synthesis from sucrose by plants or microorganisms, their use and their beneficial effects on human and animal health.

10.2.1 Structural Diversity of Fructans

Fructans, also named fructooligosaccharides (FOS, Sc-FOS), oligofructoses oligofructans or inulin, are complex carbohydrates found in several plants.

Four major classes of structurally different fructans can be distinguished: inulin, levan, mixed levan, and neoseries:

- *Inulin-type fructans* consist of linear (2-1)-linked β-D-fructofuranosyl units in which the fructofuranosyl units (F) are bound to the β-2,1 position of sucrose or GFn (1^F(1-β-D-fructofuranosyl)$_{n-1}$ sucrose) (❂ *Figure 10.1a*). They are usually found in plant species belonging to the order Asterales, such as chicory and Jerusalem artichoke. The chain length can vary from 2 to a few hundred fructose units depending on the plant (Ritsema and Smeekens, 2003).

   Fructosyl transferases derived from fungi such as *Aureobasidium pullulans* and *Aspergillus niger* also produce 1F type FOS that are mainly composed of 1-kestose (GF2), nystose (GF3) and fructosyl-nystose (GF4). The terms fructooligosaccharides (FOS) or short chain fructooligosaccharides (Sc-FOS) have been used but only for (1^F(1-β-D-fructofuranosyl)$_{n-1}$ sucrose, GFn, $2 < n < 10$) excluding long chain fructans such as inulin.

- *Levans* consist of linear (2–6)-linked β-D-fructofuranosyl units in which fructofuranosyl units (F) are β linked to the 6 position of sucrose (GFn or 6^F(1-β-D-fructofuranosyl)$_{n-1}$ sucrose) with $n = 1$–3,000 possessing minor amounts of β-2,1 branching (❂ *Figure 10.1b*). In plants, they are usually found in grasses but can also be produced by several bacteria, notably *Bacillus polymyxa*, *Bacillus subtilis*, and *Streptococcus mutans*.

- *Mixed levans* are composed of both (2–6)- and (2-1)-linked β-D-fructofuranosyl units. This type of fructan is found in many plant species such as wheat and barley. An example of this type of fructan is the molecule bifurcose (❂ *Figure 10.1c*).

- *Neoseries* consist of linear (2-1)-linked β-D-fructofuranosyl units in which fructofuranosyl units (F) are bound to C6 of the glucose moiety of sucrose (6^G(1-β-D-fructofuranosyl)$_{n-1}$ sucrose, ❂ *Figure 10.1d*). In plants, they can be found in onions.

   In some cases, for example in *Asparagus officinalis*, Shiomi et al. (1979) described inulin neoseries consisting in linear (2-1)-linked β-D-fructofuranosyl units in which the fructofuranosyl units (F) are bound to both C1 of the fructose moiety and C6 of the glucose moiety of sucrose. This results in a fructan polymer with two

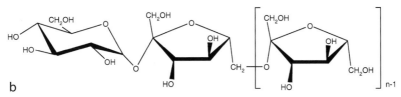

a

Inulin type fructans → 1^F (1-β-D-frutofuranosyl)$_{n-1}$ sucrose

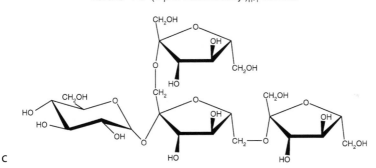

b

Levans → 6^F (1-β-D-frutofuranosyl)$_{n-1}$ sucrose

c

Mixed levan → 6^F(1-β-D-frutofuranosyl)1^F(1-β-D-frutofuranosyl) sucrose or bifurcose

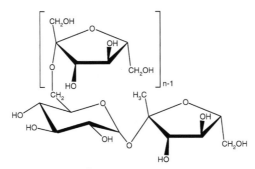

d

Neoseries → 6^G(1-β-D-frutofuranosyl)$_{n-1}$ sucrose

◘ Figure 10.1

Structural representation of different types of fructans (a) inulin type, (b) levan type, (c) Mixed levan type, (d) Neoseries type.

fructan chains on both ends of the sucrose molecule (1^F(1-β-D-fructofuranosyl)$_m$ 6^G(1-β-D-fructofuranosyl)$_{n-1}$ sucrose).

If we consider the definition of dietary fibers: prebiotics or dietary fibers are mainly non-digestible oligosaccharides of natural or synthetic origin. They are based on the monosaccharides glucose, fructose, galactose and mannose with a polymerization degree from 2 to 20 monosaccharide units. Therefore, all the types of fructans described here can be considered as dietary fibres.

The fructans can also be classified as prebiotics. They are characterized by the fact that they are neither hydrolyzed by mammalian digestive enzymes nor absorbed from the gastrointestinal tract by the host animal. Consequently, they can only be fermented in the gastrointestinal tract, which "*beneficially affects the host by selectively stimulating the growth and/or activity of one or a limited number of bacteria in the colon, and thus improves host health*" (Gibson and Roberfroid, 1995). This definition was updated by Roberfroid (2007): "A prebiotic is a selectively fermented ingredient that allows specific changes, both in the composition and/or activity in the gastrointestinal microbiota that confers benefits upon host well-being and health."

The benefits of inulin-type fructans and short chain fruto-oligosaccharides to humans have now been well studied and described for more than two decades. These molecules have been recognized as prebiotics.

Other fructans have been less studied for their prebiotic effects. It has, for instance, been demonstrated that neokestose is also a good promoter for the growth of beneficial bacteria (Kilian et al., 2002).

Levans have also been recognized as non-digestible food fibers. *In vitro* investigation showed that bifidobacteria can use levans as a source of carbon. However, this depends on the degree of polymerization – various researchers have suggested that the maximal molecular weight for a prebiotic effect from levans is 4,500 Da (Marx et al., 2000).

The production yield of fructans using enzymes from plants is low, and mass production of enzyme is quite limited by seasonal conditions. Therefore, industrial production chiefly depends on fungal enzymes, principally from *Aspergillus niger* or *Aureobasidium pullulans*.

10.2.2   Fructans From Plants

Fructans of various types can be found as natural components in honey, fruits and vegetables such as Jerusalem artichoke, banana, chicory, onion, leek, garlic, rye, barley, yacon and salsify. For most of these sources, concentration ranges are between 0.3 and 6% (w/w). For chicory and salsify, they are between 5 and 10% (w/w), while in Jerusalem artichoke and yacon they can reach up to 20% (w/w) (Mussato, 2007).

In plants, fructans are synthesized by the action of two or more fructosyl transferases.

Fructan synthesis starts with the conversion of sucrose to 1-kestose. The enzyme that performs this reaction, sucrose-sucrose 1-fructosyl transferase (1-SST, EC 2.4.1.99) is found in all fructooligosaccharide producing plants. The resulting 1-kestose is, in general, used as a substrate by more species specific fructosyl transferases to synthesize longer and/or more complex fructans.

One of the simplest fructans in plants is linear inulin. This fructan is present in plants belonging to Asterales. In members of the *Liliaceae*, inulin neoseries have been found.

Fructans of the levan type have been found in grasses, where mixed fructan types, also called, Gramminans, can also be present. Gramminans sometimes consist of even more complex structures in which neosugars are combined with branched fructan chains (Ritsema and Smeekens, 2003).

The reason for the variety of structures in plants is unknown; it might be based on different physiological needs or could be the consequence of different evolutionary origins of fructan biosynthesis in different families.

10.2.3   Fructans From Microorganisms

The nomenclature of fructan-producing enzymes remains polemical. Some workers still use the term of β-D-fructofuranosidase (hydrolase numbered EC.3.2.1.26), whereas others designate it as fructosyl transferase (EC.2.4.1.9), concentrating on the nature of transfructosylation of the reaction catalyzed by the enzyme. The reason why many authors have used the name β-D-fructofuranosidase is probably because the transfructosylating activity was first found in an invertase preparation used at high sucrose concentrations. Transfer and hydrolysis reactions are very similar; as hydrolysis is a transfer reaction onto water as acceptor. In each case fructose is transferred from sucrose donor to (i) water acceptor for hydrolysis reaction or (ii) fructan / sucrose for transfer reaction.

Recently different enzymes designated as β-D-fructofuranosidases were compared for their transferase activity (Arrojo et al, 2005). The transferase activity of these enzymes was analysed in detail and the maximum Sc-FOS synthesis yield, expressed as weight percentage of the total amount of carbohydrates in the final mixture, was determined. While the β-fructofuranosidase from *Saccharomyces cerevisiae* gives a maximum yield of 8% fructooligosaccharides, the enzyme from *Aspergillus aculeatus* produces 61% of fructooligosaccharides, close to the value obtained with the enzyme from *Aspergillus niger* ATCC 20611. Some other strains such as *Xanthophyllomyces dendrorhous*, *Schwanniomyces occidentalis and Rhodotorula gracilis* produced 17–19% yield fructooligosaccharide syrups. These results clearly demonstrate that there is no obvious relationship between enzyme name in literature and the corresponding enzyme activity.

These results greatly underline the necessity to characterize an enzyme preparation not only by its name, but also experimentally, and to determine the ratio transferase/hydrolase activity at 50 Brix sucrose (concentration above which this transferase/hydrolase ratio generally does not change).

10.2.3.1 Fructans From Fungi

Many articles on fructan synthesis have been published and a lot of fructooligosaccharide producing fungi studied for their potential application to short-chain fructooligosaccharide (Sc-FOS) production.

Generally, fructosyl transferases from fungi produce 1F-type FOS with a high regiospecificity. Using sucrose as substrate, these enzymes are able to transfer fructofuranosyl groups from sucrose as donor to yield the corresponding series of Sc-FOS: 1-kestose (GF2), nystose (GF3) and fructosyl-nystose (GF4).

The first fungus reported to achieve a high yield of Sc-FOS production was described in by Hidaka et al. (1987). Using this enzyme, from *Aspergillus niger* ATCC 20611, the maximum Sc-FOS conversion from sucrose reached 55–60% (w/w) with respect to total sugar. Rapidly Hidaka et al. (1988) and Hirayama et al. (1988) fully characterized this enzyme and developed it in Japan for an industrial production of Sc-FOS syrup Neosugar®, Meiji Seika Co.).

At the same time, Jung et al. (1987) found another fungal enzyme for Sc-FOS production using *Aureobasidium pullulans*. This enzyme gave rise to another industrial process (Cheil Foods and Chemicals Co. in Korea), well described by Yun et al. (1990, 1992).

Rapidly (Su et al. 1991) discovered another fungus producing a transfructosylating enzyme for fructooligosaccharide synthesis: *Aspergillus japonicus* isolated from a natural habitat was found to produce a significant amount of fructosyl transferase with a good potential for industrial production of fructooligosaccharides. The fructooligosaccharide yield from initial sucrose was about 60%.

Fructosyl transferase activity was also found in a commercial fruit juice processing enzyme preparation (Pectinex Ultra SPL, Novo Nordisk) from *Aspergillus foetidus* (Hang and Woodmans, 1996). The conditions affecting the enzymatic production of high fructooligosaccharide content syrup from sucrose were investigated. As for the enzymes previously described, Pectinex Ultra SPL showed a high transferase/hydrolase activity ratio. The enzyme can convert a 50 Brix sucrose syrup to a 55–60% (on total sugar) fructooligosaccharide syrup conferring a great potential for industrial production.

The synthesis of fructooligosaccharides using whole *Penicillium citrinum* at high sucrose concentration has also been investigated (Lee and Shinohara, 2001). In this case, both 1-kestose and neokestose were produced. However, no reaction product was obtained from neofructooligosaccharides. The oligosaccharides produced were 1-kestose, neokestose, nystose and fructosyl-nystose. Based on these experimental results, a hypothetical reaction pathway was proposed to illustrate how neofructooligosaccharides are formed from 1-kestose.

Penicillium citrinum KCTC 10225BP of soil origin was also described to produce both 1-kestose and neokestose oligosaccharide series, in which the degree of polymerization of each type was reported to be 3–6 (Han et al., 2003).

In Katapodis et al. (2003), production of β-fructofuranosidase from the thermophilic fungus *Sporotrichum thermophile* was studied. This enzyme was optimally active at 60°C. The optimal pH described for transfructosylation was 6.0. The major sugar produced by the transfructosylating activity of the enzyme was 1-kestose. The optimum initial sucrose concentration of 250 g/l allowed the production of only 12.5 g FOS/l.

More recently, Sangeetha et al. (2004), described the production of 1-F type fructooligosaccharides from *Aspergillus oryzae* CFR 202 and *Aureobasidium pulullans* CFR 77. The conversion yield was limited to 55–60% as the glucose released during the enzymatic reaction acts as a competitive inhibitor. These studies gave rise to a US patent application (Ramesh et al., 2005) for the preparation of prebiotics.

The use of recombinant *Aspergillus niger* enzyme was also described (Zuccaro et al., 2008). The combination of sucrose analogs as novel substrates with highly active recombinant *Aspergillus niger* enzyme provides a new and powerful tool for the efficient preparative synthesis of tailor-made saccharides. Molecules of the important 1-kestose and nystose types, headed with different modified monosaccharides of interest were prepared.

10.2.3.2   Fructans From Yeast

The recent increase in the use of prebiotic oligosaccharides in the food industry has lead to the search for new microorganisms and enzymes for their production.

Maugeri and Hernalsteens et al. (2007) screened yeast obtained from fruits and flowers from the Brazilian tropical forest, capable of secreting extracellular enzymes with high fructosyl transferase activity. The screening and isolation procedure resulted in the isolation of one potentially interesting yeast strain, *Rhodotorula spp.* (LEB-V10). This enzyme showed no hydrolytic activity with respect to the Sc-FOS produced and a conversion yield from sucrose of 50% (w/w).

Inulinase from *Kluyveromyces marxianus* was also studied for fructo-oligosaccharide production (Santos and Maugeri, 2007). This enzyme is able to specifically produce 1-F fructooligosaccharides. In this work, experimental factorial design was applied to optimize the fructooligosaccharide synthesis. The variables studied were temperature, pH, sucrose concentration and enzyme concentration. Only temperature and sucrose concentration were shown to be significant parameters. In this case, the maximum conversion yield from sucrose was only 10%.

Other yeast such as *Schanniomyces occidentalis* has been described for fructosyl transferase activity yielding the prebiotic 6-kestose (Álvaro-Benito et al., 2007). However, the main reaction catalyzed by this enzyme is sucrose hydrolysis.

10.2.3.3   Fructans From Bacteria

Bacterial Levansucrase

Fructan-producing bacteria can be found in a wide range of taxa, including plant pathogens and bacteria present in oral and gut flora of animals and humans. In general, bacteria produce levan-type fructan molecules consisting mainly of β, 2–6 linked fructosyl residues, occasionally containing β, 2–1 linked branches.

These polymers are found in many plants and microbial products and are useful as emulsifying and thickening agents in the food industry. Plant levans have a shorter chain length (about 100 fructofuranosyl residues) than microbial levans that contain up to three million residues.

Bacterial levans are synthesized extracellularly by a single enzyme, levansucrase, which produces levan directly from sucrose. Examples of bacterial genera in which fructan-producing strains can be found are *Bacillus, Aerobacter, Streptococcus and Zymomonas.*

Levansucrase (sucrose:2,6-β-D-fructan 6-β-D-fructosyl transferase; E.C. 2.4.1.10) secreted by *Bacillus subtilis* after induction by sucrose was purified by Dedonder (1966). Since, it has been extensively investigated. Detailed enzymological studies by Chambert and Gonzy-Treboul (1976) showed that the enzyme obeys a ping-pong mechanism, which involves a covalent fructofuranosyl-enzyme intermediate that has been isolated.

Production of short-chain levans (fructooligosaccharides) by levansucrase from *Bacillus subtilis* C4 was also investigated (Euzenat et al., 1991).

More recently, according to the potential of short-chain levans to be used by bifidobacteria (Marx et al., 2000), several studies on the use of levansucrase for fructooligosaccharide production have been reported.

As in previous work by Euzenat, Ahmed et al. (2005) described the production and use of extracellular levansucrase from *Bacillus subtillis*. With partially purified enzyme, in optimal conditions, conversion of sucrose to levan reached 84%. Sucrose concentration is the most effective factor controlling the molecular weight of the synthesized levan. The molecular weight of levan decreased from 60 to 0.5 KDa with increasing sucrose concentration from 2.5 to 40%.

Byun et al. (2007) optimized the conditions for the formulation of fructooligosaccharides from sucrose by a transfructosylation reaction using levansucrase from *Pseudomonas aurantiaca* and *Zymomonas mobilis*. As with the other bacteria, formation of fructooligosaccharides by levansucrases of both origins was sucrose concentration dependent. The optimum initial sucrose concentration for the formation of short-chain fructooligosaccharides (DP 3–6) was 70% (w/w). Under these conditions, yields of levan-type fructooligosaccharides were 24–26%. These effects of sucrose concentration on molecular weight are in agreement with those reported by numerous authors.

Modification of ionic strength can also modify chain length. Under conditions of high ionic strength, only levan with an average DP of 120 was synthesized with a reasonable yield by Tanaka et al. (1979). It was suggested in this work that the DP of the levan is generally regulated by ionic strength.

Another method for producing short-chain levan from long-chain levan, consists of hydrolyzing the molecule by levanase from fungus to give levan octaose (DP8, Lizuka et al., 1995).

Bacterial Fructosyl Transferase

Some bacteria such as *Bacillus macerans* produce fructosyl transferase. The crude enzyme unexpectedly selectively synthesized GF5 and GF6 fructooligosaccharides whereas purified enzyme produced mainly 1-kestose and nystose as do other transfructosylating enzymes (Park et al., 2001).

A process for fructooligosaccharide production from sucrose using extra-cellular enzyme from *Zymomonas mobilis* has also been developed (Bekers et al., 2004). The fructan syrup obtained contained 45–48% (w/w) of fructans with respect to total sugar (1-kestose, 6 kestose, neokestose and nystose).

10.2.4   Kinetic Modeling of Fructooligosaccharide Synthesis

Fructosyl transferases catalyze the transfer of fructofuranosyl residues from sucrose to various acceptor substrates. The synthesis involves group transfer without intervention of a cofactor. Two different reactions area catalyzed:

- Transglycosylation using the growing fructan chain (oligosaccharide synthesis) as the acceptor substrate.
- Hydrolysis of sucrose and/or oligosaccharides, when water is used as an acceptor.

Fructosyl transferases belong to Glycoside Hydrolase family 68. They are β-retaining enzymes, employing a double-displacement mechanism that involves formation and subsequent hydrolysis of a covalent glycosyl-enzyme intermediate (a ping-pong type mechanism, Chambert et al., 1974).

For short chain fructooligosaccharides (Sc-FOS), different kinetic models of fructooligosaccharide synthesis have been studied for *Aureobasidium pullulans* (Jung et al., 1989), *Aspergillus niger* (Ouarne and Guibert, 1992) and more recently, *Rhodotorula spp.* (Alvarado and Maugeri, 2007). For these enzymes, during the time course of the reaction, the medium contains unreacted sucrose, glucose from the sucrose used as fructosyl donor and synthesis products, GF2, GF3 and GF4. The presence of fructose in very limited concentrations can also be noted.

In these studies, the initial reaction rate was determined independently by varying the concentrations of sucrose, GF2, GF3 and GF4. The initial inhibitory effect was also studied by adding glucose to pure substrate solutions (sucrose, GF2, GF3 and GF4) at different concentrations.

Different mathematical models have been proposed and found to fit well with experimental data. All authors agree to describe the chain of reactions below:

$$GF + GF \rightarrow GF_2 + G$$

$$GF_2 + GF_2 \rightarrow GF_3 + GF$$

$$GF_3 + GF_3 \rightarrow GF_4 + GF2$$

$$GF + G \rightarrow G + GF$$

The formation of fructooligosaccharides resulted from a consecutive set of transfer reactions. For sucrose, the enzyme kinetics was characteristic of a Michaelis-Menten type with inhibition at high substrate concentrations.

In addition to these basic equations, the fructosyl transfer (equation below) from GFn to G should also be considered. For all oligosaccharides including sucrose, transfructosylation was found to be inhibited competitively by glucose:

$$GF_{n(1<n<4)} + G \rightarrow GF_{n-1} + GF$$

GFn hydrolysis was also described:

$$GF_{n(1<n<4)} + H_2O \rightarrow GF_{n-1} + G$$

In particular, the enzymes from *Aureobasidium pullulans* and *Aspergillus niger* have a high regiospecificity. They selectively transfer the fructosyl moiety of sucrose to the 1-OH fructofuranoside of the other sucrose molecules resulting in the formation of very predominantly 1-kestose based Sc-FOS.

10.2.5 Process

Recent developments in industrial enzymology have made possible the large-scale production of Sc-FOS by enzymatic synthesis. It appears that industrial production of Sc-FOS by enzymatic synthesis can be divided into two types of processes. The first one is the batch system using soluble or immobilized enzyme and the second one the continuous process using immobilized enzyme or whole cells.

Industrial processes for the synthesis of fructooligosaccharides were studied using enzymes with high transfructosylating activity. The best enzymes described were from fungi such as *Aspergillus niger, Aspergillus japonicus* and *Aureobasidium pullulans*. All produced Sc-FOS.

The industrial processes for fructooligosaccharide synthesis are schematized in ❯ *Figure 10.2*. The steps are described in detail below.

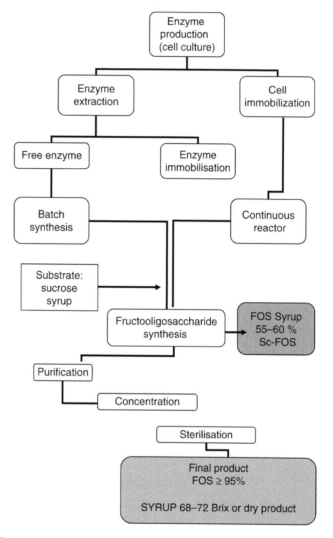

■ Figure 10.2

Flow chart of typical process of Sc-FOS production, by free enzyme, immobilized enzyme or immobilized cells.

10.2.5.1   Enzyme Production

For some fungal strains, production of fructosyl transferase can be by aerobic submerged culture, by fluid-bed culture, culture broth or semi-solid culture medium. Fermentation parameters (temperature, pH, aeration, agitation) should be established for each microorganism. Generally, sucrose is the best carbon source for both cell growth and enzyme activity production. The optimum temperature for growth is often around 30°C and the pH above 5.5.

The enzymes thus obtained are cell bound and require complex operations for separating the enzyme from the mycelium. At the end of the culture period, cells are easily harvested by a basket or continuous centrifuge, and the enzyme extracted.

Cells can also be directly immobilized for Sc-FOS production.

Non cell-bound enzyme can be obtained using semi-solid *Aspergillus niger* 489 culture medium, with the addition of cereal bran. The enzyme thus obtained was easily separated from the cells without any extraction step (Park and Pastore, 1998).

The example of *Aspergillus oryzae* grown in liquid medium with sucrose and yeast extract was also described (Rao et al., 2005) for soluble extracellular enzyme production – a process that is easy to carry out and that is of low cost.

Using an optimized medium, *Aspergillus oryzae* produces round pellets which were stable during the fermentation period (Sangeetha et al., 2005a).

For industrial scale fermentation, these pellets consisting of compact masses of hyphae are advantageous, since the filamentous form of fungi may wrap around the impeller and damage the agitator blade. Also pellet formation makes downstream processing easier in industrial fermentation.

10.2.5.2 Enzyme Extraction

At the end of the log-phase period, cells are collected by centrifugation and washed twice with deionized water, physiological saline solution or buffer. For mycelium-bound enzyme, washed cells are resuspended in water or buffer before enzyme extraction.

Different methods can be used for enzyme extraction:

- Ultrasonication
- Lysozyme
- Cell grinding

An additional centrifugation or filtration step is then needed to produce clear supernatant containing free enzyme. In some cases (Yun and Song, 1999), the enzyme solution thus obtained required an additional concentration step by ultrafiltration or dialysis (molecular weight cutoff 10 KDa). The resulting enzyme is used crude without further purification.

10.2.5.3 Substrates

Sucrose is the natural substrate of fructosyl transferase.

The effect of the sucrose concentration was investigated to minimize the hydrolysis reaction. The maximum yield of Sc-FOS is currently obtained with an initial sucrose concentration of 55 Brix.

Moreover, there is a consensus that a high concentration of commercial food grade sucrose syrup (60–70 Brix) should be used for both the batch and continuous process of fructooligosaccharide production. At such high concentrations, sucrose syrups have the benefit of a low water activity reducing the risk of contamination during the enzymatic synthesis step and reducing the final evaporation costs.

Commercial food grade Sc-FOS is being produced from pure food grade sucrose.

In order to use Sc-FOS as animal feed additive, it would be necessary to reduce production costs. With batch production based on free enzyme or cells, the use of molasses was described to produce 166 g/l fructooligosaccharides from 360 g/l molasses sugar (Shin et al., 2004). Raw sugar can also be used.

10.2.5.4 Cell Immobilization

When enzymes are cell-bound, a method in which microorganisms are directly immobilized on a carrier can be used for Sc-FOS synthesis. This method does not require separation of enzyme from cells and can prevent the reduction of enzyme activity during the extraction step.

Meiji Seika Co., which was the first to produce Sc-FOSs, used initially a continuous process involving immobilized *Aspergillus niger* cells entrapped in calcium alginate beads.

Yun and Song (1999) described *Aureobasidium pullulans* cell immobilization with calcium alginate. Cells were suspended with 3% sodium alginate solution and extruded through syringe needles to form small beads dropping into calcium chloride solution.

Cheil Foods and Chemicals Co. also developed a continuous process using immobilized cells of *Aureobasidium pullulans* entrapped in calcium alginate beads. Two one-cubic meter packed bed reactors have been in commercial operation since 1990. With concentrated sucrose as substrate, the stability of the immobilized cells is about 3 months at 50°C (Yun, 1996).

β-fructofuranosidase from *Aspergillus japonicus* mycelium was also immobilized by entrapment in calcium alginate gel (Cheng et al., 1996). A packed bed reactor was employed for the production of fructooligosaccharides at 42°C. The reaction was continued for 35 days and only 17% of enzyme activity was lost during this period.

10.2.5.5 Enzyme Immobilization

A column reactor filled with immobilized cells, is hard to operate at high flow rate due to a diffusional limitation of substrate and products within the calcium alginate gel matrix. The immobilized enzyme column is essentially superior to the immobilized cell column from the practical point of view:

- The immobilization process is quite simple.
- The unit volumetric immobilized activity is higher.
- There are fewer diffusional limitations.

For these reasons, a lot of studies on fructosyl transferase immobilization have been done. Several of adsorbents, ion exchangers and carriers for covalent enzyme linkage have been studied for the immobilization of fructosyl transferase.

The following examples are illustrative of some of the relevant studies for industrial immobilization of fungal fructosyl transferase.

Yun and Song (1999) described *Aureobasidium pullulans* fructosyl transferase immobilization process on a highly porous anion exchange resin (Diaion HPA25, Samyang Co., Korea). The support consisted of styrene-divinylbenzene polymer with quaternary alkylamine as functional group. Immobilization was very simple: after intracellular enzyme extraction with lysozyme, soluble enzyme was passed through a column containing the porous ion exchange resin (10 h at room temperature). After immobilization, only washing was done. In order to simplify the immobilization procedure, no activation of the support was carried out before or after immobilization.

Kono et al. (1994) patented for Meiji Seika Co. a fructose transferring enzyme immobilization on a granular carrier with or without crosslinking agent. Fourteen carriers were tested and in accordance with the results obtained, anion exchange resin which carrying primary to quaternary amines as functional groups were defined to be essential for immobilization. Immobilization resulted in 63–100% relative activity.

More recently, another industrial immobilization process was described for *Aureobasidium pullulans* fructosyl-transferase (Vankova et al., 2008).

As in the previous case, an anion-exchanger suitable for food applications (Dowex Marathon MSA resin, Dow Chemical, USA) which is a styrene-divinyl-benzene matrix with quaternary amine as a functional group was used. For enzyme extraction, the cells were disrupted in a bead mill and the debris removed in a centrifuge. In order to increase the efficiency of the immobilization process, the low molecular weight solutes present in the enzyme solution were removed by ultra-filtration. For immobilization, resin particles were simply mixed gently with free purified enzyme, for 12 h at 12°C and washed to eliminate unbound compounds. Immobilization resulted in 65% relative activity.

Pectinex ultra SP-L (a commercial preparation from Novozymes) containing fructosyl transferase activity, was covalently immobilized onto Eupergit C (Tanriseven and Asla, 2005). Immobilization resulted in 96% relative activity and immobilized enzyme retained its activity for 20 days without any decrease. This enzyme was also covalently immobilized on an industrial polymer polymethacrylate-based (Sepabeads EC) activated with epoxy groups. The influence of pore volume and average pore size on biocatalyst performance was studied. Several parameters that affect immobilization, such as buffer concentration, pH and amount (mg) of protein added per gram of support were analyzed. Authors found that Pectinex Ultra SP-L can be efficiently immobilized on these supports without adding any external salt or buffer (Ghazi et al., 2005).

10.2.5.6   Fructooligosaccharide Synthesis

Two enzymatic reaction systems were described for industrial fructooligosaccharide (Sc-FOS) production:

- A batch system with soluble fructosyl transferase or cells containing fructosyl transferase activity.
- A continuous system using immobilized cells entrapped in calcium alginate gel or immobilized enzyme on insoluble carrier.

Batch Production

For batch production of fructooligosaccharides, enzymes are suspended in high sucrose concentration syrup (55–70 Brix) and incubated at 55–60°C for

20–25 h with gentle agitation. From a practical standpoint, the batch production with free enzyme needs an additional process to destroy and remove the residual enzymes or cells after conversion. This process results in high costs for the production of a unit amount of Sc-FOS, therefore, a continuous process employing immobilized enzyme or cells is a better alternative.

The use of *Aspergillus oryzae* mycelium pellets was also described (Sangeetha et al., 2005a) for batch production of fructooligosaccharides. The pellets, incubated with 60% sucrose solution at 55°C, gave a Sc-FOS yield of 53%. This Sc-FOS yield was maintained up to the sixth cycle. However, it was not possible to recycle the pellets beyond the sixth cycle due to their disintegration. Compared to conventional batch production, the system was described as advantageous and economical because it does not require supplementation of any additional nutrients or the use of fresh inocula.

In the process of Sc-FOS production with fructosyl transferase, the main problem is the generation of by-products, principally glucose and residual sucrose. At the end of the production step, the maximum Sc-FOS content is known to be 55–60% per unit dry substance. Some authors have focused on searching for an economical method for the production of higher Sc-FOS contents.

- Jung et al. (1993) examined a mixed-enzyme system of fructosyl transferase and glucose oxidase. The mixed enzyme reaction was carried out in a stirred tank reactor containing, sucrose, fructosyl transferase and glucose oxidase. The net Sc-FOS content in the commercial syrup was limited to only 55–60% due to the accumulation of glucose which acts as an inhibitor of the fructosyl transferase. By supplementing glucose oxidase to the conventional Sc-FOS reaction system, it was possible to convert the glucose into gluconic acid readily separable from neutral sugars by simple ion exchange operation. By reducing or eliminating the glucose produced during fructooligosaccharide synthesis, the conversion of fructooligo-saccharides was enhanced to reach up to 90%. This process was only described for batch reactions.
- Another process with simultaneous removal of glucose was described with free purified enzyme (Nizikawa et al., 2001). A membrane reactor system was developed using a nanofiltration membrane through which glucose permeated but sucrose and fructooligosaccharides did not. Percentage fructooligosaccharide reaction product was thus increased to above 90%, a yield comparable to that of the product obtained following chromatographic separation.

Continuous Production

From the standpoint of production cost, a continuous process using an immo-bilized enzyme or an immobilized microorganism is preferred to a batch process with native enzyme.

For continuous production of fructooligosaccharides, particles of immobi-lized biocatalyst were packed into fixed bed reactors. Industrial reactor volumes are conventionally between 1 and 2 m^3. To avoid channeling problems in the reactor, column operation is facilitated by the upward flow of substrate, in this way, compression of the immobilized enzyme is minimized (Yun and Song, 1999). The reactor temperature is kept constant by circulating hot water through a jacket.

In case of fructosyl transferase from *Aureobasidium pullulans*, for cell immo-bilization fixed bed reactors, the initial flow rate at 50°C is about 0.15–0.3 BVH and 1.5 to 2 BVH for immobilized enzyme fixed bed reactors (Yun and Song, 1999). Authors concluded that immobilized enzyme column is essentially superi-or to the immobilized cell column

For immobilized fructosyl transferase from *Aureobasidium pullulans* CCY 27-1–94, the initial flow rate described (Vankova et al., 2008) was about 0.8 BVH for a constant temperature of 50°C.

10.2.5.7   Fructooligosaccharide Purification

Considering a maximum Sc-FOS content for fructosyl transferase, enzyme pro-cesses are known to yield 55–60% on a total sugar basis. Numerous studies have focused on searching for an economical method for the production of high Sc-FOS levels.

In 1984, the separation of Sc-FOS from residual sucrose and monosacchar-ides was performed by Meiji Seika Co by simulated moving bed chromatography (SMB). But as would be expected, the recovery yield of this process was far too low to be applicable to the production scale (Yun, 1996).

Over the last two decades there has been remarkable progress toward the understanding of the phenomenon of chromatographic separation. This has expanded the field of application of this technology and led to major improve-ments in process control and column design. More recently, Vankova et al. (2008) described an optimized simulated moving bed separation for fructooligosacchar-ide purification. A cation exchange adsorbent tailored for saccharide separation (Amberlite™ 1320 Ca, Rhom and Hass Company, France) which is based on a poly(styrene-codivinylbenzene) matrix with a functional group – $(SO_3)_2Ca^{2+}$, was used. The Sc-FOS solution obtained contains less than 5% monosaccharides and disaccharides.

The separation yield of Sc-FOSs in recent SMB processes is 95%. This type of separation allows high yield at low capital investment and running costs.

The potential of cross flow nanofiltration has been assessed in the purification of oligosaccharide from mixtures containing contaminant mono and disaccharides (Nobre et al., 2006). The work showed the concentration of the high molecular weight sugar fraction from a Sc-FOS mixture obtained by fermentation. Fractionation of sugar was studied using 0.5 KDa membranes. The retention yields obtained for the oligosaccharides were 27 and 8% for mono and disaccharides with a volume concentration ratio of 3.6 as compared to the initial conditions. The fraction of Sc-FOS was concentrated during the separation process from 52 to 72%. To be applied for fractionation of oligosaccharides on a large scale, this process requires further optimization.

10.2.5.8 Concentration

For concentration, evaporation of the reaction products is performed by a traditional process for sucrose concentration.

For the final dry product, spray drying of purified syrup can also be carried out.

10.2.5.9 Sterilization

In the sterilization procedure, heat, ultraviolet radiation or sterilizing filtration can be used. To avoid coloring the reaction products, high-temperature sterilization is not recommended.

10.2.6 Physicochemical Properties

Short chain fructooligosaccharides are water soluble, non-viscous and easy-to-use ingredients. Indeed, their technological properties are similar to those of sucrose. They also enhance aromas and mask the off-flavors coming from intense sweeteners.

Fructooligosaccharides have a nice, clean sweet taste typically 0.3–0.6 times as sweet as sucrose depending on the chain length – sweetness decreases with increasing chain length.

The relatively low sweetness makes oligosaccharides useful in food production when a bulking agent with reduced sweetness is desirable to enhance other food flavors.

They are stable in the pH range of 5.0–10.0 and thermally stable at food processing temperatures. Nevertheless, as a whole, when the pH falls below 4.0 and treatment temperature is high, they can become hydrolyzed.

Oligosaccharides can also be used to alter the freezing temperature of frozen food and to control the intensity of browning due to the Maillard reaction in heat-processed food. They also provide high moisture retaining activity preventing excessive drying (Mussato and Mancilha, 2007).

The caloric value of value of purified fructooligosaccharides (%Sc-FOS > 95%) has been estimated to be 1.5–2.0 kCal/g. This is approximately 40–50% the caloric value of digestible carbohydrates such as sucrose.

10.2.7 Functional Properties of Fructooligosaccharides

Fructooligosaccharides are prebiotic agents. The dietary modulation of the gut microbiota by prebiotics is designed to improve health by stimulating the numbers and/or activities of bifidobacteria and lactobacilli.

For a dietary substance to be classified as a prebiotic, at least three criteria are required:

- It must not be hydrolyzed or absorbed in the stomach or the small intestine.
- It must be selective for beneficial commensal bacteria in the colon such as bifido-bacteria.
- Fermentation of the substrate should give beneficial luminal/systemic effects within the host.

The human digestive system plays a crucial role in overall body health. Far from being a purely energy absorbing system, the gut is sometimes compared to a second brain for is high number of nerve cells, whose operating system is very similar to the brain. The colonic ecosystem with billions of bacteria is also recognized to have major, immunological, hormonal, and physiological roles influencing many aspects of general health.

In the past decades, numerous *in vitro*, clinical works have done and reviews have detailed the beneficial effects of non-digestible oligosaccharides on animal and human health (Gibson, 2004; Grizard and Bartomeuf, 1999; Hirayama, 2002;

Mussato and Mancilha, 2007; Sangeetha et al., 2005b; Swennen et al., 2006). The interactions between dietary carbohydrates reaching the colon and microbiota have proved to be of a great importance.

After few days, the use of fructooligosaccharides in human or animal diets leads to changes in the balance of colonic bacteria in a favorable manner. Interestingly, one of the main potential benefits of increasing fecal bifidobacteria is that the bifidobacteria could maintain potentially pathogenic bacteria such as *Escherichia coli* and clostridia at low level.

For the metabolism of non-digestible oligosaccharides, bacteria produce glycolytic enzymes which hydrolyze them into mono and disaccharides. The sugars can then be transported into the cell where they can be further metabolized to short chain fatty acids (SCFAs), in particular, acetate, butyrate and propionate with a resulting acidification of the colon environment.

This pH decrease is beneficial for the development of bacteria able to metabolize Sc-FOS such as bifidobacteria or lactobacilli and detrimental to the growth of the species potentially involved in bowel diseases.

The SCFAs produced are rapidly absorbed by the hindgut and provide energy to colon cells, but are also important regulators of many physiological processes influencing gut health.

As hypothesized by various authors:

- Butyrate particularly plays a key role in the large intestine mucosa maintenance and the regulation of the epithelial cell renewal.
- Propionate is metabolized in the liver influencing hepatic gluconeogenesis, inhibiting ureagenesis and influencing cholesterol production.
- Acetate, the major SCFA in the gut, is directly used by muscle and influences cholesterol production.

The relative proportions of the different SCFAs depends on the type of carbohydrate entering into the colon. It was described, from *in vitro* cultures, animal experiments or gut models that Sc-FOS will lead to the production of high levels of butyrate. This influences the mucosa depth by increasing cell differentiation at the bottom of the villi and reducing apoptosis at their apex. As mentioned above, this mucosa is the first barrier against pathogens arriving from the gut content, and thus its integrity and thickness are of great importance.

This colonic ecosystem and gut component modifications lead to several benefits for the host. Physiological functions are classified into three types:

- The primary function, approved by FOSHU in Japan (Food of Specified Health Use) in 1993, encourages good gastrointestinal conditions including normal stool frequency, less constipation, healthy intestinal microbiota.
- The second function, approved by FOSHU in 2000, is related to better mineral absorption, including an increase in bone density and relief of anemia.
- The third concern involves immunomodulation, such as allergy, cancer prevention, effect on lipid metabolism. Studies on these functions are in progress, and FOSHU should approve some of them in the near future.

Unfortunately, some undesirable effects have also been described but the purpose of this section is to summarize the role of non-digestible oligosaccharides in promoting health and treating diseases so these negative aspects will not be dealt with here.

10.2.7.1   Improved Gastrointestinal Condition

As described previously, the metabolic end products of non-digestible oligosaccharides (NDOs) in the gut are SCFAs. SCFAs are efficiently absorbed by colonic epithelial cells, stimulating their growth as well as water and salt absorption, increasing the humidity of the fecal mass through osmotic pressure. This consequently improves the intestinal mobility allowing reduction in constipation.

Inhibition of diarrhea was also observed with fructooligosaccharide diets especially when diarrhea is associated with intestinal infection. *In vitro* studies confirm the potential indirect protective effect of fructans via bifidobacteria. It has been reported that bifidobacteria may secrete a bacteriocin-type substance against *E. coli*, *Salmonella*, *Listeria*, *Camylobacter*, *Shigella* and *Vibrio cholerae*. These results are in accordance with the fact that Sc-FOS reduces susceptibility to *Salmonella* colonization in chickens.

A decrease in putrefactive substances with an increase of bifidobacterial ratio and SCFA amount was also measured. As a practical application, Sc-FOS has been added to the feed of livestock and pet animals to decrease fecal odor.

10.2.7.2   Effect on Mineral Absorption

Several mechanisms have been postulated for increased mineral absorption induced by Sc-FOS.

SCFAs produced by bifidobacteria lead to a reduction of lumen colonic pH, increasing calcium and magnesium solubility, mucosal surface production and absorptive capacity of the epithelium.

It is also likely that NDOs increase the water content in the colon and thus increase the solubility of some minerals.

This has been confirmed by a long-term beneficial effect of bone mineral content in growing rats or prevention of bone loss in ovarectomized rats. The addition of 5% fructooligosaccharides prevented bone loss significantly in femur and lumbar vertebra in the presence of 1% calcium. At 0.5% calcium in diet, 10% Sc-FOS was necessary to significantly increase bone mineralization.

Improved calcium and magnesium absorption in adolescents and/or menopausal women has been demonstrated for Sc-FOS.

A better absorption of iron was also demonstrated in rats, this supports the claim that Sc-FOS diet can relieve anemia caused by malabsorption of iron.

10.2.7.3   Modulation of the Immune System

The mucosal barrier is a complex structure that physically separates the internal milieu from the environment in the digestive tract. It includes an epithelium covered in a vascular system and also mucus inhabited by intestinal bacteria, which all play a role in host defense. To protect against infection, the human intestine can benefit from three lines of defence, namely the gut mucosa itself, the gut lymphoid-associated tissue and the gut microbiota.

Feeding mice with diets supplemented with Sc-FOS increased the activities of natural killer cells and phagocytes and enhanced lymphocyte functions compared to mice fed a diet without Sc-FOS (Sangeetha et al., 2005b).

Decrease in the Risk of Colon Cancer

Diets are well known to be a major factor influencing colon cancer prevalence. Diets containing high concentrations of animal protein have been associated with greater risk.

In agreement with the beneficial effect of fruits and vegetables and the positive effect on colon cancer preservation, research suggested that NDOs such as fructooligosaccharides can decrease the entire process of colon cancer development (Swennen, 2006).

This study showed that Sc-FOSs induced a decrease or even a suppression of colon tumor in Min mice. This effect may have been related to stimulation of gut-associated lymphoid tissue, owing to the development of Peyer's patches.

This anticarcinogenic effect appears to be relative to cellular immunity

More generally FOSs were described as fermentable fibers providing protection against earlier stages of colon carcinogenesis.

- Butyrate produced in the colon is known to stimulate apoptosis in colonic cancer cell lines.
- Health promoting bacteria inhibit the growth of pathogenic bacteria and thus decrease the production of carcinogenic substances and bacterial enzymes that play a role in colon carcinogenesis (e.g., bile acid, p-cresol).
- At the same time, bacterial growth accelerates the colonic transit time. As a result, the duration of exposure to potential carcinogens decreases.

These are still results on animal studies, but now several clinical studies on preventing cancer by dietary fibers are under throughout the world.

10.2.7.4   Effect on Lipid Metabolism

The first data on the effect of NDOs on serum cholesterol and lipid level suggested that they have hypocholesterolemic and hypotriglyceridemic effects (Grizard and Bartomeuf, 1999). Besides its effect on the intestinal tract, FOS was described to exert a systemic effect, namely by modifying the metabolism of lipids in several animal models.

FOS were described to reduce postprandial triglyceridemia by 50% and avoid the increase in serum free cholesterol levels occurring in rats fed with a western-type high-fat diet. Changes in the concentration of serum cholesterol have been related with changes in the intestinal microbiota. Some strains of *Lactobacillus* spp. have been described to assimilate cholesterol while others appear to inhibit the absorption of cholesterol through the intestinal wall (Mussato and Mancilha, 2007).

The change in lipid metabolism was also suggested to be a consequence of the action of propionate on the liver.

However, these results were modulated by Giacco et al. (2004). In individuals with mild hypercholesterolemia, they showed that moderate intake of Sc-FOS (10.8 g/kg) had no major effects on lipid metabolism (either in the fasting or in the postprandial period). However, a small but significant increase in lipoprotein

A concentration was observed with Sc-FOS consumption together with a reduction of the postprandial insulin response. The clinical relevance of these small effects is unclear.

Human studies on the lipid-lowering properties of prebiotics when consumed at a moderate dose are sometimes contradictory, and the role in lipid reduction remains to be confirmed.

In order to quantify the effects in humans of dietary 1F-type fructans on serum triacylglycerols, Brighenti (2007) conducted a meta-analysis of the available literature. Fifteen eligible randomized, controlled trials published from 1995 to 2005, for a total of 16 comparisons, were identified from databases. Standardized mean effect sizes were calculated for net changes in serum triacylglycerol concentrations using a random-effect model.

The intake of fructans was associated with significant decreases in serum triacylglycerols by 0.17 mmol/L or 7.5%. Given the limited number of studies, no specific effects for gender, amount fed, duration of the study, background diet, overweight, hyperlipidemia, or diabetes were further formally investigated. But, from the test for heterogeneity it appears that the effect of inulin-type fructans on circulating triacylglycerols is consistent across conditions. In conclusion, dietary 1F-type fructans significantly reduced serum triacylglycerols. The mechanisms, possibly related to colonic fermentation and/or incretin release from the distal gut, warrant further studies.

Role of FOS in Blood Glucose Control

It was found (Luo et al., 2000) that the daily consumption of 20 g FOS decreased basal hepatic glucose production in healthy subjects without any effect on insulin-stimulated glucose metabolism.

This study allowed authors to evaluate the effects of the chronic ingestion of FOS on, plasma, lipid and glucose concentrations but also hepatic glucose production and insulin resistance in type 2 diabetics. Contrary to other studies, the results showed that 4 weeks of 20 g/d of FOS had no effect on glucose or lipid metabolism in type 2 diabetics. The subjects had had type-2 diabetes for a long time, so their metabolism might not be modified as easily as that of healthy subjects.

10.2.7.5   Undesirable Effects

In spite of the above, Sc-FOS have some undesirable effects. As other non-digestible carbohydrates, they have osmotic properties and are extensively fermented in the large bowel. As consequence, at high daily doses (25–30 g/day), they may cause flatulence and or function as a laxative. These effects are clearly dose related and may vary significantly from one individual to another.

10.2.8   Applications of Fructooligosaccharides

Considering all the claims detailed above combined and the physiochemical properties of fructooligosaccharides, their use as food ingredients has increased rapidly.

Development of synbiotics combining FOS with targeted probiotic strains or other NDOs, are increasingly described. These synbiotics promote the growth and/or activate the metabolism of various health promoting bacteria.

Fructooligosaccharides are used in dairy products for sugar reduction and prebiotic effects. They are also associated with probiotic bacteria in yoghurts and fermented drinks. In dairy products, FOS improves the mouth feel by increasing unctuousness, which is appreciated in low-fat reduction creams, sorbets, chocolate, mousse, milk, etc.

Other applications in food industries include desserts such as, fruits preparations, confectionary, cereals, bakery products, biscuits, sweet, gums, soup, salad dressing.

Fructooligosaccharides also find uses in processed meat, canned fish, sauces and pet food with health and nutrition claims.

In addition, fructooligosaccharides are used for health benefits in feed for many animals including: broilers, layers, rabbits, calves, lambs, kids, piglets, sows, fattening pigs, pets (dog, cat), horses and fish.

10.2.9   Market

Due to differing definitions, there are specific difficulties to analyze the development of the Functional Foods market, resulting in strongly varying estimations concerning the market volume of such products.

The global market for Functional Foods was estimated in 2000 at around US $ 33 billion. The largest and moat dynamic market is the USA with an estimated market share of around 50%.

Japan is another important market for functional foods, with a specific health related food category called FOSHU. An estimated market value of FOSHU was US$ three billion in 2000 with an estimated turnover of 14 billion US$.

Sales of functional food and drink in Europe have experienced considerable growth, with a doubling of sales between 2000 and 2005. In 2005 the value of European sales of functional foods was US$ eight billion and likely to reach in excess of ten billion euros by 2010.

Prebiotics concern a very small part of the functional food market but this is growing rapidly. Prebiotics are associated with breakfast cereals, baked goods, cereal bars, baby food as well as some dairy products. The main commercial food oligosaccharides (DP 2–10), used as prebiotic agents, are essentially obtained by enzymatic technology. They include non-digestible oligosaccharides such as inulin, fructooligosaccharides, galacto-oligosaccharides, lactulose, isomalto-oligosaccharides, soybean oligosaccharides, xylo-oligosaccharides.

In 2007 an analysis by Frost and Sullivan found that the US prebiotics markets, earned revenues of US$ 69 million, estimated to reach 198 million in 2014.

In Japan the demand for prebiotic oligosaccharides was 69,000 tons/year. Among these oligosaccharides the demand for starch oligosaccharides was the largest. The fructooligosaccharide market was only 6.5% of the global demand of prebiotics (Nakakuki, 2002). FOS products were first approved as food for specified health uses in 1993 and are sold as functional ingredients in Japan.

In Europe, total sales of functional food was estimated in 2005 at around eight billion euros while the prebiotics sector was valued at 87 million euros, with expectations to reach 180 million euros in 2010 (Source: Frost and Sullivan). Currently, the European market is believed to be dominated by fructans and galacto-oligosaccharides. Resistant starch products are at the development stage. The production of prebiotic fibers in Europe is estimated of about 30,000 tons, including inulin, oligofructose, FOS, GOS and lactulose.

FOS are available worldwide, mainly in East Asia, North America and Europe as prebiotics and/or dietary fiber like oligosaccharides. Market growth of oligosaccharides is about 15% per year.

In 1984, Meiji Seika Co. first succeeded in the commercial production of FOS using *Aspergillus niger*. The product was marketed in Japan under the name "Meioligo." This company later established a joint venture with Beghin-Say from France called Beghin-Meiji-Industry. They produced FOS under the trademark "Actilight."

Some other companies produced fructooligosaccharides from sucrose, Golden Technology (USA) under the trade mark NutaFlora and Cheil Food and Chemical Co (Korea) with Oligo-Sugar.

The global market for fructooligosaccharides from sucrose was estimated in 1995 to be 20,000 tons.

10.3 Lactosucrose From Sucrose

Lactosucrose is a functional trisaccharide consisting of D-galactose, D-glucose and D-fructose (O-β-D-galactopyranosyl-(1–4)-O-α-Dglucopyranosyl-(1–2)-O-β-D-frucofuranoside) used as an artificial sweetener. It is 30% as sweet as sucrose. It is produced enzymatically by the enzymatic transfer of the fructofuranosylsyl residue of sucrose to lactose. This fructofuranosyl transfer was principally described with bacterial enzymes from *Arthobacter sp, Bacillus sp, Klebsiella pneumonia* and *Rahnella aquatilis*. Lactosucrose can be obtained either by transfructosylation reaction catalyzed by β-D-fructofuranosidase or levansucrase in which lactose is used as a fructofuranosyl acceptor and sucrose as fructofuranosyl donor.

10.3.1 Lactosucrose Production

10.3.1.1 Lactosucrose Synthesis Using Bacterial β-D-Fructofuranosidase

Production of lactosucrose (4G-β-D-galactosylsucrose) using β-fructofuranosidase from *Arthrobacter* sp K-1 was first described by Fujita et al. (1990) who investigated the transfructosylation reaction with sucrose and various acceptors such as lactose or xylose. When incubated with sucrose alone, this enzyme catalyzed both transfructosylation and hydrolysis. But in the presence of a suitable acceptor such as lactose, the enzyme catalyzed mostly transfructosylation and transferred the fructose residue preferentially to the C1-position of the glucose moiety of lactose, producing a non-reducing oligosaccharide.

Lactosucrose transfer ratio with 10% (w/v) sucrose and 20% (w/v) lactose concentrations was around 50% with lactosucrose yield close to 27% (w/w) of total sugar.

More recently, Pilgrim et al. (2001) once more studied lactosucrose synthesis from sucrose and lactose by using β-fructofuranosidase from *Arthrobacter* spp. K-1. The transfructosylation mechanism was studied and reaction rate constants, equilibrium constants, dissociation Michaelis constants, glucose and sucrose inhibition were determined at 35°C and 50°C. The transfer reaction was found to be of an ordered bi-bi type in which sucrose first bound to the enzyme and lactosucrose was released last. The maximum yield attainable in a batch process with equimolar concentrations of substrates was reported to be around 50%.

This group also investigated (Kawase et al., 2001) lactosucrose production using a simulated moving bed. Simulation of the batch process showed that product removal could increase the lactosucrose yield from 50 to over 80%. Unfortunately, the expected optimum was not attained due to strong product hydrolysis around the raffinate outlet port.

10.3.1.2 Lactosucrose Synthesis Using From Bacterial Levansucrase

Production of lactosucrose by levansucrase from *Aerobacter levanicum* was first described by Avigad (1957) who reported the synthesis of a new trisaccharide, lactosucrose (α-lactosyl-β-fructofuranoside), by levansucrase acting on a system comprised of sucrose as fructosyl donor and lactose as acceptor.

Later, production of lactosucrose, by various microorganisms containing levansucrase activity was also investigated (Park et al., 2005). Among the tested bacteria, *Bacillus subtilis* was the most effective producer using lactose as an acceptor and sucrose as a fructosyl donor. Lactosucrose production by this strain was optimal at pH 6.0 and 55°C where from 225 g/l lactose and 225 g/l sucrose in 10 h lactosucrose yield was 32.9% (w/w).

Another way to produce lactosucrose was described by Choi et al. (2004). In this case, whole cells harboring transfructosylation activity of levansucrase was used. Levansucrase-induced cells of *Paenibacillus polymyxa* were obtained in the medium containing sucrose, and the transfructosylation activity in the whole cell was optimized for lactosucrose production. The optimal cell concentration was 2.0% (w/v) with substrate ratio of lactose and sucrose 22.5% (w/v) each, 55°C, and pH 6.0. Under these conditions, lactosucrose yield was 30.9% of total sugar with a productivity of 2.8% (w/v)/h.

10.3.2   Functional Properties of Lactosucrose

Lactosucrose was described to be resistant to digestion in the stomach and small intestine.

It was reported to be selectively utilized by intestinal *Bifidobacterium* species resulting in significant induction in the growth of these bacteria in the colon. It is widely used in Japan as a dietary supplement and in functional foods, including yoghurt. Lactosucrose is being developed in the United States for similar uses (Park et al., 2005).

However, Park et al. (2008) concluded, with regard to *in vitro* and clinical works, "the evidence for prebiotic status of lactosucrose is still not sufficient and lactosucrose cannot, at present, be classified as a prebiotic."

10.4   Glucooligosaccharides

A wide range of non-digestible glucooligosaccharides can be synthesized from sucrose using a variety of glucosyltransferases with different transglucosylating specificities. Such enzymes, named glucansucrases (E.C. 2.4.1.X), catalyze the synthesis of high molecular weight glucan polysaccharides (Monsan et al., 2000) following the reaction:

$$n \text{ G} - \text{F} \xrightarrow{\text{Glucansucrase}} (\text{G})_n + n \text{ F}$$

The specificity of the main glucansucrases is described in ❷ *Table 10.1*. The structure of corresponding glucan polymers is given in ❷ *Figure 10.3*.

This reaction involves the formation of an intermediary D-glucopyranosyl-enzyme covalent intermediate (❷ *Figure 10.4*). In the presence of an efficient

❏ Table 10.1

Glucansucrases and their regiospecificity

Glucansucrase	Glucan	Osidic linkages
Dextransucrase (E.C. 2.4.1.5)	Dextran	α-1,6 (>50%)
Mutansucrase (E.C. 2.4.1.5)	Mutan	α-1,3 (>50%)
Alternansucrase (E.C. 2.4.1.140)	Alternan	Alternating α-1,3/ α-1,6
Amylosucrase (E.C. 2.4.1.4)	Amylose	α-1,4 (100%)

🔳 Figure 10.3
**Structural representation of different types of glucans synthesized by glucansucrases
(a) dextran, (b) mutan, (c) alternan, (d) amylose.**

carbohydrate acceptor, like maltose, the polymerization reaction is limited, and low molecular weight glucooligosaccharides (◉ *Figure 10.4*) are obtained (Koepsell et al., 1952). Very interestingly, in such acceptor reactions, glucansucrases present the same regiospecificity as for glucan polymer synthesis.

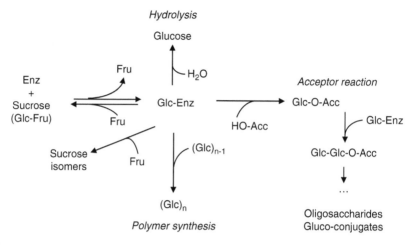

◘ Figure 10.4
Different reactions catalyzed by glucansucrases.

This results in the possibility to use sucrose to directly synthesize several types of glucooligosaccharides of interest for prebiotic applications.

In addition, the D-glucopyranosyl residue can be transferred onto water (sucrose hydrolysis) or onto a non-carbohydrate acceptor to yield a gluco-conjugate (● *Figure 10.4*).

10.4.1   Dextransucrases

Dextransucrases are extracellular enzymes produced by lactic acid bacteria of the genera *Leuconostoc, Streptococcus, Lactococcus* and *Lactobacillus* (Monsan et al., 2001). They belong to Family 70 of glycoside-hydrolases (Moulis et al., 2006). They share a general common structure consisting of (● *Figure 10.5*):

- An N-terminal signal peptide involved in their excretion
- A variable region of unknown role
- A highly conserved N-terminal catalytic domain which contains the amino acids involved in the catalytic mechanism and presents a permutated $(\beta/\alpha)_8$ barrel structure (Moulis et al., 2006)
- A C-terminal glucan binding domain, involved both in polysaccharide and oligo-saccharide synthesis, and containing a series of repeating amino acid sequences (Moulis et al., 2008)

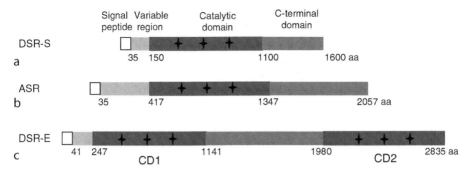

Structural organization of glucansucrases (a) dextransucrase from *Leuconostoc mesenteroides* NRRL B-512F, (b) alternansucrase from *Leuconostoc mesenteroides* NRRL B-1355, (c) dextransucrase from *Leuconostoc mesenteroides* NRRL B-1299. The position of catalytic amino-acids is indicated.

The dextransucrase from *Leuconostoc mesenteroides* NRRL B-512F is widely used for producing industrial dextran. It is a large protein of 1,527 amino acids. After acidic hydrolysis, dextran is mainly used for chromatograpy support (Sephadex®) production, as well as blood plasma substitutes and iron carriers (dextran-sulphate).

The *in vitro* utilization by human gut microbiota of the following carbohydrates:

* Industrial-grade dextran
* Oligodextran fractions produced by controlled enzymatic hydrolysis of dextran (Mountzouris et al., 1999)
* Maltodextrin

was studied using anaerobic batch culture fermenters (Olano-Martin et al., 2000). Glucose and fructooligosaccharides were used as reference carbohydrates. Fructooligosaccharides selectively increased numbers of bifidobacteria in the early stages of fermentation. Dextran and oligodextran resulted in an enrichment of bacteria with high levels of persistence up to 48 h, with production of elevated levels of butyrate ranging from 5 to 14.85 mmol/l. A three-stage continuous culture cascade system was used for a more effective simulation of the conditions that prevail in different regions of the large intestine. A low-molecular-mass oligodextran fraction was then shown to be the best substrate for bifidobacteria and lactobacilli, when compared to dextran and maltodextrin. In addition, dextran and oligodextran stimulate butyrate production more efficiently, which has been shown to present potentially very interesting anti-neoplastic properties (Olano-Martin et al., 2000). These results underline the interest of oligodextrans as modulators of the gut microbiota.

In the presence of glucose or isomaltose as acceptors, dextransucrase synthesizes a series of glucooligosaccharides of the isomaltosaccharide type (IMOs). Such partially non-digestible oligosaccharides are presently the main prebiotics produced for the Japanese market (Monsan and Auriol, 2004). They are obtained from starch hydrolysates (maltooligosacharides) by the action of an α-transglucosidase from *Aspergillus* spp. They only present short chain oligosaccharides: commercial preparations contain (Mountzouris et al., 1999) a mixture of isomaltose (DP 2: 23%), isomaltotriose (DP 3: 17%), oligosaccharides (DP 4–6: 26%), non-isomalto-oligosaccharides (panose, maltose, maltotriose, nigerose, kojibiose: 30%).

Recently, the possibility to engineer dextransucrase at the molecular level by truncating its C-terminal domain has resulted in variants able to directly synthesize isomaltooligosaccharides of controlled molecular weight. Synthesis yields of up to 69 and 75% have been obtained for 40 and 10 kDa dextran, respectively (Moulis et al., 2008). This opens the way to the direct production of prebiotic IMOs from sucrose, and particularly with increased chain lengths to limit their digestibility.

The dextransucrase from *Leuconostoc mesenteroides* NRRL B-1299 produces a dextran containing about 30% α-1,2 linkages (Monsan et al., 2000). This type of glycosidic bond is still synthesized by the enzyme when maltose is used as an acceptor (Remaud-Siméon et al., 1994). A mixture of three families of oligosaccharides is then obtained, which contain a maltose residue at the reducing end and either (1) only α-1,6 linkages (series OD), (2) α-1,6 linkages and one α-1,2 linkage at the non-reducing end (series R), or (3) α-1,6 linkages and one α-1,2 linkage on the penultimate D-glucosyl residue (series R′, ❍ *Figure 10.6*). The corresponding glucooligosaccharides (GOS) are highly resistant to attack by digestive enzymes from humans and many other animals (Valette et al., 1993), and are not metabolized by germ-free rats (Djouzi et al., 1995). They induce a broad range of glycolytic enzymes without increased production of gases (Djouzi and Andrieux, 1997). Bifidobacteria, lactobacilli and particularly bacteroides specifically utilize such GOS as carbon source (Monsan and Auriol, 2004).

Dextransucrase is obtained by fed-batch culture of *L. mesenteroides* NRRL B-1299, using sucrose as carbon source and specific enzyme inducer (Dols et al., 1997a). As sucrose is also the substrate of the extracellularly

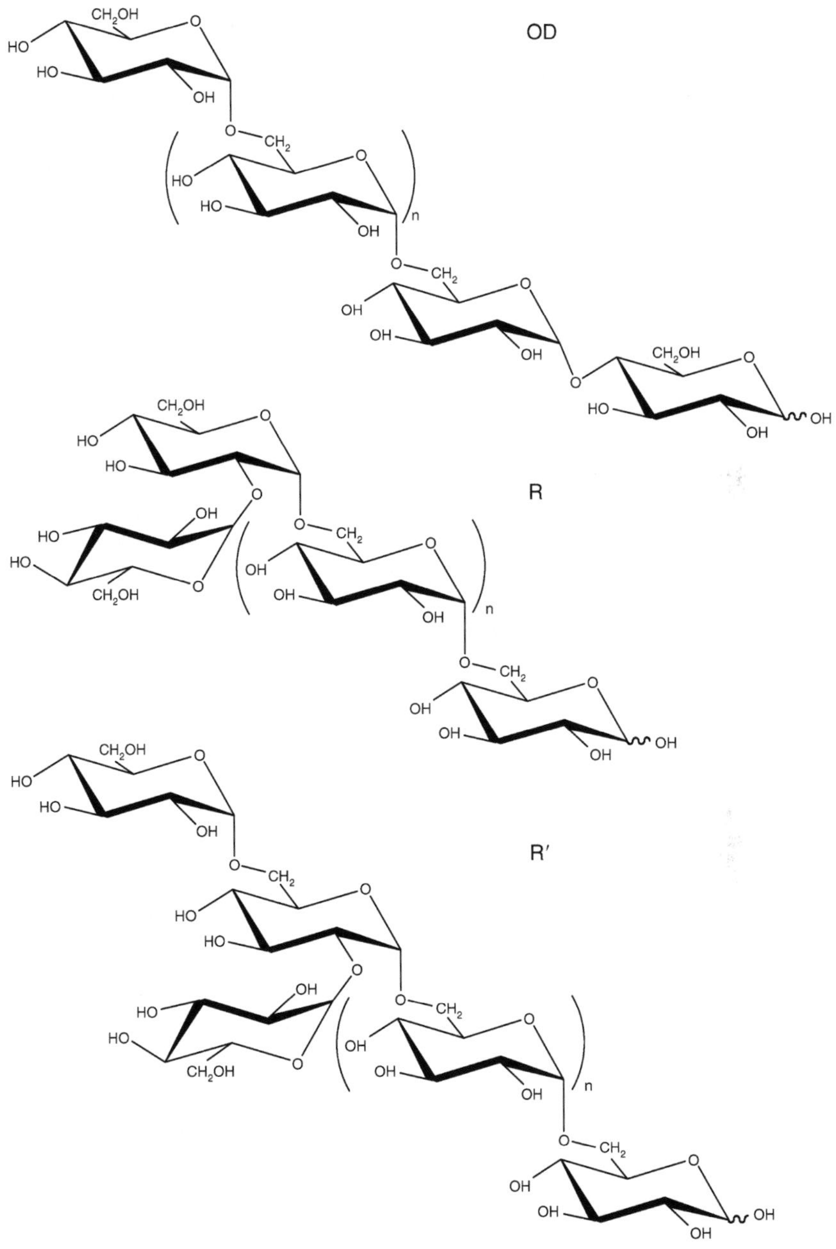

Figure 10.6
Structure of the glucooligosaccharides synthesized by the dextransucrase from *Leuconostoc mesenteroides* NRRL B-1299 from sucrose when using maltose as acceptor: series OD, series R and series R'.

obtained dextransucrase, dextran is synthesized during microbial growth and D-fructose is produced. The presence of the D-fructose results in the repression of dextransucrase production. This repression effect can be suppressed in the presence of D-glucose. The simultaneous fed-batch addition of sucrose and D-glucose thus increased the dextransucrase activity obtained in the culture medium by 100% to reach a final activity of 9.7 U/ml (Dols et al., 1997b). Very fortunately, more than 90% of this dextransucrase activity is associated to the cells and the surrounding dextran slime. It can thus be very easily recovered from the culture medium by centrifugation.

To develop a continuous process for glucooligosaccharide synthesis, the dextransucrase must be immobilized. This can be easily done, thanks to the strong association of dextransucrase with dextran slime and cells, by simply entrapping the catalytic activity within calcium alginate beads. The immobilization yield is 93%, and a specific activity of 4.1 U/ml of gel can be obtained (Dols et al., 1997b). The present industrial production is of 60 tons per year, and is obtained by operating the immobilized dextransucrase in a 1-m^3 continuous packed-bed reactor. The key parameter for efficient operation is the sucrose-to-maltose ratio, which controls both the glucooligosaccharide yield and the size distribution (Dols et al., 1997b). Sucrose conversion at the outlet of the reactor is 100%.

Roberfroid (2008a,b) stressed the fact that this type of oligosaccharide cannot be classified as a prebiotic, due to the lack of evaluation in humans. In fact, these GOS are produced on the industrial scale essentially for cosmetic application, to stimulate the growth of beneficial skin microorganisms (Lamothe et al., 1994).

It is increasingly accepted that the concept of prebiotic need not only apply to the gastrointestinal microbiota. It could also be applied to any other microbiota, like the skin microbiota or the vaginal microbiota, for example. Different oligosaccharides have been evaluated for their prebiotic effect using selected human vaginal lactobacilli presenting a high level of hydrogen peroxide production, and pathogenic microorganisms (Rousseau et al, 2005).

In any case, the evidence of a prebiotic effect is clear in animal trials, e.g., the administration of 0.15% (w/w) GOS to young calves (two populations of 1,300 animals) resulted in 20% decrease in veterinary costs (Monsan and Auriol., 2004), without any side effects.

In addition, it was demonstrated that GOS prevent the installation of type II diabetes in over-fed mice. Female C57/Bl6/J mice were fed with a high-fat diet (45% fat, 35% carbohydrate, 20% protein) supplemented or not with 1.5 g/kg/day of GOS. The GOS supplementation did not change body weight nor fat pad mass, nor any of the blood parameters measured (glucose, insulin, leptin, triglycerides, and free fatty acids). Mice which received the GOS supplemented diet showed increased glucose utilization after a 1 g/kg load of glucose compared with the mice fed the high-fat diet alone (Boucher et al., 2003).

It must be underlined that the effect of the glycosidic linkage involved in a wide range of disaccharides upon selectivity of fermentation by intestinal bacteria has been recently determined, showing that α-1,2 linkages are particularly selective (Sanz et al., 2005).

The originality of the dextransucrase from *Leuconostoc mesenteroides* NRRL B-1299 is to present not only one, but two catalytic sites (⊘ *Figure 10.3*): one is specific for α-1,6 linkage synthesis and is located at the N-terminal end, while the second one is specific for α-1,2 linkage synthesis and is located at the C-terminal end (Bozonnet et al., 2002). The N-terminal truncation of the native corresponding gene results in the design of a new enzyme able to catalyze the controlled synthesis of α-1,2 branched oligodextrans, with different sizes and branching degrees (Fabre et al., 2005). These products are presently under evaluation for their prebiotic properties.

10.4.2 Alternansucrase

When grown on sucrose, *Leuconostoc mesenteroides* NRRL B-1355 produces a glucosyltransferase, alternansucrase, which catalyses the synthesis from sucrose of a glucan containing alternating α-1,3 and α-1,6 linkages, alternan (Côté and Robyt, 1982a; Jeanes et al., 1954; Lopez-Munguia et al., 1993; Seymour et al., 1979).

The corresponding encoding gene was independently isolated by two groups (Argüello-Morales et al., 2000; Kosmann et al., 1999). Alternansucrase is a 2,057 amino acid enzyme, with a molecular weight of 245 kDa. Its structural organization (⊘ *Figure 10.5*), with an N-terminal catalytic domain and a C-terminal glucan binding domain containing amino-acid repeats, is similar to that of most of the glucansucrases of Family GH-70 of the glycoside-hydrolases (Argüello-Morales et al., 2000).

In the presence of acceptor carbohydrates, such as maltose, alternansucrase presents the same specificity as for alternan synthesis and catalyses the production of oligoalternans (Argüello-Morales et al., 2001; Côté and Dunlap, 2003; Côté and Robyt, 1982b; Pelenc et al., 1991). Such oligosaccharides have been demonstrated to efficiently control enteric bacterial pathogens (Côté and Holt, 2007; Holt et al., 2005).

Gentiobiose was also used as an acceptor for the synthesis of oligosaccharides with alternansucrase. These products were compared to gentio-oligosaccharides (containing D-glucopyranosyl units linked with β-1,6 linkages) to study their effect on the growth of predominant gut bacteria. A prebiotic index (PI) was calculated to obtain a general quantitative comparative measurement of the selectivity of fermentation and to compare the influence of size and structure in the selective fermentation. This index sets a relationship between changes in the "beneficial" and "undesirable" elements within the microbiota (Sanz et al., 2006a). DP4–5 alternansucrase gentiobiose acceptor products generated the highest PI values (PI of 5.87). Regarding the production of short-chain fatty acids, the mixture of DP6–10 gave the highest levels of butyric acid but the lowest levels of lactic acid. From a general point of view, for similar molecular weights, alternansucrase gentiobiose acceptor products gave higher PI values than gentio-oligosaccharides (Sanz et al., 2006a).

The influence of glucosidic linkages and molecular weight on the fermentation by gut bacteria of maltose-based oligosaccharides synthesized by alternansucrase and dextransucrase has been determined (Sanz et al., 2006b). When using an anaerobic *in vitro* fermentation method, DP3 oligosaccharides showed the highest selectivity. Oligosaccharides with higher molecular weight (DP6–7) also resulted in selective fermentation. Oligosaccharides with DP above seven did not promote the growth of beneficial bacteria.

The C-terminal domain of alternansucrase is composed of two different series of homologous repeating units, the CW-like repeats and the APY-repeats (Joucla et al., 2006). The CW-like repeats are 20-amino-acid-long motifs with a high representation of conserved glycine and aromatic residues. APY-repeats are 79-amino-acid-long motifs, with a high number of conserved glycine and aromatic residues, specifically characterized by the presence of the three consecutive residues alanine, proline and tyrosine. Seven APY repeats are present within the last 550 C-terminal amino acids. Fully active variants of alternansucrase, truncated of its C-terminal glucan-binding domain have been successfully designed (Joucla et al., 2006): the truncation of the APY repeats, keeping four CW-like repeats, results in a variant with a molecular weight of 175 kDa. This variant presents a high specific activity and the same specificity of product synthesis as the native enzyme. In addition, it is more soluble and suffers less degradation when the corresponding gene is expressed in *E. coli* than full length alternansucrase (Joucla et al., 2006).

Alternansucrase is used by Cargill (Carlsson et al., 2006) to produce from sucrose and maltose a sweetening prebiotic mixture, Xtend™ Sucromalt, containing residual fructose (37%), leucrose (13%) and alternan oligosaccharides (48%).

10.4.3 Amylosucrase

Amylosucrase catalyses the transfer of D-glucopyranosyl units from sucrose to synthesize only α-1,4 linkages. This reaction yields highly pure amylose chains. The gene coding for the amylosucrase from *N. polysaccharea* has been isolated, cloned and sequenced (Potocki de Montalk et al., 1999). Amylosucrase has been purified to homogeneity as a fusion protein with glutathion-S-transferase, using glutathion-Sepharose-4-B affinity chromatography. Its molecular mass is 70 kDa.

In the presence of sucrose alone, amylosucrase simultaneously catalyses sucrose hydrolysis, and maltose and maltotriose synthesis (using D-glucose as acceptor), and high molecular weight glucan synthesis (containing only α-1,4 linkages). Very surprisingly, amylosucrase is activated by high sucrose concentrations: while Vmax and Km values are equal to 510 U/g and 2 mM respectively when sucrose concentrations are lower than 20 mM, they are equal to 906 U/g and 26 mM respectively when sucrose concentrations are higher than 20 mM (Potocki de Montalk et al., 1999).

In the presence of sucrose and glycogen, amylosucrase is highly activated: at 105 mM sucrose, the addition of 30 g/l glycogen results in a 100-fold increase in amylosucrase activity (Potocki de Montalk et al., 1999). This activator effect decreases when increasing sucrose concentration, suggesting a competition between sucrose and glycogen. This reaction results in the obtention of modified glycogen, which contains extended linear chains with an average DP of 75 D-glucosyl units according to iodine-complex characterization.

This enzyme is the only glucansucrase for which the 3D structure has been solved (Skov et al., 2001). It belongs to Family 13 of glycoside-hydrolases, which is the family of α-amylases and CGTases. The amino-acid residues involved in the catalytic mechanism of D-glucopyranosyl unit transfer have been identified (Albenne et al., 2003). The structure of the α-glucan (Rolland-Sabaté et al., 2004) and amylose elongation products has been determined. Such amylose products present a nanoparticle size and are highly resistant to the attack of digestive enzymes, which corresponds to a new range of resistant nano-starches (Potocki-Véronèse et al., 2005).

The German company Südzucker is presently developing the production of such resistant amyloses, obtained by amylosucrase action, under the brand name NEO-amylose® (Südzucker 2005/2006).

10.5   Summary

- Sucrose is not a simple carbohydrate, but a protected and activated form of D-glucose and D-fructose
- Fructosyl transferases catalyze the synthesis of a variety of fructans and fructooligosaccharides using sucrose as a D-fructofuranosyl unit donor
- When using a fungal fructosyl transferase as catalyst, non-digestible prebiotic short-chain fructooligosaccharides (Sc-FOS) containing β-2,1 linkages are mainly obtained: kestose (GF2), nystose (GF3) and fructosyl-nystose (GF4)
- Sc-FOS are very efficient prebiotics which are produced on the industrial scale in continuous plug-flow reactors containing immobilized fructosyl transferase from *Aspergillus oryzae*
- Lactosucrose is a functional trisaccharide obtained by transfer of a β-D-fructofuranosyl residue from sucrose onto the reducing group of lactose, catalysed by a fructofuranosidase or a levansucrase
- Glucosyltransferases catalyze the synthesis of a variety of glucans and glucooligosaccharides using sucrose as D-glucopyranosyl unit donor
- In the presence of efficient carbohydrate acceptors, like maltose, glucosyltransferases catalyze the synthesis of low-molecular-weight oligosaccharides instead of high molecular weight polymers (acceptor reaction)
- Dextransucrase, alternansucrase and amylosucrase are microbial glucosyltransferases (transglucosidases) which can be efficiently used to synthesize non-digestible prebiotic carbohydrates, glucooligosaccharides (GOS), oligoalternans, and amylose respectively.

List of Abbreviations

Brix	degrees brix, sucrose concentration in % (w/w)
BVH	bed volume per hour
Da	daltons
DP	degree of polymerization
F	fructose
FOS	fructooligosaccharides
FOSHU	food of specified health use in Japan
G	glucose
GF	sucrose

GF2	1-kestose
GF3	nystose
GF4	fructosyl-nystose
G(n)	dextrans
Kda	kilo daltons in g/mole
NDOs	non-digestible oligosaccharides
SCFAs	short chain fatty acids
Sc-FOS	short chain fructooligosaccharides
w/w	weight per weight
w/v	weight per volume

References

Ahmed FAF, Doaa ARM, Mona ATE (2005) Current Microbiol 51:402–407

Albenne C, Skov LK, Mirza O, Gajhede M, Feller G, D'Amico S, André G, Potocki-Véronèse G, van der Veen B, Monsan P, Remaud-Siméon M (2003) J Biol Chem 279:726–734

Alvarado M, Maugeri F (2007) J Biotechnol 131:S91–S92

Álvaro-Benito M, De Abreu M, Fernández-Arrojo L, Plou FJ, Jiménez-Barbero J, Ballesteros A, Polaina J, Fernández-Lobato M (2007) J Biotechnol 132:75–81

Argüello-Morales MA, Remaud-Siméon M, Pizzut S, Sarçabal P, Willemot RM, Monsan P (2000) FEMS Microbiol Lett 182:81–85

Argüello-Morales MA, Remaud-Siméon M, Willemot RM, Vignon MR, Monsan P (2001) Carbohydr Res 331:403–411

Arrojo LF, Alvaro M, Ghazi I, De Abreu M, Linde D, Gutierrez-Alonso P, Alcalde M, Jimenez-Barbero J, Jimenez A, Ballesteros A, Fernandez-Lobato M, Plou FJ (2007) J Biotechnol 131S:S107

Avigad G (1957) J Biol Chem 229:121–129

Bekers M, Marauska M, Grube M, Karklina D, Duma M (2004) Acta Alimentaria 33:31–37

Boucher J, Daviaud D, Remaud-Siméon M, Carpéné C, Saulnier-Blache JS, Monsan P,

Valet P (2003) J Physiol Biochem 59:169–174

Bozonnet S, Dols-Laffargue M, Fabre E, Pizzut S, Remaud-Siméon M, Monsan P, Willemot RM (2002) J Bacteriol 184: 5753–5761

Brighenti F (2007) J Nutr 137:2552S–2556S

Byun SH, Han WC, Soon AK, Chul HK, Jang KH (2007) J Biotechnol 131:S112

Carlsson TL, Woo A, Zheng GH (2006) WO 088884

Chambert R, Treboul G, Dedonder R (1974) Eur J Biochem 41:285–300

Cheng CY, Duan KJ, Sheu DC, Lin CT, Li SY (1996) J Chem Tech Biotechnol 66:135–138

Choi HJ, Kim CS, Kim P, Jung HC, Oh DK (2004) Biotechnol Prog 20:1876–1879

Côté GL, Dunlap CA (2003) Carbohydr Res 338:1961–1967

Côté GL, Robyt JF (1982a) Carbohydr Res 101:57–74

Côté GL, Robyt JF (1982b) Carbohydr Res 111:127–142

Dedonder R (1966) Meth Enzymol 8:500–505

Djouzi Z, Andrieux C (1997) Br J Nutr 78:313–324

Djouzi Z, Andrieux C, Pelenc V, Somarriba S, Popot F, Paul F, Monsan P, Szylit O (1995) J Appl Bacteriol 79:117–127

Dols M, Remaud-Siméon M, Monsan P (1997a) Enzyme Microb Technol 20:523–530

Dols M, Remaud-Siméon M, Monsan P (1997b) Appl Biochem Biotechnol 62: 47–59

Euzenat O, Guibert A, Combes D (1997) Process Biochem 32:237–243

Fabre E, Bozonnet S, Arcache A, Willemot RM, Vignon V, Monsan P, Remaud-Siméon M (2005) J Bacteriol 187:296–230

Fujita K, Hara K, Hashimoto H, Kitahata S (1990) Agric Biol Chem 54:2655–2661

Ghazi I, De Segura AG, Femandez-Arrojo L, Alcade M, Yates M, Rojas-Cervantes ML, Plou FJ, Ballesteros A (2005) J Mol Catal B Enzym 35:19–27

Giacco R, Clemente G, Luongo D, Lasorella G, Fiume I, Brouns F, Bornet F, Patti L, Cipriano P, Rivellese AA, Riccardi G (2004) Clin Nutr 23:331–340

Gibson GR (2004) Best Pract Res Clin Gastroenterol 18:287–298

Gibson GR, Roberfroid MB (1995) J Nutr 125:1401–1412

Grizard D, Bartomeuf C (1999) Reprod Nutr Dev 39:563–588

Han JS, Park KJ, Shin DS, Kim JH, Kim JC, Lee KC, Kim SW, Park SW (2003) US Patent Application, 2003/0054499 A1

Hang YD, Woodmans EE (1996) Lebnsm Wiss Technol 29:578–580

Hidaka H, Eida T, Saitoh Y (1987) Nippon Nogeikagaku Kaishi 61:915–923

Hidaka H, Hirayama M, Sumi N (1988) Agric Biol Chem 52:1181–1187

Hirayama M (2002) Pure Appl Chem 74:1271–1279

Hirayama M, Sumi N, Hidaka H (1988) Agric Biol Chem 53:667–673

Holt SM, Miller-Fosmore CM, Côté GL (2005) Lett Appl Microbiol 40:385–390

Jeanes AR, Haynaes WC, Wilham CA, Rankin JC, Melvin EH, Austin MJ, Cluskey JE, Fisher BE, Tsuchiya HM, Rist CE (1954) J Am Chem Soc 76:5041–5052

Joucla G, Pizzut S, Monsan P, Remaud-Siméon M (2006) FEBS Lett 580:763–768

Jung KH, Kim JH, Jeon YJ, Lee JH (1993) Biotechnol Lett 15:65–70

Jung KH, Jong WY, Kyung RK, Jai YL, Jae HL (1989) Enz Microbiol Technol 11:491–494

Jung KH, Lim JY, Yoo SJ, Lee JH, Yoo MY (1987) Biotechnol Lett 9:703–708

Katopodis P, Kalogeris E, Kekos D, Macris BJ, Christakopoulos P (2003) Food Biotechnol 17:1–14

Kawase M, Pilgrim A, Araki T, Hashimoto K (2001) Chem Eng Sci 56:453–458

Kilian S, Kritzinger S, Rycroft C, Gibson GR (2002) World J Microbiol Biotechnol 18:637–644

Koepsell HJ, Tsuchiya HM, Hellman NN, Kasenko A, Hoffman CA, Sharpe ES, Jackson RW (1952) J Biol Chem 200: 793–801

Kono T, Yamaguchi G, Hidaka H (1994) US Patent 5314810

Kosmann J, Welsh T, Quanz M, Knuth K (1999) European Patent 1 151 085:B1

Lamothe JP, Marchenay Y, Monsan P, Paul F, Pelenc V (1992) French Patent 2678166

Lee JH, Shinohara S (2001) J Microbiol 12:331–333

Lizuka M, Minamiura Y, Kojima I (1995) Japan Patent 71155986

Lopez-Munguia A, Pelenc V, Remaud M, Biton J, Michel J, Lang C, Monsan P (1993) Enzyme Microb Technol 15:77–85

Luo J, Van Yperselle M, Salwa WR, Rossi F, Bornet FRJ, Slama G (2000) J Nutr 130:1572–1577

Marx SP, Winkler S, Hartmeier W (2000) FEMS Microbiol Lett 44:647–649

Maugeri F, Hernalsteens S (2007) J Mol Catal B Enzym 49:43–49

Monsan P, Auriol D (2004) In: Neeser JR, German JB (eds) Bioprocesses and biotechnology for functional foods and nutraceuticals. Marcel Dekker Inc., New York, pp. 135–149

Monsan P, Bozonnet S, Albenne C, Joucla G, Willemot RM, Remaud-Siméon M (2001)

Monsan P, Paul F (1995) In: Wallace RJ, Chesson A (eds) Biotechnology in animal feed and animal feeding. VCH, Weinheim, pp. 233–245

Monsan P, Potocki de Montalk G, Sarçabal P, Remaud-Siméon M, Willemot RM (2000) In: Bielecki S, Tramper J, Polak J (eds) Food biotechnology. Elsevier, Amsterdam, pp. 115–122

Moulis C, Joucla G, Harrison D, Fabre E, Potocki-Véronèse G, Monsan P, Remaud-Siméon M (2006) J Biol Chem 281:31254–31267

Moulis C, Vaca-Medina G, Suwannarangsee S, Monsan P, Potocki-Véronèse G, Remaud-Siméon M (2008) Biocatal Biotransformation 24:141–151

Mountzouris KC, Gilmour SG, Grandison AS, Rastall RA (1999) Enzyme Microb Technol 24:75–85

Mussato SI, Mancilha IM (2007) Carbohydr Polym 68:587–597

Nakakuki T (2002) Pure Appl Chem 74:1245–1251

Nizikawa K, Nakajima M, Natabeni H (2001) Food Sci Technol Res 7:39–44

Nobre C, Dominguez A, Torres D, Rocha O, Rodriguez L, Rocha I, Teixeira J, Ferreira E (2006) World Congress of Food Science and Tecnology: Food is life, France, September 21

Olano-Martin E, Mountzouris KC, Gibson GR, Rastall RA (2000) Br J Nutr 83:247–255

Ouarne F, Guibert A (1995) Zuckerind 120:793–798

Park JP, Oh TK, Yun JW (2001) Process Biochem 37:471–476

Park NH, Choi HJ, Oh DK (2005) Biotechnol Lett 27:495–497

Park YK, Pastore GM (1998) PCT patent BR98/00022

Pelenc V, Lopez-Munguia A, Remaud M, Biton J, Michel J, Paul F, Monsan P (1991) Sci Alim 11:465–476

Pilgrim A, Kawase M, Ohashi M, Fujita K, Murakami K, Hashimoto K (2001) Biosci Biotechnol Biochem 65:758–765

Potocki de Montalk G, Remaud-Siméon M, Willemot RM, Planchot V, Monsan P (1999) J Bacteriol 181:375–381

Potocki-Véronèse G, Puteaux JL, Dupeyre D, Albenne C, Remaud-Siméon M, Monsan P, Buléon A (2005) Biomacromolecules 6:1000–1011

Ramesh MN, Shivakumara M, Sangeetha PT, Gurudutt P, Prakash M (2005) US patent application 0069627 A1

Remaud-Siméon M, López-Munguía A, Pelenc V, Paul F, Monsan P (1994) Appl Biochem Biotechnol 44:101–117

Ritsema T, Smeekens (2003) Curr Opin Plant Biol 6:223–230

Roberfroid M (2008a) In: Gibson GR, Roberfroid MB (eds) Handbook of prebiotics.CRC Press, Boca Raton, pp. 39–68

Roberfroid MB (2007) J Nutr 137:830S–837S

Roberfroid MB (2008b) Handbook of prebiotics. Taylor & Francis group, Boca Raton, pp. 39–68

Rolland-Sabaté A, Planchot V, Potocki-Véronèse V, Monsan P, Colonna P (2004) J Cereal Sci 40:17–30

Rousseau V, Lepargneur JP, Roques C, Remaud-Siméon M, Paul F (2005) Anaerobe 11:145–153

Sangeetha PT, Ramesh MN, Prapulla SG (2004) Process Biochem 39:753–758

Sangeetha PT, Ramesh MN, Prapulla SG (2005a) Process Biochem 40:1085–1088

Sangeetha PT, Ramesh MN, Prapulla SG (2005b) Trend Food Sci Technol 16:442–457

Santos AMP, Maugeri F (2007) Food Technol Biotechnol 45:181–186

Sanz ML, Côté GL, Gibson GR, Rastall RA (2006a) FEMS Microbiol Ecol 56:383–388

Sanz ML, Côté GL, Gibson GR, Rastall RA (2006b) J Agric Food Chem 54:9779–9784

Sanz ML, Gibson GR, Rastall RA (2005) J Agric Food Chem 53:5192–5199

Seymour FR, Knapp RD, Chen ECM, Bishop H, Jeanes A (1979) Carbohydr Res 74:41–62

Shin HT, Baig SY, Lee SW, Suh DS, Kwon ST, Lim YB, Lee JH (2004) Biores Technol 93:59–62

Shiomi N, Yamada J, Izawa M (1979) Agric Biol Chem 43:2233–2244

Skov LK, Mirza O, Henriksen A, Potocki de Montalk G, Remaud-Siméon M, Willemot RM, Monsan P, Gajhede M (2001) J Biol Chem 276:25273–25278

Su YC, Sheu CS, Chien JY, Tzan TK (1991) Proc Natl Sci Counc Repub China B 15:131–139

Südzucker AG (2005/2006) Annual Report

Swennen K, Courtin CM, Delcour JA (2006) Crit Rev Food Sci Nutr 46:459–471

Tanaka T, Oi S, Yamamoto T (1979) J Biochem 85:287–293

Tanriseven A, Asla Y (2005) Enzyme Microbial Technol 36:550–554

Valette P, Pelenc V, Djouzi Z, Andrieux C, Paul F, Monsan P, Szylit O (1993) J Sci Food Agric 62:121–127

Vankova K, Onderkova Z, Antosova M, Polkovic M (2008) Chem Papers 4:375–381

Yun JW (1996) Enzyme Microbial Technol 19:107–117

Yun JW, Jung KH, Jeon YJ, Lee JH (1992) J Microbiol Biotechnol 2:98–101

Yun JW, Jung KH, Oh JW, Lee JH (1990) Appl Biochem Biotechnol 24/25:299–308

Yun JW, Song KS (1999) Methods in biotechnology, In: Bucke C (ed) Carbohydrate biotechnology protocols. vol. 10: Human Press Inc., Totowa, pp. 141–151

Zuccaro A, Goetze S, Kneip S, Dersch P, Seibel J (2008) Chem Bio Chem 9:143–149

11 Prebiotic Potential of Polydextrose

Julian D. Stowell

11.1 Introduction

Polydextrose is a randomly bonded polymer of glucose with some sorbitol end-groups. It was originally developed by scientists at Pfizer seeking a low calorie bulking agent to be used in conjunction with intense sweeteners. Polydextrose has been used for more than 25 years in human food and beverage products around the world. It is currently marketed by Danisco A/S as Litesse®Two and Litesse®Ultra™ and by A E Staley as Sta-Lite III.

11.2 Characteristics of Polydextrose

11.2.1 Manufacture and Structure

Polydextrose is produced by the bulk melt polycondensation of glucose and sorbitol in conjunction with small amounts of food grade acid in vacuo. Typically, corn glucose is used. Further purification steps are then involved to generate a range of products suited to different applications. A representative structure of polydextrose is given in ● *Figure 11.1*. Recent studies have shown that polydextrose has a highly branched structure with a wide spectrum of glycosidic linkages represented, 1–6 linkages predominate. The average degree of polymerization (DP) is 12 glucose units. However, a complete spectrum of DPs up to 30 and above have been detected in the polydextrose mix (Stowell 2009).

Polydextrose is manufactured and marketed in accordance with the Food Chemical Codex Specification (Anon 2004). AOAC Method 2000.11 is used to measure polydextrose in foods (Anon 2002).

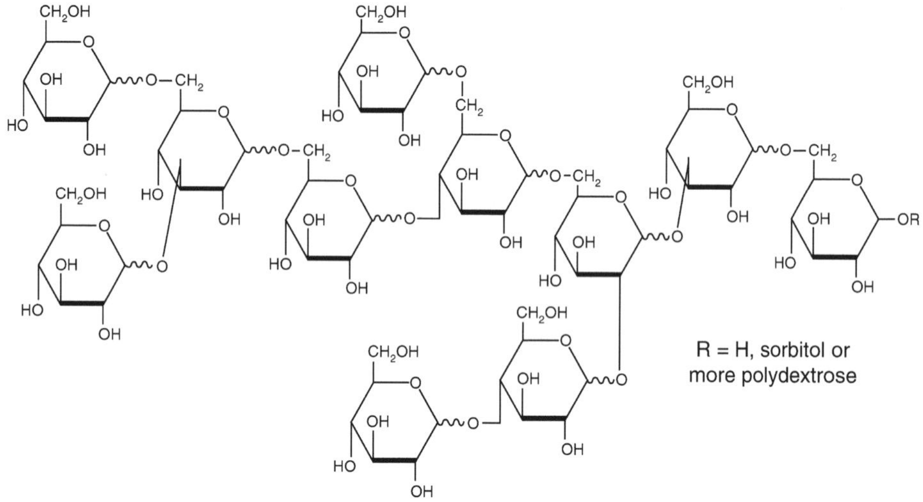

■ Figure 11.1

Representative structure for polydextrose. R = H, sorbitol, sorbitol bridge, or more poly-dextrose.

11.2.2 Safety

The safety of polydextrose in the human diet has been comprehensively demon-strated (Burdock and Flamm 1999). The data have been examined by many expert national and supra-national panels. Both the Joint Food and Agriculture Office of the United Nations (FAO) and World Health Organization (WHO) Expert Committee on Food Additives (JECFA) and the European Commission, Scientific Committee on Food (EC/SCF) have assigned an acceptable daily intake (ADI) "not specified", meaning that polydextrose can be added to foods at the level required to achieve the desired functionality (JECFA 1987; EC/SCF 1990).

11.2.3 Tolerance

Nine clinical studies have been conducted with polydextrose to evaluate the extent of gastrointestinal symptoms in the lower gastrointestinal tract. These studies showed that polydextrose is better tolerated than most other low diges-tible carbohydrates. This is to be expected as polydextrose has a higher molecular weight, leading to a lower risk of osmotic diarrhea and also it is only partially fermented (see below). Having evaluated the data JECFA and EC/SCF concluded that polydextrose has a mean laxative threshold of about 90 g/day (1.3 g/Kg bodyweight) or 50 g as a single dose (Flood et al. 2004).

11.2.4 Polydextrose as a Bulking Agent

Polydextrose is resistant to digestion in the upper gastrointestinal tract and is partially fermented in the large intestine producing volatile fatty acids. Approximately 50% of the glucose equivalents of polydextrose are excreted. A number of energy balance and isotope-label disposition studies have been conducted in animals and man to evaluate the energy contribution of polydextrose. A review of the data from 14 studies concluded that polydextrose contributes an energy value of approximately 1 Kc/g (Auerbach et al. 2007).

Polydextrose can be incorporated into a wide variety of foods and beverages in place of fully caloric carbohydrates and, to some extent, dietary fats. Polydextrose itself is only slightly sweet and intense sweeteners can be used to adjust the overall sweetness of the product. In this way polydextrose facilitates the production of reduced energy foods and beverages. Indeed, this application as a low calorie bulking agent was the main *raison d'être* for polydextrose during the initial commercialization phase in the 1980's. Since then a diverse programme of investigations has been undertaken to evaluate the physiological implications of polydextrose and today the emphasis has shifted towards the use of polydextrose for its physiological benefits. These include:

- Oral health – polydextrose has been shown to be non-cariogenic
- Dietary fiber properties
- Reducing glycaemic impact – polydextrose can be used to replace glycaemic carbohydrates to reduce the overall glycaemic response to foods and diets, and
- Prebiotic properties

Only the latter will be considered in detail here. However, it is appropriate to summarize the fiber attributes of polydextrose as these overlap with its consideration as a prebiotic. Danisco currently markets its Litesse® polydextrose as the "*Single ingredient with multiple benefits*" (www.litesse.com). In this way polydextrose differs from most other commercially available fibers in that it is typically added to foods to facilitate a number of claims and not just for its potential prebiotic properties.

11.2.5 Polydextrose as Dietary Fiber

The lack of a globally accepted definition of dietary fiber has led to confusion. The WHO/FAO Codex Alimentarius Committee on Nutrition and Foods for Special Dietary Uses (CCNFSDU) has been debating the subject since 1992. In 2005 CCNFSDU proposed the following definition of dietary fiber:

"Dietary fiber means carbohydrate polymers with a degree of polymerizations (DP) not lower than three, which are neither digested nor absorbed in the small intestine. A degree of polymerization not lower than three is intended to exclude mono- and disaccharides. It is not intended to reflect the average DP of a mixture. Dietary fiber consists of one or more of:

* *Edible carbohydrate polymers naturally occurring in the food as consumed*
* *Carbohydrate polymers, which have been obtained from food raw material by physical, enzymatic or chemical means*
* *Synthetic carbohydrate polymers"*

With the exception of non-digestible edible carbohydrate polymers naturally occurring in foods a physiological effect would need to be scientifically demonstrated (*http://www.ccnfsdu.de/fileadmin/user_upload/PDF/ReportCCNFSDU2005.pdf*).

There is widespread international support for this definition of dietary fiber. It takes into account the latest science and ingredient developments and relates best to potential consumer benefits. It is hoped that a definition along these lines will soon be finalized. On this basis polydextrose would be considered as dietary fiber.

Human feeding studies have consistently reported improved bowel function in diets supplemented with polydextrose. Animal and human data indicate that polydextrose has a moderate beneficial effect on cholesterol metabolism, similar to other soluble fermentable dietary fibers. Polydextrose elicits a negligible glycaemic and insulinaemic response (◉ *Figure 11.2*) and, in addition, has been shown to attenuate the blood glucose raising potential of glucose itself (Stowell 2009).

Even in the absence of a harmonized definition, the fiber status of polydextrose is widely accepted in most countries around the world. The Fibe Mini produced and marketed by Otsuka Pharmaceutical Company since the 1980s (http://www.fibemini.hk/) is an excellent example of a functional food targeted at improving bowel function. This product is based on polydextrose as the active ingredient.

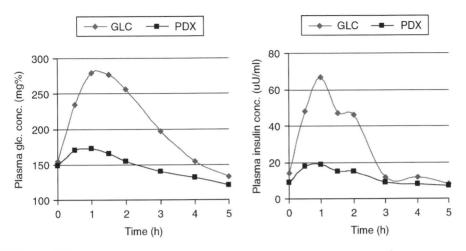

■ Figure 11.2
A comparison of postprandial plasma glucose and insulin response to either glucose or polydextrose ingestion. The mean response has been calculated from ten subjects (McMahon 1978).

11.3 Application of Polydextrose as a Food Ingredient

Polydextrose is a highly versatile food ingredient with excellent stability (both heat and acid) and compatibility with most food and beverage matrices. In some systems it can act as a humectant, facilitating moisture management, and it may also form a stable glass structure which can be used to positive effect in a variety of applications. It is used on a global basis in a bewildering array of commercial food and beverage products, to facilitate the production of foods with:

- A reduced or low energy content and/or energy density
- A reduced or low glycaemic impact
- An enhanced fiber content, and
- A prebiotic effect where accepted

Applications include:

- Confectionery – both chocolate and sugar confectionery
- Baked goods
- Frozen dairy desserts
- Cultured dairy products
- Beverages and dairy drinks

- Fruit spreads and fruit fillings
- Meat applications
- Pasta and noodles, and
- Pharmaceuticals
- Among others

The technological aspects of polydextrose and its food applications have been well described elsewhere (Auerbach et al. 2006; Mitchell et al. 2001; Stowell 2009).

11.4   Evaluation of Polydextrose as a Prebiotic

11.4.1   Introduction

In considering the potential prebiotic properties of polydextrose it is important to refer to the accepted definition of a "prebiotic." Until recently it appeared that consensus existed with regard to this definition. Gibson and Roberfroid (1995) first defined a prebiotic as "*a non-digestible food ingredient that beneficially affects the host by selectively stimulating the growth/and or activity of one or a limited number of bacteria in the colon, and thus improves host health*". The concept was then updated and it was proposed that the ability of a food component to function as a prebiotic should be assessed against the following criteria (Gibson et al. 2004):

- Resistance to gastric acidity, hydrolysis by mammalian enzymes and gastrointestinal absorption
- Fermentation by the intestinal microbiota, and
- Selective stimulation and/or activity intestinal bacteria associated with health and wellbeing (currently agreed at the present time to include bacteria of the genera *Bifidobacterium* and *Lactobacillus*)

Recently the definition of a "prebiotic" has been revisited both by the Food and Agriculture Organization of the United Nations (FAO – www.fao.org/ag/agn/agns/files/Prebiotics_Tech_Meeting_Report.pdf) by the International Life Sciences Institute (ILSI – http://europe.ilsi.org/activities/taskforces/diet/PrebioticsTF.htm). And by the International Scientific Association for Probiotics and Prebiotics (ISAPP) who state that:

"Prebiotics are selectively fermented, dietary ingredients that result in specific changes in the composition and/or activity of the gastrointestinal microbiota, thus conferring benefit(s) upon host health." (http://www.isapp.net/docs/ Consumer Guidelines-prebiotic.pdf)

A range of studies have been undertaken on polydextrose to evaluate its prebiotic properties. These include:

- *In vitro* studies
- Metabolism studies in animals, and
- Human intervention studies.

11.4.2 Resistance to Gastric Acidity, Hydrolysis by Mammalian Enzymes and Gastrointestinal Absorption

Because of its range of glycosidic linkages, polydextrose has a compact structure that resists hydrolysis by digestive enzymes found in the upper gastrointestinal tract of mammals.

Oku et al. (1991) studied the hydrolysis of polydextrose *in vitro* using a homogenate of rat intestinal mucosa and found it to be very weakly hydrolyzed. It was concluded that polydextrose contains only a small proportion of α-1,4 and/or α-1,6 glycosidic linkages that are accessible to hydrolysis by mammalian GI enzymes. The low frequency of linkages susceptible to enzymatic hydrolysis was confirmed by Kobayashi and Yoshino (1989) and Stumm and Baltes (1997). In metabolic studies in rats Figdor and Rennhard (1981) concluded that approximately 60% of the orally administered polydextrose was eliminated in the feces. In a study by Fava et al. (2007) in which pigs were fed diets supplemented with polydextrose, residual polydextrose was found to be present in the distal colon.

In a study in humans by Figdor and Bianchine (1983) in which polydextrose was consumed daily for 10 days, 50% of a radiolabel incorporated in the polydextrose consumed on day 8 was recovered in the feces over the subsequent 2 days. In a study by Achour et al. (1994) in which humans consumed polydextrose over a 30-day period, 33% and 32% of a radiolabel incorporated in the polydextrose consumed on days 5 and 27 respectively was recovered in the feces and considered to represent intact polydextrose.

The glycaemic response to polydextrose is given in the International table of glycaemic index and glycaemic load values as 7 ± 2 compared to glucose at 100. Litesse®UltraTM, Danisco's premium branded polydextrose, mediates a glycaemic response of 4 ± 2 (Foster-Powell et al. 2002).

These studies confirm that polydextrose is poorly digested and poorly absorbed from the GI tract in humans.

11.4.3 Fermentation by the Intestinal Microbiota

Animal data (Figdor and Rennhard, 1981; Juhr and Franke, 1992) and human data (Figdor and Bianchine, 1983; Achour et al., 1994; Endo et al. 1991) indicate that the proportion of polydextrose fermented is of the order of 30 to 50%, which is similar to that for other types of dietary fibre such as cellulose. In an *in vitro* study in which the fermentation of polydextrose by human stool homogenates was assessed by hydrogen evolution, polydextrose was found to have a fermentability of 24.8% compared to a glucose reference (Solomons and Rosenthal 1985). When the fermentability of polydextrose was studied in an *in vitro* four-stage model using human fecal inocula and simulating digestion from the proximal to the distal colon, progressive degradation of polydextrose throughout the four stages was observed, accompanied by a progressive increase in short chain fatty acids (SCFAs) (Mäkivuokko et al. 2005). Fava et al. (2007) studied the effect of polydextrose on intestinal microbes and immune function in pigs. They demonstrated that polydextrose is fermented gradually throughout the colon and total concentrations of SCFAs increased in the colon.

Another *in vitro* study using human fecal homogenates to compare the fermentation profiles of a range of carbohydrates found that the quantities of SCFAs produced by polydextrose were similar to those produced by other nondigestible carbohydrate substrates while the quantity of gas produced was lower (Arrigoni et al. 1999). Within the SCFAs, the molecular ratio of acetate/propionate/butyrate produced by the fermentation of polydextrose was found to be similar to that produced by fermentations of fructo-oligosaccharides (FOS) and xylo-oligosaccharides. A further *in vitro* study using human fecal homogenates compared the compounds resulting from fermentation of polydextrose with those produced by fermentation of other carbohydrate substrates (oligofructose, inulin, glucose, arabinose, galactose, fructose, lactose, sucrose, lactulose, cellobiose, sorbitol, lactitol, Litner starch, pectins, maltitol and arabinogalactan) and concluded that the quantities of SCFAs produced in 48 h were comparable to those seen with the other carbohydrates (Wang and Gibson 1993). These studies confirm that polydextrose is partially fermentable by human colonic bacteria. The extent of this fermentation varies but is typically in the region of 50% as measured by glucose equivalents.

One positive impact of selectively enhanced saccharolytic (carbohydrate) versus putrefactive (protein) fermentation is a reduction in colonic pH which discourages the growth of pathogens which favor a higher pH.

Recent colon simulator studies, summarized in ❯ *Figure 11.3*, have shown that polydextrose enhances the production of butyrate at each stage in the colon. Acetate and propionate were also enhanced (data not shown). It is well established that butyrate acts as a substrate for the colonic mucosa and enhanced butyrate would be expected to contribute positively to mucosal integrity. The enhancement of butyrate was achieved without any accumulation of lactic acid. Hence, a balance was established between lactic acid generation and its subsequent fermentation to short chain fatty acids.

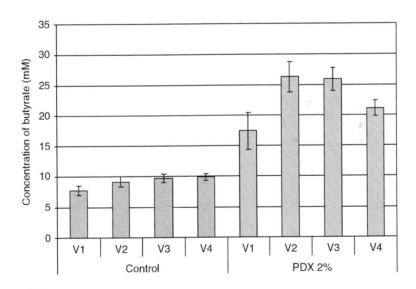

◙ Figure 11.3
Butyrate concentrations in colon simulator vessels after 48 h simulations (average of 7 control and 8 polydextrose [PDX] simulations). V1 represents the ascending part of the colonic model, V2 represents the transverse colon, V3 the descending colon, and V4 the sigmoid and rectum area. Different parts of the model (V1–V4) differ from each other in pH levels and volume of the media and hence in microbial numbers (Mäkivuokko et al., 2005; Mäkeläinen et al., 2007).

11.4.4   Selective Stimulation and/or Activity of Intestinal Bacteria Associated with Health and Wellbeing

In a randomized double-blind, placebo-controlled study by Jie et al. (2000), in which 120 subjects consumed polydextrose over a 28-day period at different dose levels, increases in the numbers of viable bacteria of the *Lactobacillus* and *Bifidobacterium* species were detected in the subjects' stools. The increases in numbers of these bacteria were dose-dependent and were statistically significant from a dose of 4g/day (the lowest dose tested) upwards. At the same time, the number of *Bacteroides spp.* organisms in subjects' stools was reduced. In a study in human subjects by Endo et al. (1991), cited above, in which eight healthy volunteers were fed a diet rich in cholesterol and had a daily intake of 15 g of polydextrose for 6 weeks, changes in colonic flora were accompanied by a decrease in fecal concentrations of *Clostridium spp.* In a study in humans by Tiihonen et al. (2008), supplementation with polydextrose at 5 g/day and a probiotic mixture together was found to increase culturable fecal bifidobacteria over supplementation with the probiotic mixture alone when compared over a 2 week period in twenty subjects (summarized in ◉ *Table 11.1*). Consumption of the probiotic mixture increased the faecal levels of culturable lactobacilli and propionibacteria. These levels were not further increased by the addition of polydextrose to the probiotic mixture.

In an *in vitro* model simulating bacterial activity in different segments of the human colon, Probert et al. (2004) investigated the effect on bacterial populations of a range of fermentable substrates (polydextrose, lactitol and FOS).

◻ Table 11.1

A probiotic mixture supplemented with polydextrose increased cultured bifidobacteria (Tiihonen et al. 2007)

Period	Bifidobacteria (Log_{10}cfu/g wet weight feces)		
	Mean	S.D.	p
Run-in	7.0	2.2	NS
Probiotic	7.0	2.0	NS
PDX + Probiotic	8.9	2.5	$<.001$
Follow-up	8.5	1.5	<0.05

Study involved 22 young adults (15 female/5 male). Polydextrose was fed at 5 g/day. The probiotic mixture included: *Lactobacillus* GG, *L. rhamnosus* LC705, *Propionibacterium shermannii* JS and *Bifidobacterium breve* Bbi. Fecal samples were analyzed after each 2 week period. Total bifidobacteria counts were measured by plating

Polydextrose was found to stimulate the number of bifidobacteria and to support the growth of a wide variety of bifidobacteria species in all three of the colon segments modelled. In an *in vitro* study using human faecal homogenates Wang and Gibson, 1993, previously referenced above, found that polydextrose stimulated bacterial growth but less selectively so than other substrates studied.

Taken together, these studies suggest that administration of polydextrose can favorably influence the balance of the microbial genera present in the colonic microbiota.

11.4.5 Other Studies

Several additional studies have assessed the prebiotic potential of polydextrose. Some of the findings are in line with those for other prebiotics whilst some point to interesting and possibly unique effects of polydextrose.

Hara et al. (2000) showed that dietary polydextrose increased calcium absorption and bone mineralization in rats. The effect may partly have been due to colonic acidification which is known to increase calcium solubility. However, the main positive effect was, unexpectedly, seen in the small intestine.

Ishizuka et al. (2003), using a rat model, showed that polydextrose considerably reduced the formation of aberrant crypt foci (ACF) in the presence of a carcinogen. The effect was most pronounced in the rectum where the reduction was up to 65%. The authors concluded that ingestion of polydextrose may prevent colorectal carcinogenesis. This study highlights the fact that polydextrose mediates its effect into the distal colon and colorectal area (at least in the rat model).

As seen above, fermentation of polydextrose has beneficial effects for mucosal functions. Enhanced butyrate production serves as an important energy source, not only for epithelial cells, but also for mucosal immune cells. Polydextrose has been shown to increase production of immunoglobulin A (IgA) in the large intestine of rats and a synergistic effect was seen with a polydextrose: lactitol combination (Peuranen et al. 2004).

Balancing immune responses in the large intestine is especially important for reducing the risk of colon cancer development. A possible mechanism for reduction in cancer development involves the regulation of mucosal gene expression. Over-expression of the cyclo-oxygenase 2 (cox-2) gene is related to early stages of colon cancer development and chronic inflammatory diseases in the intestine. Mäkivuokko et al. (2005) combined two different *in vitro* systems, namely a four-stage simulator of colonic fermentation and a cell-culture based model of human intestinal epithelial function, in order to study the effects of polydextrose on colon cancer development. A dose-dependent decreasing effect on cox-2 expression was observed in Caco-2 cells (a human colon cancer cell line). This reduction of cox-2 expression associated with the colonic fermentation of polydextrose further suggests a protective role of polydextrose against colon cancer. Fava et al. (2007) showed a reduction in the cox-2 inflammatory bio-marker in association with polydextrose consumption in a pig model. They did, however, not observe a change in the microbiota composition as determined by fluorescent *in situ* hybridisation, this may relate to the relatively high fibre content of the test feed, and the young age of the animals. The reduction in cox-2 is nevertheless a good example of the application of the emerging science of nutrigenomics in the evaluation of a prebiotic.

Prebiotics have recently been proposed as ingredients to encourage satiety or, more specifically, to enhance satiety. It is not clear whether this effect is mediated via the microbiota or some other mechanism so this may or may not be a prebiotic effect *sensu stricto*. King et al. (2005) studied the impact of xylitol and polydextrose on satiety in 20 human subjects. Xylitol (25 g), xylitol:polydextrose (50:50; 25 g) or polydextrose (25 g) in yoghurt ingested as a preload consumed 90 min before lunch suppressed combined calorie intake by 5–8% versus a sucrose control. This study demonstrated that xylitol and/or polydextrose, under these specific conditions, mediates a satiety effect. Whilst the dose of poly-dextrose and/or xylitol was relatively high the effect was sustained over a 10 day period, which is encouraging.

Vasankari and Ahotupa (2005) found that the ingestion of 12.5 g of poly-dextrose along with a hamburger meal reduced the total postprandial hypertri-glyceridemia by 25%. It has been postulated that the polydextrose passing through the lumen of the intestine interferes with fat uptake in some way. Further investigations are required to confirm the effect and clarify the mechanism.

11.5   Regulatory Considerations

As noted above, there is no universally adopted definition of a "prebiotic". Although further studies are certainly needed, polydextrose seems to comply with most current scientific opinion on what constitutes a "prebiotic" and, where formal opinions have been given they have largely been positive. In the absence of a formal opinion polydextrose may be considered as a prebiotic on the basis that this informs and does not mislead consumers. In Europe, under the auspices of the recently adopted Regulation on Nutrition and Health Claims made on foods (EC1924/2006), two relevant claims for polydextrose have been forwarded by EU Member States to Brussels and are now on the consolidated list under consideration by the European Food Safety Authority (EFSA). These are:

Polydextrose improves bowel function
Conditions of use: 4g/day
Nature of evidence: Authoritative bodies, reviews, human studies, animal studies
Examples of wording: Polydextrose promotes good intestinal health; Polydextose improves bowel function and gut comfort; Polydextrose stimulates metabolic activity
Polydextrose is a prebiotic with bifidogenic properties
Conditions of use: 4g/day
Nature of evidence: Human studies, *in vitro* studies
Examples of wording: Polydextrose stimulates the growth of beneficial bacteria in the gut; Polydextrose stimulates the growth of bifidobacteria in the colon; Polydextrose stimulates the growth of lactobacilli bacteria in the gut; Prebiotics promote healthy/well-balanced gut bacteria/flora;

Time will tell which claims will finally be consolidated into the "Article 13" health claims list due for publication in January 2010. Local authorities should always be consulted before finalizing a food or beverage claim.

11.6   Future Developments

Further studies are currently underway to elucidate the prebiotic properties of polydextrose. The emphasis is mainly on human intervention studies and subgroups of the population will be targeted in some of these. Mechanistic data are also considered important to explain observations in humans and to support claims. The impact on novel biomarkers will be included in the evaluation where relevant.

11.7   Conclusion

A wide variety of studies have been undertaken to evaluate the prebiotic properties of polydextrose. The following attributes have been demonstrated:

- Slow fermentation rate
- Sustained fermentation throughout the colon

- Enhancement of bifidobacteria
- Increase in butyrate production
- No lactic acid accumulation
- Decrease in putrefaction (protein fermentation)
- Stimulation of the gastric immune system – IgA (Rat model)
- Reduction of colon cancer risk (based on rat and *in vitro* models)
- Attenuation of postprandial triglycerides

The technological versatility of polydextrose, combined with the interesting range of physiological benefits it mediates, makes polydextrose an excellent choice for those seeking to enhance the nutritional profile of foods and beverages. However, it should be noted that the prebiotic status of polydextrose is still under debate, with further studies in progress and questions over data generated using more traditional techniques. It is important when intending to label a food or beverage as 'prebiotic' to consider the totality of the evidence and to always seek local opinions in respect of legislative requirements.

References

Achour L, Flourie B, Briet F, Pellier P, Marteau P, Rambaud J (1994) Gastro-intestinal effects and energy value of polydextrose in healthy non-obese men. Am J Clin Nutr 59:1362

Anon (2002) AOAC official method 2000.11 Polydextrose in foods. Official methods of analysis of AOAC International, 17th edition, revision 2, 2003, Chap. 45, p 78C

Anon (2004) Food chemicals codex. Fifth Edition, The National Academies Press, Washington, D.C. ISBN 0-309-08866-6, 336–339

Arrigoni E, Jann A, Rochat F, Amado R (1999) In vitro fermentability of indigestible oligo- and polysaccharides. Swiss Federal Institute of Technology, Institute of Food Science, Zurich, Switzerland

Auerbach M, Craig S, Mitchell H (2006) Bulking agents: multifunctional ingredients. In: Mitchell H (ed) Sweeteners and sugar alternatives in food technology Blackwell Publishing Ltd, Oxford, ISBN-13: 978-14051-3434-7

Auerbach MH, Craig SAS, Howlett JF, Hayes KC (2007) Caloric availability of polydextrose. Nutr Rev 65(12):544–549

Burdock GA, Flamm WG (1999) A review of the studies of the safety of polydextrose in food. Food Chem Toxicol 37(2–3): 233–264

EC/SCF (1990) 4.2 Polydextrose. Excerpt from the minutes of the 71st meeting of the Scientific Committee for Foods held on 25–26 January

Endo K, Kumemura M, Nakamura K, Fujisawa T, Suzuki K, Benno Y, Mitsuoka T (1991) Effect of high cholesterol diet and polydextrose supplementation on the microflora, bacterial enzyme activity, putrefactive products, volatile fatty acid (VFA) profile, weight, and pH of the feces in healthy volunteers. Bifidobact Microflora 10:53

Fava F, Mäkivuokko H, Siljander-Rasi H, Putaala H, Tiihonen K, Stowell J, Tuohy K, Gibson G, Rautonen N (2007) Effect of polydextrose and intestinal microbes and immune functions in pigs. B J Nutr 98:123

Figdor SK, Bianchine JR (1983) Caloric utilization and disposition of [^{14}C]polydextrose in man. J Agric Food Chem 31:389

Figdor SK, Rennhard HH (1981) Caloric utilization and disposition of [^{14}C]polydextrose in the rat. J Agric Food Chem 29:1181

Flood MT, Auerbach MH, Craig SAS (2004) A review of the clinical toleration studies of polydextrose in food. Food Chem Toxicol 42:1531–1542

Foster-Powell K, Holt SHA, Brand-Miller JC (2002) International table of glycemic index and glycemic load values. Am J Clin Nutr 76(1):5–56

Gibson GR, Roberfroid MB (1995) Dietary modulation of the human colonic microbiota: introducing the concept of prebiotics. J Nutr 125:1401–1412

Gibson GR, Probert HM, Van Loo J, Rastall RA, Roberfroid MB (2004) Dietary modulation of the human colonic microbiota: updating the concept of prebiotics. Nut Res Rev 17:259–275

Hara H, Suzuki T, Aoyama Y (2000) Ingestion of the soluble dietary fibre, polydextrose, increases calcium absorption and bone mineralization in normal and total-gastrectomized rats. B J Nutr 84:655–661

Ishizuka S, Nagai T, Hara H (2003) Reduction of aberrant crypt foci by ingestion of polydextrose in the rat colorectum. Nutr Res 23:117–122

JECFA (1987) Evaluation of certain food additives and contaminants. Thirty-first report of the Joint FAO/WHO Expert Committee on Food Additives, World Health Organization Technical Report Series 759, Geneva

Jie Z, Luo B, Xiang M, Liu H, Zhai Z, Wang T, Craig SAS (2000) Studies on the effects of polydextrose on physiological function in Chinese people. Am J Clin Nutr 72:1503–1509

Juhr N, Franke J (1992) A method for estimating the available energy of incompletely digested carbohydrates in rats. J Nutr 122:1425

King NA, Craig SAS, Pepper T, Blundell JE (2005) Evaluation of the independent and combined effects of xylitol and polydextrose consumed as a snack on hunger and energy intake over 10d. Br J Nutr 93:911–915

Knapp BK, Parsons CM, Swanson KS, Fahey GC, Jr. (2008), "Physiological responses to novel carbohydrates as assessed using canine and avian models". J Agric Food Chem 56(17):7999–8006

Kobayashi T, Yoshino H (1989). Enzymatic hydrolysis of "Polydextrose". Denpun Kagaku 36:283–286

Mäkeläinen HS, Mäkivuokko HA, Salminen SJ, Rautonen NE, Ouwehand AC (2007) The effects of polydextrose and xylitol on microbial community and activity in a 4-stage colon simulator. J Food Sci 72(5):M153–M159

Mäkivuokko H, Kettunen H, Saarinen M, Kamiwaki T, Yokoyama Y, Stowell J, Rautonen N (2007) "The effect of cocoa and polydextrose on bacterial fermentation in gastrointestinal tract simulations". Biosci Biotechnol Biochem 71(8):1834–1843

Mäkivuokko H, Nurmi J, Nurminen P, Stowell J, Rautonen N (2005) In vitro effects on polydextrose by colonic bacteria and Caco-2 cell cyclooxygenase gene expression. Nutr Cancer 52:94

McMahon FG (1978) Polydextrose food additive petition. FDA Petition 9A3441 Pfizer Inc., New York

Mineo H, Hara H, Kikuchi H, Sakura H, Tomita F (2001) "Various indigestible saccharides enhance net calcium transport from the epithelium of the small and large intestine of rats in vitro". J Nutr 131:3243–3246

Mitchell H, Auerbach MH, Moppett FK (2001) Polydextrose. In: Nabors L, O'Brien (eds) Alternative Sweeteners, Third Edition Marcel Dekker Inc, New York, ISBN: 0-8247-0437-1

Oku T, Fuji Y, Okamatsu H (1991) Polydextrose as dietary fiber: hydrolysis by digestive enzymes and its effect on gastrointestineal transit time in rats. J Clin Biochem Nutr 11:31–40

Peuranen S, Tiihonen K, Apajalahti J, Kettunen A, Saarinen M, Rautonen N (2004) Combination of polydextrose and lactitol affects microbial ecosystem and immune responses in rat gastrointestinal tract. B J Nutr 91:905–914

Probert HM, Apajalahti JHA, Rautonen N, Stowell J, Gibson GR (2004) Polydextrose, lactitol and fructo-oligosaccharide fermentation by colonic bacteria in a three-stage continuous culture system. Appl Environ Microbiol 70:4505–4511

Solomons NW, Rosenthal A (1985) Intestinal metabolism of random-bonded polyglucose bulking agent in humans: in vitro and in vivo studies of hydrogen evolution. J Lab Clin Med 105:585–592

Stowell JD (2009) Polydextrose. In: Cho SS (ed) Fiber ingredients: food applications and health benefits. Taylor & Francis, ISBN10 1420043846, pp. 173–204

Stumm I, Baltes W (1997) Analysis of the linkage positions in polydextrose by the reductive cleavage method. Food Chem 59:291–297

Tiihonen K, Suomalainen T, Tynkkynen S, Rautonen N (2008) Effect of prebiotic supplementation on a probiotic bacteria mixture: comparison between a rat model and clinical trials. B J Nutr 99:826–831

Vasankari TJ, Ahotupa M (2005) Supplementation of polydextrose reduced a hamburger meal induced postprandial hypertrigly-ceridemia. AHA Congress Proceedings, vol. 112. American Heart Association (AHA) Congress, Dallas, 16th November, 2005, pp. II-833

Wang X, Gibson GR (1993) Effects of the in vitro fermentation of oligofructose and inulin by bacteria growing in the human large intestine. J Appl Bacteriol 75:373

12 Prebiotics in Companion and Livestock Animal Nutrition

Kathleen A. Barry · Brittany M. Vester · George C. Fahey, Jr.

12.1 Introduction

Prebiotic supplementation of animal diets began in an attempt to increase concentrations of beneficial intestinal microbiota. It was understood that prebiotics inhibited growth of intestinal pathogens and decreased concentrations of stool odor-causing metabolites. Since the use of prebiotics began, several countries have banned the use of antimicrobials in livestock animal feeds, and several more have placed restrictions on the quantity of antimicrobials that can be used. Prebiotic supplementation has become increasingly popular as the body of evidence supporting its use continues to grow. As this literature expands, the number of potential prebiotic substances has grown beyond those that are naturally occurring, such as those found in chicory and yeast products, to include a large number of synthetic or chemically/enzymatically manufactured prebiotics.

Definition and types of prebiotics. To be considered a prebiotic, a compound must conform to the following guidelines: first, it must be resistant to digestion in the upper gastrointestinal tract (remain unaltered through hydrolytic-enzymatic digestion); second, it must selectively stimulate one or a limited number of beneficial microbiota; and third, it must benefit host health by improving colonic microbiota composition (Roberfroid et al., 1998). Fructans are one of the most popular prebiotic supplements available, comprising 61% of the publications on the topic of prebiotic supplementation. While fructans occur naturally in feeds, hydrolytic and enzymatic methodologies have produced fructans of varying chain lengths. Examples of fructans include fructooligosaccharides (FOS), short-chain FOS (scFOS), oligofructose (OF), and inulin, as well as several blends of varying chain lengths. Generally, fructans contain a linear chain of fructose, which can vary in length from 2 to 60 units, with or without a glucose unit on the end of the chain (Flickinger et al., 2003). Fructans escape hydrolytic digestion due to their β-(2,1) bonds (Flickinger et al., 2003).

Mannanoligosaccharide (MOS) research comprises 33% of the reports about prebiotics. Similar to fructans, several forms of MOS are available. These forms can differ greatly in their ability to influence intestinal microbiota. The crudest form of MOS is unmodified yeast (*Saccharomyces cervisiae*), which appears to act as a prebiotic due in part to the MOS contained in its cell wall. In addition, purified yeast cell wall has been extracted and utilized in animal feeds. Manna-noligosaccharides are able to bind with certain intestinal microbiota due to the mannan component in its structure and thereby prevent attachment of micro-biota to the intestinal cells (Spring et al., 2000).

Soy oligosaccharides also have received attention as prebiotic supplements, comprising 9% of prebiotic literature. These oligosaccharides have been fed intact as a component of the whole soybean, or as purified raffinose and stachyose to determine their individual and synergistic effects. Raffinose and stachyose are noted for gas production which occurs when these oligosaccharides are fed intact in soybeans or as purified oligosaccharides. Additionally, several guar fractions, namely gum, germ, hull, and meal, have been investigated as potential prebiotics due to their fermentative characteristics.

Galactooligosaccharide (GOS) and trans-galactooligosaccharide (TGOS/TOS) research comprises 9% of prebiotic literature. Galactooligosaccharides are produced by subjecting lactose to microbial galactosyltransferases (Tzortis et al., 2005). Trans-galactooligosaccharides are water-soluble, contain galactose units with β-linkages that prevent hydrolytic digestion, and are rarely found in feeds (Houdijk et al., 1998). Gluconic acid, a GOS-like prebiotic, also has been inves-tigated (Biagi et al., 2006).

In addition to naturally occurring prebiotic substances, several synthetic pre-biotics have been investigated. Lactosucrose, also called 4G-β-D-galactosylsucrose or 0-β-D-galactopyranosyl-(1–4)-0-α-D-glucopyranosyl-(1–2)-β-D-fructofuranoside, contains galactose, glucose, and fructose and is produced by *Arthrobacter* spp. K-1 (Fujita et al., 1990). Isomaltooligosaccharide (IMO) is produced by the enzymatic transgalactosylation of glucose and contains isomaltose, panose, isomaltotriose, and other oligosaccharides with four or five glucose residues (Zhang et al., 2003). Pullulan consists of glucose units linked as α-1,6-maltotriose subunits (Spears et al., 2005). γ-Cyclodextrin is a ring-shaped oligosaccharide composed of α-1,4-linked glucose units (Spears et al., 2005). Both pullulan and γ-cyclodextrin are partially hydrolyzed in the upper digestive tract of non-ruminants, but the majority of each compound is believed to pass into the large intestine.

Digestive strategies of companion and livestock species. Prebiotic efficacy in any animal species is closely correlated to its digestive strategy. Non-ruminant animals, such as the dog and cat, employ post-gastric fermentation. The dog and cat have a very small or nonexistent cecum and a tubular colon with little sacculation (NRC, 2006). Due to these features, they typically have a short (approximately 12 h) intestinal residence time and digest approximately 8% of food in the colon (NRC, 2006). Other non-ruminants, such as poultry and swine, also employ post-gastric fermentation but are capable of a higher level of colonic fermentation than cats and dogs. These animals have one cecum (swine) or two ceca (poultry) adapted to microbial fermentation. In addition, the colon of swine is highly sacculated and allows for an increased residence time. Fermentation in the colon of swine accounts for 5–28% of their maintenance energy requirement (NRC, 1998). Herbivorous non-ruminants like the horse have a very large cecum and highly sacculated colon that allows for up to 60% of the animal's energy needs to be met through fermentation (NRC, 2007a). Unlike other non-ruminants, herbivorous non-ruminants are able to consume and utilize cellulose for energy.

Similar to non-ruminant animals, the young ruminant, also referred to as a pre-ruminant, employs post-gastric fermentation until the ruminal microbiota are established. Ruminal microbiota stabilize after 6–8 wk, and these animals are fed as ruminants after this point (Damron, 2006). Adult ruminant animals are pre-gastric fermenters. The majority of fermentation in adult ruminants occurs in the reticulorumen, which can provide 50–70% of the energy requirements of the animal through short-chain fatty acid (SCFA) production (Damron, 2006). While these animals are both pre- and post-gastric fermenters, hindgut fermentation has not been the focus of prebiotic research in ruminants.

Microbiota methodologies. One of the key outcomes of prebiotic supplementation is a beneficial alteration in the microbiota composition of the host animal, so identification of specific microbial communities is necessary to determine efficacy. For many years, researchers have investigated alterations in microbiota using culturing techniques (e.g., plating on selective media). While plating can quantify microbial communities, there are several drawbacks to its use. First, only viable microbiota can be cultured. If the species of interest is an anaerobic colony, this task can be quite challenging as viable members will begin to expire shortly after the sample is collected, even under conditions of proper storage and transport. Second, selective media must be employed to ensure that only a specific microorganism is cultured. However, several microbes can be grown on the same media, which can lead to inaccurate quantification (Rastall, 2004). Additionally, it is now known that several microbial species are dependent upon one another for growth and maintenance of the colony in question (NRC, 2007b). To culture such microbial species individually is impossible using current plating methods.

Fluorescence *in situ* hybridization (FISH) initially was developed as a qualitative tool to identify two or more species in a growth medium. With FISH, a fluorescent marker is hybridized to a specific region of microbial DNA. The microbe then is fixed to a microscope slide (Harmsen and Welling, 2002). The accuracy of FISH counting varies greatly, as general practice is to count the number of microbiota in a specific area and multiply by the entire area to arrive at a conclusion. Further advancements in FISH technology allowed separation and, thus, quantification of microbiota by joining FISH with flow cytometry.

To address the problems associated with plating, molecular methods have been developed to more clearly understand the makeup of the intestinal microbiome. Quantitative PCR (qPCR) has provided insight into the microbial community of the intestine. Primers and probes specific to a microbial genus or species are prepared, and then added to a mixture of DNA and buffer to quantify the genus or species based on their DNA. This method, like the others, is not without its problems. First, the numbers of microbes quantified do not strictly represent viable microbiota, as the primers and probes amplify DNA of all the cells present. This may lead to an inaccurate count of changes in microbiota due to prebiotic activity (Middelbos et al., 2007a). Second, the primers and probes may not be specific enough to the species in question. While the primers and probes selected for each organism strive to specify an exact microorganism, there can be cross-reactivity among species and genera of microorganisms with similar genetic codes. Therefore, each primer/probe pairing should be checked against a "negative control" or a pure culture of microbial DNA that should not react with the primer/probe pairing. While several issues remain unsolved as regards current methods in quantifying microbial communities, technologies continue to improve.

12.2   Prebiotic Use in Companion Animal and Livestock Species

Canines. Several types of prebiotics have been utilized in canine nutrition research. Details of the experiments are presented in ❯ *Table 12.1*. As microbiota

○ Table 12.1

In vivo experiments, listed in chronological order beginning in 2005, reporting effects of prebiotics in dogs[a,b,c] (Cont'd p. 358)

Reference	Outcome variables quantified	Animals/ treatment (age, initial BW)	Dietary information; time on treatment	Daily prebiotic dose; source	Major findings
Jeusette et al. (2005)	Plasma leptin	12 Obese dogs (between 1 and 9 yr; 21.9 ± 0.8 kg)	Basal diet (Obesity veterinary diet, Royal Canin, France) fed to promote weight loss	Control (0%)	No effect of scFOS supplementation on leptin, insulin, ghrelin, or gluose
	Plasma insulin		Chemical composition 34% CP, 10% fat, 19.8% dietary fiber, 1% scFOS	2% scFOS (Beghin-Meiji Industrie)	
	Plasma ghrelin		Study continued until ideal BCS obtained (5 out of 9)		
	Blood glucose				
Spears et al. (2005)	Food intake	Five purpose-bred adult dogs (3.7 yr; 28.9 kg)	Basal diet: Hill's prescription diet d/d- Rice and Duck	No supplement (0 g)	↑ γ-cyclodextrin:
	Apparent ileal and total tract digestibility			1 g high molecular weight pullulan	Linear ↓ food intake (control: 352 g; 4 g/d treatment: 305 g)***

◘ Table 12.1 (Cont'd p. 360)

Reference	Outcome variables quantified	Animals/treatment (age, initial BW)	Dietary information; time on treatment	Daily prebiotic dose; source	Major findings
Spears et al. (2005) (con't.)	Microbial populations		Chemical composition 17% CP, 14% fat, 4.0% TDF	2 g high molecular weight pullulan	Quadratic ↑ ileal bifidobacteria (control: 9.46 CFU/g DM feces; 2 g/d: 10.20 CFU/g DM feces; 4 g/d: 9.83 CFU/g DM feces) and lactobacilli (control: 9.12 CFU/g DM feces; 2 g/d: 10.11 CFU/g DM feces; 4 g/d: 9.42 CFU/g DM feces)***
	Fecal characteristics		14 d study	1 g γ-cyclodextrin	Quadratic ↓ fecal *Clostridium perfringens* (control: 9.75; 2 g/d: 9.44; 4 g/d: 9.76 CFU/g DM feces)**
				2 g γ-cyclodextrin	↑ pullulan:
					Linear ↑ ileal bifidobacteria (9.46 CFU/g DM feces, control; 10.12 CFU/g DM feces, treatment) and lactobacilli (9.12 CFU/g DM feces, control; 10.02 CFU/g DM feces, 4 g/d treatment)***
Vanhoutte et al. (2005)	Fecal population banding patterns (DGGE)	Seven adult dogs	Basal diet: see Propst et al. (2003)	Baseline	Inulin:
			Chemical composition not provided	4.5 g/d OF	↑ *Streptococcus lutetiensis*
			10 d study	5.6 g/d inulin	

Reference	Parameter	Animals	Basal diet	Control (0 g)	Results
Gouveia et al. (2006)	Number of leukocytes, neutrophils, and lymphocytes	Eight dogs (2–6 mo) all suffering from gastroenteritis	Basal diet not provided	Control (0 g)	MOS:
	Fecal microbiota		Chemical composition not provided	2 g MOS (Bio-MOS)	Elimination of E. coli in 85.71% of animals
			10 d study		
Verlinden et al. (2006)	Nutrient digestibility	Four adult beagle dogs (2–11 yr; 6–15 kg)	Basal diets: hydrolyzed protein diet (Hill's z/d ultra, allergen-free); intact protein source (Hill's d/d with duck and rice)	0% Supplementation	Inulin:
	Fecal characteristics		Chemical composition:	3% Inulin (Raftifeed IPS, DP 2–60; Orafti, Tienen, Belgium)	↓ Fecal DM (12%)***
	Hematology		Hydrolyzed-18% CP, 3% TDF		↓ Apparent CP digestibility in intact protein + inulin diet (4.6%)***
	Serum and fecal IgA, IgG, IgE, IgM				↑ Estimated bacterial protein content in feces (% of fecal DM, 16%; and % of CP intake, 33%) in intact protein + inulin diet*
			Intact protein-14% CP, 5% TDF		
			21 d study		

◘ Table 12.1   (Cont'd p. 362)

Reference	Outcome variables quantified	Animals/ treatment (age, initial BW)	Dietary information; time on treatment	Daily prebiotic dose; source	Major findings
Adogony et al. (2007)	Mammary, nasal, and blood immunoglobulin concentrations	Eight primiparous female beagles (10–12 kg)	Control diet: 31% CP, 16% crude fat, 7.8% TDF	0% Supplementation	scFOS:
	Diarrhea incidence		Test diet: 30% CP, 15% crude fat, 6.6% TDF	1% scFOS (Profeed, Beghin-Meiji, France)	↑ IgM in colostrum and milk (40%)**
			Gestation d 35 through weaning		↑ Blood IgM concentrations (60%)***
Apanavicius et al. (2007)	Food intake	Six hound-cross puppies (12 wk)	Control diet: 32% CP, 19% fat, 3% TDF	0% Supplementation	scFOS and Inulin:
	Gastrointestinal tract histopathology		scFOS diet: 32% CP, 19% fat, 3% TDF	1% scFOS	↓ Change in food intake following infection (26%)**
	Body temperature after infection		Inulin diet: 30% CP, 19% fat, 5% TDF	1% Inulin	↓ Enterocyte sloughing severity (9%)**
	Ileal and colonic nutrient and ion transport				Maintenance of ileal Na + -dependent glucose transport (400% ↓ in control, no change in supplemented puppies)
			14 d study		Inulin:

Apanavicius et al. (2007) (con't.)	Microbial populations				↑ Change from baseline in fecal acetate (control: −37.7 μmol/g; inulin 85.5 μmol/g) and SCFA concentrations (control: −53.5 μmol/g; inulin: 145.5 μmol/g)**
					↑ Change in fecal lactobacilli concentrations (7%)**
Middelbos et al. (2007a)	Nutrient digestibility	Six purpose-bred adult female dogs (4.5 yr; 23 kg)	350 g/d	Control – no supplemental fermentable carbohydrate	Supplemented diets:
			14 d study	Control + 2.5% cellulose	↓ CP digestibility (15%)**
	Fecal microbial populations			Control + 2.5% beet pulp	↑ Fecal bifidobacteria concentrations (14%)**
	Fecal fermentative end-products				↑ Fecal lactobacilli concentrations (8%)***
	Immunological indices			Control + 1.0% cellulose + 1.5% scFOS (Nutraflora P-95)	
				Control + 1.0% cellulose + 1.2% scFOS + 0.3% yeast cell wall (YCW, Safmannan)	↑ Fecal butyrate concentrations (67%)**
				Control + 1.0% cellulose + 0.9% scFOS + 0.6% YCW	

◻ Table 12.1

Reference	Outcome variables quantified	Animals/ treatment (age, initial BW)	Dietary information; time on treatment	Daily prebiotic dose; source	Major findings
Middelbos et al. (2007b)	Apparent ileal and total tract nutrient digestibility	Five purpose-bred adult female dogs (4 yr; 23 kg BW)	280 g basal diet	0 g supplementation	YCW:
	Serum IgA, IgM, and IgG		Chemical composition 30% CP, 21% fat, 4% TDF	Control+0.07 g YCW (Safmannan)	↑ ileal nutrient digestibility (10% DM, 11% CP)***
	Fecal microbial populations		14 d study	Control+0.35 g YCW	Linear ↓ in monocyte counts (control: 1.0 thousands/µL; 0.65% supplementation: 0.7 thousands/µL)**
				Control+0.63 g YCW	Linear ↓ in fecal *E. coli* populations (control: 9.1 CFU/g DM feces; 0.65% supplementation: 8.2 CFU/g DM feces)**
				Control+0.91 g YCW	

[a]For research published prior to 2004, please refer to the review of Swanson and Fahey (2006)

[b]*BW* body weight, *CFU* colony forming units, *CP* crude protein, *d* day, *DGGE* denaturing gradient gel electrophoresis, *DM* dry matter, *g* gram, *IgA* immunoglobulin A, *IgE* immunoglobulin E, *IgG* immunoglobulin G, *IgM* immunoglobulin M, *kg* kilogram, *mo* month, *MOS* mannanoligosaccharide, *OF* oligofructose, *SCFA* short-chain fatty acid, *scFOS* short-chain fructooligosaccharide, *TDF* total dietary fiber, *yr* year, *YCW* yeast cell wall, *wk* week

[c]*P < 0.001, **P < 0.05, ***P < 0.10

factor significantly into the general health and well being of all animals, the majority of canine prebiotic research has focused on monitoring changes in gut microbiota. Microbiota reside in all areas of the intestinal tract, but are generally studied in feces rather than using invasive techniques to obtain colonic contents. Fifty-five percent of authors observing changes in microbiota reported increased bifidobacteria (Middelbos et al., 2007a; Swanson and Fahey, 2006), while 32% reported increased lactobacilli (Apanavicius et al., 2007; Middelbos et al., 2007a; Swanson and Fahey, 2006). Potentially pathogenic microbiota generally decreased in response to prebiotic supplementation. Thirty-six percent of authors reported decreased clostridia (Spears et al., 2005; Swanson and Fahey, 2006); however, two authors reported increased clostridia in response to scFOS and OF supplementation (Swanson and Fahey, 2006). *Escherichia coli* was reported to decrease in response to MOS-supplementation (Gouveia et al., 2006; Middelbos et al., 2007b; Swanson and Fahey, 2006), while streptococci increased in response to fructan supplementation (Swanson and Fahey, 2006; Vanhoutte et al., 2005). Overall, prebiotic supplementation appears to benefit the microbial ecology of the dog.

Commensurate with changes in microbiota populations, changes in fermentative end-products occur with prebiotic supplementation. Increased concentrations of fecal SCFA, particularly butyrate, were reported in 46% of studies observing changes in fecal metabolites (Apanavicius et al., 2007; Middelbos et al., 2007a; Swanson and Fahey, 2006). Concentrations of acetate and propionate increased in 23 and 31% of studies, respectively (Apanavicius et al., 2007; Swanson and Fahey, 2006). Most SCFA are generated via carbohydrate fermentation. Conversely, branched-chain fatty acids (BCFA), ammonia, phenols, and indoles are generated via protein fermentation. In general, concentrations of BCFA, phenol, and indole decreased in response to prebiotic supplementation (Swanson and Fahey, 2006). Ammonia concentrations responded inconsistently to prebiotic supplementation: 23% of studies reported increased ammonia (Swanson and Fahey, 2006) while 15% of studies reported decreased ammonia in feces (Swanson and Fahey, 2006).

Prebiotics have demonstrated clear effects on nutrient digestibility. Crude protein digestibility is negatively impacted by inulin consumption (Middelbos et al., 2007a; Verlinden et al., 2006). However, ileal dry matter and crude protein digestibilities increased in dogs fed diets supplemented with yeast cell wall (Middelbos et al., 2007b). Although no differences in nutrient digestibility were observed, dogs consumed less food in a linear fashion when provided diets with increasing concentrations of γ-cyclodextrin (Spears et al., 2005).

Closely tied to nutrient digestibility, fecal quality and consistency are affected by prebiotics. As reported in 50% of published studies, higher concentrations of supplemental prebiotic compounds led to decreased fecal dry matter (Swanson and Fahey, 2006; Verlinden et al., 2006). Three percent supplementation of inulin resulted in a decrease in fecal dry matter in dogs consuming both hydrolyzed and intact protein diets (Verlinden et al., 2006). Also affected by prebiotic supplementation are fecal volume and fecal score. Thirty-five percent of studies reported an increase in fecal volume with prebiotic supplementation (Swanson and Fahey, 2006). Fecal scores were reported to increase in three studies (decreased consistency) while decreased fecal score was reported in two studies (Swanson and Fahey, 2006).

Increased immune responses also have been observed in dogs consuming prebiotics. When fed to primiparous beagles, scFOS increased IgM concentrations in colostrum, milk, and blood (Adogony et al., 2007). Under a *Salmonella typhimurium* challenge, puppies consuming either scFOS or inulin experienced decreased enterocyte sloughing and maintenance of sodium-dependent glucose transport (Apanavicius et al., 2007). While feed intake decreased, it appeared that supplementation of prebiotic fructans allowed for normal metabolism to occur. When fed diets supplemented with increasing concentrations of yeast cell wall, a linear decrease in monocyte concentration was observed (Middelbos et al., 2007b).

Felines. Five studies involving prebiotic supplementation of felines have been published (⊙ *Table 12.2*). Increased bifidobacteria and lactobacilli, along with decreased clostridia, staphylococci, and *E. coli* have been reported (Swanson and Fahey, 2006). Supplementation with fructans also increased fecal nitrogen of microbial origin in supplemented cats (Hesta et al., 2005; Swanson and Fahey, 2006). Fecal fermentative end-products – ammonia, phenol, and indole – decreased with lactosucrose supplementation (Swanson and Fahey, 2006), whereas inulin increased total SCFA concentrations (Swanson and Fahey, 2006).

Nitrogen and crude protein digestibilities decreased in response to supplementation with fructans (Swanson and Fahey, 2006). This likely would be observed in response to other prebiotics; however, these effects remain open to investigation in the cat. Fructan supplementation also increased fecal nitrogen excretion and decreased fat digestibility (Hesta et al., 2005; Swanson and Fahey, 2006). With respect to fecal characteristics, fructan supplementation decreased fecal dry matter and fecal score while increasing dry fecal output, fecal volume, and number of defecations per day (Hesta et al., 2005; Swanson and Fahey, 2006).

◘ Table 12.2

In vivo experiments, listed in chronological order beginning in 2005, reporting effects of prebiotics in cats[a,b,c]

Reference	Outcome variables quantified	Animals/ treatment (age, initial BW)	Dietary information; time on treatment	Daily prebiotic dose; source	Major findings
Hesta et al. (2005)	Urea metabolism	Four female cats (>7 yr; 2.2 and 4 kg)	Basal diet fed for ME requirement of ideal BW	0% Supplementation	FOS:
	Fecal odor components		Chemical composition 29% CP, 37% crude fat, 1% CF	3.11% FOS (Raftilose), DMB	↑ Fecal moisture (6%)***
			3 wk study		↑ DM fecal output (27%)***
					↑ Fecal N excretion (36%)***
					↓ Urinary N excretion (48%)***
					↑ Fecal bacterial N (% of N intake; 125%)***

[a]For research published prior to 2004, please refer to the review of Swanson and Fahey (2006)
[b]*BW* body weight, *CF* crude fiber, *CP* crude protein, *DM* dry matter, *DMB* dry matter basis, *FOS* fructooligosaccharide, *kg* kilogram, *ME* metabolic energy, *N* nitrogen, *yr* year, *wk* week
[c]*P < 0.001, **P < 0.05, ***P < 0.10

Poultry. Prebiotic research has been conducted in poultry since 1990 and, as a result, a large database of research is available in this area (◉ *Table 12.3*). Increased bifidobacteria (Cao et al., 2005; Jiang et al., 2006; Sims et al., 2004; Terada et al., 1994; Thitaram et al., 2005; Xu et al., 2003) and decreased clostridia (Biggs et al., 2007; Butel et al., 2001; Cao et al., 2005; Sims et al., 2004; Terada et al., 1994) have been reported in 30% of studies that investigated the microbial effects of prebiotic supplementation. Fifteen percent of studies also reported increased lactobacilli (Baurhoo et al., 2007a; Jiang et al., 2006; Xu et al., 2003); however, one author reported decreased ileal and cecal lactobacilli with MOS

12

◻ Table 12.3

In vivo experiments, listed in chronological order, reporting effects of prebiotics in poultry[a,b] *(Cont'd p. 368)*

Reference	Outcome variables quantified	Animals/treatment (age, initial BW)	Dietary information; time on treatment	Daily prebiotic dose; source	Major findings
Coon et al. (1990)	Study 1	Study 1	Study 1	Study 1	Study 1
	TME_n	Eight Leghorn roosters	30 g (precision-fed) test diet	SBM	ESBM:
			Chemical composition 44% CP	ESBM	↑ TME_n (21%)**
	Study 2	Study 2	Study 2	Study 2	Study 2
	Ileal and total excreta CHO digestibilities	30 Leghorn roosters	Same as study 1	Same as study 1	ESBM:
	Intestinal transit time				↓ Ileal sucrose (42%), cellulose (0.5%), total water-soluble CHO (78%) digestibilities**
					↓ Excreta sucrose (54%), raffinose (35%), stachyose (37%), total water-soluble CHO (90%) digestibilities**
					↑ Excreta hemicellulose (52%), cellulose (35%) digestibilities**

Reference	Study	Details	Treatment	Results
Coon et al. (1990) (con't.)				↑ Excreta true DM digestibility (25%)**
				↑ Intestinal transit time (62%)**
				↑ Cecal pH (9%)***
Leske et al. (1993)	Study 1	Study 1	Study 1	Study 1
	Nutrient digestibility	6–9 Leghorn roosters	Control (0%)	1% raffinose:
	TME_n	30 g (precision-fed) soy protein concentrate	Control + 1% raffinose	↓ TME_n (10%)**
		Chemical composition 73.1% CP, 0.36% stachyose, 0.05% raffinose	Control + 5% stachyose	5% stachyose:
		48 h study	Control + 1% raffinose + 5% stachyose	↓ TME_n (12%)**
			Control + 6% sucrose	1% raffinose + 5% stachyose:
			Control + 1% raffinose + 5% stachyose + Control + 6% sucrose	↓ TME_n (12%)**
	Study 2	Study 2	Study 2	Study 2
	Same as study 1	Same as study 1	Control (0%)	All concentrations stachyose:
			Control + 1% stachyose	↓ TME_n (13–18%)**
			Control + 2% stachyose	↓ DM digestibility (12–16%)

Table 12.3 (Cont'd p. 370)

Reference	Outcome variables quantified	Animals/treatment (age, initial BW)	Dietary information; time on treatment	Daily prebiotic dose; source	Major findings
Leske et al. (1993) (con't.)				Control + 3% stachyose	
				Control + 4% stachyose	
				Control + 5% stachyose	
	Study 3	Study 3	Study 3	Study 3	Study 3
	Same as study 1	Same as study 1	Same as study 1	Control (0%)	0.6% raffinose:
				Control + 0.4% raffinose	↓ TME_n (12%)**
				Control + 0.6% raffinose	↓ DM digestibility (11%)**
					0.8% raffinose:
				Control + 0.8% raffinose	↓ TME_n (15%)**
				Control + 1.0% raffinose	↓ DM digestibility (12%)**
					1.0% raffinose:
					↓ TME_n (16%)**
					↓ DM digestibility (13%)**

Terada et al. (1994)	Performance responses (BW, mortality)	6,360 Cobb broiler chicks (1 d)	Corn-SBM diet	Control (0%)	Lactosucrose:
	Cecal microbiota		Chemical composition (starter–finisher) 23.3–28.3% CP, 1.20–0.91% lys, 0.98–0.87% met + cys, 1.07–1.03% Ca, 0.69–0.75% P	Control + 0.15% lactosucrose	↓ Lecithinase-negative clostridia on d 20 (9%)**
			62 d study		↓ Lecithinase-positive clostridia (31%), bacterioidaceae (2%), total anaerobes (1%) on d 62**
	Cecal metabolites				↑ Bifidobacteria (4%) on d 62**
					↓ Pseudomonad occurence (63%) on d 62**
	Pen ammonia				↓ Staphylococci on d 62 (20%)**
					↓ Cecal ammonia on d 62 (50%)***
					↓ Cecal phenol (38%), p-cresol on d 62 (34%)**
					↑ Cecal acetate on d 62 (98%)*
					↑ Cecal butyrate on d 62 (100%)**

■ Table 12.3 *(Cont'd* p. 372)

Reference	Outcome variables quantified	Animals/treatment (age, initial BW)	Dietary information; time on treatment	Daily prebiotic dose; source	Major findings
Leske and Coon (1999a)	Study 1	Study 1	Study 1	Study 1	Study 1
	TME_n AA digestibility	Seven adult Leghorn roosters	25 g (precision-fed) test diet	SBM	ESBM:
	NSP digestibility		48 h study	ESBM	↑ DM (11%), cellulose (7%), noncellulosic NSP (22%), arabinose (27%), fucose (29%), galactose (25%), glucose (16%), mannose (13%), xylose (39%), uronic acid (18%) digestibility**
					↑ TME_n (14%)**
					↑ Digestible cellulose (414%), noncellulosic NSP (519%)**
	Study 2	Study 2	Study 2	Study 2	Study 2
	TME_n	Nine roosters	Roosters: 25 g (precision-fed) test diet	47% CP SBM	47% CP SBM (10:1 vs. control):
	AME_n	20 Male Ross × Ross broiler chicks	48 h study	47% CP SBM extracted with 80% ethanol, then water (10 parts ethanol/water: 1 part SBM)	↑ TME_n (9%)**
			Chicks: unlimited access to test diet	44% CP SBM	44% CP SBM (8:1 vs. control):
			12 d study		↑ AME_n (14%)**

Reference	Response	Animal	Diet	Treatment	Results
Leske and Coon (1999a) (con't.)	Hydrogen gas production	Two male Ross × Ross broiler chicks (10 d, 156 g)	6 g (precision-fed) test diet	44% CP SBM extracted with 80% ethanol, then water (8:1)	↑ TME_n (8%)**
					44% CP SBM (water: ethanol vs. control):
				44% CP SBM extracted with water, then 80% ethanol (water:ethanol)	↑ AME_n (14%)**
					↑ TME_n (21%)**
Leske and Coon (1999b)	Hydrogen gas production		28 h study	47% CP SBM	No differences among treatments
				ESBM (α-galactoside-free)	
				ESBM with α-galactoside concentrations of standard SBM	
Fernandez et al. (2000)	Study 1	Study 1	Study 1	Study 1	Study 1
	Salmonella challenge	6 or 12 Ross-1 broiler chicks (1 d)	Mash diet (ingredients, chemical compostion not provided)	Control (0%)	All dilutions of HCC dosed prevented cecal S. enteritidis colonization (12/12 control birds colonized)
	Cecal microbiota		9 d study – Chicks primed with hen cecal contents (HCC) adapted to designated treatment diet	Control + 2.5% D-mannose	
				Control + 2.5% MOS	
				Control + palm kernel meal	

12

◘ Table 12.3 (Cont'd p. 374)

Reference	Outcome variables quantified	Animals/treatment (age, initial BW)	Dietary information; time on treatment	Daily prebiotic dose; source	Major findings
Fernandez et al. (2000) (con't.)	Study 2	Study 2	Study 2	Study 2	Study 2
	Salmonella challenge	12 or 24 Ross-1 broiler chicks (1 d)	Same as study 1	Same as study 1	MOS:
	Cecal microbiota				↓ Cecal S. enteritidis with 10^{-4} HCC (51% vs. mannose group)**
	Study 3	Study 3	Study 3	Study 3	Study 3
	Salmonella challenge	12 or 24 Ross-1 broiler chicks (1 d)	Same as study 1	Same as study 1	MOS:
	Cecal microbiota				↓ Cecal S. enteritidis with 10^{-3} (51%), 10^{-4} (67%), and $10^{-3} + 10^{-4}$ HCC (58%; mannose also decreased 37%, 50%, and 42%, respectively)**
	Study 4	Study 4	Study 4	Study 4	Study 4
	Salmonella challenge	12 or 24 Ross-1 broiler chicks (1 d)	Same as study 1	Same as study 1	MOS:
	Cecal microbiota				↓ Cecal S. enteritidis with 10^{-3} (38%), 10^{-5} (33%), and $10^{-3} + 10^{-6}$ HCC (24%; mannose also decreased 19% with $10^{-3} + 10^{-6}$ HCC)**

	Study 1	Study1 (*S. typhimurium* groups)	Study 1	Study 1	Study 1
Spring et al. (2000)	Salmonella challenge	Ten line 24 broiler chicks (hatched, groups 1 and 2) plus 26 chicks (hatched, group 3)	Unmedicated corn-SBM diet, all birds' microbiota standardized prior to initial feeding	Control (0 ppm)	MOS:
				Control + 4,000 ppm MOS (Bio-Mos)	↓ Salmonellae (26%)**
	Cecal microbiota		Chemical composition not provided		
	Cecal metabolites		10 d study		↓ Coliforms (3%)***
	Study 2	Study 2 (*S. dublin* groups)	Study 2	Study 2	Study 2
	Salmonella challenge	Nine chicks (hatched, groups 4 and 5) plus an unspecified number of chicks (hatched, group 6)	Same as study 1	Same as study 1	MOS:
	Cecal microbiota				↓ Salmonella-positive birds (38%)**
	Cecal metabolites				
Butel et al. (2001)	Cecal metabolites	Three "flora" groups (inoculated with human infant fecal flora)	Semi-synthetic diet (ingredients and chemical composition not provided)	Control (6% w/w lactose)	Lactose + OF:
	Cecal microbiota	(1) 11 Germ-free quail (*Coturnix coturnix* subsp. *japonica*, 2 wk) (2) Nine quail (2 wk) (3) 14 quail (2 wk)	28 d study	OF [3% lactose + 3% OF (Raftilose)]	(1) No difference in infection or cecal morphology, metabolites, or microbiota

■ Table 12.3 (Cont'd p. 376)

Reference	Outcome variables quantified	Animals/treatment (age, initial BW)	Dietary information; time on treatment	Daily prebiotic dose; source	Major findings
Butel et al. (2001) (con't.)	Cecal morphology associated with necrotizing enterocolitis				(2) ↓ Cecal wall weight in ill quails (20%)**
					(3) ↓ Cecal C. perfringens (19%) and C. paraputrificum (31%)**
Fairchild et al. (2001)	Performance responses (BW, feed conversion, [F:G])	56 Young (33 wk) and 56 old (58 wk) male BIG-6 turkey poults (1 d)	Corn-SBM diet	Control (0 g/kg)	MOS with E. coli:
			Chemical composition 27.9% CP, 5.2% fat, 0.68% met, 1.13% met + cys, 1.71% lys, 1.4% Ca, 0.7% P	1 g/kg MOS (Bio-Mos)	↑ BW in young (13%) and old (15%) chicks at wk 1**
	E. coli challenge		3 wk study	2.2 mg/kg flavomycin	MOS with E. coli:
	Intestinal microbiota			1 g/kg MOS + 2.2 mg/kg flavomycin	
					↑ BW in young chicks at wk 3 (5%)**
					↓ Liver total coliforms on d 7 (42%)**

Reference	Performance responses (BW, feed consumption, feed conversion)	160 Male hybrid large White turkey poults	Corn-SBM diets	Control (0 g/kg)	MOS:
Parks et al. (2001)	Bird mortality		Chemical composition (initial–final) 28.14–16.50% CP, 3.16–5.90% crude fat, 0.59–0.31% met, 1.05–0.61% met + cys, 1.60–0.80% lys, 1.20–0.65% Ca, 0.60–0.32% P	1 g/kg MOS (reduced to 0.5 kg/ton at 6 wk)	↑ BW at wk 20 (3%)**
			140 d study	2 mg/kg flavomycin	↓ F:G from 0–3 wk (4%)**
				20 mg/kg Stafac	MOS + flavomycin: ↑ BW at wk 12 (4%), 15 (4%), 20 (2%)**
				1 g/kg MOS (reduced to 0.5 g/kg at 6 wk) + 2 mg/kg flavomycin	↓ F:G from wk 0–3 (4%), 0–6 (4%), 0–12 (3%), 0–18 (2%)**
				1 g/kg MOS (reduced to 0.5 g/kg at 6 wk) + 20 mg/kg Stafac	MOS + Stafac: ↑ BW at wk 15 (3%)**
					↓ F:G from wk 0–3 (6%), 0–6 (4%), 0–12 (3%), 0–18 (3%)**

● Table 12.3 (*Cont'd* p. 378)

Reference	Outcome variables quantified	Animals/treatment (age, initial BW)	Dietary information; time on treatment	Daily prebiotic dose; source	Major findings
Fernandez et al. (2002)	Study 1	Study 1	Study 1	Study 1	Study 1
	Salmonella challenge	Five ross-1 broiler chicks (1 d)	Wheat-SBM mash diet (ingredients and chemical composition not provided)	Control (0%)	MOS:
	Cecal Microbiota		2 wk study - chicks (except control) dosed with HCC adapted to designated treatment diet	Control + 2.5% (w/w) MOS	↑ Total anaerobes (4%)**
				Control + 2.5% (w/w) palm kernel meal	↓ Coliforms (17%) and S. enteritidis colonization (98%)**
	Study 2	Study 2	Study 2	Study 2	Study 2
	Salmonella challenge	15 Ross-1 broiler chicks (1 d)	Same as study 1 except 4 wk study	Same as study 1	No difference in S. enteritidis colonization with MOS
	Cecal microbiota				

Samarasinghe et al. (2003)	Study 1	Study 1	Study 1	Study 1	Study 1
	Performance responses (BW, feed intake, feed conversion)	30 Mixed-sex broilers (unspecified strain, 19 d)	Corn-SBM-rice polish mash diet	Control (0%)	MOS:
	Carcass composition		Chemical composition 24.1% CP, 9.3% crude fat, 5.4% CF, 7.9% ash	Control + 500 ppm virginiamycin	↑ Net protein utilization (31%)**
	Intestinal microbiology		4 wk study	Control + 200 ppm MOS	
				Control + 1 g/kg turmeric root powder	
				Control + 2 g/kg turmeric root powder	
				Control + 3 g/kg turmeric root powder	
	Study 2	Study 2	Study 2	Study 2	Study 2
	Performance responses (BW, feed intake, feed conversion)	36 Mixed-sex broilers (unspecified strain, 21 d)	Same as study 1	Control (0 ppm)	MOS:
	Carcass composition			Control + 500 ppm virginiamycin	↓ Feed intake (5%)**
				Control + 200 ppm MOS	↑ Daily weight gain (8%), feed conversion (13%), energy utilization (6%), protein utilization (5%)**

■ Table 12.3 (Cont'd p. 380)

Reference	Outcome variables quantified	Animals/treatment (age, initial BW)	Dietary information; time on treatment	Daily prebiotic dose; source	Major findings
Samarasinghe et al. (2003) (con't.)	Intestinal microbiology			Control + 1 g/kg turmeric root powder	↑ % Live carcass (2%), liver (25%), abdominal fat weight (55%)**
					↓ Coliforms (58%), yeasts and molds (84%), total viable microbial counts (51%) in intestinal contents**
Shashidhara and Devegowda (2003)	Study 1	Study 1	Study 1	Study 1	Study 1
	Production responses (hatchability, sperm count)	1,560 Cobb broiler breeders (1,440 female, 120 male)	Corn-SBM-rice bran diet Chemical composition (Males) 15.7% CP, 7.6% fiber, 1.0% Ca, 0.34% P	Control (0 g/kg)	MOS: ↑ Egg production on wk 60–62 (3–6%)**
	Immune response		(Females) 16.1% CP, 5.6% fiber, 3.3% Ca, 0.33%P	Control + 0.5 g/kg MOS (Bio-Mos)	↑ Hatchability [total (0.6–5.2%) and fertile egg sets (0.8–5.3%)]**
			8 wk study		↓ Infertile (5–30%) and dead-in-shell eggs (18, 26% on wk 64 and 66)**
					↑ Sperm count (18%)**
					↑ IBDV antibody titer in breeder females (17%)**

		Study 2	Study 2	Study 2	Study 2
Shashidhara and Devegowda (2003) (con't.)	Production responses (hatchability, sperm count)	1,440 Cobb breeders (1,284 female, 156 male)	Same as study 1 except 12 wk study	Control (0 g/kg)	MOS:
	Immune response			Control + 1 g/kg MOS	↑ Hatchability (fertile egg set only, 1–3%)** ↓ Infertile eggs (9–27%)** ↑ IBDV antibody titer in breeder females (15%) and chicks (47%)**
Xu et al. (2003)	Performance responses (Feed intake, ADG, F:G)	60 Male Avian farms broiler chicks (1 d)	Corn-SBM diet	Control (0 g/kg)	2 g/kg FOS:
	Intestinal microbiota		Chemical composition (initial–final) 22.8–18.2% CP, 0.94–0.81% Ca, 0.85–0.67% P	Control + 2 g/kg FOS (Meioligo-P)	↓ F:G (5%)**
			49 d study	Control + 4 g/kg FOS	↓ Cecal E. coli (8%)**
	Enzyme activity			Control + 8 g/kg FOS	↑ Amylase activity (52%)**
	Intestinal morphology				↑ Ileal villus height:crypt depth (31%), microvillus height (25%)**

■ Table 12.3 (Cont'd p. 382)

Reference	Outcome variables quantified	Animals/treatment (age, initial BW)	Dietary information; time on treatment	Daily prebiotic dose; source	Major findings
Xu et al. (2003) (con't.)					4 g/kg FOS:
					↑ ADG (11%)**
					↓ F:G (9%)**
					↑ Intestinal bifidobacteria (12%), lactobacilli (14%)**
					↓ Intestinal *E. coli* (12%)**
					↑ Cecal bifidobacteria (7%), lactobacilli (8%), total anaerobes (3%)**
					↓ Cecal *E. coli* (7%)**
					↑ Protease (27%), amylase (75%) activities**
					↑ Jejunal villus height: crypt depth (26%), microvillus height (21%)**

Reference	Measurement	Animal	Diet/Study	Treatment	Results
Xu et al. (2003) (con't.)					↓ Jejunal (17%), ileal crypt depth (25%)**
					↑ Ileal villus height (16%), villus height: crypt depth (55%), microvillus height (39%)**
Yusrizal and Chen (2003)	Performance responses (BW, feed intake, F:G)	32 Ross × Ross broiler chicks (16 male, 16 female; 1 d)	Corn–SBM diet	Control (0%)	Inulin:
	Carcass composition		Chemical composition (initial-final) 21.3–19.8% CP	Control + 1% inulin (Raftifeed IPF)	Males:
	Serum profile		6 wk study	Control + 1% OF (Raftifeed OPS)	↑ Carcass % (2%)**
					↑ BW at wk 2 (10%)**
					↓ F:G at wk 6 (18%)**
	Intestinal morphology				↓ Abdominal fat as % carcass (32%), % live weight (30%)**
					Females:
					↓ Serum cholesterol (17%)**
					↓ Abdominal fat as % carcass (30%), % live weight (30%)**

Table 12.3 (Cont'd p. 384)

Reference	Outcome variables quantified	Animals/treatment (age, initial BW)	Dietary information; time on treatment	Daily prebiotic dose; source	Major findings
Yusrizal and Chen (2003) (con't.)					OF:
					Males:
					↑ BW at wk 2 (5%)**
					Females:
					↑ BW (10%)**
					↑ Carcass weight (13%), carcass % (3%)**
					↑ Gut length (8%)**
					↓ F:G at wk 2 (4%), 5 (13%), 6 (18%)**
					↓ Serum cholesterol (20%)**
Zhang et al. (2003)	Performance responses (BW, feed intake, feed efficiency)	72 Male Arbor acres broiler chicks (1 d)	Corn–SBM diet	Control (0%)	0.3% IMO:
	Immune organ weight		Chemical composition (initial–final) 22.1–18.0% CP, 1.1–0.8% Ca, 0.52–0.40% P	Control + 0.3% IMO (IMO-900)	↑ ADG (7%)***
			7 wk study	Control + 0.6% IMO	↑ Thymus weight (49%)**
	Intestinal and cecal microbiota			Control + 0.9% IMO	↓ Crop isobutyrate (61%)**

Zhang et al. (2003) (con't.)					
	Intestinal metabolites			Control + 1.2% IMO	↓ Duodenal isovalerate (61%)*
					↑ Jejunal butyrate (102%), isobutyrate (85%)**
					0.6% IMO:
					↓ Duodenal isovalerate (59%)*
					All concentrations IMO:
					↑ Feed efficiency (3–6%)***
					↓ Crop isobutyrate (35–61%)**
					↓ Duodenal acetate (30–64%)**
Sims et al. (2004)	Performance responses (BW, feed consumption, conversion)	180 tom Hybrid turkey poults (1 d)	Ingredient composition not provided	Control (0%)	MOS:
	Intestinal microbiota		Chemical composition (initial-final %) 29–17% CP, 6.1–10.2% crude fat, 1.4–0.9% Ca, 0.75–0.48% P	Control + bacitracin (BAC; 55 mg/kg until 6 wk, 27.5 mg/kg after)	↑ BW at 18 wk (6%)**
			18 wk study	Control + MOS (Bio-Mos; 0.1% until 6 wk, 0.05% after)	↑ Feed conversion at 12 (6%), 15 wk (6%)**

▣ Table 12.3 (Cont'd p. 386)

Reference	Outcome variables quantified	Animals/treatment (age, initial BW)	Dietary information; time on treatment	Daily prebiotic dose; source	Major findings
Sims et al. (2004) (con't.)	Intestinal histology			Control + BAC + MOS (same doses)	↑ Bifidobacteria (21%), total anaerobes (9%)**
					↓ Clostridia (29%)**
					MOS + BAC:
					↑ BW at 15 (5%), 18 wk (8%)**
					↑ Feed conversion at 18 wk (11%)**
Zdunczyk et al. (2004)	Performance responses (BW, feed consumption, conversion)	45 BUT-9 turkey chickens (3 d)	Wheat-corn-SBM mash diet with or without an antibiotic	Control (0.0%)	0.1% MOS:
	Cecal enzyme activity		Chemical composition (initial-final) 28.8–25.9% CP, 4.4–5.5% crude fat, 3.4–3.7% CF, 1.3–1.1% Ca, 0.73–0.64% P	Control + 0.1% MOS (Bio-Mos)	↓ Cecal tissue weight (19%)**
	Cecal metabolites		8 wk study	Control + 0.25% MOS	↓ Molar ratio of cecal propionate (60%)**
				Control + 0.5% MOS	0.25% MOS:

Reference	Measured parameter	Study details	Diet	Treatment	Results
Zdunczyk et al. (2004) (con't.)					↑ Cecal DM (19%), pH (11%)**
					↑ Cecal propionate (72%)**
					↑ Cecal α-glucosidase (68%), β-glucosidase (200%), α-galactosidase (89%), β-galactosidase (219%), and β-glucuronidase (65%)**
					0.5% MOS:
					↑ Cecal acetate (40%), butyrate (72%), total SCFA (42%)**
					↓ Molar ratio of cecal propionate (60%)**
Cao et al. (2005)	Performance responses (BW, feed intake, mortality)	1,500 Arbor acres commercial broiler hens (28 d)	Isolated soy protein-cornstarch diet	Control (0%)	FOS:
	Cecal microbiota		Chemical composition 18% CP, 79.8 g/kg Ca, 40.0 g/kg P	Control + 4.1 g/kg green tea polyphenols	↓ Mortality (42%)**
	Cecal metabolites	14 d study		Control + 4.1 g/kg FOS	↑ Cecal bifidobacteria (9%), eubacteria (5%)**

■ Table 12.3 (Cont'd p. 388)

Reference	Outcome variables quantified	Animals/treatment (age, initial BW)	Dietary information; time on treatment	Daily prebiotic dose; source	Major findings
Cao et al. (2005) (con't.)					↓ Cecal bacilli (28%), lecithinase-positive clostridia (30%), Peptococcaceae spp. (11%), Streptococci spp. (25%), Staphylococci spp. (43%)**
					↑ Cecal valerate (54%)**
					↓ Cecal phenol (31%), cresol (48%), ethyl phenol (35%), and indole (36%)**
Cetin et al. (2005)	Blood metabolites	24 White hybrid converter turkey poults (15 d)	Corn-SBM diets	Control (0 g/kg)	MOS:
	IgG and IgM		Chemical composition (initial–final) 27.52–19.45% CP, 4.89–5.25% CF	1 g/kg MOS (Bio-Mos; reduced to 0.5 g/kg after wk 8)	↑ IgG (37%), IgM (44%)**
			15 wk study	1 g/kg Probiotic (Primalac 454; reduced to 0.5 g/kg after wk 8)	↓ α-naphthyl acetate esterase positive T lymphocytes (21%)**

Lee et al. (2005)	90 Mixed-sex Ross × Ross broiler chicks	Corn-SBM diet containing no guar	0, 2.5, 5, 7.5, or 10% of Guar gum, Guar germ, OR Guar hull	Guar gum:
Performance responses (BW, feed consumption, F:G, mortality)				↑ BW (8–9%), F:G (10–14%)**
Carcass and deboned breast muscle yield		Chemical composition (starter-finisher) 22.8–17.7% CP, 0.96–0.77% Ca, 0.46–0.36% P		5% treatments (vs. 2.5%):
		6 wk study		↓ Feed consumption (9%)**
				↓ Carcass weight (16%), deboned breast weight (14%) and yield (5%)**
				7.5% treatments (vs. 2.5%):
				↓ Carcass weight (16%), carcass yield (2%), deboned breast weight (18%) and yield (8%)*
				10% Treatments (vs. 2.5%):
				↓ Feed consumption (12%)**

■ Table 12.3 (*Cont'd* p. 390)

Reference	Outcome variables quantified	Animals/treatment (age, initial BW)	Dietary information; time on treatment	Daily prebiotic dose; source	Major findings
Lee et al. (2005) (con't.)					↓ Carcass weight (35%), carcass yield (5%), deboned breast weight (43%) and yield (19%)**
	Study 2	Study 2	Study 2	Study 2	Study 2
	Performance responses (BW, feed consumption, F:G, mortality)	100 Mixed-sex Ross × Ross broiler chicks	Same as study 1	All treatments with or without Hemicell (endo-beta-mannanase):	Guar meal:
				Control (0%)	↓ BW (14%)**
				Control + 5% guar meal	↑ F:G with (8%) and without Hemicell (20%)**
					Guar hull:
				Control + 5% guar germ	↓ BW (14%)**
				Control + 5% guar hull	↓ feed consumption (6%)**
					↑ F:G with (5%) and without Hemicell (15%)**
					Guar germ:
					↑ F:G with Hemicell (6%)**

Parks et al. (2005)	Performance responses (BW, feed consumption, feed conversion)	160 Female large white hybrid converter turkey poults (1 d)	Corn-SBM diets	Control (0 mg/kg)	MOS + virginiamycin:
	Mortality		Chemical composition (initial–final) 28.00–20.45% CP, 4.61–7.22% crude fat, 0.71–0.35% met, 1.20–0.73% met + cys, 1.80–1.03% lys, 1.60–0.85% Ca, 0.75–0.38% P	22 mg/kg Virginiamycin	↓ BW at 3 wk (5%)**
	Production responses		98 d study	500 mg/kg MOS (reduced to 22 mg/kg after 6 wk) + 22 mg/kg virginiamycin	↑ BW at 9 (4%), 12 wk (4%)**
					↑ BW gain at 6–9 wk (8%)**
					↓ Feed conversion rates at 6–9 wk (7%), overall (1%)**
Thitaram et al. (2005)	Performance responses (BW, feed conversion, feed efficiency)	Ten Ross × Ross broiler chicks (1 d)	Standard unmedicated corn-soy diet	Control (0%)	All doses IMO: ↑ Bifidobacteria (10–11%)** 1% IMO:
				Control + 1% IMO	↓ BW gain (10%)** ↓ Salmonella enterica ser. Typhimurium (31%)**

■ Table 12.3 (Cont'd p. 392)

Reference	Outcome variables quantified	Animals/treatment (age, initial BW)	Dietary information; time on treatment	Daily prebiotic dose; source	Major findings
Thitaram et al. (2005) (con't.)	Cecal microbiota		Chemical composition 22.5% CP, 5.3% crude fat, 2.5% CF, 0.45% P	Control + 2% IMO	
			21 d study	Control + 4% IMO	
Zaghini et al. (2005)	Performance responses (feed consumption)	24 Warren laying hens (44 wk, 2.2 kg)	Corn-SBM diet	Control (0%)	MOS:
	Egg production and quality		Chemical composition 14.79% CP, 3.23% crude fat, 4.54% CF, 3.53% Ca, 0.86% P	2.5 ppm aflatoxin B1	↓ Egg weight at wk 2 (6%), 3 (4%)**
	Liver histology		4 wk study	0.11% MOS (Bio-Mos)	↓ Yolk a* color (6%)**
				0.11% MOS + 2.5 ppm aflatoxin B1	↑ Albumen CP at wk 2 (3%)*, 4 (3%)**
					↑ Albumen ash at wk 4 (14%)*
					MOS + aflatoxin:
					↓ Egg weight at wk 2 (8%)**
					↑ Albumen ash at wk 4 (12%)*

Reference	Measure	Animal	Diet	Treatment	Results
Zaghini et al. (2005) (con't.)					↑ Yolk L (5%)*, b* (6%)** color at wk 4
					↑ Albumen CP at wk 2 (3%)*, 4 (3%)**
Zdunczyk et al. (2005)	Cecal microbiota	39 Male BUT-9 turkey poults (3 d)	SBM-wheat-corn diet	Control (0%)	All concentrations MOS:
	Cecal metabolites		Chemical composition (initial–final) 28.77–18.78% CP, 3.4–3.0% CF, 4.4–6.6% crude fat, 1.8–1.2% lys, 1.16–0.73% met + cys, 1.3–1.0% Ca, 0.72–0.50% P	0.1% MOS (Bio-Mos)	↓ Colon weight (16–30%)**
	Cecal enzyme activity		16 wk study	0.4% MOS (decreased to 0.2% after 8 wk)	↑ Cecal ammonia as % BW (49–64%)**
				1.0% MOS (decreased to 0.4% after 8 wk)	0.1% MOS:
					↓ Cecal pH (14%)**
					↑ Cecal ammonia (30%)**
					0.4/0.2% MOS:
					↑ BW at 16 wk (5%)**
					↓ Cecal acetate (45%), total SCFA (34%)**
					↓ Cecal E. coli (13%)**
					1.0/0.4% MOS:
					↑ BW at 16 wk (3%)**

Table 12.3 (Cont'd p. 394)

Reference	Outcome variables quantified	Animals/treatment (age, initial BW)	Dietary information; time on treatment	Daily prebiotic dose; source	Major findings
Zdunczyk et al. (2005) (con't.)					↓ Cecal acetate (46%), valerate (35%), total SCFA (34%)**
					↓ Cecal E. coli (15%)**
Jiang et al. (2006)	Performance responses (BW, mortality, feed consumption, feed efficiency)	72 Male Arbor acres broiler chicks (1 d)	Corn-soy protein isolate diet	Control (0%)	4 g/kg stachyose:
	Cecal metabolites		Chemical composition (starter-grower) 220.5–200.0 g/kg CP, 10.2–9.6 g/kg Ca, 6.7–6.4 g/kg P, 12.3–10.9 g/kg lys, 7.0–5.7 g/kg met, 10.7–9.1 g/kg met + cys	Control + 4 g/kg stachyose	↑ Cecal butyrate on d 21 (23%)**
	Cecal microbiota		42 d study	Control + 8 g/kg stachyose	12 g/kg stachyose:
				Control + 12 g/kg stachyose	↓ 6 wk BW (4%)*
					↓ Overall ADG (4%)*
	Nutrient digestibility			Control + 16 g/kg stachyose	16 g/kg stachyose:
				Control + SBM (10.5 g/kg raffinose, 32.1 g/kg stachyose; no soy protein isolate)	↓ 3 (5%), 6 wk BW (6%)*

Jiang et al. (2006) (con't.)	↑ Stachyose concentrations:	Quadratic ↓ ADG (control: 49.4 g/d; 4 g/kg: 49.4 g/d; 8 g/kg: 48.8 g/d; 12 g/kg: 47.4 g/d; 16 g/kg: 46.1 g/d)*	Quadratic ↓ feed intake (control: 92.9 g/d; 4 g/kg: 93.5 g/d; 8 g/kg: 93.3 g/d; 12 g/kg: 91.0 g/d; 16 g/kg: 90.1 g/d)*	Linear ↓ DM digestibility on d 42 (control: 76.32%; 4 g/kg: 76.36%; 8 g/kg: 74.44%; 12 g/kg: 74.65%; 16 g/kg: 72.17%)*	Linear ↓ OM digestibility on d 21(control: 81.36%; 4 g/kg: 81.09%; 8 g/kg: 79.69%; 12 g/kg: 79.02%; 16 g/kg: 78.70%)***	Linear ↓ OM digestibility on d 42 (control: 81.76%; 4 g/kg: 81.57%; 8 g/kg: 80.30%; 12 g/kg: 80.51%; 16 g/kg: 78.46%)**

↓ 3 (5%) and 6 (7%) wk, overall ADG (7%)*

Table 12.3 (Cont'd p. 396)

Reference	Outcome variables quantified	Animals/treatment (age, initial BW)	Dietary information; time on treatment	Daily prebiotic dose; source	Major findings
Jiang et al. (2006) (con't.)					Linear ↓ CP digestibility on d 42 (control: 63.22%; 4 g/kg: 62.48%; 8 g/kg: 60.21%; 12 g/kg: 62.21%; 16 g/kg: 57.69%)***
					Quadratic ↑ bifidobacteria on d 21 (control: 8.09 log CFU/g; 4 g/kg: 8.58 log CFU/g; 8 g/kg: 8.18 log CFU/g; 12 g/kg: 8.02 log CFU/g; 16 g/kg: 7.72 log CFU/g)***
					Quadratic ↑ lactobacilli on d 21 (control: 7.63 log CFU/g; 4 g/kg: 8.69 log CFU/g; 8 g/kg: 8.20 log CFU/g; 12 g/kg: 8.00 log CFU/g; 16 g/kg: 7.67 log CFU/g)***
					Quadratic ↑ cecal butyrate on d 21 (control: 0.48 mmol/g; 4 g/kg: 0.59 mmol/g; 8 g/kg: 0.52 mmol/g; 12 g/kg: 0.54 mmol/g; 16 g/kg: 0.43 mmol/g)**

Reference	Measurement	Diet / Composition	Treatment	Results
Jiang et al. (2006) (con't.)				SBM:
				↓ 3 wk (4%), 6 wk BW (4%)*
				↓ 3 wk (4%), overall ADG (4%)*
Kim et al. (2006)	Egg production	Complete layer ration	Control (0%)	0.75% FOS:
	Bone measurements	Chemical composition not provided	Full feed	↓ First day out of egg production (20%)**
		51 d study	Feed withdrawal	
			100% Alfalfa	
			0.375% FOS with alfalfa	
			0.75% FOS with alfalfa	
Ma et al. (2006)	Performance responses (BW, feed consumption)	Corn-SBM diet	Control (0 g/kg)	MOS + Ligustrum lucidium:
	Enzyme activity	Chemical composition (initial–final) 220–200 g/kg CP, 11–10 g/kg lys, 4.5–4.4 g/kg met, 10–9.5 g/kg Ca, 7.2–7.0 g/kg P	10 g/kg Ligustrum lucidium	↓ Serum (21%), thigh malondialdehyde concentration (35%)**
			10 g/kg Schisandra chinensis	↑ Serum (41%), heart glutathione reductase activity (39%)**

● Table 12.3 (Cont'd p. 398)

Reference	Outcome variables quantified	Animals/treatment (age, initial BW)	Dietary information; time on treatment	Daily prebiotic dose; source	Major findings
Ma et al. (2006) (con't.)	Serum antibodies		49 d study	50 g/kg MOS + 10 g/kg *Ligustrum lucidium*	↑ Antibody titer against Newcastle disease virus on d 21 (21%), 28 (25%)**
	Spleen lymphocyte proliferation				↑ Spleen lymphocyte proliferation (33%)**
				50 g/kg MOS + 10 g/kg *Schisandra chinensis*	MOS + *Schisandra chinensis*:
					↓ Serum (21%), thigh malondialdehyde concentration (36%)**
					↑ Serum glutathione reductase activity (50%)**
					↑ Antibody titer against Newcastle disease virus on d 21 (18%), 28 (24%)**
					↑ Spleen lymphocyte proliferation (28%)**

Reference	Response	Animals	Diet	Treatment	Findings
Baurhoo et al. (2007a)	Performance responses (BW; feed intake, conversion)	160 Male Cobb broiler chicks (1 d)	Corn-SBM based diet	Control	MOS:
	Cecal microbiota		Chemical composition (starter-grower) 22.0–20.0% CP, 1.0–0.9% Ca, 0.50–0.47% P	Control with antibiotic	↓ Feed intake (9%), BW (9%)**
	Intestinal morphology and histology		42 d study	Control + 0.2% MOS (Bio-Mos; 0.1% in grower diet)	↑ Villus height on d 28 (7%)**
				Control + 1.25% lignin	↑ Goblet cells per villus on d 28 (79%), 42 (94%)**
					↑ Cecal lactobacilli on d 42 (5%)**
				Control + 2.5% lignin	↓ E. coli in litter on d 28 (76%), 42 (30%)**
Baurhoo et al. (2007b)	Performance responses (BW; feed intake, conversion)	208 Male Cobb broiler chicks (1 d)	Corn-SBM based diet	Control	MOS:
				Control with antibiotic	↑ Cecal lactobacilli on d 38 (3%)**
				Control + 0.2% MOS (Bio-Mos; 0.1% in grower diet)	↓ E. coli on d 9 post-challenge (18%)**
	Cecal microbiota		Chemical composition not provided	Control + 1.25% lignin	
	E. coli challenge		38 d study (E. coli challenge at d 21–12 birds/treatment)	Control + 2.5% lignin	

■ Table 12.3 (Cont'd p. 400)

Reference	Outcome variables quantified	Animals/treatment (age, initial BW)	Dietary information; time on treatment	Daily prebiotic dose; source	Major findings
Biggs and Parsons (2007)	AA digestibility	Four cecectomized roosters plus four conventional roosters	30 g Corn-isolated soy protein diet (precision-fed)	Control (0 g/kg)	Cecectomized roosters:
	TME_n		Chemical composition 160 g/kg CP	4 g/kg inulin	8 g/kg OF:
			48 h study	4 g/kg OF	↑ met digestibility (3%)**
				4 g/kg scFOS	4 g/kg scFOS:
				4 g/kg MOS	↑ met digestibility (3%)**
				4 g/kg TOS	4 g/kg MOS:
				8 g/kg Inulin	↑ met digestibility (5%)**
				8 g/kg OF	8 g/kg MOS:
				8 g/kg scFOS	↑ ile (6%), lys digestibility (6%)**
				8 g/kg MOS	4 g/kg TOS:
				8 g/kg TOS	↑ ile digestibility (5%)**
					8 g/kg TOS:
					↑ ile (8%), lys (8%), met (5%), val digestibility (9%)**

Reference	Measurement	Treatment	Diet / Chemical composition	Conventional roosters:
Biggs and Parsons (2007) (con't.)				4 g/kg Inulin: ↓ met digestibility (5%)**
				8 g/kg Inulin: ↓ met digestibility (3%)**
Biggs et al. (2007)	Study 1	Study 1	Study 1	Study 1
	Performance responses (BW, feed intake, G:F)	Control (0 g/kg)	Corn-SBM diet	Inulin:
			40 Male Hampshire × Columbian chicks (1 d)	
	Nutrient digestibility	8 g/kg Inulin	Chemical composition 23% CP, 1.27% lys, 0.53% met, 0.90% met + cys	↓ Metabolizable energy on d 7 (8%), 14 (4%), 21 (4%)**
	Metabolizable energy	8 g/kg OF	21 d study	↓ ile (4%), met (4%) digestibility on d 3**
		8 g/kg scFOS		↓ arg (2%), cys (7%), his (4%), ile (4%), leu (4%), lys (3%), met (4%), phe (3%), thr (6%), val (7%) digestibility on d 7**
		8 g/kg MOS		
		8 g/kg TOS		↓ arg (5%), cys (13%), his (7%), ile (8%), leu (6%), lys (5%), met (5%), phe (6%), thr (10%), val (8%) digestibility on d 21**

■ Table 12.3 (Cont'd p. 402)

Reference	Outcome variables quantified	Animals/treatment (age, initial BW)	Dietary information; time on treatment	Daily prebiotic dose; source	Major findings
Biggs et al. (2007) (con't.)					scFOS:
					↓ Metabolizable energy on d 7 (10%), 14 (3%)**
					↓ ile (4%), met (4%) digestibility on d 3*
					↓ arg (3%), cys (9%), his (5%), ile (4%), leu (4%), lys (4%), met (5%), phe (4%), thr (8%), val (10%) digestibility on d 7**
					↓ arg (4%), cys (11%), his (5%), ile (7%), leu (5%), lys (5%), met (5%), phe (5%), thr (8%), val (6%) digestibility on d 21**
					MOS:
					↓ Metabolizable energy on d 14 (4%)**
					↓ met digestibility on d 3 (5%)**

Reference	Response criteria	Animal/design	Design	Treatment	Results
Biggs et al. (2007) (con't.)					↓ arg (2%), cys (5%), his (3%), ile (3%), leu (3%), lys (2%), met (4%), phe (3%), thr (5%), val (3%) digestibility on d 7**
					TOS:
					↓ arg (2%), cys (4%), his (2%), ile (3%), leu (3%), lys (2%), met (3%), phe (3%), thr (5%), val (4%) digestibility on d 7**
					↓ arg (2%), cys (5%), his (2%), ile (2%), leu (2%), lys (2%), met (2%), phe (2%), thr (4%), val (3%) digestibility on d 21**
	Study 2	Study 2	Study 2	Study 2	Study 2
	Performance responses (BW, feed intake, G:F)	40 Male Hampshire × Columbian chicks (1 d)	Same as study 1	Control (0 g/kg)	Inulin:
	Nutrient digestibility		21 d study	4 g/kg inulin	↑ Metabolizable energy on d 7 (7%), 14 (3%), 21 (5%)**
				4 g/kg OF	↓ phe (3%), thr (7%) digestibility on d 3–4**
	Metabolizable energy			4 g/kg scFOS	↓ cys (3%), lys (2%), thr (2%) digestibility on d 21**

◧ Table 12.3 (*Cont'd* p. 404)

Reference	Outcome variables quantified	Animals/treatment (age, initial BW)	Dietary information; time on treatment	Daily prebiotic dose; source	Major findings
Biggs et al. (2007) (con't.)	Cecal microbiota			4 g/kg MOS	OF:
				4 g/kg TOS	↑ Metabolizable energy on d 7 (6%), 21 (3%)**
					↓ ile (4%), leu (4%), phe (3%), thr (7%), val (6%) digestibility on d 3–4**
					↓ cys (2%), met (1%) digestibility on d 21 **
					scFOS:
					↓ Metabolizable energy on d 3–4 (8%)**
					↑ Metabolizable energy on d 21 (2%)**
					↓ arg (5%), cys (8%), his (7%), ile (7%), leu (7%), lys (8%), met (10%), phe (6%), thr (13%), val (11%) digestibility on d 3–4**
					↓ arg (2%), his (1%), ile (2%), leu (1%), lys (3%), met (3%), phe (1%), thr (4%), val (2%) digestibility on d 21**

Biggs et al. (2007) (con't.)	MOS:			TOS:			
	↓ Metabolizable energy on d 3–4 (8%)**	↑ Metabolizable energy on d 7 (2%), 14 (2%), 21 (5%)**	↑ arg (1%), cys (6%), his (2%), ile (2%), leu (2%), lys (1%), met (1%), phe (2%), thr (3%), val (4%) digestibility on d 21**	↓ Metabolizable energy on d 3–4 (12%)**	↑ Metabolizable energy on d 21 (2%)**	↓ arg (6%), cys (14%), his (7%), ile (8%), leu (7%), lys (8%), met (5%), phe (7%), thr (13%), val (10%) digestibility on d 3–4**	↓ arg (2%), cys (1%), his (1%), ile (3%), leu (1%), lys (4%), met (2%), thr (3%) digestibility on d 21**

■ Table 12.3 *(Cont'd p. 406)*

Reference	Outcome variables quantified	Animals/treatment (age, initial BW)	Dietary information; time on treatment	Daily prebiotic dose; source	Major findings
Biggs et al. (2007) (con't.)	Study 3	Study 3	Study 3	Study 3	Study 3
	Cecal microbiota	20 Male Hampshire × Columbian chicks (1 d)	Dextrose-isolated soy protein diet	Same as study	scFOS:
			Chemical composition 23% CP, 1.41% lys, 0.54% met, 0.90% met + cys		↓ Cecal *Clostridium perfringens* (5%)**
			21 d study		MOS:
					↓ Cecal *Clostridium perfringens* (6%)**
Elmusharaf et al. (2007)	Performance responses (BW; feed intake, conversion)	64 Male Ross 308 broiler chicks (1 d)	Corn-SBM-wheat based pelleted diet	Control (0 g kg)	MOS:
	Coccidiosis challenge (oocyst count, lesion scoring)		Chemical composition 215.6 g/kg CP	Control + 10 g/kg MOS (Bio-Mos) with or without coccidial challenge	↓ oocyst shedding on d 5 (55%), 12 (70%)**
			19 d study		↓ Lesions from *Eimeria acervulina* (72%)**

Reference	Parameter	Subjects	Diet	Dietary composition / study	Control groups	Results
Lan et al. (2007)	Cecal lactic acid bacteria	60 Male Arbor acres broiler chicks (1 d)	Corn-SBM diet	Dietary composition 202.1 g/kg CP, 10.0 g/kg Ca, 6.4 g/kg P	Negative control (0% SBM, 0% antibiotics)	SBM oligosaccharides:
	Intestinal morphology				Positive control (SBM, antibiotics)	↑ Lactic acid bacteria (determined via threshold cycle values, 17%)**
				15 d study	Negative control + 10 g/kg SBM oligosaccharides	
Rehman et al. (2007)	Performance response (BW)	20 Ross 308 broiler chicks (1 d)	Corn-SBM-wheat diet		Control (0%)	Inulin:
	Intestinal histomorpho-logy			Dietary composition 24.9% CP, 3.5% crude fat, 1.5% CF, 1.1% Ca, 0.87% P	Control + 1% inulin (Raftiline-GR)	↑ Jejunal villus height (20%)**
	Nutrient transport activity			35 d study		↑ Crypt depth (31%)*
						↑ Short-circuit current with glucose (669%) and glutamine (1,571%)*
						↓ Basal values of transmucosal tissue resistance with glucose (47%) and glutamine (37%)*

● Table 12.3 (*Cont'd* p. 408)

Reference	Outcome variables quantified	Animals/treatment (age, initial BW)	Dietary information; time on treatment	Daily prebiotic dose; source	Major findings
Solis de los Santos et al. (2007)	Performance response (BW)	18 Hybrid converter tom turkey poults (1 d)	Standard unmedicated corn-soybean starter diet	Control (0%)	All concentrations MOS:
			21 d study	Control + 0.455 g/kg MOS (Alphamune, yeast extract)	↑ BW on d 7 (8%)**
	Intestinal morphology			Control + 0.91 g/kg MOS	↑ Duodenal neutral (47–71%), sialomucin (56–64%), sulfomucin (44–56%) goblet cells on d 7**
					↑ Jejunal crypt depth on d 7 (40–48%), 21 (46–76%)**
					↑ Jejunal villus height (57–79%), villus surface area on d 21 (89–127%)**
					↑ Jejunal neutral (40–51%), sulfomucin (23–66%) goblet cells on d 21**
					↑ Ileal villus height (31–33%), villus surface area (56–57%) on d 21**

Solis de los Santos et al. (2007) (con't.)	↑ Ileal crypt depth on d 7 (30–49%), 21 (41–57%)** ↑ ileal neutral (60–92%), sialomucin (72–116%), sulfomucin (32–72%) goblet cells on d 7** 0.455 g/kg MOS: ↑ Duodenal villus height (21%), surface area (38%) on d 7** ↑ Duodenal crypt depth on d 21 (3%)** ↑ Duodenal sulfomucin goblet cells on d 21 (53%)** ↑ Jejunal villus surface area on d 7 (89%)** ↑ Jejunal lamina propria thickness on d 21 (64%)** ↑ Jejunal sialomucin (31%), sulfomucin (50%) goblet cells on d 7** ↑ Jejunal sialomucin goblet cells on d 21 (84%)** ↑ Ileal lamina propria thickness on d 7 (44%), 21 (38%)**			

Table 12.3 (Cont'd p. 410)

Reference	Outcome variables quantified	Animals/treatment (age, initial BW)	Dietary information; time on treatment	Daily prebiotic dose; source	Major findings
Solis de los Santos et al. (2007) (con't.)					↑ Ileal villus height on d 7 (49%)**
					↑ Ileal villus surface area (86%)**
					↑ Ileal neutral (120%), sialomucin (58%), sulfomucin (36%) goblet cells on d 21**
Donaldson et al. (2008)	Study 1	Study 1	Study 1	Study 1	Study 1
	Salmonella enterica serovar Enteritidis challenge	12 Laying hens	Corn-soybean mash	Control (nonmolted)	0.75% FOS:
	Performance responses (feed intake, BW, ovarian weight)		Chemical composition 16.5% CP, 3.5% Ca, 0.48% P	Feed withdrawal (molted)	↓ Number of hens positive for Salmonella enteritidis in crop (17%)**
	Crop and cecal metabolites		28 d study	90% alfalfa + 10% layer ration	

				90% alfalfa + 10% layer ration + 0.375% FOS	
Donaldson et al. (2008) (con't.)	Crop, cecal, and organ *Salmonella enteritidis* colonization	Study 2	Study 2	Study 2	Study 2
		Same as study 1	Same as study 1	Same as study 1	No differences observed
		Study 3	Study 3	Study 3	Study 3
		Same as study 1 except no feed intake data	Same as study 1	Same as study 1	0.375% FOS:
	Intestinal *Salmonella enteritidis* shedding			90% alfalfa + 10% layer ration + 0.75% FOS	↓ Number of hens positive for *Salmonella enteritidis* in intestines (60%)**
					↓ Intestinal *Salmonella enteritidis* colonization (118%)**
					↓ Crop lactic acid (28%)**

■ Table 12.3   (Cont'd p. 412)

Reference	Outcome variables quantified	Animals/treatment (age, initial BW)	Dietary information; time on treatment	Daily prebiotic dose; source	Major findings
Donaldson et al. (2008) (con't.)					0.75% FOS:
					↑ Crop lactic acid (320%)**
					↑ Crop *Salmonella enteritidis* colonization (127%)**
					↓ Number of hens positive for *Salmonella enteritidis* in intestines (60%)**
					↓ Intestinal *Salmonella enteritidis* colonization (118%)**
	Study 4	Study 4	Study 4	Study 4	Study 4
	Same as study 3	Same as study 1	Same as study 1	Same as study 1	0.375% FOS:
					↑ Crop lactic acid (295%)**
					0.75% FOS:
					↑ Cecal isobutyrate (250%)**

Yang et al. (2008)	Study 1	Study 1	Study 1	Study 1	Study 1
		96 Male Cobb broiler chicks (1 d)	SBM-sorghum diet	Control (0 g/kg)	Study 1
Performance responses (BW, feed intake, feed conversion)					1 g/kg MOS:
Metabolizable energy			Chemical composition (starter-grower) 230.0–210.0 g/kg CP, 40.8–44.0 g/kg CF, 57.3–52.1 g/kg crude fat, 12.5–11.0 g/kg lys, 9.0–8.3 g/kg met + cys, 10.0 g/kg, 4.0–3.7 g/kg P	Control + 50 ppm zinc bacitracin	↑ AME (2%)**
Intestinal microbiota			5 wk study	1 g/kg MOS	↑ Ileal pH (4%)**
Intestinal metabolites				2 g/kg MOS	↓ Ileal total SCFA (66%)**
					↓ Cecal lactobacilli (5%)**
					↓ Cecal propionate (50%)**
					↑ Cecal butyrate (92%)**
					2 g/kg MOS:
					↓ Feed conversion during wk 1–3 (4%)***

◘ Table 12.3

Reference	Outcome variables quantified	Animals/treatment (age, initial BW)	Dietary information; time on treatment	Daily prebiotic dose; source	Major findings
Yang et al. (2008)					↑ AME (2%)**
					↑ Ileal pH (5%)**
					↓ Ileal lactobacilli (8%), coliforms (8%)**
					↓ Ileal lactic acid (47%), total SCFA (42%)**
					↓ Cecal propionate (59%)**
	Study 2	Study 2	Study 2	Study 2	Study 2
	Metabolizable energy	Eight male Cobb broiler chicks (1 d)	Same as study 1	Same as study 1	MOS:
	Nutrient digestibility				↓ Soluble NSP digestibility (20%)**

[a]AA amino acid, ADG average daily gain, AME_n metabolizable energy corrected for nitrogen, BAC bacitracin, BW body weight, Ca calcium, CF crude fiber, CFU colony forming unit, CHO carbohydrate, CP crude protein, cys cysteine, d day, DM dry matter, ESBM ethanol-extracted soybean meal, F:G feed:gain ratio, FOS fructooligosaccharide, g gram, h hour, HCC hen cecal contents, IBDV infectious bursal disease virus, IgG immunoglobulin G, IgM immunoglobulin M, IMO isomaltooligosaccharide, kg kilogram, lys lysine, met methionine, mmol millimole, MOS mannanoligosaccharide, NSP non-starch polysaccharide, OF oligofructose, P phosphorus, ppm parts per million, SBM soybean meal, SCFA short-chain fatty acid, TME_n true metabolizable energy corrected for nitrogen, TOS transgalactooligosaccharide, val valine, wk week

[b]*$P < 0.01$, **$P < 0.05$, ***$P < 0.10$

supplementation (Yang et al., 2008). Pathogenic microbiota also decreased with prebiotic supplementation. Five authors reported decreased salmonellae and coliforms (Donaldson et al., 2008; Fernandez et al., 2000, 2002; Spring et al., 2000; Thitaram et al., 2005), while four authors reported decreased *E. coli* (Baurhoo et al., 2007a, b; Xu et al., 2003; Zdunczyk et al., 2005). Streptococci, peptococci, bacilli, staphylococci, bacteriodaeceae, pseudomonad, yeast, and mold populations also have been reported to decrease with prebiotic supplementation (Cao et al., 2005; Samarasinghe et al., 2003; Terada et al., 1994).

Changes in fermentation profiles also occur with prebiotic supplementation of poultry. Thirty-six percent of studies observed increased butyrate concentrations with prebiotic supplementation (Jiang et al., 2006; Terada et al., 1994; Yang et al., 2008; Zdunczyk et al., 2004; Zhang et al., 2003). When supplemented with lactosucrose and FOS, decreased cecal phenol and indole concentrations were observed (Cao et al., 2005; Terada et al., 1994). Although inconclusive, total SCFA and lactic acid concentrations generally decreased while intestinal pH increased with prebiotic supplementation (Donaldson et al., 2008; Yang et al., 2008; Zdunczyk et al., 2005). As fermentative end-products are volatile, they can pose a significant problem not only to the health of the animals but also to the people who work with them on a daily basis. Reducing emissions of volatile components such as ammonia improves air quality in production settings and has the potential to improve the environment in general. However, mixed results have been observed with regard to cecal ammonia concentrations: ammonia decreased in response to lactosucrose supplementation (Terada et al., 1994), but increased in response to MOS supplementation (Zdunczyk et al., 2005). Mixed results also have been reported for acetate, propionate, isobutyrate, and valerate concentrations with prebiotic supplementation (Cao et al., 2005; Donaldson et al., 2008; Terada et al., 1994; Yang et al., 2008; Zdunczyk et al., 2004, 2005; Zhang et al., 2003).

Prebiotic supplementation also affects cell proliferation in poultry. Increased intestinal villus height was reported in 40% of studies measuring intestinal morphology (Baurhoo et al., 2007a; Rehman et al., 2007; Solis de los Santos et al., 2007; Xu et al., 2003). Two studies investigating the effects of MOS supplementation reported increased numbers of goblet cells per villus, as well as increases in several types of goblet cells along the intestinal tract (Baurhoo et al., 2007a; Solis de los Santos et al., 2007). Although inconclusive, crypt depth generally increased with prebiotic supplementation (Rehman et al., 2007; Solis de los Santos et al., 2007). Other intestinal characteristics have been observed in several studies, including increased gut length (Yusrizal and Chen, 2003) and microvillus height (Xu et al., 2003).

Changes in nutrient digestibility are of utmost concern whenever a feed supplement is added to a diet. For young chicks, amino acid (AA) digestibility and metabolizable energy generally decreased in response to prebiotic supplementation (Biggs et al., 2007). Decreased AA digestibility also was observed in intact adult roosters; however, when observed in cecectomized roosters, increased AA digestibility was observed (Biggs and Parsons, 2007). This implies that the observed changes in digestibility are caused by microbial fermentation in the ceca. Soy oligosaccharides decreased dry matter digestibility in four studies (Coon et al., 1990; Jiang et al., 2006; Leske et al., 1993; Leske and Coon, 1999a) and true metabolizable energy in three studies (Coon et al., 1990; Leske et al., 1993; Leske and Coon, 1999a), and decreased digestibilities of other nutrients in one study (Leske and Coon, 1999a). Mixed results in apparent metabolizable energy data were observed by three authors with select prebiotic supplements (Biggs et al., 2007; Leske and Coon, 1999a; Yang et al., 2008).

Performance and production responses have been evaluated as regards prebiotic supplementation. Body weight was measured in 57% of studies and, although inconclusive, increased in the majority of those studies (Fairchild et al., 2001; Lee et al., 2005; Parks et al., 2001; Sims et al., 2004; Solis de los Santos et al., 2007; Yusrizal and Chen, 2003; Zdunczyk et al., 2005). Similarly, body weight gain increased (Parks et al., 2005; Samarasinghe et al., 2003), while average daily gain responses were variable with prebiotic supplementation (Parks et al., 2005; Sims et al., 2004; Xu et al., 2003; Yang et al., 2008). Feed intake and feed:gain ratios (F:G) generally decreased with supplementation of fructans and MOS (Baurhoo et al., 2007a; Parks et al., 2001; Samarasinghe et al., 2003; Xu et al., 2003; Yusrizal and Chen, 2003). Increased carcass weight and abdominal fat weight were observed with MOS and inulin supplementation (Samarasinghe et al., 2003; Yusrizal and Chen, 2003). Egg production and hatchability increased in one study (Shashidhara and Devegowda, 2003), while egg weight and yolk color decreased in another study with MOS supplementation (Zaghini et al., 2005).

Swine. As has been observed in poultry, major changes in intestinal microbiota occur in swine in response to prebiotic consumption (◉ *Table 12.4*). Increased bifidobacteria (Estrada et al., 2001; Howard et al., 1995; Loh et al., 2006; Lynch et al., 2007; Pierce et al., 2006; Smiricky-Tjardes et al., 2003; Tako et al., 2008; Tzortis et al., 2005; Xu et al., 2002) and lactobacilli (Lynch et al., 2007; Oli et al., 1998; Smiricky-Tjardes et al., 2003; Tako et al., 2008; Tzortis et al., 2005; White et al., 2002; Xu et al., 2002) were reported in 65 and 53% of studies observing changes in microbial ecology, respectively. Generally, decreased populations of enterobacteria, clostridia, coliforms, and *E. coli* were observed with

■ Table 12.4

In vivo experiments, listed in chronological order, reporting effects of prebiotics in swine (*Cont'd* p. 416)

Reference	Outcome variables quantified	Animals/treatment (age, initial BW)	Dietary information; time on treatment	Daily prebiotic dose; source	Major findings
Howard et al. (1995)	Study 1	Study 1	Study 1	Study 1	Study 1
	Performance responses (BW, feed intake)	Ten newborn male pigs (36 h)	Infant formula	Control (0%)	FOS:
	Cecal and colonic microbiota		Chemical composition: 59.4 g/L CP, 56.2 g/L fat, 50.8 g/L CHO	Control + 3 g/L FOS	↑ Cecal cell density (11%)*
	Cecal metabolites		15 d study		↑ Cecal labeled cells (17%)**
	Cecal and colonic morphology				↑ Proximal colonic crypt height (10%), leading edge (21%), labeled cells (42%), proliferation zone (12%), and labeling index (34%)*
					↑ Proximal colonic cell density (5%)***
	Carcass morphology				↑ Distal colonic crypt height (20%), leading edge (42%), cell density (26%), labeled cells (82%), and labeling index (43%)*
					↑ Distal colonic proliferation zone (13%)***
	Study 2	Study 2	Study 2	Study 2	Study 2
	Fecal microbiota	Six newborn male pigs (36 h)	Same as study 1 except 6 d study	Same as study 1	FOS: ↑ Bifidobacteria on d 6 (247%)***

◪ Table 12.4   *(Cont'd p. 418)*

Reference	Outcome variables quantified	Animals/treatment (age, initial BW)	Dietary information; time on treatment	Daily prebiotic dose; source	Major findings
Houdijk et al. (1998)	Performance responses (BW, feed intake, feed conversion)	Ten castrated male Great Yorkshire × Great Yorkshire × Landrace × Great Yorkshire piglets (57 d, 15.6 kg)	Corn-casein-fishmeal-meat meal diet	Control (0%)	FOS:
			Chemical composition not provided	Control + 7.5 g/kg FOS (Raftilose P95)	↓ ADG (low FOS) during wk 2 (7%)***
			6 wk study	Control + 15 g/kg FOS	↓ ADG during wk 1 (10–21%)*
				Control + 10 g/kg TOS (Oligostroop)	↓ ADG during wk 1–3 (11%)**
				Control + 20 g/kg TOS	↓ DMI during wk 1 (4–11%)***
					↓ DMI during wk 1–3 (8%)**
					↑ Feed conversion during wk 1 (6–16%)*
					↓ Feed conversion (high FOS) during wk 2 (11%)**, 3 (2%)***
					↓ Fecal DM from d 0–35 (4%)***
	Fecal microbiota				TOS:
					↓ ADG during wk 1 (19–20%)*
					↓ ADG (low TOS) during wk 2 (16%)***
					↓ ADG during wk 1–3 (11–14%)**
					↓ DMI during wk 1 (6–13%)***
					↓ DMI during wk 1–3 (7–11%)**
					↑ Feed conversion during wk 1 (7–21%)*
					↓ Feed conversion during wk 2 (22%)*, 3 (16%)***
					↓ Fecal DM from d 0–35 (7–13%)***

Reference	Measurement	Animals	Diet/Conditions	Control/Treatment	Results
Oli et al. (1998)	Clinical symptoms of infection	Five crossbred pigs (21 d)	Antibiotic-free early-wean pig pellets plus 15 µg/kg BW cholera toxin to induce diarrhea	Oral electrolyte solution	FOS at 24 h:
	Stool consistency		Chemical composition not provided	Oral electrolyte solution + 0.5% FOS	↑ Redox potential (29%)**
	Intestinal metabolites		72 h study		↑ Small intestinal (26%), cecal (21%) lactobacilli**
					↓ Small intestine mucosal enterobacteria (25%)**
	Intestinal microbiology				↓ Cecal (16%), colonic (27%) enterobacteria**
					FOS at 72 h:
					↓ Cecal enterobacteria (18%)**
Estrada et al. (2001)	Study 1	Study 1	Study 1	Study 1	Study
	Performance responses (BW, ADFI, F:G)	20 Weanling Canabrid × Camborough 15 pigs (8 d, 6.2 kg)	Wheat-SBM diets	Control (0.0%)	FOS:
	Fecal microbiota		Chemical composition not provided	Control + 0.5% FOS (Raftilose P95) + 10^{10} Bifidobacterium longum str. 75119 (given on d 1, 3)	↑ ADG from d 0–7 (16%)*
			21 d periods		↑ F:G from d 0–7 (20%)**
					↓ Enterobacteria (7%), clostridia (11%), and total anaerobes (12%) on d 7**
					↑ Bifidobacteria on d 7 (16%)**

Table 12.4   (Cont'd p. 420)

Reference	Outcome variables quantified	Animals/treatment (age, initial BW)	Dietary information; time on treatment	Daily prebiotic dose; source	Major findings
Estrada et al. (2001) (con't.)	Study 2	Study 2	Study 2	Study 2	Study 2
	Performance responses (BW, ADFI)	20 Weanling Canabrid × Camborough 15 pigs (8 d, 6.2 kg)	Wheat-SBM diets	Control (0.0%)	FOS:
	Fecal microbiota		Chemical composition not provided	Control + 10^7 B. longum str. 75119	↓ BW on d 12 (5%), 19 (9%)*
	Serum IGF-I			Control + 0.5% FOS	↓ ADG on d 5–12 (22%), 12–19 (15%), 0–19 (18%)*
				Control + 0.5% FOS + 10^{10} B. longum str. 75119 (given on d 1, 3)	↓ ADFI from 0–5 (12%)***
					↓ Serum IGF-I on d 12 (56%), 19 (42%)**
Zhang et al. (2001)	Performance responses (ADG, ADFI, F:G)	8 Landrace × Large white × Duroc barrows (12.5 kg)	Corn-soy protein isolate diet	Control (0%)	SBM:
	Nutrient digestibilities		Chemical composition 19.75% CP, 0.97% Ca, 0.50% P, 1.09% lys, 0.37% met, 0.75% met + cys	Control + 1% stachyose	↓ Nitrogen retention (6%)**
	Nitrogen retention		7 d periods	Control + 2% stachyose	
				Control + 23.5% SBM (0.16% raffinose, 0.75% stachyose; no soy protein isolate)	

					Study 1
Davis et al. (2002)	Performance responses (ADG, ADFI, G:F)	Study 1	54 Weanling Hampshire × Duroc × Yorkshire x Landrace barrows (18 d-old, 6 kg)	Study 1	Study 1	Control (0%)	MOS:
		Corn-soy diet with supplemental Cu at 0 or 175 ppm	Control + 0.2% MOS	↑ ADG with (76%) and without (47%) Cu during phase 1**
		Chemical composition (initial-final) 1.5–1.2% lys, 0.98–0.77% thr, 0.27–0.24% trp, 0.90–0.72% met + cys		↑ G:F with (62%) and without (38%) Cu during phase 1**
		38 d study		↑ ADG during phase 3 (8%), overall (6%)**
				↑ G:F during phase 3 (7%)***, overall (6%)**
	Study 2	Study 2	Study 2	Study 2
	Performance responses (ADG, ADFI, G:F)	36 Crossbred barrows and gilts (20 kg)	Corn-soy diet with supplemental Cu at 0 or 125 ppm in starter and grower phases and 0 or 175 ppm in finisher phase	Starter:	MOS:
			Chemical composition (grower-finisher) 20.2–16.7% CP, 1.10–0.85% lys, 0.78–0.64% thr, 0.24–0.19% trp, 0.67–0.57% met + cys	Control (0%)	↑ ADG without Cu in finisher phase (7%)**
			Study end at 106 kg average BW	Control + 0.2% MOS	↓ ADG with Cu in finisher phase (5%)**
				Grower:
				Control (0%)
				Control + 0.1% MOS
				Finisher:
				Control (0%)
				Control + 0.05% MOS

■ Table 12.4 (Cont'd p. 422)

Reference	Outcome variables quantified	Animals/treatment (age, initial BW)	Dietary information; time on treatment	Daily prebiotic dose; source	Major findings
Houdijk et al. (2002)	Intestinal and fecal metabolites	Four castrated Great Yorkshire-Landrace × Great Yorkshire pigs (38 d, 10.4 kg)	Cornstarch-casein liquid diet	Control (0 g/kg)	40 g/kg FOS:
			Chemical composition 168.4 g/kg CP, 18.7 g/kg crude fat, 17.7 g/kg hemicelluloses, 49.1 g/kg cellulose, 5.4 g/kg lignin	Control + 10 g/kg FOS (Raftilose P95)	↓ Ileal pH (6%)**
			37 d study	Control + 40 g/kg FOS	↓ Ileal acetate (32%)**
				Control + 10 g/kg TOS (Oligostroop)	↑ Ileal isovalerate (405%)**
	Intestinal and fecal microbiota			Control + 40 g/kg TOS	↑ Fecal pH (11%)**
					↑ Fecal isobutyrate (100%)**
					40 g/kg TOS:
					↑ Fecal pH (11%)**
					↑ Fecal isobutyrate (143%)**
White et al. (2002)	Study 1	Study 1	Study 1	Study 1	Study 1
	Performance responses (ADG, ADFI, G:F)	35 Hampshire × Landrace-Yorkshire barrows and gilts (21.8 d, 6.6 kg)	Corn-SBM-whey diet	Control (0%)	Yeast diet:
	Fecal microbiota		28 d study	Control + antibiotic	↓ Feed intake on wk 1 (22%), 3 (13%), overall (11%)**
	Circulating antibodies			Control + 3% brewer's yeast	↓ Overall daily gain (9%)**
	Intestinal morphology			Control + 3% brewer's yeast + 2% citric acid	↑ Lactobacilli on d 28 (4%)**
	Fecal metabolites				↓ Total coliforms on d 14 (3%), 28 (3%)**

	Study 2	Study 2	Study 2	Study 2
White et al. (2002) (con't.)				Yeast + acid diet:
				↓ Feed intake on wk 3 (10%), 4 (14%), overall (10%)**
				↓ Daily gain on wk 1 (23%), 4 (11%), overall (9%)**
				↓ Bifidobacteria (6%), aerotolerant aerobes (7%) on d 28**
				↑ IgG (42%)*
				↓ Isovalerate (32%)**
Study 2	Eight Hampshire × Landrace-Yorkshire pigs (11 d old, 4.1 kg)	Same as study 1 except 39 d study	Control (0%)	Study 2
Performance responses (ADG, ADFI, G:F)				Yeast diets:
Fecal microbiota		E. coli K88 challenge on 29	Control + antibiotic	↓ Coliform shedding (10%)**
E. coli challenge			Control + 3% brewer's yeast	↓ Coliforms in jejunum (23%)*
Circulating antibodies				↓ Coliforms in cecum (9%)**

■ Table 12.4   (Cont'd p. 424)

Reference	Outcome variables quantified	Animals/treatment (age, initial BW)	Dietary information; time on treatment	Daily prebiotic dose; source	Major findings
Xu et al. (2002)	Performance responses (ADG, ADFI, F:G)	32 Jiaxing Black × Duroc × Landrace barrows (20.8 kg)	Corn-SBM diet without antibiotic supplementation	Control (0%)	4 g/kg FOS:
				Control + 2 g/kg FOS (Meioligo-P)	↑ ADG (8%)**
					↓ F:G (7%)**
				Control + 4 g/kg FOS	↑ Bifidobacteria (9%) and Lactobacilli (11%) in small intestine**
	Digestive enzyme activity		Chemical composition 17.8% CP, 4.8% crude fat, 2.0% CF, 0.94% lys, 0.56% met + cys, 0.7% Ca, 0.55% P	Control + 6 g/kg FOS	↓ clostridia in small intestine (19%)**
	Intestinal and colonic microbiota		42 d study		↑ Bifidobacteria (8%) and lactobacilli (11%) in proximal colon**
	Intestinal histomorphology				↓ E. coli (8%) in proximal colon**
					↑ Jejunal villus height (17%)**
					↑ Jejunal villus height:crypt depth (40%)**
					↑ Protease (56%), trypsin (43%), and amylase (30%) activities in small intestinal contents**
					6 g/kg FOS:
					↑ ADG (7%)**
					↓ F:G (8%)**

Reference	Study	Animals	Measurement	Treatment	Results
Xu et al. (2002) (con't.)					↑ Bifidobacteria (8%) and lactobacilli (10%) in small intestine**
					↓ Clostridia in small intestine (22%)**
					↑ Bifidobacteria in proximal colon (8%)**
					↑ Trypsin (50%) and amylase (40%) activities in small intestinal contents**
					↑ Jejunal villus height (15%)** All concentrations FOS:
					↓ Clostridia in proximal colon (12–21%)**
Correa-Matos et al. (2003)	Salmonella typhimurium challenge	48 piglets (2 d)		Control (0 g/L)	Soy polysaccharide:
	Sow's milk replacer formula (Advance Baby Pig Liqui-Wean)			Control + 7.5 g/L methylcellulose	↑ Ileal sucrase activity (131%)**
	Chemical composition not provided		Performance responses (BW, feed intake)	Control + 7.5 g/L soy polysaccharide	↓ Ileal transmucosal resistance (39%)** ↑ Colonic total SCFA (70%)**
	14 d study		Physical activity	Control + 7.5 g/L FOS	↑ Fecal moisture (100%)**
			Stool consistency		FOS:
			Histomorphology		↑ Ileal sucrase activity (152%)**
			Disaccharidase activity Intestinal electro-physiology		↑ Colonic total SCFA (50%)**
			Intestinal metabolites		

□ Table 12.4 (Cont'd p. 426)

Reference	Outcome variables quantified	Animals/treatment (age, initial BW)	Dietary information; time on treatment	Daily prebiotic dose; source	Major findings
LeMieux et al. (2003)	Study 1	Study 1	Study 1	Study 1	Study 1
	Performance responses (ADG, ADFI, G:F) over 3 phases and overall	35 Weanling Yorkshire × Landrace or Yorkshire × Landrace × Duroc barrows, boars, or gilts (20 d, 4.8 kg)	Corn-soy diet with no additional Zn (contained antibiotic)	Control (0%)	No effect of supplementation
				Control + 0.2% MOS (Bio-Mos)	
				Control + 0.3% MOS	
			Chemical composition (phase 1–3) 23.8–18.9% CP, 1.6–1.1% lys, 0.9% Ca, 0.8% P		
			28 d study		
	Study 2	Study 2	Study 2	Study 2	Study 2
	Performance responses (ADG, ADFI, G:F) over 3 phases and overall	25 Weanling Yorkshire × Landrace or Yorkshire × Landrace × Duroc barrows, boars, or gilts (17 d, 5.4 kg)	Corn-soy diet with supplemental Zn at 0 or 3,000 ppm (contained antibiotic)	Control (0%)	0.2% MOS:
				Control + 0.2% MOS	↑ ADG without Zn in phase 2 (11%), phase 3 (14%), overall (18%)***
			Chemical composition same as study 1	Control + 0.3% MOS	↑ ADFI without Zn in phase 2 (22%), overall (14%)***
			28 d study		↓ G:F without Zn in phase 2 (9%)***
					↑ G:F with Zn in phase 3 (17%)***
					↑ G:F with Zn (8%)***
	Study 3	Study 3	Study 3	Study 3	Study 3
	Performance responses (ADG, ADFI, G:F) over 3 phases and overall	25 Weanling Yorkshire × Landrace or Yorkshire × Landrace × Duroc barrows, boars, or gilts (16 d, 4.9 kg)	Corn-soy diet with supplemental Zn at 0 or 3,000 ppm (contained antibiotic)	Control (0.0%)	0.2% MOS:
				Control + 0.2% MOS	↓ ADFI without Zn in phase 2 (5%)***, overall (9%)**
			Chemical composition same as study 1		↑ ADG without Zn in phase 2 (17%), overall (6%)***
			21 d study		↑ G:F without Zn in phase 2 (24%)***

Reference	Measurement	Animals	Diets	Control	Prebiotic	Results
LeMieux et al. (2003) (con't.)						↓ ADG with Zn in phase 2 (14%), overall (16%)***
						↓ ADFI with Zn in phase 2 (10%)***, overall (11%)**
						↓ G:F with Zn in phase 2 (5%)**
	Study 4	Study 4	Study 4	Study 4	Study 4	Study 4
	Performance responses (ADG, ADFI, G:F) over 3 phases and overall	20 weanling Yorkshire × Landrace or Yorkshire × Landrace × Duroc barrows, boars, or gilts (18 d, 4.7 kg)	Corn-soy diet with supplemental antibiotic at 0.00 or 0.75% (no supplemental Zn), or with supplemental Zn at 0 or 3,000 ppm (no supplemental antibiotic)	Control (0%)	0.2% MOS:	
			Chemical composition same as study 1	Control + 0.2% MOS		↓ ADFI in phase 2 with (4%) and without antibiotic (10%)***
			27 d study			↓ ADG in phase 2 without antibiotic (15%)**
						↓ G:F in phase 2 without antibiotic (6%)*
						↑ G:F in phase 2 with antibiotic (7%)**
						↑ ADG in phase 2 with antibiotic (2%)**
						↓ G:F in phase 2 without Zn (6%)**
						↑ G:F in phase 2 with Zn (7%)**
Mikkelsen et al. (2003)	Performance responses (BW, feed consumption)	28 Mixed-sex Landrace × Yorkshire piglets (4 wk)	Cornstarch-wheat-fishmeal-casein diets Chemical composition not provided	Control	FOS:	
	Fecal scores		4 wk study	Control + 4% FOS (Raffilene ST)		↑ Daily feed intake on 14–28 (12%)***
	Fecal microbiota			Control + 4% TOS (Elix'or)		↓ Fecal valerate (30%) and isobutyrate + isovalerate (23%) on d 3**
	Fecal metabolites					↑ Fecal yeast on d 14 (24%)**
					TOS:	
						↑ Daily feed intake on d 14–28 (22%)***
						↑ Fecal DM on d 3 (19%)**
						↑ Fecal valerate on d 28 (51%)**
						↑ Fecal yeast on d 7 (67%), 14 (42%), and 28 (38%)**

■ Table 12.4   (Cont'd p. 428)

Reference	Outcome variables quantified	Animals/treatment (age, initial BW)	Dietary information; time on treatment	Daily prebiotic dose; source	Major findings
Petkevicius et al. (2003)	Oesophagostomum dentatum challenge	Eight mixed-sex Landrace × Yorkshire × Duroc pigs (10 wk, 20.4 kg)	Barley flour-SBM diet	Basal + 300 g/kg oat hull meal (Control)	Inulin:
	Performance response (BW)		Chemical composition 17% CP, 2% fat, 19.8% dietary fiber	Basal + 160 g/kg inulin (Raftiline)	↓ Fecal DM (23%), NSP (53%)*
	Fecal egg counts		13 wk study: all pigs maintained on control for 10 wk, but dosed with 6,000 O. dentatum infective larvae at 3 wk; treatment diets fed for 3 wk	Basal + 210 g/kg sugar beet fiber	↑ Fecal ash (93%), protein (177%), and fat (129%)*
				Basal + 150 g/kg sugar beet fiber + 60 g/kg inulin	↓ Fecal acetate (13%), butyrate (27%)*
	Fecal metabolites				↑ Fecal propionate (35%), valerate (155%), and total SCFA (49%)*
					↓ Intestinal O. dentatum egg counts (100%), worm burdens (100%), and lower worm concentrations in female pigs compared to male pigs (60%)** Sugar beet fiber:
					↓ Fecal DM (27%), NSP (64%)*
					↑ Fecal ash (104%), protein (146%), and fat (167%)*
					↓ Fecal propionate (13%)*
					↑ Fecal total SCFA (35%)*
					↓ Intestinal O. dentatum egg counts (38%), worm burdens (72%)**
					Inulin + sugar beet fiber:
					↓ Fecal DM (26%), NSP (60%)*

Reference	Measure	Animals	Diet / Methods	Control	Treatment	Results
Petkevicius et al. (2003) (con't.)						↑ Fecal ash (94%), protein (140%), and fat (173%)*; ↓ Fecal propionate (16%)*; ↑ Fecal total SCFA (37%)*; ↓ Intestinal *O. dentatum* egg counts (84%), worm burdens (89%)**
Smiricky-Tjardes et al. (2003)	Nutrient digestibility	12 PIC 326 × C22 pigs (30 kg)	Casein-cornstarch diet	Control (0%)	17% soy solubles (6.9 g raffinose, 27.7 g stachyose, 1.2 g verbascose)	Soy solubles: ↓ Ileal DM (4%), OM (6%), N (4%) digestibility**
	Intestinal microbiota		Chemical composition 3.02% N	6% TOS		↑ Ileal GOS digestibility (77%)**; ↑ Ileal propionate (52%), butyrate (195%)**; ↓ Total tract N digestibility (4%)**
	Fecal microbiota		7 d periods			↑Fecal bifidobacteria (21%), lactobacilli (7%)**; TOS: ↓ Ileal DM (4%), OM (4%) digestibility**; ↑ Ileal GOS digestibility (100%)**; ↓ Total tract DM (2%), OM (3%) digestibility**; ↑ Fecal bifidobacteria (13%), lactobacilli (4%)**
Tsukahara et al. (2003)	Intestinal metabolites	Three castrated male Landrace × Large white × Duroc piglets (40 d, 12 kg)	Standard swine diet No. 1 (Nippon Formula Feed, Yokohama, Japan) without antibiotics or prebiotics	Standard diet (0%)	Standard diet + 10% (w/w) FOS	FOS: ↑ Digesta water content along large intestine (8–14%)**
	Intestinal morphology		Chemical composition 22% CP, 4.6% crude fat, 0.9% CF, 6.3% ash			↓ pH of digesta samples along large intestine (4–12%)***; ↑ Butyrate (110–243%), total SCFA (13–70%) along large intestine*
			10d study			↑ Acetate (12–54%), valerate (3–444%) along large intestine***

■ Table 12.4 (Cont'd p. 430)

Reference	Outcome variables quantified	Animals/treatment (age, initial BW)	Dietary information; time on treatment	Daily prebiotic dose; source	Major findings
Tsukahara et al. (2003) (con't.)					↑ Crypt depth (22–43%), crypt density (8–26%), epithelial cell count (24–45%), mitotic cell count (113–275%), mitotic zone (48–98%), mitotic index (50–157%), mucin-containing cell count (39–53%) along large intestine*
Davis et al. (2004a)	Study 1	Study 1	Study 1	Study 1	Study 1
	Performance responses (ADG, ADFI, G:F)	54 Weanling Hampshire × Duroc × Yorkshire × Landrace barrows (19 d, 6.2 kg)	Corn–whey-SBM diets supplemented with 200 or 2,500 ppm ZnO	Control (0%)	No performance responses to MOS
			38 d study	Control + 0.2% MOS	
	Study 2	Study 2	Study 2	Study 2	Study 2
	Performance responses (ADG, ADFI, G:F)	36 Weanling Hampshire × Duroc × Yorkshire × Landrace barrows (19 d, 4.6 kg)	Corn–whey-SBM diets supplemented with 200 or 2,500 ppm ZnO	Control (0%)	0.2% MOS:
				Control + 0.2% MOS	↑ ADG during d 10–24 with high Zn (15%)**
					0.3% MOS:
				Control + 0.3% MOS	↑ ADG during d 10–24 with high Zn (13%)**
			38 d study		

	Study 3	Study 3	Study 3	Study 3	Study 3
Davis et al. (2004a) (con't.)	Performance responses (ADG, ADFI, G:F)	36 Weanling Hampshire × Duroc × Yorkshire × Landrace barrows (19 d, 5.6 kg)	Corn-whey-SBM diets supplemented with 200, 500, or 2,500 ppm ZnO 35 d study	Control (0%) Control + 0.3% MOS	0.3% MOS: ↑ G:F during d 7–21 (4%)** ↑ Overall G:F (3%)** ↓ Lymphocyte proliferation with 200 ppm Zn (33%)** ↑ Lymphocyte proliferation with 500 ppm Zn (29%)** ↓ Unstimulated lymphocyte proliferation (32%)** ↓ Phytohemagglutinin-stimulated lymphocyte proliferation (19%)**
Davis et al. (2004b)	Performance responses (ADG, ADFI, G:F)	16 Yorkshire × Landrace × DeKalb EB barrows and gilts (19 d, 5.7 kg)	Corn-whey-SBM diet Chemical composition 1.5% lys, 0.90% met + cys, 0.9% Ca, 0.8% P 26 d study	Control (0%) Control + 0.3% MOS (Bio-MOS)	0.3% MOS: ↑ ADG during d 0–14 (64%), 0-final (31%)** ↑ G:F during d 0-final (20%)** ↑ BW on d 14 (20%)**, 21 (17%)*
	Hematologic responses (α1-acid glycoprotein, macrophages, monocytes)				↓ % Neutrophils (14%)*** ↓ α1-acid glycoprotein on d 0 (18%)***
	Intestinal cell immunity				↑ % Lymphocytes (18%)** ↑ CD14+ in lamina propria on d 19 (96%)*** ↓ CD14+ in blood on d 21 (25%)** ↓ CD14+MCHII+ in lamina propria on d 21 (82%)*** ↓ CD3+CD4+:CD3+CD8+ in jejunal lamina propria on d 21 (76%)**

Table 12.4 (Cont'd p. 432)

Reference	Outcome variables quantified	Animals/treatment (age, initial BW)	Dietary information; time on treatment	Daily prebiotic dose; source	Major findings
Mikkelsen and Jensen (2004)	Intestinal metabolites	Ten piglets (4 wk)	Same as Mikkelsen et al. (2003)	Control (0%)	FOS:
				Control + 40 g/kg FOS (Raftiline ST)	↑ Cecal butyrate (35%)**
	Intestinal microbiota		28 d study	Control + 40 g/kg TOS (Elix' or)	↑ Proximal colonic acetate (11%), butyrate (21%), valerate (136%), and capronic acid (80%)**
					↑ Intestinal (17%), cecal (28%), and colonic yeast (30%)**
	Microbial activity				TOS:
					↑ Distal colonic isobutyrate (29%)**
					↑ Stomach (31%), intestinal (31%), cecal (31%), and colonic yeast (38%)**
Rideout and Fan (2004)	Nutrient digestibility	Six Yorkshire barrows (30 kg)	Corn-SBM diet	Control (0%)	Inulin:
			Chemical composition 207.3 g/kg CP, 118.3 g/kg NDF, 41.5 g/kg ADF, 6.3 g/kg Ca, 7.6 g/kg P	Control + 50 g/kg inulin	↓ Apparent digestible CP intake (9%)**
	Ca and P digestibility		14 d periods		↓ Apparent CP retention (9%), digestibility (5%)***
					↑ Apparent fecal CP loss (29%)***
					↓ Total fecal (8%), overall Ca loss (8%)***
					↓ Total P loss (9%)***
					↓ Urine P loss (61%)*
					↑ TCA-insoluble CP output (9%)***
					↑ Water-soluble Ca fecal output (13%)**
					↓ Water-soluble P fecal output (18%)**

Reference	Response	Animals	Diet	Treatment groups	Results
Rideout et al. (2004)	Performance responses (feed intake, water intake)	Six Yorkshire barrows (30 kg)	Corn-SBM diets	Control (0%)	Inulin:
	Fecal metabolites		Chemical composition 20.7% CP, 11.8% NDF, 4.15% ADF, 0.63% Ca, 0.76% P	Control + 5% chicory inulin	↓ Fecal pH (3%)**
			14 d periods		↑ Total fecal N (9%)**; ↓ Fecal skatole (57%)**
Bohmer et al. (2005)	Prececal and total tract nutrient digestibilities	Eight male German Landrace × Pietrain pigs (36 kg)	Corn-wheat-barley-SBM diets	Control (0% inulin, 0 CFU Enterococcus faecium)	Inulin: ↑ Crude ash (19%), CF digestibility (6%) in IRA pigs**
	Intestinal and fecal microbiota	Four normal, four surgically modified with an ileo-rectal-anastomosis (IRA)	Chemical composition 16.6% CP, 5.6% crude fat, 3.6% CF	Control + 8 × 10^9 CFU E. faecium	↑ Crude ash digestibility (22%)**
			12 d periods	Control + 2% inulin (Raftifeed IPS)	E. faecium + inulin:
	Intestinal and fecal metabolites			Control + 2% inulin + 8 × 10^9 CFU E. faecium	↑ Crude ash (19%), CF digestibility (12%) in IRA pigs**
					↑ Crude ash (19%),CF digestibility (8%)**
Rozeboom et al. (2005)	Performance responses (ADG, ADFI, G:F)	481 Mixed-sex crossbred piglets (19 d, 6.15 kg)	Corn-whey-SBM diets	Control (0%)	MOS:
			Chemical composition (phase 1-4) 1.7-1.15% lys, 0.9-0.8% Ca, 0.8-0.7% P	Control + 0.3% MOS (Bio-Mos; reduced to 0.2% in phase 2)	↑ ADG on d 11-42 (8%), overall (7%)**
				Control + 110 mg each of tylosin and sulfamethazine	↑ G:F on d 0-11 (2%)**; MOS + antimicrobial: ↑ ADG on d 0-11 (20%), 11-42 (10%), overall (11%)**
			42 d study	Control + 0.3% MOS (Bio-Mos; reduced to 0.2% in phase 2) + 110 mg each of tylosin and sulfamethazine	↑ ADFI overall (7%)**; ↑ G:F on d 0-11 (18%)**

■ Table 12.4 (Cont'd p. 434)

Reference	Outcome variables quantified	Animals/treatment (age, initial BW)	Dietary information; time on treatment	Daily prebiotic dose; source	Major findings
Shim et al. (2005)	Performance responses (BW, feed intake, feed conversion)	Four male Large white × Landrace pigs (24 d, 8.2 kg)	Wheat-SBM diets	Control (0%)	0.25% FOS:
				Control + 0.25% FOS (Neosugar)	↓ Large intestinal acetate (16%)*
	Intestinal metabolites		Chemical composition 21.9% CP, 1.3% lys, 0.8% Ca, 0.68% P	Control + 3.0% FOS	↑ Large intestinal isobutyrate (140%) and isovalerate (220%)*
	Intestinal morphology		21 d study		↑ Large intestinal butyrate (46%)**
					3.0% FOS:
					↓ pH in cecum (7%)*
	Disaccharidase activity				↓ pH in proximal colon (7%)*
					↓ Large intestinal acetate (9%)*
					↑ Large intestinal valerate (75%)*
					↑ Large intestinal isovalerate (98%)*
Tzortis et al. (2005)	Performance responses (BW, feed intake)	Ten weaned male pigs (35 d, 14.7 kg)	Commercial pelleted diet (Deltawean 15 NGP)	Control (0%)	1.6% GOS:
				Control + 1.6% GOS	↑ Lactobacilli in feces (9%)**
				Control + 4.0% GOS	↑ Lactate in proximal colon (153%)**
				Control + 1.6% inulin	
	Colonic microbiota		Chemical composition 192 g/kg CP, 33 g/kg oil, 13.2 g/kg lys		↓ Propionate in proximal colon (23%) **
	Colonic metabolites		34 d study		4% GOS:
	Fecal microbiota				↑ Bifidobacteria in proximal (10%), distal colon (5%) and feces (13%)**
					↑ Lactobacilli in proximal (6%), distal colon (8%) and feces (13%)**

Reference	Response measured	Animal	Diet	Treatment	Results
Tzortis et al. (2005) (con't.)					↑ Lactate (68%), acetate (25%) in proximal colon** ↓ pH in proximal colon (4%)** Inulin: ↑ Bacteroides in proximal colon (5%) and feces (11%)** ↑ Bifidobacteria in distal colon (8%) and feces (15%)** ↑ Lactobacilli in distal colon (7%) and feces (10%)** ↑ Acetate in distal colon (22%)**
Xu et al. (2005)	Performance responses (BW, ADFI, ADG, F:G)	30 Mixed-sex Duroc × Landrace × Large white pigs (33 d, 7.9 kg)	Corn-SBM diet Chemical composition 18% CP, 1.24% lys, 0.32% met, 0.68% met + cys, 0.81% Ca, 0.70% P	Control (0%) Control + 75 mg/kg antibiotic Control + 0.4% FOS	FOS: ↑ ADG (32%)* ↑ F:G (24%)* ↓ Diarrhea rate (71%)**
	Cecal and fecal metabolites				↑ Serum glucose (143%)*
	Intestinal morphology		21 d study		↑ Cecal propionate (56%), butyrate (44%), and total SCFA (31%)* ↑ Fecal acetate (43%), isovalerate (35%), and total SFCA (38%)**
	Serum chemistry				↑ Jejunal (45%) and cecum villus height (100%)* ↑ Jejunal (4%)**, ileal (7%)*, and cecal (11%)* goblet cell percentages

■ Table 12.4 (Cont'd p. 436)

Reference	Outcome variables quantified	Animals/treatment (age, initial BW)	Dietary information; time on treatment	Daily prebiotic dose; source	Major findings
Awati et al. (2006)	Performance responses (ADFI)	16 Mixed-sex Hypor × Pietrain pigs (4 wk)	Cornstarch-fishmeal diet	Control (0 g/g)	Inulin + lactulose: ↓ Fecal DM (9%)** ↓ Cecal (60%), colonic (43%), Fecal (18%) ammonia** ↓ Proximal small intestinal (117%), fecal (15%) BCFA**
	Intestinal and fecal metabolites		Chemical composition 179 g/kg CP, 11.5 g/kg lys, 8.3 g/kg Ca, 5.7 g/kg P 10 d periods	Control + 7.5 g/kg inulin + 20 g/kg lactulose	
Biagi et al. (2006)	Performance responses (BW, feed intake, ADG)	12 German Landrace × Pietrain piglets (28 d, 7.4 kg)	Chemical composition (weeks 1–3): 20.7% CP, 3.1% fat, 4.2% CF; (weeks 4–6): 18.5% CP, 3.9% fat, 4.9% CF	Control (0 ppm) Control + 3,000 ppm 50% Gluconic acid Control + 6,000 ppm 50% Gluconic acid	All concentrations Gluconic acid: Quadratic ↑ final BW (0 ppm: 24.5 kg; 3,000 ppm: 26.7 kg; 6,000 ppm: 26.9 kg; 12,000 ppm: 25.3 kg)**
	Intestinal and fecal microbiota			Control + 12,000 ppm 50% Gluconic acid	Quadratic ↑ ADG (0 ppm: 409 g; 3,000 ppm: 464 g; 6,000 ppm: 466 g; 12,000 ppm: 428 g)**
	Intestinal morphology and metabolites		6 wk study		Quadratic ↑ jejunal acetate (0 ppm: 14.6 ppm; 3,000 ppm: 83.2 ppm; 6,000 ppm: 65.4 ppm; 12,000 ppm: 42.7 ppm)** Quadratic ↑ total SCFA (0 ppm: 64.5 ppm; 3,000 ppm: 177.1 ppm; 6,000 ppm: 120.7 ppm; 12,000 ppm: 112.1 ppm)***

Krag et al. (2006)	Trichuris suis challenge	Seven mixed-sex Landrace × Yorkshire × Duroc pigs	Barley flour–SBM diets	Basal + 300 g/kg oat husk (control)	Inulin groups: ↓ Number of recovered worms on weeks 7 (69%)* and 9 (52%)**
	Performance response (BW)		Nine wk study: all pigs infected with 2,000 infective *T. suis* eggs at 3 wk; groups N-7, N-9 and N/I-9 fed oat hull meal, group I-7 fed inulin until wk 7; N-7 and I-7 Slaughtered at wk 7, N/I-9 fed inulin for 2 wk	Basal + 160 g/kg inulin (Raftiline)	↓ IgA+ cells in lamina propria on week 7 (25%)*
	Intestinal parasitology				↓ IgG+cells in lamina propria on week 7 (64%)*
	Intestinal histology				↓ CD3+ cells in lamina propria on week 7 (45%)*
					↓ Mast cells in lamina propria on wk 9 (37%)
					↓ Eosinophils in tela submucosa on wk 9 (55%)**
Loh et al. (2006)	Performance responses (BW, feed intake)	16 German Landrace pigs (42 d)	Wheat-barley–SBM or Corn-wheat gluten meal diets	Control (0%)	Wheat + Inulin:
				Control + 3% inulin (Raftilene ST)	↑ Colonic bifidobacteria (192%)**
	Intestinal microbiota		Chemical composition		↓ Colonic acetate as molar % (7%)*
	Intestinal metabolites				↑ Colonic butyrate as molar % (26%)*, propionate as molar % (8%)***
			Wheat: 20.3% CP, 4.0% crude fat, 48.0% starch, 0.8% inulin		↓ Colonic pH (5%)*
			Corn: 20.9% CP, 4.6% crude fat, 49.9% starch, 0.9% inulin		Corn + Inulin:
					↑ Colonic bifidobacteria (238%)**
					↓ Colonic acetate as molar % (5%)*
					↑ Colonic butyrate as molar % (28%)*, propionate as molar % (6%)***
					↑ Colonic pH (4%)*

■ Table 12.4   (Cont'd p. 438)

Reference	Outcome variables quantified	Animals/treatment (age, initial BW)	Dietary information; time on treatment	Daily prebiotic dose; source	Major findings
Mountzouris et al. (2006)	Intestinal microbiota	Four castrated growing Duroc × (Large white × Landrace) pigs (12 kg)	Corn-SBM diet	Control (0 g/kg)	FOS: ↑ Cecal β-galacturonidase activity (161%)**
	Intestinal metabolites		Chemical composition not provided	10 g/kg FOS (Raftifeed OPS)	TOS: ↑ Cecal (199%), colonic (62%), fecal (119%) β-galacturonidase activity**
	Microbial enzyme activities		30 d periods	10 g/kg TOS (Vivinal GOS 10)	
Pierce et al. (2006)	Intestinal morphology	Five large white × Large white × Landrace piglets (24 d, 7.8 kg)	Wheat-SBM-whey protein isolate diet	Control (0 g/kg)	Low lactose + Inulin:
	Intestinal microbiota		Chemical composition	Control + 15 g/kg inulin	↑ Food intake (24%)**
	Intestinal metabolies		Low lactose: 191.4 g/kg CP, 91.6 g/kg NDF, 14.8 g/kg lys, 5.3 g/kg met		↑ Duodenal (29%), jejunal villus:crypt ratio (53%)***
			High lactose: 191.7 g/kg CP, 63.7 g/kg NDF, 15.2 g/kg lys, 5.2 g/kg met		↓ Ileal pH (1%)**
					↑ Jejunal villus height (50%)**
					↑ Colonic bifidobacteria (8%)***
					↑ Cecal propionate (4%)***, total SCFA (23%)**
					↑ Cecal isobutyrate (50%), isovalerate (53%)**
					↑ Colonic total SCFA (42%)**
					High Lactose + Inulin:
					↓ Ileal pH (3%)**

Reference	Measurement	Animals	Conditions	Treatment	Results
Pierce et al. (2006) (con't.)					↑ Jejunal villus height (4%)**
					↓ Ileal lactobacilli (50%)**
					↓ Colonic bifidobacteria (9%)**
					↑ Cecal propionate (27%)***
					↓ Cecal isobutyrate (33%), isovalerate (11%), total SCFA (25%)**
					↑ Colonic propionate (27%)**
					↓ Colonic total SCFA (27%)**
Lynch et al. (2007)	Study 1	Study 1	Study 1	Study 1	Study 1
	Nutrient digestibility	Four meat-line boars × large white × Landrace finishing boars (74 kg)	Wheat-SBM diets with high (200 g/kg) CP or low (140 g/kg) CP	Control (0%)	High CP + Inulin: ↓ Hemicellulose (2%)**, N (3%)* digestibility
	N balance		Chemical composition	Control + 12.5% inulin (Raftiline ST)	↑ ADF digestibility (3%)**
	Ammonia emissions		High CP: 202.4 g/kg CP, 127.8 g/kg NDF, 47.4 g/kg ADF, 20.0 g/kg crude fat		↑ Fecal N excretion (33%)**
					↓ Urinary N:fecal N (25%)**
					Low CP + Inulin:
					↓ NDF (16%), ADF (13%), hemicellulose (18%) digestibility**
			Low CP: 148.2 g/kg CP, 103.1 g/kg NDF, 53.7 g/kg ADF, 14.4 g/kg crude fat		↓ N digestibility (3%)*
					↑ Fecal N excretion (22%)**
			6 d study		↓ Urinary N:fecal N (26%)**
	Study 2	Study 2	Study 2	Study 2	Study 2
	Cecal and colonic microbiota	Six Meat-line boars × Large white × Landrace finishing boars (74 kg)	Same as study 1	Same as study 1	High CP + Inulin: ↑ Cecal bifidobacteria (6%)*
	Cecal and fecal metabolites				↑ Cecal lactobacilli (9%)***
					↑ Cecal enterobacteria (8%)**
					↓ Colonic propionate (3%)**
					↑ Fresh feces:urine ratio (66%)**

Table 12.4

Reference	Outcome variables quantified	Animals/treatment (age; initial BW)	Dietary information; time on treatment	Daily prebiotic dose; source	Major findings
Lynch et al. (2007) (con't.)					Low CP + Inulin: ↑ Cecal bifidobacteria (3%)* ↑ Cecal enterobacteria (10%)** ↓ Colonic propionate (6%)** ↓ Fresh feces:urine ratio (22%)**
Yasuda et al. (2007)	Study 1 Digesta inulin content	Study 1 Six weanling Yorkshire × Hampshire × Landrace pigs (7.7 kg)	Study 1 Corn-SBM diet	Study 1 Control (0%)	Study 1 Inulin:
	Intestinal saccharide concentrations		Chemical composition 18.5% CP, 4.3% CF 6 wk study	Control + 4% inulin (Synergy 1)	↑ Inulin in stomach (0.5%), upper jejunum (3%), and lower jejunum (5%)** ↑ Fructose in stomach (151%), lower jejunum (156%), and ileum (3,175%)** ↑ Raffinose in lower jejunum (300%)** ↑ Stachyose in lower jejunum (6,100%)**
	Study 2 Digesta inulin content	Study 2 Six weanling Yorkshire × Hampshire × Landrace pigs (11.2 kg)	Study 2 Same as study 1 except 8 wk study	Study 2 Same as study 1	Study 2 Inulin: ↑ Inulin in stomach (1%), lower jejunum (1%), and ileum (2%)** ↓ Glucose in stomach (150%)** ↑ Fructose in stomach (400%) and lower jejunum (600%)** ↑ Sucrose in ileum (700%)**

Tako et al. (2008)	Six weanling Yorkshire × Hampshire × Landrace pigs	Corn-SBM diets	Control (0 g/kg)	Inulin:
	Blood hemoglobin			
	Cecal microbiota	Chemical composition 18.1% CP, 4.5% CF, 54.4 mg/kg Fe	Control + 32 g/kg inulin (Synergy 1)	↑ Duodenal expression of ferritin (63%), divalent metal transporter 1 (300%), transferring receptor (67%), cytochrome b reductase (75%), mucin (122%), ferroportin (150%)**
	Intestinal transporter gene expression	6 wk study		↑ Colonic expression of ferritin (50%), divalent metal transporter 1 (133%), and transferring receptor (83%)**
				↑ Cecal lactobacilli (171%) and bifidobacteria (100%)**

[a]ADF acid detergent fiber, ADG average daily gain, ADFI average daily feed intake, BW body weight, Ca calcium, CHO carbohydrate, CF crude fiber, CFU colony forming unit, CP crude protein, Cu copper, cys cysteine, d day, DM dry matter, DMI dry matter intake, Fe iron, F:G feed:gain ratio, FOS fructooligosaccharide, g gram, G:F gain:feed ratio, GOS galactooligosaccharide, h hour, IgA immunoglobulin A, IgG immunoglobulin G, IGF-I insulin-like growth factor-I, IRA ileo-rectal anastomosis, kg kilogram, L liter, lys lysine, met methionine, MOS mannanoligosaccharide, N nitrogen, NDF neutral detergent fiber, NSP nonstarch polysaccharide, OF oligofructose, OM organic matter, P phosphorus, ppm parts per million, SBM soybean meal, SCFA short-chain fatty acid, thr threonine, trp tryptophan, TOS trans-galactooligosaccharide, μg microgram, wk week, Zn(O) zinc (oxide)

[b]*P < 0.01, **P < 0.05, ***P < 0.10

prebiotic supplementation (Estrada et al., 2001; Lynch et al., 2007; Oli et al., 1998; White et al., 2002; Xu et al., 2002). Concentrations of intestinal and fecal yeast increased in response to FOS and TOS supplementation (Mikkelsen et al., 2003; Mikkelsen and Jensen, 2004). When challenged with *Oesophagostomum dentatum*, pigs supplemented with inulin experienced decreased intestinal *Oesophagostomum dentatum* eggs and worms (Petkevicius et al., 2003).

Changes in fermentative end-products as a result of prebiotic supplementation of swine follow patterns similar to that of other non-ruminant species. While some inconsistencies occur within the literature, acetate, butyrate, and total SCFA increase with prebiotic supplementation (Biagi et al., 2006; Correa-Matos et al., 2003; Loh et al., 2006; Mikkelsen and Jensen, 2004; Petkevicius et al., 2003; Pierce et al., 2006; Shim et al., 2005; Tsukahara et al., 2003; Tzortis et al., 2005; Xu et al., 2005). Fecal propionate concentrations increased in 28% of studies (Loh et al., 2006; Petkevicius et al., 2003; Pierce et al., 2006; Smiricky-Tjardes et al., 2003; Xu et al., 2005), whereas decreased propionate was observed in 22% of studies (Lynch et al., 2007; Petkevicius et al., 2003; Tzortis et al., 2005). Isobutyrate and valerate concentrations also increased with prebiotic supplementation (Houdijk et al., 2002; Mikkelsen and Jensen, 2004; Shim et al., 2005; Tsukahara et al., 2003), whereas isovalerate concentrations were variable (Houdijk et al., 1998; Mikkelsen et al., 2003; Pierce et al., 2006; Shim et al., 2005; White et al., 2002; Xu et al., 2005). Lactate and cupronic acid increased with GOS and FOS supplementation, respectively (Mikkelsen and Jensen, 2004; Tzortis et al., 2005).

Changes in intestinal morphology with respect to prebiotic supplementation were observed by ten authors. Villus height and villus height:crypt depth increased in response to fructan supplementation (Pierce et al., 2006; Xu et al., 2002; Xu et al., 2005). Fructan supplementation increased cell density, crypt height, crypt depth, crypt density, epithelial cell count, mitotic cell count, mucin-containing cell count, and number of goblet cells (Howard et al., 1995; Tsukahara et al., 2003; Xu et al., 2005). Cell death (CD) markers, mast cells, eosinophils, IgA, and IgG decreased in the lamina propria when pigs were supplemented with inulin (Krag et al., 2006). While observed in only one study, inulin increased intestinal expression of ferritin, divalent metal transporter 1, transferring receptor, cytochrome b reductase, mucin, and ferroportin (Tako et al., 2008).

Changes in nutrient digestibility occurred in response to prebiotic-supplemented diets. Dry matter digestibility, organic matter digestibility, nitrogen digestibility, and nitrogen retention decreased in response to prebiotic supplementation (Lynch et al., 2007; Rideout and Fan, 2004; Smiricky-Tjardes et al., 2003; Zhang et al., 2001). Fiber digestibility also decreased in response to prebiotic supplementation (Lynch et al., 2007). In contrast, GOS digestibility increased with prebiotic supplementation (Smiricky-Tjardes et al., 2003). Several enzyme activities increased in response to prebiotic supplementation, including protease (Xu et al., 2002), trypsin (Xu et al., 2002), amylase (Xu et al., 2002), sucrase (Correa-Matos et al., 2003), and β-galacturonidase (Mountzouris et al., 2006).

Growth performance response to prebiotics was the most commonly reported outcome variable in the swine. Excluding MOS (discussed below), ADG increased in 21% of studies (Biagi et al., 2006; Estrada et al., 2001; Xu et al., 2002; Xu et al., 2005), while it decreased in 16% of studies (Estrada et al., 2001; Houdijk et al., 1998). Dry matter intake and body weight were variable (Estrada et al., 2001; Houdijk et al., 1998; Mikkelsen et al., 2003; Pierce et al., 2006). Feed:gain ratios generally increased in response to FOS supplementation (Estrada et al., 2001; Xu et al., 2005).

Mannanoligosaccharides increased ADG, body weight, and G:F of swine (Davis et al., 2004b). When measured in response to MOS plus supplemental copper, ADG increased without supplemental copper but the response was variable with supplemental copper (Davis et al., 2002). These responses appear to depend on the age and production phase of the pig, as a greater increase in ADG occurred in younger pigs with copper while a greater increase in ADG occurred in finisher pigs without copper. Gain:feed increased with MOS with and without supplemental copper (Davis et al., 2002). Average daily gain increased as often with MOS and supplemental zinc (Davis et al., 2004a) as without (LeMieux et al., 2003), although supplemental zinc appeared to be less beneficial to older pigs. Average daily feed intake was variable with MOS but without supplemental zinc (Davis et al., 2004a; LeMieux et al., 2003). Mixed results appear in the literature for G:F; however, diets supplemented with MOS and zinc appeared to increase G:F in older pigs (Davis et al., 2004a; LeMieux et al., 2003). Average daily gain and G:F increased in response to dietary supplementation of MOS and antibiotics while decreasing in response to supplementation with MOS without antibiotics (LeMieux et al., 2003). However, ADFI decreased with MOS consumption regardless of antibiotic supplementation (LeMieux et al., 2003).

Horses. Prebiotic supplements are not well researched in horses, as only five studies have been published in this area (● *Table 12.5*). Generally, studies involving prebiotic supplementation have been performed to simulate or provoke laminitis symptoms. Decreased lactobacilli, lactic acid-utilizing bacteria, *E. coli*, and streptococci were observed in two studies that measured changes in microbial populations in response to fructan supplementation (Berg et al., 2005; Respondek et al., 2008). Fermentative responses varied greatly among studies, as acetate and lactate concentrations both increased and decreased in response to fructan supplementation (Berg et al., 2005; Respondek et al., 2007). However, butyrate and total SCFA concentrations increased consistently with fructan supplementation in these studies (Berg et al., 2005; Respondek et al., 2007). Fecal pH also decreased consistently with fructan supplementation (Berg et al., 2005; Crawford et al., 2007; van Eps and Pollet, 2006). Two studies observed increased insulin concentrations with increased fructan consumption (Bailey et al., 2007; van Eps and Pollet, 2006).

Pre-ruminants and adult ruminants. As the rumen of a pre-ruminant is sterile at birth and, thus, easily influenced by feed and environmental microbiota and because pre-ruminants are under continued stress from the standpoint of health maintenance, prebiotic supplementation may be an excellent nutritional intervention for them. Two studies have been conducted (● *Table 12.6*). Fructooligosaccharides increased ADG, BW, insulin concentrations, leukocytes, and eosinophils while decreasing lactate, NEFA, and post-prandial glucose concentrations (Kaufhold et al., 2000). Mannanoligosaccharides increased fecal scores and rate of recovery from diet-induced scours (Heinrichs et al., 2003). Clearly, more research is needed in this area.

Prebiotic supplementation of adult ruminants is not commonly performed as they have a relatively stable microbial population; however, several studies have been conducted (● *Table 12.7*). Mannanoligosaccharide supplementation of dry dairy cows increased serum rotavirus neutralization titers, while increasing serum protein concentrations and IgA concentrations, and rotavirus neutralization titers were observed in the calves of dams that had consumed MOS (Franklin et al., 2005). However, supplementation with GOS, a yeast culture, or a combination negatively impacted the microbial nitrogen concentration and nitrogen supply available to the host. These supplements also decreased ruminal propionate, which is used for gluconeogenesis, while increasing acetate, which is used for *de novo* fatty acid synthesis (Mwenya et al., 2005). Together, these results can negatively impact the health of the cow by decreasing the blood glucose concentration on which the cow relies for energy and creating fat mass.

12.3 Special Considerations for Prebiotic Use

Age. While the factor of age remains mainly undeveloped with respect to prebiotic supplementation, certain considerations given to dietary fibers also must be considered when adding prebiotics to animal feedstuffs. Concentrations of

■ Table 12.5

In vivo experiments, listed in chronological order, reporting effects of prebiotics in horses[a,b] *(Cont'd p. 444)*

Reference	Outcome variables quantified	Animals/treatment (age, initial BW)	Dietary information; time on treatment	Daily prebiotic dose; source	Major findings
Berg et al. (2005)	Fecal metabolites	Nine quarter horses: six male, three female (489–539 d, 400.6 kg)	Pasture grass ad libitum with corn-oat concentrate supplement at 1% BW	Control (0 g/d)	All concentrations FOS:
	Fecal microbiota		Chemical composition (concentrate) 13.44% CP, 3.93% fat, 0.61%Ca, 0.56% P	8 g/d FOS	Linear ↓ pH (control: 6.48; 8 g/d: 6.44; 24 g/d: 6.38)*
			10 d blocks	24 g/d FOS	Quadratic ↓ fecal *E. coli* (control 4.90 $\log_{10}$ population; 8 g/d: 4.75 $\log_{10}$ population; 24 g/d: 4.93 $\log_{10}$ population)*
					Linear ↑ fecal lactate (control: 0.36 mg/g; 8 g/d: 0.41 mg/g; 24 g/d: 0.47 mg/g)**
					Linear ↑ fecal acetate (control: 2.13 mg/g; 8 g/d: 2.18 mg/g; 24 g/d: 2.52 mg/g)**
					Linear ↑ fecal propionate (control: 0.58 mg/g; 8 g/d: 0.64 mg/g; 24 g/d: 0.73 mg/g)*
					Linear ↑ fecal butyrate (control: 0.40 mg/g; 8 g/d: 0.46 mg/g; 24 g/d: 0.54 mg/g)**

■ Table 12.5   (Cont'd p. 446)

Reference	Outcome variables quantified	Animals/treatment (age, initial BW)	Dietary information; time on treatment	Daily prebiotic dose; source	Major findings
Berg et al. (2005) (con't.)					Linear ↑ fecal total VFA (control: 3.47 mg/g; 8 g/d: 3.69 mg/g; 24 g/d: 4.25 mg/g)*
van Eps and Pollet (2006)	Performance responses (temperature, heart rate, fecal pH)	18 Mature Standardbred horses (no age, BW given)	60% Lucerne chaff, 15% micronized barley, 25% commercial feed (Cool Command)	Control (0 g/kg)	All concentrations OF:
	Haematological and biochemical data	Control: 6 horses 7.5 g/kg: 2 horses 10.0 g/kg: 8 horses 12.5 g/kg: 2 horses	Chemical composition: not provided	7.5 g/kg BW OF (Raftilose P95)	Lameness induced
	Laminitis score		48 h study	10 g/kg BW OF	Distal limb edema induced
				12.5 g/kg BW OF	↓ Lymphocytes after 12 h (28%)**
					↑ Plasma L-lactate (188%), D-lactate (275%)**
					↓ Fecal pH (3–42%)**
					↓ Plasma bicarbonate (37%)**
					↑ Plasma glucose (26–80%), cortisol (100–950%)**

Reference	Measure	Subjects	Diet	Composition	Results
van Eps and Pollet (2006) (con't.)					10 g/kg BW OF: ↑ Heart rate after 4 h (24–70%)** ↑ Rectal temperature after 8 h (1–6%)** ↓ Plasma sodium, chloride after 24 h** ↑ Plasma potassium at 16–32 h (54%)** ↑ Neutrophils after (56–100%)** ↑ Plasma fibrinogen (63%)** ↑ Plasma insulin at 12, 40, 44 h** 12.5 g/kg BW OF: ↑ Plasma sodium, chloride at 8 h**
Bailey et al. (2007)	Glucose tolerance	Study 1 Ten native-breed horses (15.1 y)	Study 1 Pasture (mixed sward with clover) for 2 mo, then mature Timothy hay for 7 d	Study 1 Sward: 138 g/kg DM fructans (DP ≥ 3), Timothy hay: 34 g/kg DM	Study 1 Grass: ↑ serum insulin concentration (107%)**
	Insulin sensitivity				Hay: ↓ serum insulin vs. grass serum insulin concentrations (34%; similar to hay-fed control horses)**
	Plasma TG concentrations				

Table 12.5

Reference	Outcome variables quantified	Animals/treatment (age, initial BW)	Dietary information; time on treatment	Daily prebiotic dose; source	Major findings
Bailey et al. (2007) (con't.)	Study 2	Study 2	Study 2	Study 2	Study 2
	Glucose tolerance	10 laminitis-prone, 11 control native-breed horses (15.1 y)	Hay for 2 wk, then high-protein grass (15% protein)	Hay: 4.1 g/kg/d; Grass: 6.1 g/kg/d (includes 3 g/kg inulin top-dressed on grass)	Laminitis-prone horses:
	Insulin sensitivity				↑ Serum insulin concentrations with fructan addition (206%)*
	Plasma TG concentrations				
Crawford et al. (2007)	Plasma amine concentrations Hoof wall and coronary band temperature	Six laminitis-prone, 6 control mixed, adult native-breed horses (13.2 y, 337 kg)	Hay for 2 wk, then high-CHO diet (2/3 hay plus 1/3 dried grass)	Control (0 g/kg)	Inulin: ↓ Fecal pH (10%)*
	Fecal metabolites		3 wk study	3 g/kg inulin	↑ fecal tyramine (2.5-fold), tryptamine (2-fold)**

Respondek et al. (2007)	Cecal and colonic metabolites	4 cross-bred, cecal cannulated geldings (7 y, 425 kg)	Commercial pelleted feed (HIPPO 122) and wheat straw; barley fed on d 21 in place of morning concentrate meal	Control (0 g/d)	scFOS:
	Cecal and colonic microbiota		Chemical composition: 11.16% CP, 1.91% fat, 17.24% CF, 40.4% NDF, 20.3% starch	30 g/d scFOS (Profeed P95)	↓ Streptococci 5 h post-barley consumption (6%)**
					↓ Lactobacilli (7%), lactate-utilizing microbiota (6%), streptococci (9%) 29 h post-barley consumption**
					↑ Cecal L-lactate concentrations 5 h post-barley consumption (111%)**
					↓ Cecal L-lactate concentrations 29 h post-barley consumption (61%)**
					↑ Cecal D-lactate concentrations 29 h post-barley consumption (121%)**
			24 d periods		↑ Cecal acetate (21–25%), butyrate (30–49%), total VFA (24–31%)***
					↓ Colonic acetate proportion post-barley consumption (7%)***

[a]*BW* body weight, *Ca* calcium, *CF* crude fiber, *CHO* carbohydrate, *CP* crude protein, *d* day, *DM* dry matter, *DP* degree of polymerization, *FOS* fructooligosaccharide, *g* gram, *h* hour, *kg* kilogram, *mg* milligram, *NDF* neutral detergent fiber, *OF* oligofructose, *P* phosphorus, *scFOS* short-chain fructooligosaccharide, *TG* triglyceride, *VFA* volatile fatty acid, *wk* week, *yr* year

[b]*P < 0.01, **P < 0.05, ***P < 0.10

■ Table 12.6

In vivo experiments, listed in chronological order, reporting effects of prebiotics in pre-ruminants[a,b]

Reference	Outcome variables quantified	Animals/ treatment (age, initial BW)	Dietary information; time on treatment	Daily prebiotic dose; source	Major findings
Kaufhold et al. (2000)	Performance responses (BW, feed intake, ADG)	Seven Simmental × Red Holstein veal calves (10 wk old)	11 L/d whole milk	Control (0 g/d)	FOS:
	Hematological profile		Chemical composition: 122 g DM/kg, 274 g CP/kg DM, 273 g Crude fat/ kg DM, 379 g Lactose/kg DM with supplemental milk replace and vitamin premix to allow ADG of 1.4–1.5 kg	Control + 10 g/d FOS (Profeed P95) with morning meal	↑ ADG (10%)***
					↑ Leukocytes on d 14 (45%)**
					↑ Absolute BW gain (10%)***
					↑ Eosinophils on d 21 (560%)**
					↓ Post-prandial plasma glucose concentrations on d 21 (12–27%)**

					↓ Lactate (37%) and NEFA concentrations (46%) on d 14**
					↑ Insulin concentrations on d 14 (58–63%)**
Heinrichs et al. (2003)	Performance responses (BW, skeletal growth)	24 Holstein calves (2 d)	Milk replacer and calf starter feed	Control (0 g/d)	MOS:
	Blood metabolites		Chemical composition: Milk: 20% CP, 20% fat, 0.75% Ca, 0.70% P Starter: 20.9% CP, 3.03% fat, 1.65% Ca, 0.88% P	1.4 g/kg antibiotic	↑ Rate of recovery from diet-induced scours (41%)*
	Fecal scores		6 wk study	4 g/d MOS (Bio-MOS)	↓ Scour severity score in category 2 (38%), 3 (50%)*
					↑ probability of normal fecal score after wk 3 (13%), overall (86%)*

[a]ADG average daily gain, BW body weight, CP crude protein, d day, DM dry matter, FOS fructooligosaccharide, g gram, kg kilogram, L liter, MOS mannanoligosaccharide, NEFA non-esterified fatty acid, wk week
[b]*P < 0.01, **P < 0.05, ***P < 0.10

◘ Table 12.7

In vivo experiments, listed in chronological order, reporting effects of prebiotics in adult ruminants[a,b] (Cont'd p. 452)

Reference	Outcome variables quantified	Animals/treatment (age, initial BW)	Dietary information; time on treatment	Daily prebiotic dose; source	Major findings
Franklin et al. (2005)	Performance responses (BW-calves)	19 Multiparous dairy cows (14 Holstein, 5 Jersey)	Alfalfa silage-concentrate TMR	Control (0 g/d)	MOS:
	Blood metabolites (cows and calves)		Chemical composition 18% CP, 3.1% fat, 37% NDF, 24% ADF	Control + 10 g/d MOS	Dams:
	Colostrum analysis		4 wk study		↑ Serum rotavirus neutralization titer (2%)**
	Serum Ig analysis				Calves:
					↑ Serum protein concentration at 24 h (29%)***
					↓ Serum IgA (13%)**
					↑ Serum rotavirus neutralization titer (5%)***
Mwenya et al. (2005)	Total tract digestibility	Four nonlactating, ruminally cannulated Holstein cows (697 kg)	Alfalfa hay cube-oat straw TMR	Control (0%)	GOS:
	Microbial N supply		Chemical composition 13.9% CP, 3.1% crude fat, 38.6% NDF, 25.6% ADF, 4.8% ADL	Control + 10 g/d yeast culture	↓ Uric acid (33%), allantoin (28%), total urinary excretion (29%)*
	Ruminal metabolites	Rumen emptied and switched over at end of each period	27 d periods	Control + 2% GOS control + 10 g/d yeast culture + 2% GOS	↓ Absorption of purine derivatives (39%)*

Mwenya et al. (2005) (con't.)	Protozoa enumeration	↓ Microbial N (38%), N supply (39%)*
		↓ Ruminal pH (2%)*
		↑ Ruminal acetate (5%)*
		↓ Ruminal propionate (22%)*
		↑ Ruminal acetate:propionate (19%)*
		Yeast:
		↓ Uric acid (35%), allantoin (21%), total urinary excretion (23%)*
		↓ Absorption of purine derivatives (31%)*
		↓ Microbial N (31%), N supply (32%)*
		↑ Ruminal acetate (8%)*
		↓ Ruminal propionate (25%), total VFA (11%)*
		↑ Ruminal acetate:propionate (23%)*
		Yeast + GOS:
		↑ Urine N (18%)**
		↑ N retained (144%)**
		↓ Uric acid (23%), allantoin (19%), total urinary excretion (20%)*

● Table 12.7

Reference	Outcome variables quantified	Animals/treatment (age, initial BW)	Dietary information; time on treatment	Daily prebiotic dose; source	Major findings
Mwenya et al. (2005) (con't.)					↓ Absorption of purine derivatives (26%)*
					↓ Microbial N (24%), N supply (27%)*
					↑ Ruminal acetate (9%)*
					↓ Ruminal propionate (29%), total VFA (9%)*
					↑ Ruminal acetate:propionate (34%)*
					↓ Ruminal isoacids (26%)*

[a]ADF acid detergent fiber, ADL acid detergent lignin, BW body weight, CP crude protein, d day, GOS galactooligosaccharide, Ig immunoglobulin, IgA immunoglobulin A, kg kilogram, MOS mannanoligosaccharide, N nitrogen, NDF neutral detergent fiber, TMR total mixed ration, VFA volatile fatty acids

[b]*P < 0.01, **P < 0.05, ***P < 0.10

prebiotics in the diet of very young and very old animals should remain low so as not to drastically change diet digestibility. Very young animals require highly digestible diets to maintain rapid growth rates. While this is important for all species, it is of particular importance for livestock species as profit is closely tied to optimal animal growth and health. While not as great a concern for production livestock species, some aging animals experience a decline in diet digestibility as they approach senior and geriatric status. A dilution of nutrients at this life stage can lead to loss of muscle mass and stimulate other health concerns.

Health status. As demonstrated by several studies, animals with challenged immunity respond better to prebiotic supplementation as compared to healthy animals. When exposed to pathogenic microorganisms and parasites, poultry and swine exhibited more consistent responses to treatment with prebiotic supplements than control groups. While still important in disease prevention, use of prebiotics may not appear as efficacious in healthy, non-challenged animals. Prophylactic use of prebiotic supplements is important, however, in non-challenged animals as the changes that they induce in intestinal microbial ecology lead to an overall healthier animal.

Species. While supplementation appears to have clear benefits to most animal species, certain species respond poorly to prebiotic compounds. For example, the majority of research published in horses links fructan consumption to laminitis. Indeed, fructans have been used to induce laminitis and rapid intake of carbohydrate has been shown to induce laminitis. In contrast, pre-ruminants appear to benefit from the addition of prebiotic compounds to their diets. These animals do not have a strong microbial population established in their rumen and are highly affected by microorganisms in their environment, particularly due to susceptibility to diet-induced shifts in microbiota. Prebiotic supplementation of these animals can influence the populations of microbiota that establish within the rumen and, potentially, can lead to an improved ruminal environment for the animal.

Animal housing. The way in which an animal is housed can greatly affect its response to prebiotic supplementation. Companion animals, generally accepted as indoor pets, are less likely to exhibit a dramatic response to a prebiotic compound unless ill when compared with some livestock species. Livestock species also respond very differently, which generally can be related to the type of living conditions in which they reside. One would expect a greater response to prebiotic supplementation of poultry housed in a floor pen where the birds have free access to their environment, including excreta of other birds, than would be observed in group-housed swine or possibly even caged poultry.

12.4   Conclusion

As evidenced by the body of literature covered in this essay, prebiotic supplementation of companion and livestock animals demonstrates many benefits. By increasing beneficial microbiota while decreasing populations of potentially pathogenic microbiota, animal owners and producers can improve the health and well being of their animals. Prebiotic compounds also improve fermentation in the intestine of animals, as has been observed in all species to which prebiotics have been fed. Increasing SCFA concentrations provides energy to the colon, allows for proper cell maintenance and turnover, and can contribute to the beneficial microbial environment by decreasing intestinal pH. Fermentation associated with odor- and disease-producing compounds such as phenols and branched-chain fatty acids generally decreased with prebiotic supplementation, adding to the list of benefits of prebiotic supplementation. While some conflicting information exists, benefits generally outweigh costs of supplementation. With proper consideration of health status, living conditions, species to be supplemented, and age, prebiotic supplementation as a nutritional intervention strategy has the potential to improve overall health status of many species.

12.4.1 Future Research

Several areas remain open for further discovery in the area of prebiotics as regards animal nutrition. Prebiotic compounds are used in some companion animal foods currently, but further use might be warranted for young, old, and health-compromised pets. Poultry and swine prebiotic research demonstrates clear benefits to these species in the area of intestinal health. While production and performance responses are inconsistent, the animal industry, particularly in the European Union, has shifted or will shift away from antibiotic supplementation. This opens an avenue for further research into the use of prebiotics as antibiotic proxies. Use of prebiotics by horses, pre-ruminants, and ruminant animals can be expanded. New potential prebiotic compounds are created as new production technologies become more advanced, leading to new and potentially more active compounds that stimulate beneficial microbial population growth and fermentative profiles in the intestinal tracts of companion and livestock animal species.

12.5   Summary

- Prebiotics have been investigated in companion animals and many livestock species with varied, but generally positive, results.
- When prebiotics are fed to non-ruminant companion and livestock animals, similar responses are observed across species. Prebiotics can aid the host animal in fighting microbial infections, improving growth performance responses, improving microbial ecology, and optimizing animal health.
- Prebiotic studies in the equine generally have dealt with induction of laminitis with fructans, implying that horses do not tolerate large concentrations of fructans. However, there are no published studies investigating the effects of other prebiotics or low concentration supplementation strategies.
- Ruminant and pre-ruminant studies with prebiotics are sparse, but measurable changes in fermentative end-products, nutrient digestibilities, and immunological indices have been noted.
- Several considerations must be taken into account prior to prebiotic supplementation with regard to age, species, housing conditions, and health status to obtain optimal results.
- While prebiotics have been widely researched, much research remains to be performed as results with some species are sorely lacking. Improved technologies allow for the creation of novel prebiotics that should be investigated in select animal species.
- Overall, prebiotics have the potential to improve animal health by altering gastrointestinal events that impact host animal metabolism.

List of Abbreviations

AA	amino acid
ADF	acid detergent fiber
ADFI	average daily feed intake
ADG	average daily gain
ADL	acid detergent lignin
AME_n	metabolizable energy corrected for nitrogen
BAC	bacitracin
BW	body weight
Ca	calcium

CF	crude fiber
CFU	colony forming units
CHO	carbohydrate
CP	crude protein
Cu	copper
cys	cysteine
d	day
DGGE	denaturing gradient gel electrophoresis
DM	dry matter
DMB	dry matter basis
DMI	dry matter intake
DP	degree of polymerization
ESBM	ethanol-extracted soybean meal
Fe	iron
F:G	feed:gain ratio
FISH	fluorescence *in situ* hybridization
FOS	fructooligosaccharide
g	gram
G:F	gain:feed ratio
GOS	galactooligosaccharide
h	hour
HCC	hen cecal contents
IBDV	infectious bursal disease virus
IgA	immunoglobulin A
IgE	immunoglobulin E
IgG	immunoglobulin G
IGF-I	insulin-like growth factor-I
IgM	immunoglobulin M
IMO	isomaltooligosaccharide
IRA	ileo-rectal anastomosis
kg	kilogram
L	liter
lys	lysine
ME	metabolic energy
met	methionine
μg	microgram
mmol	millimole
mo	month

MOS	mannanoligosaccharide
N	nitrogen
NDF	neutral detergent fiber
NEFA	non-esterified fatty acid
NSP	non-starch polysaccharide
OF	oligofructose
OM	organic matter
P	phosphorus
ppm	parts per million
qPCR	quantitative polymerase chain reaction
SBM	soybean meal
SCFA	short-chain fatty acid
scFOS	short-chain fructooligosaccharide
TDF	total dietary fiber
TG	triglyceride
thr	threonine
TME_n	true metabolizable energy corrected for nitrogen
TMR	total mixed ration
TOS	trans-galactooligosaccharide
trp	tryptophan
val	valine
VFA	volatile fatty acid
wk	week
YCW	yeast cell wall
yr	year
Zn(O)	zinc (oxide)

References

Adogony V, Respondek F, Biourge V, Rudeaux F, Delaval J, Bind J-L, Salmon H (2007) Effects of dietary scFOS on immunoglobulins in colostrums and milk of bitches. J Anim Physiol Anim Nutr 91:169–174

Apanavicius CJ, Powell KL, Vester BM, Karr-Lilienthal LK, Pope LL, Fastinger ND, Wallig MA, Tappenden KA, Swanson KS (2007) Fructan supplementation and infection affect food intake, fever, and epithelial sloughing from salmonella challenge in weanling puppies. J Nutr 137:1923–1930

Awati A, Williams BA, Bosch MW, Gerrits WJJ, Verstegen MWA (2006) Effect of inclusion of fermentable carbohydrates in the diet on fermentation end-products on feces of weanling pigs. J Anim Sci 84:2133–2140

Bailey SR, Menzies-Gow NJ, Harris PA, Habershon-Butcher JL, Crawford C, Berhane Y, Boston RC, Elliot J (2007) Effect of dietary fructans and dexamethasone administration on the insulin response of ponies predisposed to laminitis. J Am Vet Med Assoc 231:1365–1373

Baurhoo B, Phillip L, Ruiz-Feria CA (2007a) Effects of purified lignin and mannan oligosaccharides on intestinal integrity and microbial populations in the ceca and litter of broiler chickens. Poult Sci 86:1070–1078

Baurhoo B, Letellier A, Zhao X, Ruiz-Feria CA (2007b) Cecal populations of lactobacilli and bifidobacteria and *Escherichia coli* populations after *in vivo Escherichia coli* challenge in birds fed diets with purified lignin or mannanoligosaccharides. Poult Sci 86:2509–2516

Berg EL, Fu CJ, Porter JH, Kerley MS (2005) Fructooligosaccharide supplementation in the yearling horse: Effects on fecal pH, microbial content, and volatile fatty acid concentrations. J Anim Sci 83:1549–1553

Biagi G, Piva A, Moschini M, Vezzali E, Roth FX (2006) Effect of gluconic acid on piglet growth performance, intestinal microflora, and intestinal wall morphology. J Anim Sci 84:370–378

Biggs P, Parsons CM (2007a) The effects of several oligosaccharides on true amino acid digestibility and true metabolizable energy in cecectomized and conventional roosters. Poult Sci 86:1161–1165

Biggs P, Parsons CM, Fahey GC (2007b) The effects of several oligosaccharides on growth performance, nutrient digestibilities, and cecal microbial populations in young chicks. Poult Sci 86:2327–2336

Bohmer BM, Branner GR, Roth-Maier DA (2005) Precaecal and faecal digestibility of inulin (DP 10-12) or an inulin/*Enterococcus faecium* mix and effects on nutrient digestibility and microbial gut flora. J Anim Physiol Anim Nutr 89:388–396

Butel M-J, Catala I, Waligora-Dupriet A-J, Taper H, Tessedre A-C, Durao J, Szylit O (2001) Protective effect of dietary oligofructose against cecitis induced by clostridia in gnotobiotic quails. Microb Ecol Health Dis 13:166–172

Cao BH, Karasawa Y, Guo YM (2005) Effects of green tea polyphenols and fructooligosaccharides in semi-purified diets in broilers' performance and caecal microflora and their metabolites. Asian-Aust J Anim Sci 18:85–89

Cetin N, Guclu BK, Cetin E (2005) The effects of probiotic and mannanoligosaccharide on some haematological and immunological parameters in turkeys. J Vet Med A 52:263–267

Coon CN, Leske KL, Akavanichan O, Cheng TK (1990) Effect of oligosaccharide-free soybean meal on true metabolizable energy and fiber digestion in adult roosters. Poult Sci 69:787–793

Correa-Matos NJ, Donovan SM, Isaacson RE, Gaskins HR, White BA, Tappenden KA (2003) Fermentable fiber reduces recovery time and improves intestinal function in piglets following *Salmonella typhimurium* infection. J Nutr 133:1845–1852

Crawford C, Sepulveda MF, Elliot J, Harris PA, Bailey SR (2007) Dietary fructan carbohydrate increases amine production in the equine large intestine: Implications for pasture-associated laminitis. J Anim Sci 85:2949–2958

Damron WS (2006) The gastrointestinal tract and nutrition. In: Yarnell D (ed) Introduction to animal science: Global, biological, social, and industry perspectives. Upper Saddle River, New Jersey, Pearson Education, Inc., pp. 97–115

Davis ME, Maxwell CV, Brown DC, de Rodas BZ, Johnson ZB, Kegley EB, Hellwig DH, Dvorak RA (2002) Effect of dietary mannan oligosaccharides and(or) pharmacological additions of copper sulfate on growth performance and immunocompetence of weanling and growing/finishing pigs. J Anim Sci 80:2887–2894

Davis ME, Brown DC, Maxwell CV, Johnson ZB, Kegley EB, Dvorak RA (2004a) Effect of phosphorylated mannans and pharmacological addition of zinc oxide on growth and immunocompetence of weanling pigs. J Anim Sci 82:581–587

Davis ME, Maxwell CV, Erf GF, Brown DC, Wistuba TJ (2004b) Dietary supplementation with phosporylated mannans improves growth response and modulates immune function of weanling pigs. J Anim Sci 82:1882–1891

Donaldson LM, McReynolds JL, Kim WK, Chalova VI, Woodward CL, Kubena LF, Nisbet DJ, Ricke SC (2008) The influence of a fructooligosaccharide prebiotic combined with alfalfa molt diets on the gastrointestinal tract fermentation, *Salmonella enteriditis* infection, and intestinal shedding in laying hens. Poult Sci 87:1253–1262

Elmusharaf MA, Peek HW, Nollet L, Beynen AC (2007) The effect of an in-feed mannanoligosaccharide preparation (MOS) on a coccidiosis infection in broilers. Anim Sci Feed Technol 134:347–354

Estrada A, Drew MD, Van Kessel A (2001) Effect of the dietary supplementation of fructooligosaccharides and *Bifidobacterium longum* to early-weaned pigs on performance and fecal bacterial populations. Can J Anim Sci 81:141–148

Fairchild AS, Grimes JL, Jones FT, Wineland MJ, Edens FW, Sefton AE (2001) Effects of hen age, Bio-Mos, and flavomycin on poult susceptibility to oral *Escherichia coli* challenge. Poult Sci 80:562–571

Fernandez F, Hinton M, Van Gils B (2000) Evaluation of the effect of mannanoligosaccharides on the competitive exclusion of *Salmonella enteritidis* colonization in broiler chicks. Avian Pathol 29:575–581

Fernandez F, Hinton M, Van Gils B (2002) Dietary mannan-oligosaccharides and their effect on chicken caecal microflora in relation to *Salmonella eneritidis* colonization. Avian Pathol 31:49–58

Flickinger EA, Van Loo J, Fahey GC (2003) Nutritional responses to the presence of inulin and oligofructose in the diets of domesticated animals: A review. Crit Rev Food Sci Nutr 43:19–60

Franklin ST, Newman MC, Newman KE, Meek KI (2005) Immune parameters of dry cows fed mannan oligosaccharide and subsequent transfer of immunity to calves. J Dairy Sci 88:766–775

Fujita K, Hara K, Hashimoto H, Kitahta S (1990) Transfructosylation catalyzed by β-fructofuranoside I from *Arthrobacter* sp. K-1. Agric Biol Chem 54:2655–2661

Gouveia EMF, Silva IS, Van Onselem VJ, Correa RAC, Silva CJ (2006) Use of mannanoligosaccharides as an adjuvant treatment for gastrointestinal diseases and this effects on *E. coli* inactivated in dogs. Acta Cir Bras 21: [serial on the Internet]

Harmsen HJM, Welling GW (2002) Fluorescence *in situ* hybridization as a tool in intestinal bacteriology. In: Tannock GW (ed) Probiotics and prebiotics: Where are we going? Wymondham, UK, Caiser Academic Press, pp. 41–58

Heinrichs AJ, Jones CM, Heinrichs BS (2003) Effects of mannan oligosaccharide or antibiotics in neonatal diets on health and growth of dairy calves. J Dairy Sci 86:4064–4069

Hesta M, Hoornaert E, Verlinden A, Janssens GPJ (2005) The effect of oligofructose on urea metabolism and faecal odour components in cats. J Anim Physiol Anim Nutr 89:208–214

Houdijk JGM, Bosch MW, Verstegen MWA, Berenpas HJ (1998) Effects of dietary oligosaccharides on the growth performance and faecal characteristics of young growing pigs. Anim Sci Feed Technol 71:35–48

Houdijk JGM, Hartemink R, Verstegen MWA, Bosch MW (2002) Effects of dietary non-digestible oligosaccharides on microbial characteristics of ileal chime and faeces in weaner pigs. Arch Anim Nutr 56:297–307

Howard MD, Gordon DT, Pace LW, Garleb KA, Kerley MS (1995) Effects of dietary supplementation with fructooligosaccharides on colonic microbiota populations and epithelial cell proliferation in neonatal pigs. J Pediatr Gastroenterol Nutr 21:297–303

Jeusette IC, Detilleux J, Shibata H, Saito M, Honjoh T, Delobel A, Istasse L, Diez M (2005) Effects of chronic obesity and weight loss on plasma ghrelin and leptin concentrations in dogs. Res Vet Sci 79:169–175

Jiang HQ, Gong LM, Ma YX, He YH, Li DF, Zhai HX (2006) Effect of stachyose supplementation on growth performance, nutrient digestibility, and caecal fermentation characteristics in broilers. Br Poult Sci 47:516–522

Kaufhold J, Hammon HM, Blum JW (2000) Fructo-oligosaccharide supplementation: Effects on metabolic, endocrine, and hematological traits in veal calves. J Vet Med A 47:17–29

Kim WK, Donalson LM, Mitchell AD, Kubena LF, Nisbet DJ, Ricke SC (2006) Effects of alfalfa and fructooligosaccharide on molting parameters and bone qualities using dual X-ray absorptiometry and conventional bone assays. Poult Sci 85:15–20

Krag L, Thomsen LE, Iburg T (2006) Pathology of Trichuris suis infection in pigs fed an inulin- and a non-inulin-containing diet. J Vet Med 53:405–409

Lan Y, Williams BA, Verstegen MWA, Patterson R, Tamminga S (2007) Soy oligosaccharides in vitro fermentation characteristics and its effects on caecal microorganisms of young broiler chickens. Anim Sci Feed Technol 133:286–297

LeMieux FM, Southern LL, Bidner TD (2003) Effect of mannan oligosaccharides on growth performance of weanling pigs. J Anim Sci 81:2482–2487

Lee JT, Connor-Appleton S, Bailey CA, Cartwright AL (2005) Effects of guar meal by-product with and without β-mannanase Hemicell on broiler performance. Poult Sci 84:1261–1267

Leske KL, Jevne CJ, Coon CN (1993) Effect of oligosaccharide additions on nitrogen-corrected true metabolizable energy of soy protein concentrate. Poult Sci 72:664–668

Leske KL, Coon CN (1999a) Nutrient content and protein and energy digestibilities of ethanol-extracted, low α-galactoside soybean meal as compared to intact soybean meal. Poult Sci 78:1177–1183

Leske KL, Coon CN (1999b) Hydrogen gas production of broiler chicks in response to soybean meal and α-galactoside free, ethanol-extracted soybean meal. Poult Sci 78:1313–1316

Loh G, Eberhard M, Brunner RM, Hennig U, Kuhla S, Klessen B, Metges CG (2006) Inulin alters the intestinal microbiota and short-chain fatty acid concentrations in growing pigs regardless of their basal diet. J Nutr 136:1198–1202

Lynch MB, Sweeney T, Callan JJ, Flynn B, O'Doherty JV (2007) The effect of high and low dietary crude protein and inulin supplementation on nutrient digestibility, nitrogen excretion, intestinal microflora and manure ammonia emissions from finisher pigs. Animal 18:1112–1121

Ma D, Li Q, Du J, Liu Y, Liu S, Shan A (2006) Influence of mannan oligosaccharide, Ligustrum lucidum, and Schisandra chinensis on parameters of antioxidative and immunological status of broilers. Arch Anim Nutr 60:467–476

Middelbos IS, Fastinger ND, Fahey GC (2007a) Evaulation of fermentable oligosaccharides in diets fed to dogs in comparison to fiber standards. J Anim Sci 85:3033–3044

Middelbos IS, Godoy MR, Fastinger ND, Fahey GC (2007b) A dose-response evaluation of spray-dried yeast cell wall supplementation of diets fed to adult dogs: Effects on butrient digestibility, immune indices, and fecal microbial populations. J Anim Sci 85:3022–3032

Mikkelsen LL, Jakobsen M, Jensen BB (2003) Effects of dietary oligosaccharides on microbial diversity and fructo-oligosaccharide degrading bacteria in faeces of piglets postweaning. Anim Sci Feed Technol 109:133–150

Mikkelsen LL, Jensen BB (2004) Effect of fructo-oligosaccharides and transgalacto-oligosaccharides on microbial populations and microbial activity in the gastro-intestinal tract of piglets post-weaning. Anim Sci Feed Technol 117:107–119

Mountzouris KC, Balaskas C, Fava F, Tuohy KM, Gibson GR, Fegeros K (2006) Profiling of composition and metabolic activities of the colonic microflora of growing pigs fed diets supplemented with prebiotic oligosaccharides. Anaerobe 12:178–185

Mwenya B, Santoso B, Sar C, Pen B, Morikawa R, Takaura K, Umetsu K, Kimura K, Takahashi J (2005) Effects of yeast culture and galacto-oligosaccharides on ruminal fermentation in Holstein cows. J Dairy Sci 88:1404–1412

NRC (1998) Nutrient requirements of swine. Washington, DC: National Academies Press

NRC (2006) Nutrient requirements of dogs and cats. Washington, DC: National Academies Press

NRC (2007a) Nutrient requirements of horses. Washington, DC: National Academies Press

NRC (2007b) The new science of metagenomics: Revealing the secrets of our microbial planet. Washington, DC: National Academies Press

Oli MW, Petschow BW, Buddington RK (1998) Evalutation of fructooligosaccharide supplementation of oral electrolyte solutions for treatment of diarrhea: Recovery of the intestinal bacteria. Dig Dis Sci 43:138–147

Parks CW, Grimes JL, Ferket PR, Fairchild AS (2001) The effect of mannanoligosaccharides, bambermycins, and virginiamycin on performance of large white male market turkeys. Poult Sci 80:718–723

Parks CW, Grimes JL, Ferket PR (2005) Effects of virginiamycin and a mannanoligosaccharide-virginiamycin shuttle program on the growth and performance of large white female turkeys. Poult Sci 84:1967–1973

Petkevicius S, Bach Knudsen KE, Murrell KD, Wachmann H (2003) The effect of inulin and sugar beet fiber on *Oesophagostomum dentatum* infection in pigs. Parasitol 127:61–68

Pierce KM, Sweeney T, Brophy PO, Callan JJ, Fitzpatrick E, McCarthy P, O'Dougherty JV (2006) The effect of lactose and inulin on intestinal morphology, selected microbial populations and volatile fatty acid concentrations in the gastro-intestinal tract of the weanling pig. Anim Sci 82:311–318

Propst EL, Flickinger EA, Bauer LL, Merchen NR, Fahey GC (2003) A dose-response experiment evaluating the effects of oligofructose and inulin on nutrient digestibility, stool quality, and fecal protein catabolites in healthy adult dogs. J Anim Sci 81:3057–3066

Rastall RA (2004) Bacteria in the gut: Friends and foes and how to alter the balance. J Nutr 134:2022S–2026S

Rehman H, Rosenkrantz C, Bohm J, Zentek J (2007) Dietary inulin affects the morphology but not the sodium-dependent glucose and glutamine transport in the jejunum of broilers. Poult Sci 86:118–122

Respondek F, Goachet AG, Julliand V (2008) Effects of dietary short-chain fructo-oligosaccharides on the intestinal microflora of horses subjected to a sudden change in diet. J Anim Sci 86:316–323

Rideout TC, Fan MZ (2004) Nutrient utilization in response to dietary supplementation of chicory inulin in growing pigs. J Sci Food Agric 84:1005–1012

Rideout TC, Fan MZ, Cant JP, Wagner-Riddle C, Stonehouse P (2004) Excretion of major odor-causing and acidifying compounds in response to dietary supplementation of chicory inulin in growing pigs. J Anim Sci 82:1678–1684

Roberfroid MB, Van Loo JA, Gibson GR (1998) The bifidogenic nature of chicory inulin and its hydrolysis products. J Nutr 128:11–19

Rozeboom DW, Shaw DT, Tempelman RJ, Miguel JC, Pettigrew JE, Connolly A

(2005) Effects of mannan oligosaccharide and an antimicrobial product in nursery diets on performance of pigs reared on three different farms. J Anim Sci 83:2637–2644

Samarasinghe K, Wenk C, Silva KFST, Gunasekera JMDM (2003) Turmeric (*Curcuma longa*) root powder and mannanoligosaccharides as alternatives to antibiotics in broiler chicken diets. Asian-Aust J Anim Sci 10:1495–1500

Shashidhara RG, Devegowda G (2003) Effect of dietary mannan oligosaccharide on broiler breeder production traits and immunity. Poult Sci 82:1319–1325

Shim SB, Williams IH, Verstegen MWA (2005) Effects of dietary fructo-oligosaccharide on villous height and disaccharidase activity of the small intestine, pH, VFA, and ammonia concentrations in the large intestine of weaned pigs. Acta Agric Scand A 55:91–97

Sims MD, Dawson KA, Newman KE, Spring P, Hooge DM (2004) Effects of dietary mannan oligosaccharide, bacitracin methylene disalicylate, or both on the live performance and intestinal microbiology of turkeys. Poult Sci 83:1148–1154

Smiricky-Tjardes MR, Grieshop CM, Flickinger EA, Bauer LL, Fahey GC (2003) Dietary galactooligosaccharides affect ileal and total-tract nutrient digestibility, ileal and fecal bacterial concentrations, and ileal fermentative characteristics of growing pigs. J Anim Sci 81:2535–2545

Solis de los Santos F, Donoghue AM, Farnell MB, Huff GR, Huff WE, Donoghue DJ (2007) Gastrointestinal maturation is accelerated in turkey poults supplemented with a mannan-oligosaccharide yeast extract (Alphamune). Poult Sci 86:921–930

Spears JK, Karr-Lilienthal LK, Fahey GC (2005) Influence of supplemental high molecular weight pullulan or γ-cyclodextrin on ileal and total tract nutrient digestibility, fecal characteristics, and microbial populations in the dog. Arch Anim Nutr 59:257–270

Spring P, Wenk C, Dawson KA, Newman KE (2000) The effects of dietary mannanoligosaccharides on cecal parameters and the concentrations of enteric bacteria in the ceca of salmonella-challenged broiler chicks. Poult Sci 79:205–211

Swanson KS, Fahey GC (2006) Prebiotic impacts on companion animals. In: Gibson GR, Rastall RA (eds) Prebiotics: Development and Application. New York, Wiley, pp. 213–236

Tako E, Glahn RP, Welch RM, Lei X, Yasuda K, Miller DD (2008) Dietary inulin affects the expression of intestinal enterocyte iron transporters, receptors and storage protein and alters the microbiota in the pig intestine. Br J Nutr 99:472–480

Terada A, Hara H, Sakamoto J, Sato N, Takagi S, Mitsuoka T (1994) Effects of dietary supplementation with lactosucrose (4G-β-D-Galactosylsucrose) on cecal flora, cecal metabolites, and performance in broiler chickens. Poult Sci 73:1663–1672

Thitaram SN, Chung C-H, Day DF, Hinton A, Bailey S, Siragusa GR (2005) Isomaltooligosaccharide increases cecal *Bifidobacterium* population in young broiler chickens. Poult Sci 84:998–1003

Tsukahara T, Iwasaki Y, Nakayama K, Ushida K (2003) Stimulation of butyrate production in the large intestine of weanling piglets by dietary fructooligosaccharides and its influence on the histological variables of the large intestinal mucosa. J Nutr Sci Vitaminol 49:414–421

Tzortis G, Goulas AK, Gee JM, Gibson GR (2005) A novel galactooligosaccharide mixture increases the bifidobacterial population numbers in a continuous *in vitro* fermentation system and in the proximal colonic contents of pigs *in vivo*. J Nutr 135:1726–1731

van Eps AW, Pollitt CC (2006) Equine laminitis induced with oligofructose. Equine Vet J 38:203–208

Vanhoutte T, Huys G, De Brandt E, Fahey GC, Swings J (2005) Molecular monitoring and characterization of the faecal microbiota of healthy dogs during fructan supplementation. FEMS Microbiol Letters 249:65–71

Verlinden A, Hesta M, Hermans JM, Janssens GPJ (2006) The effects of inulin supplementation of diets with or without hydrolyzed protein sources on digestibility, faecal characteristics, haematology, and immunoglobulins in dogs. Br J Nutr 96:936–944

White LA, Newman MC, Cromwell GL, Lindeman MD (2002) Brewers dried yeast as a source of mannan oligosaccharides for weanling pigs. J Anim Sci 80:2619–2628

Xu C, Chen X, Ji C, Ma Q, Hao K (2005) Study of the application of fructooligosaccharides in piglets. Asian-Aust J Anim Sci 18:1011–1016

Xu ZR, Hu CH, Xia MS, Zhan XA, Wang MQ (2003) Effects of dietary fructooligosaccharide on digestive enzyme activities, intestinal microflora and morphology of male broilers. Poult Sci 82:1030–1036

Xu ZR, Zuo XT, Hu CH, Xia MS, Zhan XA, Wang MQ (2002) Effects of dietary fructooligosaccharide on digestive enzyme activities, intestinal microflora and morphology of growing pigs. Asian-Aust J Anim Sci 15:1784–1789

Yang Y, Iji PA, Kocher A, Thomson E, Mikkelsen LL, Choct M (2008) Effects of mannanoligosaccharide in broiler chicken diets on growth performance, energy, energy utilization, nutrient digestibility and intestinal microflora. Br Poult Sci 49:186–194

Yasuda K, Maiorano R, Welch RM, Miller DD, Lei XG (2007) Cecum is the major degradation site of ingested inulin in young pigs. J Nutr 137:2399–2404

Yuzrial X, Chen TC (2003) Effect of adding chicory fructans in feed on broiler growth performance, serum cholesterol, and intestinal length. Int J Poult Sci 2:214–219

Zaghini A, Martelli G, Roncada P, Simioli M, Rizzi L (2005) Mannanoligosaccharides and aflatoxin B1 in feed for laying hens: Effects on egg quality, aflatoxins B1 and M1 residues in eggs, and aflatoxin B1 levels in liver. Poult Sci 84:825–832

Zdunczyk Z, Juskiewicz J, Jankowski J, Koncicki A (2004) Performance and caecal adaptation of turkeys to diets without or with antibiotic and with different levels of mannan-oligosaccharide. Arch Anim Nutr 58:367–378

Zdunczyk Z, Juskiewicz J, Jankowski J, Bierdrzycka E, Koncicki A (2005) Metabolic response of the gastrointestinal tract of turkeys to diets with different levels of mannan-oligosaccharide. Poult Sci 84:903–909

Zhang L, Li D, Qiao S, Wang J, Bai L, Wang A, Han IK (2001) The effect of soybean galactooligosaccharides on nutrient and energy digestibility and digesta transit time in weanling piglets. Asian-Aust J Anim Sci 14:1598–1604

Zhang WF, Li DF, Lu WQ, Yi GF (2003) Effects of isomaltooligosaccharides on broiler performance and intestinal microflora. Poult Sci 82:657–663

13 Analysis of Prebiotic Oligosaccharides

M. L. Sanz · A. I. Ruiz-Matute · N. Corzo · I. Martínez-Castro

13.1   Introduction

Carbohydrates and more specifically prebiotics, are complex mixtures of isomers with different degrees of polymerization (DP), monosaccharide units and/or glycosidic linkages. Many efforts are focused on the search for new products and the determination of their biological activity. However, the study of their chemical structure is fundamental to both acquire a basic knowledge of the carbohydrate and to increase the understanding of the mechanisms for their metabolic effect.

Both the identification of their constituents (qualitative) and the determination of their concentrations (quantitative) are the aims of an analytical process. Selection of an appropriate analytical technique, sample preparation (purification, fractionation, etc.) and optimization of the methodology are necessary for determining prebiotic structure. These steps are clearly dependent on the analytes of interest and the type of sample.

There are two main groups of analytical techniques used for the analysis of prebiotics: separation and spectroscopic techniques. Separation techniques (chromatographic and electrophoretic) give rise to the resolution of the constituents of a sample allowing the obtainment of quantitative information; however, the structural knowledge afforded is usually limited. Spectroscopic techniques are frequently necessary to provide detailed structural data of an isolated compound or a simple mixture. Combination of several techniques is often necessary to achieve all the required information about composition of complex mixtures.

Detailed information about the different analytical techniques required for the characterization of prebiotics as well as the state of the art of their applications has been included in this chapter. Although colorimetric methods such as determination of total carbohydrate or reducing sugar contents are still in use for oligosaccharide characterization, the separation techniques such as planar

13

chromatography, gas chromatography (GC), high performance liquid chromatography (HPLC) and capillary electrophoresis (CE), which provide qualitative and quantitative information of independent oligosaccharides, are the most widely used and therefore the main aim of this section. These techniques can be coupled to spectroscopic instruments in order to obtain structural information. Moreover, nuclear magnetic resonance (NMR) and mass spectrometry (MS) are directly used for prebiotic structural analysis. These techniques have experienced exceptional advances in recent years, although application to prebiotics is still in progress.

13.2   Analytical Techniques

13.2.1   Planar Chromatography

This term was proposed in 1983 by the Chromatographic Society. It covers both paper (PC) and thin-layer chromatography (TLC). PC was one of the earliest chromatographic techniques used for carbohydrate analysis, but at present it is scarcely utilised and mainly combined with other techniques to give supporting information.

A combination of PC, HPLC and high performance anion exchange chromatography (HPAEC) was used for the isolation of two octasaccharides, two dodecasaccharides and a tridecasaccharide from human milk (Haeuwfievre et al., 1993). Paper chromatography of milk oligosaccharides was used to purify some fractions eluted from an anion exchanger; migration time was 5 days (Guerardel et al., 1999). It has also been used to isolate the transglycosylation product of sucrose with beta-glycosidase from recombinant *Sulfolobus shibatae*, which was identified as α-D-galp-(1→6)-α-D-glucp-β-D-fructofuranoside (Park et al., 2005).

Planar chromatography also covers modern techniques derived from TLC such as HPTLC (High Performance TLC), OPTLC (Over Pressured TLC) and UTLC (Ultra TLC). These techniques are relatively low-cost, easy to perform and they display simultaneously in the chromatogram the overall components present in the sample.

Especially for saving time, thin-layer plates replaced paper, but the earliest methods have evolved and new modes have been introduced. At present, TLC is a well established technique which offers several additional advantages: it is relatively cheap, automated, allows satisfactory quantification and it can even be coupled to spectroscopic techniques such as MS.

13.2.1.1 Pretreatment of Samples

As TLC plates are not reused, the careful pretreatments necessary to keep the integrity of HPLC columns can be omitted. A carbohydrate solution not excessively turbid can be applied to plates, only avoiding the presence (in high amounts) of those compounds which can interfere the elution of analytes, such as proteins, lipids, certain salts, amines or acids.

13.2.1.2 Sorbents and Eluents

The preferred sorbents for carbohydrates are based on silica gel. This substance basically retains solutes by adsorption; separation thus occurs by solubility. In order to introduce new interaction mechanisms, different approaches have been proposed. Impregnation with inorganic salts allows modulation of the separation through complex formation: boric acid, sodium acetate, sodium bisulphite and phosphate buffers have been used for this purpose. Silica can also be functionalised with different organic groups in order to work in reverse-phase mode: amine and diol groups are preferred for carbohydrate analysis. Amino plates can be buffered with phosphates in order to avoid the reaction of the amino groups with the free carbonyls of the sugars.

Elution is carried out with aqueous mixtures of alcohols (methanol, ethanol, isopropanol, butanol); minor amounts of less-polar solvents (acetonitrile, ethyl acetate, acetone) are frequently added depending on the mixture to separate.

Although classic TLC is carried out in isocratic mode, at present it is feasible to use elution gradients by means of AMD (automatic mode development) which allows the formation of step-to-step gradients. AMD is performed by means of commercial equipment which allows a careful control of the process; in brief, the plate is eluted for a short time with the starting solvent mixture, then the solvent is removed and the layer is dried under vacuum; finally, another run starts in the same direction with another solvent of lower elution strength than that used before, and so on. In this way, a stepwise elution gradient is formed. Resolution is improved since spots are focused through the successive elution steps, which affords very narrow bands.

The introduction of HPTLC has improved both resolution and quantitative measurements. This technique is based on the use of special plates which have been prepared with very thin and uniform particles of silica (5–6 μm average) which allows shorter migration distances (3–6 cm) and reduced elution times

(3–20 min). Resolution can be improved by the use of plates with spherical particles such as LiChrospher™ (Merck).

OPTLC is based on a pressured chamber in which the vapor phase above the sorbent is almost eliminated. The eluent is pushed through the layer by a pump; continuous development can be performed.

UTLC was introduced in 2002 and it is based on monolithic structures created on the plate without particles; the sorbent layer is a continuous bed about 10 μm thickness containing macropores (1–2 μm) and mesopores (3–4 nm) (Hauck and Schulz, 2003). This geometry allows faster separations (1–6 min), lower limits of detection and reduced analyte and solvent volumes, but resolution has not yet been totally optimized. This technique seems to be promising for carbohydrate analysis, although applications need to be developed.

13.2.1.3   Visualization and Quantification

There is a broad range of choices for visualization of sugar spots. Both spraying and dipping modes have been used, although the latter is preferred at present, especially with high-performance methods. Chromogenic reagents based on amines and strong acids such as diphenylamine-aniline-phosphoric acid in acetone (Martinez-Castro and Olano, 1981; Reiffova and Nemcova, 2006), urea-phosphoric acid (Bonnett et al., 1997) and N-(1-naphthyl)ethylenediamine dihydrochloride dissolved in sulfuric acid-methanol (Bounias, 1980) have been sprayed on silica gel plates. This latter reagent has been shown to be very sensitive (50 ng). 4-Aminobenzoic acid has been used for visualization on diol plates, whereas α-naftol has been used for amino plates. *In situ* reaction of sugars with the amino groups of amino layers also produces visualization, by simply heating the plate at about 170°C; sugars appeared as white-blue fluorescent spots under UV light at 365 nm.

13.2.1.4   Applications

⊘ *Table 13.1* summarizes some applications of TLC to the analysis of prebiotic carbohydrates.

Chromatographic analysis of lactulose was revised in 1987 (Martínez-Castro et al., 1987). A TLC method was devised for analysis of lactulose in milk using silica gel plates charged with borate to form complexes allowing separation of lactulose from lactose (Martinez-Castro and Olano, 1981).

◻ Table 13.1

TLC applications for the analysis of prebiotics *(Cont'd p. 470)*

Analytes	Bed	Elution system	Visualization	Quantification	Reference
Lactulose in milk	Silica gel G + 0.03M H_3BO_3	ACN:water (5:2)	1 g diphenylamine + 1 mL aniline + 50 mL H_3PO_4 + 50 mL acetone	Semiquantitative detection limit: 0.02% lactulose	Martínez-Castro and Olano (1981)
14 Sugars in biol fluids; maltodextrines in infant formulas	Cellulose	Threefold development (a) butanol/EtOH/w (3: 2: 1) (b) pyridine/ethyl acetate/AcH/w (5: 5: 3: 1)	3 g $AgNO_3$ + 12 mL+ (a) 500 mL acetone (b) 50 ml of 10N NaOH in 450 mL EtOH	None	Bosch-Reig et al. (1992)
Oligosaccharides in beer	Amino + 0.4M KH_2PO_4	AMD linear gradient ACN: acetone (1:1)/w 40–20% in 15 steps	Thermal "in situ" reaction	Scanner in fluorescence mode at 366 nm	Brandolini et al. (1995)
FOS in biological fluids	Silica gel F254 + 0.02M Na Ac	Butanol:ethanol:w (5:3:2)	Diphenylamine-aniline-H_3PO_4 in acetone	Reflectance densitometry at 370 nm	Reiffova et al. (2006)
Microbiologically-produced Dextrans	Silica gel F	Three ascents EtAc: ACN: w propanol (2:7:5.5:5)	0.2%N(1-naphtyl) ethylendiamine HCl in MeOH with 3% H_2SO_4	Scanner in reflectance mode	Coté and Leathers (2005)
Hydrolyzates of polysaccharides from Linghzi	HPTLC silica gel 60 and HPTLC silica gel 50,000	AMD with 2 solutions: propanol:w (6:4) and propanol:w (83:17) in 7 steps	0.5 g 4-aminobenzoic acid, 9 mL AcH, 10 mL w, 0.5 mL H_3PO_4	Scanned at 365 in absorption mode	Di et al. (2003)
FOS in plants	Silica gel	Two developments in butanol:2-propanol: w (3:12:4)	Urea + H_3PO_4	Qualitative purposes	Bonnett et al. (1997)

■ Table 13.1

Analytes	Bed	Elution system	Visualization	Quantification	Reference
Oligosaccharides in molasses and artichoke leaves	Diol HPTLC	AMD with 3 solutions: ACN: acetone (1.1)/w (85/15), (94/6) and (95/5) in 9 steps	2 g 4-aminobenzoic acid + 36 mL AcH, 40 mL w, 2 mL H_3PO_4 + 120 mL acetone	Scanning in both fluorescence and absorbance (nanomol)	Vaccari et al. (2001)
Xylooligosaccharides	Microcristalline cellulose	EtOAc:AcOH:w (3:2:1)	Aniline + H Phthalate	Qualitative purposes	Katapodis et al. (2003)
Arabinoxylo-oligosaccharides	Silica gel 60	Butanol:EtOH:w (3.2.2)	2% orcinol in (EtOH:H_2SO_4: w 8:1:1)	Qualitative purposes	Rantanen et al. (2007)
Dietary fiber	Silica gel	n-propanol:EtOH:w (7:1:2)	5% H_2SO_4 in MeOH		Mc Cleary and Rossiter (2004)
Human milk oligosaccharides	HPTLC silica	Butanol:AcH:w (2.5:1:1) two developments	0.1% orcinol in 20% H_2SO_4	Qualitative purposes	Kunz et al. (1996)

FOS, fructooligosaccharides; ACN, acetonitrile; w, water; EtOH, ethanol; MeOH, methanol

Classic TLC has been used to analyze different mixtures of oligosaccharides. Fructans of different plants from the family *Poaceae* were separated using two successive developments, allowing the separation of fructans belonging to both 2→1 and 2→6 series from DP3 to DP10 (Bonnett et al., 1997). TLC has been successfully used to monitor the formation of dextrans and isomaltooligosaccharides by the action of glucansucrases from *Leuconostoc* (Côté and Leathers, 2005), as well as xylooligosaccharides (XOS) (Katapodis et al., 2003) and arabinooligosaccharides (AROS) (Rantanen et al., 2007) formed by the action of different xylanases.

Prebiotics have been monitored in different parts of the intestinal tract (jejunum, ileum and colon) of monogastric animals by a simple TLC method (Reiffova and Nemcova, 2006).

HPTLC with AMD achieved a good separation and quantification of several oligosaccharides in molasses (Vaccari et al., 2001). This technique was also shown to be useful for the study of fingerprint profiles of hydrolyzates of polysaccharides from some important and popular Chinese medicinal mushrooms commonly known as Lingzhi (Di et al., 2003).

Human milk oligosaccharides (HMOS) are excessively complex to be directly separated by HPTLC. Nevertheless, this technique is very useful for the analysis of fractions obtained by preparative techniques, as carried out by Kunz et al. (1996).

13.2.1.5   Coupling with MS

This field was revised by Wilson in 1999. TLC was firstly used as a preparative technique for MS, thus working off-line: spots can be easily cut, solvent-extracted and injected into the ion source of the mass spectrometer (St-Hilaire et al., 1998). Nevertheless, many efforts have been directed to achieve an effective coupling, analyzing spots on the plate (Wilson, 1999). First couplings were carried out using fast atom bombardment (FAB) and liquid secondary ion (LSI). The spot was cut from an aluminum plate and attached to the MS probe. Interfaces based on motorised probes have been designed, where a strip of the plate can be slowly moved through the ion source, enabling the analysis of all spots from a lane to give a true chromatogram.

At present the more common techniques are those based on laser desorption, as MALDI (Matrix-Assisted Laser Desorption/Ionization) and SALDI (Surface-Assisted Laser Desorption/Ionization); in both cases the time of flight (ToF) analyzer is a good option, as it will be further seen.

As an example of coupling with MS, native milk oligosaccharides were separated on 10×10 silica gel plates and developed in n-butanol/acetic/water (110/45/45). MALDI-ToF was selected as MS technique. Glycerol was used as matrix, with an infrared laser for MALDI and an orthogonal ToF (o-ToF) for achieving high mass accuracy, allowing a straightforward method with a detection limit of $\sim$10 pmol of individual compounds (Dreisewerd et al., 2006).

13.2.2   Liquid Chromatography

Liquid Chromatography (LC) is a separation technique which uses a liquid as mobile phase. Although HPLC is at present generally utilized for the analysis of prebiotic carbohydrates, the use of traditional open columns packed with ion exchange resins, carbon-celite or size exclusion gels as stationary phases is still widely practiced. Although analysis of oligosaccharides (such as malto-, isomalto-, gentio- and levan oligosaccharide series; Kennedy et al., 1989) has been carried out by traditional LC open columns coupled to mainly refractive index (RI) detectors, this technique has been mainly focused on preparative purposes. The collection of fractions of homologous oligosaccharides (different molecular weights or monosaccharide units) is in many cases, a required step for their further characterization by other analytical techniques such as MS or NMR.

13.2.3   High Performance Liquid Chromatography

HPLC is one of the most widespread techniques for oligosaccharide analysis, both for analytical and preparative purposes. A high number of methodologies for qualitative and quantitative characterization of prebiotic carbohydrates have been developed using different operation modes and detectors.

13.2.3.1   Sample Preparation

Analysis by HPLC commonly requires sample preparation methods to remove interfering compounds or impurities; the analysis of prebiotic oligosaccharides is not an exception. These methodologies are mainly based on dilution, liquid-liquid or liquid-solid extraction and filtration steps (Sanz and Martinez-Castro, 2007). Nevertheless, derivatization is in some cases necessary, mainly to enhance sensitivity in the detection during analysis.

There is a wide variety of derivatization reagents for oligosaccharides; the state the art in the preparation of derivatives being included in different reviews dedicated to sample preparation or chromatographic analysis of carbohydrates (Lamari et al., 2003; Sanz and Martinez-Castro, 2007), therefore, only a summary is mentioned here.

Most methods are based on the condensation of a carbonyl group in carbohydrates with primary amines to give a Schiff base which is then reduced to a N-substituted glycosil amine. The primary amine has to posses the desired chromophore or fluorophore substituent, usually an aromatic ring. Reductive amination has been carried out with 2-aminopyridine, different trisulphonates, esters of p-aminobenzoic acid, 2-aminoacridone, etc. (Sanz and Martinez-Castro, 2007).

Acetylation reactions of oligosaccharides overcome problems of solubility in organic solvents, whereas perbenzoylated derivatives improve the chromatographic properties on reverse phase columns (Kennedy and Pagliuca, 1994).

13.2.3.2 Chromatographic Columns

Chromatographic columns used for HPLC carbohydrate analysis can be divided according to the composition of their stationary phases and their dimensions and design.

13.2.3.2.1 Stationary Phase Composition

Both reverse phase and cation exchange chromatography have been the most common HPLC modes utilized for carbohydrate analysis till Rocklin and Pohl (1983) suggested the use of HPAEC for this aim. Most of the stationary phases used in these modes are available for both analytical and preparative purposes; this section being focused on the analysis of prebiotic oligosaccharides.

Alkyl-Bonded Silica Phases   Among the alkylated silica-based stationary phases, those of octadecyl-coated (C_{18}) sorbents are the most commonly utilized, this columns being useful for the separation of oligosaccharides with different DP. Moreover, columns can present different percentages of bonded alkyl chains which could show a wide effect on carbohydrate resolution. The operation mode used for these columns is the reverse phase (RP)-HPLC where the non-polar ligands are covalently bound to a solid support and the mobile phase is mainly composed by aqueous solutions moderately polar. The retention mechanism is based on the interaction of the packing with polar materials; the most polar compounds elute first whereas those with lower polarity are more retained. The use of low temperatures for the elution improves the resolution (Kennedy and Pagliuca, 1994). General aspects of underivatised and derivatised carbohydrate analysis by RP-HPLC have been reviewed by El-Rassi (2002).

Water is the most common mobile phase chosen for underivatised carbohydrates since these compounds require high surface tension to achieve an appropriate resolution. Nevertheless, gradients with organic solvents are used, although problems of solubility of oligosaccharides can arise.

Aminoalkyl-bonded Silica Gel Phases   Aminoalkyl-modified silica gel columns provide good resolution; however, their stability is low and can be easily degraded. The most common ones for oligosaccharide analysis are aminopropyl-bonded columns, although primary and secondary diamines and secondary and tertiary amines can be also found. Non-polar organic solvents or aqueous organic mixtures are used as mobile phases.

Different mechanisms have been proposed for this chromatography such as partition or hydrogen bonding (Herbreteau, 1992). Oligosaccharides are eluted in order of increasing their molecular weight. Carbohydrates up to DP15 can be separated by this technique, although solubility problems can be found for oligosaccharides of high molecular weights. Nevertheless, if a diamine or a polyamine is added to the eluent a dynamic equilibrium is formed and oligosaccharides up to DP25 can be separated (Kennedy and Pagliuca, 1994). In these cases a presaturation column has to be placed before the injector to avoid dissolution of the analytical column packing.

Many researchers have used amino columns to analyze fructooligosaccharides (FOS) of different DP using acetonitrile:water (75:25) as mobile phase (Sangeetha et al., 2005), however, resolution is not as good as that obtained for mono- and disaccharides and solubility problems appear (Herbreteau, 1992). Moreover, the formation of Schiff bases between reducing sugars and amino groups can reduce the lifetime of these columns.

Cyclodextrin-bonded Phases   The use of cyclodextrin-based columns for the separation of neutral prebiotic carbohydrates such as those derived from xylan, inulin or mannan, has been also proposed as a substitution of aminoalkyl modified silica gel columns. These columns are particularly useful for the analysis of oligosaccharides since the retention of these compounds is mainly based on the hydrogen bonding interactions of oligosaccharide hydroxyl groups with the stationary phase which allows the separation of the different molecular weights. Mobile phases are normally constituted by different percentages of acetonitrile and water. Carbohydrates elutes in order of increasing DP (Herbreteau, 1992).

Other Polar Bonded Phases   Several stationary phases with highly polar sorbents such as cyano, hydroxyl, diol, derivatives of poly(succinimide), sulfoalkylbetaine, etc. have been also used for carbohydrate analysis (Ikegami et al., 2008). Analysis on these columns has in common that retention increases with the hydrophilicity of the stationary phase and the analytes and with decreasing hydrophilicity of solvents from mobile phase. All of them are therefore grouped under the acronym HILIC (hydrophilic interaction chromatography). The first generations of HILIC were based on the amino-silica stationary phases and mixtures of acetonitrile:water mobile phases that has been previously described. HILIC belongs to normal phase liquid chromatographic (NPLC) modes with the hydrophilic stationary phase but with the mobile phase replaced by an aqueous/organic mixture (typically acetonitrile in water or a volatile buffer). It is very useful for the separation of polar compounds such as oligosaccharides. The hydrophilic groups of the stationary phase attract water molecules from the mobile phase to form water-enriched layers. The chromatographic mechanism is therefore mainly based on partition equilibrium between both mobile and stationary phases facilitated by the aqueous layers.

Graphitized Carbon Phases   Graphitized carbon columns (GCC) were developed as an alternative to RP columns for the analysis of polar compounds (Koizumi, 2002). Their mechanism is based on the unspecific adsorption of polar compounds such as carbohydrates and interaction is enhanced with increasing molecular size. The effect of temperature is not drastic, although high temperatures can produce an increase in the retention due to the higher adsorptive activity of carbon. Eluents for mobile phases include high percentages of organic modifiers such as acetonitrile with no ion-pairing agents; these eluents being compatible to MS detectors.

Size Exclusion Phases   As it has been indicated before, size exclusion chromatography (SEC) is widely utilized in its classical form with open columns. Nevertheless, the use of size exclusion for HPLC (HPSEC) is also commonly applied, although its inability to separate linkage isomers has limited its development.

Oligosaccharides are eluting in order of decreasing molecular size from a stationary phase constituted by cross-linked polysaccharide or polyacrylamide. These packing material are available with a range of pore volumes; separation depends on the ratio of their molecular dimensions and the average diameter of the pores (Churms, 1996). Mobile phases should be carefully chosen to avoid all types of interaction, such as electrostatic interactions; acetate buffers or pyridinium acetates being used among others.

Cation Exchange Phases　Cation exchange resins are composed by cross-linked polystyrene and silica-based ion exchangers such as calcium or silver. Oligosaccharides up to DP8 for calcium columns and DP12 for silver columns can be separated (Kennedy and Pagliuca, 1994). Carbohydrates elute in order of decreasing molecular size and the chromatographic mechanism is based on both the size exclusion and ligand-exchange.

These phases can show different disadvantages such as compressibility of the gel matrix, efficiency losses when flow rate is increased, the need of high temperature operation ($85°C$) and extended analysis times (Kennedy and Pagliuca, 1994). The use of H^+ columns and 0.01M sulfuric acid as mobile phase which allows the regeneration of the H^+ reduces the losses of efficiency.

Anion Exchange Phases　The advent of HPAEC in 1983 (Rocklin and Pohl, 1983) for carbohydrate analysis notably improved knowledge about oligosaccharide composition of a wide variety of products. Carbohydrates are negatively charged at high pH (pH > 13) and oligo- and polysaccharides up to DP50 can be separated by anion-exchange chromatography using NaOH as mobile phase; amylopectins up to DP80 have even been separated by this technique (Hanashiro et al., 1996). A gradient of increasing concentration of sodium acetate is normally used to help elution of oligosaccharides. Stationary phases are composed of polymeric, non-porous, MicroBead™ pellicular resins such as polystyrene/divinylbenzene or ethylvinylbenzene/divinylbenzene substrates agglomerated with Microbead™ quaternary amine functionalized latex, which are highly resistant to high pHs. CarboPac PA100 and more recently, CarboPac PA200 are columns mainly designed for oligosaccharide separation, although CarboPac PA1 and PA10 can be also used (Cardelle-Cobas et al., 2008; Splechna et al., 2006).

Carbohydrate elution takes place with increasing the molecular weight for oligosaccharides with the same glycosidic linkage, nevertheless, this order can change when families of oligosaccharides with different linkage variants are mixed (i.e., isomaltohexaose elutes before maltotriose; Morales et al., 2006). The combination of different effects (charge, molecular size, sugar composition and glycosidic linkages) is implied in the chromatographic separation (Gohlke and Blanchard, 2008). This retention behavior is one of the disadvantages of this technique which requires the use of standards for the identification of complex mixtures of carbohydrates with different DP and glycosidic linkages. Nevertheless, this chromatographic technique coupled to a pulse amperometric detector (PAD), as it will be shown later, presents significant advantages: fast analysis, samples do not require a previous derivatization, low to sub-picomole sensitivity, high resolution, etc.

13.2.3.2.2   Column Dimensions and Design

Recently, there is a trend to develop miniaturized systems which allow reduced solvent consumption and disposal, fast analysis and increased sensitivity. Downscaling of the column dimensions to the capillary- or nano-scale has shown several advantages over the conventional chromatography for carbohydrate analysis. Glycans at femtomole level can be analyzed without derivatization. This miniaturization has been carried out for graphitized carbon stationary phases (Ninonuevo et al., 2005), normal and reverse phases (Wuhrer et al., 2005) and HPAE columns (Bruggink et al., 2005a).

Conventional HPLC phases of between 3 and 10 µm diameter of particles are commonly used for oligosaccharide analysis; 3 µm silica columns (such as aminobonded silica phases) have been demonstrated to improve the analysis of glucooligosaccharides up to a DP of 30–35 (Herbreteau, 1992). Columns with smaller diameter of particles (sub-2 µm) have recently been introduced in the market and allow faster separations without resolution losses.

The use of columns with small particle diameter and column length produces an increase in pressure. In order to solve this backpressure problem, Ultra Performance Liquid Chromatography (UPLC) systems are employed. However, applications to oligosaccharides are scarce.

Fast separation can be also achieved using monolithic columns constituted by highly porous materials with a network of interconnecting channels. These columns allow the use of very fast flows: a mixture can be resolved at a flow of 9 mL min^{-1} reducing the elution time between 5 and 10 times. A wide range of underivatized or derivatized carbohydrates (mono- and oligosaccharides) can be successfully separated with this kind of columns achieving higher column efficiencies than with particle-packed columns (Ikegami et al., 2006; Ikegami et al., 2008).

13.2.3.3   Detectors

Not only is the separation of oligosaccharides a problem in HPLC due to their similar structures, but also to achieve a sensitive detection can be a difficult task. Carbohydrates itself do not contain a specific chromophore or fluorophore, being necessary to fall back on universal detectors such as RI detectors, or on electrochemical ones. Nevertheless, ultraviolet (UV) detectors for both derivatised and non derivatised carbohydrates or fluorimetric detectors for derivatised ones are also used.

Refractive Index Detector

RI detectors are the most common detectors used for carbohydrate analysis although a lack of sensitivity is normally associated to them. Their main drawback arises from their dependence on temperature and mobile phase composition changes. Therefore, these detectors are commonly utilized with isocratic mode or, if gradients of mobile phases are used, solvents with the same refractive index must be used (Davies and Hounsell, 1996). These detectors are mainly used in the mM-µM concentration range.

UV Detector

UV detectors at low wavelengths (below 210 nm) show similar sensitivity to RI detectors however, they allow changes in temperature and gradient elution. As has been indicated, different methods to introduce chromogenic groups in saccharide molecules have been proposed. However, post-column derivatization (not considered as sample preparation) also improves UV detection. Carbohydrates can be labeled with different reagents such as tetrazolium blue or cyanoacetamide. Hase (2002) reports a table with all the possible reagents used for post-column derivatization in HPLC.

Fluorometric Detector

Postcolumn derivatization of carbohydrates with fluorescent labels such as 2-aminopyridine or 2-aminobenzoic acid allows their detection in subpicomolar concentrations (Gohlke and Blanchard, 2008). However, most of postcolumn derivatization methods have been applied to monosaccharide analysis, while only few works have been reported about oligosaccharide analysis.

Light Scattering Detectors

Evaporative light scattering detectors (ELSD) utilize a spray which atomizes the column effluent into small droplets. These droplets are evaporated and the solutes as fine particulate matter are suspended in the atomizing gas. These particles diffuse the light originated from a monochromatic or polychromatic source. These detectors are universal, more sensitive than RI and are compatible with elution gradients (Herbreteau, 1992).

Liquid light scattering detectors differ from ELSD in that they respond to the light scattered by a polymer or large molecular weight substance present in the column eluent. The high intensity light source is achieved by the use of a laser. There are two forms of the detector: Low angle laser light scattering (LALLS) and multiple angle laser light scattering (MALLS) which provide an appropriate sensitivity and baseline stability. They have commonly been applied to SEC. Light from a laser is scattered to different degrees by the concentration and size of the analyte passing through the cell flow. The intensity of scattered light is highest at low scattering angles and also increases with the molecular weight of the carbohydrate (Davies and Hounsell, 1996). However, laser light scattering can sometimes give confusing results because of molecule-molecule interactions and associations of oligosaccharides with high DP.

Pulse Amperometric Detectors

PADs are commonly coupled to HPAEC and allow the detection of non-derivatised carbohydrates at very low picomole levels. This detection provides a high selectivity; only compounds oxidizable at the selected voltages being detected. PAD is composed by a working electrode of Au or Pt, a stainless steel auxiliary electrode and a reference electrode of Ag/AgCl or H_2. The Au electrode is able to catalyze the oxidation reactions and is the best choice for detection of carbohydrates. Carbohydrates are detected by measuring the electrical current generated by their oxidation at the surface of the working electrode at the selected potential (E_1). Next, the voltage is increased (E_2) to oxidize the gold detector which causes desorption of the carbohydrate oxidation products. Finally, the potential is lowered (E_3; negative potential) to reduce the electrode surface for the next pulse. The three potentials are applied for fixed times. More recently, this potential sequence has been modified, because although good results can be achieved the gold electrode surface is gradually lost which affects reproducibility (Rohrer, 2003). Similar to the previous method, the first potential is applied to oxidise the carbohydrate on the surface of the gold electrode. However, the second potential is in this case a reductive potential to clean the working electrode surface, whereas the third short potential reactivates the electrode oxidizing its surface. A fourth potential is necessary to achieve the initial conditions of the Au surface.

Mass Detectors

The use of MS detectors coupled to HPLC systems has considerably enriched the field of carbohydrate analysis. MS detectors have been commonly utilised with alkyl- and aminoalkyl-bonded phases, although other columns such as GCC can be also used. Currently, MS (both quadrupole (Q) and ion trap (IT) MS; Bruggink et al., 2005a,b) are also being coupled to HPAEC using automated systems for neutralization and removal of eluent salts to be compatible with electrospray ionization (ESI) MS requirements. These systems also allow the collection of different desalted fractions which are suitable for further enzymatic and chemical digestions or NMR and chromatographic analysis. A splitter to divide the effluent to the PAD and a MS detector is installed after the column and a membrane-desalting device placed before the MS system convert the sodium hydroxide into water and sodium acetate into acetic acid. Lithium or sodium chlorides are added after the membrane desalter to enhance the MS signal by the formation of lithium or sodium adducts of carbohydrates (Bruggink et al., 2005a,b). These couplings are a great advantage for carbohydrate analysis allowing the acquisition of information related not only to carbohydrate retention, but also to structural characteristics.

Electrospray ionization (ESI) is the most common ionization source coupled to HPLC for the analysis of carbohydrates. Analytes (which are in an ionic state) and eluent from HPLC (which is highly volatile) are sprayed at atmospheric pressure from a needle which is subjected to a high potential (3,000–5,000 V) giving rise to small droplets. Desolvation of ions is assisted by a heated inert gas (N_2). As the solvent evaporates, the ionic charges of analytes are closer and repel each other, breaking up the droplets. The ions free from solvents are focused in the analyzer: Q, IT, Q-ToF, etc.

13.2.3.4   Multidimensional HPLC

Multidimensional HPLC has been used to mainly separate fluorescently-labeled oligosaccharides. Combinations of amine-bonded phases which can perform as both hydrophilic interaction media and as an anion exchange phase and RP-columns have been applied for this purpose (Gohlke and Blanchard, 2008).

13.2.3.5   Applications

The number of applications of HPLC to analyze prebiotic oligosaccharides in recent years is huge and only some of them can be mentioned in this chapter, although it is good to consider that most of the methods here reported for common prebiotic oligosaccharides [FOS and galactooligosaccharides (GOS)] can be applied to the analysis of different carbohydrate sources. ❂ *Table 13.2* summarizes some of the applications described below.

The most recent applications of prebiotic analysis have been developed for HPAEC-PAD. This technique is widely used to both determine the oligosaccharide composition of prebiotic carbohydrates and to study the degradation patterns of oligosaccharides during fermentation assays to evaluate their prebiotic effect. Different methodologies, mainly based in the use of eluents indicated above (NaOH and NaOAc) and Carbo-Pac PA100 or Carbo-Pac PA200 columns, have been developed by many researchers who tried to achieve the optimum separation depending on the prebiotic source to be analyzed.

Fructooligosaccharides and Inulin

Sangeetha et al. (2005) reviewed the different methods reported in the literature for the analysis of FOS, most of them being based on the use of polar-bonded phases or resin based ion exchange columns coupled to RI detectors and HPAEC-PAD analyses.

There are many reports on the analysis of FOS by RP-HPLC coupled to a diode array detector at a wavelength of 190 nm (Grizard and Barthomeuf, 1999) or to a RI detector (Mujoo and Ng, 2003) using a C18 column and water as mobile phase. Chromatographic profiles showed coelution of glucose and fructose, whereas kestose, nystose and fructofuranosylnystose appeared as separated peaks.

HPAEC-PAD analysis of fractionated inulin allows the separation of different molecular weight fructooligosaccharides (G-F_n; α-D-gluc$p$-[β-D-fruct$f$]$_{n-1}$-D-fructofuranoside) and inulooligosaccharides (F-F_n; β-D-fruct$p$-[α-D-fruct$f$]$_{n-1}$-α-D-fructofuranoside). Assuming that retention time in HPAEC-PAD of a homologous series of carbohydrates increases with increasing DP, the assignment of these carbohydrates can be feasible. However, the coexistence of both series

◻ Table 13.2

HPLC applications for the analysis of prebiotics

Prebiotic	Chromatographic column	Detector	Mobile phase	Reference
FOS in onion	CarboPac PA1	PAD	NaOH and CH_3CO_2Na	Kaack et al. (2004)
FOS from levan	Cation exchange (Ca^{2+})	RI	Water	Kennedy et al. (1989)
FOS	CarboPac PA200	MS	NaOH and CH_3CO_2Na	Bruggink et al. (2005a)
FOS in nutraceutical and functional foods	CarboPac PA100 CarboPac PA10	PAD	NaOH and CH_3CO_2Na	Corradini (2002)
GOS	CarboPac PA10	PAD	NaOH and CH_3CO_2Na	Cardelle-Cobas et al. (2008)
GOS	Cation exchange (Ca^{2+})	RI	Water	Goulas et al. (2007)
	CarboPac PA1	PAD	NaOH and CH_3CO_2Na	
Human milk oligosaccharides	On chip GCC	o-ToF	Formic acid in acetonitrile/water	Ninonuevo et al. (2005)
Soybean oligosaccharides	HPSEC	RI	$NaNO_3$ with NaN_3	Giannoccaro et al. (2008)
	CarboPac PA10	PAD	NaOH and CH_3CO_2Na	
Sucrose derived oligosaccharides	Amino bonded	ELSD	Methanol: acetonitrile:water	Yin et al. (2006)
XOS	Cation exchange (Na^+)	RI	Water	Ohara et al. (2006)
XOS	Cation exchange (Ca^{2+})	RI	5 mM H_2SO_4 in water	Moura et al. (2008)
Lactulose	Amino bonded	RI	Acetonitrile:water	Paseephol et al. (2008)
FOS and glucooligosaccharides	C18	RI	0.1% trifluoroacetic acid (v/v) in water	Rousseau et al. (2005)

(G-F$_n$ and F-F$_n$) makes their identification more difficult and coelution of some oligosaccharides can be observed. Schütz et al. (2006) investigated the chromatographic profile of inulin up to carbohydrates of DP79 in artichoke heads and dandelion roots by this technique, although quantitative analysis was only carried out for glucose, fructose, sucrose, kestose, nystose and fructofuranosylnystose. The lack of higher molecular weight standards is one the limitations of this analysis. Ronkart et al. (2007) developed a method to obtain F-F_n standards and to identify them in a complex inulin chromatogram. F-F_n standards were isolated by semi-preparative HPSEC from inulin from globe artichoke treated with endo-inulinase, and analyzed by HPAEC-PAD. Coelution problems of some G-F_n and F-F_n oligosaccharides after HPAEC-PAD analysis can be solved by the use of a coupled ESI MS detector which allows the unveiling of both series by the extraction of the ion chromatograms at the appropriate m/z ratios (Bruggink et al., 2005a,b). ❖ *Figure 13.1a* shows the extracted ion chromatograms of FOS up to DP13 obtained by HPAEC-MS using a CarboPac PA200 capillary column (Bruggink et al., 2005a), the F-F_n series being more retained than G-F_n series. A MS spectrum of $[GF_4 + Na]^+$ is also shown in ❖ *Figure 13.1b*. As can be seen in ❖ *Figures 13.1c* and ❖ *13.1d*, MS/MS spectra of the two series (i.e., for DP5 variants) showed similar fragmentation patterns with different relative intensities of m/z ions.

Wang et al. (1999) quantitatively analyzed FOS from different food matrices and compared these results to those obtained by MALDI-ToF. The PAD response was different depending on DP and oligosaccharide linkage; the use of specific standards being necessary to avoid overestimated results.

Different authors have shown the suitability of HPAEC-PAD to evaluate fermentation properties of FOS. Hartemink et al. (1997) observed a different HPAEC degradation pattern of FOS for the different strains assayed (*Ent. cloacae, E. coli, Salm. infantis* and *Sh. flexneri*, among others). Moreover, depending on the FOS source employed, different behavior was observed; i.e., using Profeed P95 (from Nutreco, Boxeer, The Netherlands) as a substrate, *Ent. Clocae and Salm. infantis* produced very little degradation of FOS, however these bacteria showed degradation, mainly of F-F_n series, using Raftilose P95 (from Orafti). Corradini et al. (2004) optimised a gradient elution program using water, 0.6 M sodium hydroxide and 0.5 M sodium acetate as eluents to selectively separate glucose, fructose, sucrose and fructans with DP from 3 to 60 in microbial cultures, obtaining good resolution during the whole chromatogram. This method allowed the evaluation of FOS and inulin consumption by pure cultures of *Bifidobacterium* spp. and by fecal cultures.

Galactooligosaccharides

GOS are produced by transgalactosylation reactions catalyzed by β-galactosidases using mainly lactose as substrate. As consequence of these reactions, a large variety of structures can be obtained (different glycosidic bonds and

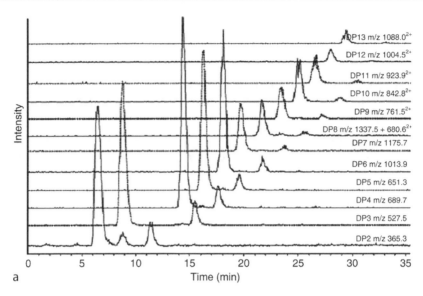

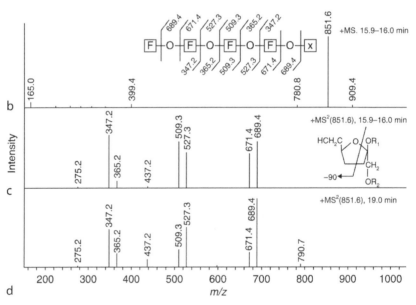

■ Figure 13.1

HPAEC-on-line-MS analysis of FOS. Extracted ion chromatograms for fructan oligosaccharides of various degrees of polymerization (DP) detected as sodium adducts by capillary HPAEC-on-line-MS (a). Mass spectra of the two isobaric sodium adducts of DP5 fructans: Part (b) shows the MS spectrum of [GF$_4$ + Na]+; parts (c) and (d) are the MS2 spectra with m/z 851.6 as precursor ion, where (c) represents GF$_4$ and (d) F$_5$. In the fragmentation scheme, F stands for fructofuranosyl and X is glucopyranosyl or fructopyranosyl, R$_1$ and R$_2$ stand for the rest part of the oligosaccharide chain and R$_2$ can also be a H. From Bruggink et al. (2005a) with permission from Elsevier.

oligosaccharides of different molecular weights are formed depending on the enzymatic source and the reaction conditions used). Therefore, it is necessary to use high resolution methods to distinguish between structural isomers of oligosaccharides consisting of monosaccharides linked together in various anomeric and positional configurations.

Splechtna et al. (2006) analyzed the composition of GOS (mainly mono-, di- and trisaccharides) by HPAEC-PAD using a Carbopac PA-1 column, although identification of all of the structures could not be completely achieved. Moreover, coelution of some carbohydrates (i.e., glucose and galactose; lactose and allolactose) was observed. These coelution problems have been avoided using a modified method with a CarboPac PA-10 column (Cardelle-Cobas et al., 2008; Martínez-Villaluenga et al., 2008a). Goulas et al. (2007) used a cation exchange column (Ca^{2+}) at 85°C with water as mobile phase and RI detector to determine the synthesis and purification of GOS. Under these conditions, two not well-resolved chromatographic peaks were obtained for oligosaccharides of DP higher than three, whereas disaccharides eluted as one peak and glucose and galactose appeared as separated peaks. HPAEC-PAD with a CarboPac PA1 allowed the separation of the disaccharides obtained in these samples.

◉ *Figure 13.2* shows an HPAEC-PAD profile of GOS before (◉ *Figure 13.2a*) and after (◉ *Figure 13.2b*) removal of mono- and disaccharides using activated charcoal (Sanz et al., 2007). The use of a CarboPac PA-100 column allowed the separation of oligosaccharides up to DP7, although a complete resolution of these carbohydrates was not achieved. The lack of standards was the main disadvantage for identification and quantification purposes, necessitating the isolation of oligosaccharides followed by ESI-MS analysis to determine their molecular weights.

Other Prebiotics

Several methods have been also developed for the analysis of other oligosaccharides such as xylooligosaccharides (XOS; Ohara et al., 2006), soybean oligosaccharides (Giannoccaro et al., 2008), lactulose (Paseephol et al., 2008) or glucooligosaccharides (Rousseaua et al., 2005), some of them only tentatively considered as prebiotics. Giannoccaro et al. (2008) have compared two methods using HPSEC-RI (using two analytical Shodex OHpak SB 802HQ columns) and HPAEC-PAD (CarboPac PA10) to analyze soybean sugars. Although both systems gave reproducible results, HPAEC-PAD was more sensitive, faster, and with higher resolution than the HPSEC-RI method.

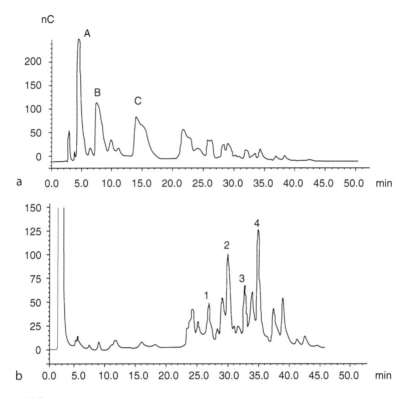

☑ Figure 13.2
HPAEC-PAD analysis of GOS. HPAEC-PAD profiles of: (a) GOS, (b) GOS previously treated with activated charcoal. (A) glucose and galactose, (B) lactose, and (C) unknown disaccharides. From Sanz et al. (2007) with permission from ACS publications.

GCC have recently been used for the separation of carbohydrates in a wide range of applications including coupling to ESI MS. Robust and rapid separations were achieved with these methods. Separation of human milk oligosaccharides has been carried out using on-chip GCC (Ninonuevo et al., 2005) coupled to a MS detector with o-ToF which allows isomeric detection. Consumption of these oligosaccharides by intestinal bacteria has also been evaluated by Ninonuevo et al. (2007) using GCC and UV detection (206 nm) obtaining different chromatographic profiles for the bacteria studied.

Recently, the use of HPLC-ESI MS has allowed the identification of 19 peptides glycated with GOS from 2 to 7 hexose units which prebiotic potential is being studied (Moreno et al., 2008).

13.2.4   Gas Chromatography

Since its first application to carbohydrate analysis by Langer et al. (1958), GC has seen widespread use for sugar determination as it is a relatively cheap, simple and powerful analytical technique. Higher oligosaccharides in foods and diets are often present at low concentrations, thus, the high resolving power, sensitivity and selectivity of GC results is extremely advantageous. The potential of this technique for carbohydrate determination was achieved with the development of capillary columns and their coupling to mass spectrometric detectors; identification and quantification of many prebiotic oligosaccharides as well as structural studies can be performed.

13.2.4.1   Sample Preparation

Purification/Fractionation

As oligosaccharides usually appear in complex matrices, a purification step is required before their analysis as with HPLC determinations. This procedure is often carried out to discard insoluble material, lipids and proteins, desalt the sample or remove impurities.

Soluble carbohydrates in foods are usually extracted with ethanolic or methanolic solutions: oligosaccharides up to DP6 are easily soluble in these solvents while other interfering substances are not, being discarded by filtration or centrifugation. Although these methods are still in use on standard protocols and regulations, other modern procedures such as membrane filtration have been introduced for more complex mixtures.

In those cases where the study of a specific carbohydrate or a group of carbohydrates is required, a fractionation step can be also necessary. This procedure provides an enrichment of the samples in carbohydrates and purifies them before their chromatographic analysis. As an example, nanofiltration, yeast (*Saccharomyces cerevisiae*) treatment, and adsorption onto activated charcoal were used by Sanz et al. (2005) prior to GC analysis in order to remove honey monosaccharides and study the potential prebiotic effect of its oligosaccharides.

Recent methods for the selective extraction of lactulose from a mixture with lactose have been developed by accelerated solvent extraction (ASE; Ruiz-Matute et al., 2007) and supercritical fluid extraction (SFE; Montañés et al., 2007), which allowed the obtainment of a high purity lactulose fraction, using rapid processes with low solvent consumption.

Derivatization

Due to the polar nature of carbohydrates, a derivatization step previous to GC analysis is required. Classical methods are based on the substitution of the polar groups in order to increase their volatility. Acetates, methyl ethers, trifluoroacetates and trimethylsilyl ethers have been the most common derivatives used for carbohydrate determination (Knapp, 1979). Among them, trimethylsilyl derivatives are the most popular, since they present good volatility and stability characteristics. Trimethylsilylation has been recognized as a quantitative rapid derivatization method for a wide range of carbohydrates and related compounds including aldoses, ketoses, aminosugars, alditols, inositols as well as oligosaccharides up to DP4 (Brobst and Lott, 1966; Sweeley et al., 1963).

However, these derivatives give multiple peaks corresponding to the different anomeric forms of carbohydrates. Even though high-resolution capillary columns can adequately resolve complex mixtures, multiple peaks may cause interferences for qualitative identification and quantitative measurement.

Sometimes, the multiple peaks obtained are not considered a disadvantage as they serve as a "fingerprint" for each sugar, aiding their identification. Nevertheless, other derivatives are preferred for analyses of mixtures containing many sugars. Usually the anomeric carbon is modified in order to reduce this effect. Several attempts have been made for this purpose, among them, some possibilities are to: (1) convert the free carbonyl group into an oxime using hydroxylamine chlorohydrate or to an O-methyloxime using O-methylhydroxylamine chlorohydrate; (2) reduce the aldehyde with sodium borohydride to the corresponding alditol or (3) convert the aldehyde into an oxime and then dehydrate into a nitrile.

Trimethylsilyl oximes can be easily obtained by a two step derivatization procedure (oximation and silylation). They have been widely used for the GC analysis of many oligosaccharides since they produce only two peaks corresponding to the syn (E) and anti (Z) forms for reducing sugars, and only one peak for non reducing carbohydrates, the derivatives formed having satisfactory GC properties (Molnar-Perl and Horváth, 1997; Sanz et al., 2002).

Alditol acetate derivatives have also been used for sugar GC analysis due to their stability and the simplicity of the resulting chromatograms. The reduction of aldoses to alditols and their conversion to alditol acetates simplifies the chromatograms by producing only one peak for each aldose. Abazia et al. (2003) used these derivatives for the simultaneous GC measurement of lactulose and other sugars in urine. Sugars were reduced with sodium borohydride and acetic acid/methanol 1:9 (v/v) was added to remove the boric acid. Then acetylation was performed by the addition of dry pyridine and acetic anhydride. Although these derivatives show a high chemical stability and low cost of reagents, they also present some disadvantages. On reduction, ketoses yield a mixture of two sugar alcohols (i.e., fructose produces glucitol and mannitol) and thus, give two chromatographic peaks. In some cases, a significant loss of information may occur in the reduction step as some aldoses and ketoses produce the same alditol and cannot be differentiated (i.e., glucose and fructose both produce glucitol). Although significant improvements have recently been made simplifying the derivatization procedure, some common versions of this method still require tedious evaporations to remove the borate before acetylation. The alditol acetate derivatization has been widely used for monomer analysis of macromolecules by GC (Fox et al., 1989).

An alternative to eliminate the anomeric center is the conversion of sugars to their aldononitrile acetates derivatives. They give a unique peak for every sugar but they cannot be applied for ketoses (Ye et al., 2006). These derivatives have been applied for structural analysis of gums and food samples (McGinnis and Biermann, 1989).

Structural Analysis

Structural analysis of complex carbohydrates requires the characterization of monomer composition and anomeric configuration, as well as the determination of the sequence of monosaccharide residues, branch position, functional groups and glycosidic linkages. The elucidation of structural chemistry of complex carbohydrates requires sophisticated instrumentation such as mass spectrometry (MS) or nuclear magnetic resonance (NMR), but the additional information that GC-MS data provides is essential for carbohydrate characterization.

GC-MS has been applied for either the determination of composition and sequence of oligosaccharides released by partial depolymerization after being converted into proper volatile derivatives or for sugar monomer analysis after complete hydrolysis and derivatization. Partial degradation of polysaccharides to oligosaccharides is achieved by means of enzymatic or mild acidic hydrolysis. The use of acids implies the optimization of conditions to achieve maximum cleavage to oligosaccharides and minimum decomposition of the liberated mono- and/or oligosaccharides.

Methylation analysis is the most widely used method for determining linkage structure of prebiotic oligosaccharides by GC. It basically consists of the following steps: Firstly, the free hydroxyl groups of polymerized sugars are completely methylated, forming their correspondent methyl ethers. Then, hydrolysis of the polymer is performed releasing the free hydroxyl groups in those places in which previously there were glycosidic linkages. Finally, these hydroxyl groups are converted into more volatile compounds, the most common derivatives being alditol or, aldonononitrile acetates. These samples are analyzed by GC-MS in order to ascertain the original linkages and to obtain quantitative linkage information on complex polysaccharides.

The major drawback of standard methylation analysis is that in certain cases the carbon involved in the cyclic hemiacetal of the monosaccharide is not distinguished from linked positions after hydrolysis of the permethylated poly-saccharide. As an alternative, the reductive-cleavage method, which yields partially methylated anhydroalditols while retaining the ring structure, has been successfully used to investigate the structure of different prebiotic oligosaccharides. Rolf and Gray (1984) studied the suitability of the reductive-cleavage method for the study of the linkage positions in D-fructofuranosyl residues of D-fructans of different sources. The same derivatization method was used by Stumm and Baltes (1997) for the structural determination of polydextrose. Previous to methylation, ultrafiltration was applied in order to yield fractions free from monomeric residues. Carbohydrate analysis was performed by GC and GC-MS using both electronic impact (EI) and chemical ionization (CI). EI MS was used to confirm the identity of the carbohydrate derivatives while CI MS with ammonia as reagent gas was useful for the determination of the molecular weight. This method resulted to be useful for the elucidation of the degree of branching, position of the linkages and the type of monomeric compounds involved in the studied samples.

Many novel prebiotic oligosaccharides synthesized either by the use of microbial cells or enzymes have been also characterized by methylation analysis. The determination of carbohydrate structures of neofructo-oligosaccharides produced by *P. citrinum* (Hayashi et al., 2000), oligosaccharides formed by a fructosyltransferase purified from asparagus (Yamamori et al., 2002), and oligosaccharides synthesized by glucosyl transfer from β-D-glucose-1-phosphate to raffinose and stachyose using *T. brockii* kojibiose phosphorylase (Okada et al., 2003), are some examples of the application of methylation analysis.

13.2.4.2   Columns and Stationary Phases

The most significant improvement in GC separation was achieved with the advances in capillary column technology. Although many separations of prebiotic oligosaccharides have been carried out on packed columns (Farhadi et al., 2003; Karoutis et al., 1992), the use of capillary columns involves an increment on resolution while the analysis time is decreased. The most common liquid stationary phases used for carbohydrate analysis by GC are those based on polysiloxanes (also called "silicones") since they present good thermal stability and high permeability towards solutes. A wide polarity range can be found and depends on the percentage of polar phenyl or cyanopropyl groups in the siloxane chain, the most apolar being 100% methylsilicone while 100% cyanopropyl silicone is the most polar. For high-temperature separations, phases based on a carborane skeleton have been proposed (Joye and Hoebergs, 2000).

Dimensions of capillary columns used for carbohydrate GC analysis can vary in the range of: 1–50 m (length); 0.1–0.5 mm (diameter), and 0.02–2 μm (df).

It has been demonstrated that oligosaccharides with up to 11 monosaccharide units can be analyzed using capillary columns with ultrathin films (<0.05 μm) of thermostable bonded stationary phases at high temperatures (Carlsson et al., 1992).

13.2.4.3   Chromatographic Conditions

As carbohydrates usually appear as a mixture of mono- and oligosaccharides, programed temperature is convenient during the chromatographic run, so that each compound can be analyzed under adequate conditions. The temperature commonly used for carbohydrate analysis range from 60 to 330°C.

GC and high temperature gas chromatography (HT-GC) have been used for carbohydrate analysis by many authors (Joye and Hoebergs, 2000; Karoutis et al., 1992; Montilla et al., 2006). The main differences between these techniques are their operating conditions: GC uses temperatures below 360–370°C whereas HT-GC works above them, but requires special apparatus, high temperature capillary columns and heat resistant fittings for the analysis. Carlsson et al. (1992) developed a HT-GC method for the determination of high oligosaccharides: the temperature program was increased from 40 to 400°C with a linear temperature program of 20°C/min. A conventional GC procedure for the determination of high DP oligosaccharides was developed by Montilla et al. (2006). A maximum temperature program of 360°C allowed the determination of oligosaccharides up to DP 7 in different foodstuff.

High carrier gas velocities and columns coated with thin layers of stationary phase can help to minimize the time spend through the column and, consequently, reduce analysis time. The use of helium as a carrier gas further decreases analysis time without any substantial loss in column efficiency.

13.2.4.4   Detectors

Flame ionization detection (FID) is the most frequently used for GC analysis of carbohydrates since it possesses good sensitivity for organic products. However, the identification by GC always requires the use of standard compounds, and only few of them are commercially available. The coupling of a MS detector to a gas chromatograph contributes to the identification and quantification of carbohydrates, especially in those cases where complex mixtures of the same DP have to be determined.

Most of GC–MS applications utilize capillary GC with Q MS and EI ionization due to its relatively low cost, high sensitivity, high information content, and the ready availability of commercial instruments. CI with ammonia as reagent gas has also been used in order to confirm the molecular weight in methylation analysis (Stumm and Baltes, 1997).

Certain carbohydrates with the same DP and only differing in the position of their hydroxyl groups show similar mass spectra. In those cases, the combination of retention information provided by GC analysis and spectral data from MS is necessary for carbohydrate identification, although it is not always enough for a proper characterization.

Spectral information of different carbohydrate derivatives such as alditol acetates (Fox et al., 1989), trimethylsilyl oximes (Molnárl-Perl and Horváth, 1997; Sanz et al., 2002), trimethylsilyl O-methyloximes (Laine and Sweeley, 1973) and methyl acetates (Stumm and Baltes, 1997) can be found in literature.

13.2.4.5   Applications

GC applications for prebiotic carbohydrate analysis have been mainly focused on both direct analysis of oligosaccharides with a previous derivatization procedure and methylation analysis for their structural determination.

Many references can be found about the analysis of lactulose by gas chromatography. Lactulose is a disaccharide obtained through lactose isomerization (through Lobry de Bruyn-Alberda Van Ekenstein rearrangement). It is not digested in the small intestine and is fermented by the colonic flora. Some examples are summarized in ❯ Table 13.3. Farhadi et al. (2003) developed two gas chromatographic methods using a packed and a capillary column respectively, for the simultaneous quantitation of urinary lactulose and other carbohydrates (sucrose and mannitol). Columns and chromatographic conditions are summarized in ⊙ *Table 13.3*. The capillary method was more sensitive, accurate and reproducible for lactulose determination. Moreover, it permitted the use of smaller volumes of urine in the analysis and did not require pretreatment of the samples.

A method was developed for the determination of lactulose in milk (Montilla et al., 2005). This method gave good chromatographic resolution, as well as precise and reproducible results when applied to commercial milk samples submitted to heat treatments of different intensity. In addition, this method provided a good separation among sucrose, lactulose and lactose peaks which allowed the suitable quantification of lactulose in samples containing a high concentration of sucrose.

The degree of polymerization as well as the presence of branches are important in inulin, since they affect to their functionality. Short chain inulin carbohydrates should be separated from their long chain analogues for prebiotic uses. Lopez-Molina et al. (2005) characterized artichoke inulin and demonstrated its health-promoting prebiotic effects. The extraction of artichoke inulin involved several physical steps (see ⊙ *Table 13.3*); hydrolysis of inulin was also performed. GC-MS analysis of their trimethylsilyl derivatives confirmed that fructose was the main monosaccharide unit in artichoke inulin and its degradation by inulinase indicated that it contained the expected β-2,1-fructan bonds.

Packed column GC was used by Sosulski et al. (1982) for the analysis of oligosaccharides in legumes but long analysis time and poor reproducibility for larger oligosaccharides were the major drawbacks of the method. Karoutis et al. (1992) optimized a methodology by GC for the analysis of raffinose, stachyose and verbascose. Different analytical parameters were assayed: carrier flow-rate, split ratio and nature of derivatization agent (trimethylimidazole or N-methyl-bis (trifluoroacetamide).

Joye and Hoebregs (2000) developed a method for the quantitative determination of oligofructose in foods. The use of high temperature chromatography with an Al-clad capillary column and oven temperatures up to 440°C allowed the determination of carbohydrates up to DP9 in complex matrices in only one chromatographic run. ⊙ *Figures 13.3a* and ⊙ *13.3b* show the chromatogram obtained for Raftilose P95 X (Orafti) and an enzymatically synthesized FOS (Actilight®). Malto-, isomalto- and galactooligosaccharides were also analyzed by this method to exclude possible interferences from other sugar compounds. ⊙ *Figure 13.3c* shows the GC profile of GOS as an example. This method was very accurate and reproducible for the study of these carbohydrates.

■ Table 13.3

GC applications for the analysis of prebiotics (Cont'd p. 496)

Carbohydrates	Derivatives	Pretreatment	Column and dimensions	Cromatographic conditions	References
Lactulose and other sugars in urine	Alditol acetates	Urine purification with Dowex mixed bed resin	ZB-1 capillary column (30 m × 0.25 mm ID)	T programme: 230–300°C	Abazia et al. (2003)
				Carrier gas: Nitrogen	
Lactulose and other sugars in urine	TMSO	Centrifugation and filtration	Glass column packed with 3% SE-30 on 80/100 chromosorb WHP (6-ft × 2 mm ID)	T programme: 220–274°C	Farhadi et al. (2003)
				Carrier gas: Nitrogen	
		None	DB1 capillary column (15 m × 0.53 mm ID × 1.5 μm)	T programme: 220–274°C	
				Carrier gas: Helium	
Lactulose from a sugar mixture in urine	TMSO	None	DB-1701 capillary column (30 m × 0.25 mm ID × 0.25 μm)	T programme: 180–250°C	Farhadi et al. (2006)
				Carrier gas: Helium	
Lactulose in dairy products	TMS	Methanol to remove proteins and fats	SPB-17 capillary column (30 m × 0.32 mm ID × 0.25 μm)	T programme: 235–270°C	Montilla et al. (2005)
				Carrier gas: Nitrogen	
Lactulose from a mixture with lactose	TMS	PLE extraction	SPB-17 capillary column (30 m × 0.25 mm ID × 0.25 μm)	T programme: 250–270°C	Ruiz–Matute et al. (2007)
				Carrier gas: Nitrogen	

	Derivatization	Extraction/clean-up	Column	T programme / Carrier gas	Reference
Inulin from artichoke	Complete hydrolysis with formic acid and TMS	Aqueous extraction, ultrafiltration, precipitation by ionic-exchange chromatography, low temperature precipitation, centrifugation and lyophilization	HP-5MS capillary column (30 m × 0.25 mm ID)	T programme: 250–280°C; Carrier gas: Helium	Lopez-Molina et al. (2005)
FOS, GOS, malto- and isomaltooligosaccharides in food products	TMSO	Extraction with water Foods containing fats: hexane and centrifugation	Al-clad capillary column coated with 5% phenyl polycarborane-siloxane (6 m × 0.25 mm ID)	T programme: 105–440°C; Carrier gas: Helium	Joye and Hoebregs (2000)
Oligosaccharides (DP up to 7) in foods (FOS and GOS)	TMSO	Dilution with methanol and centrifugation	CP-SIL 5CB capillary column (8 m × 0.25 mm ID × 0.25 μm)	T programme: 130–360°C; Carrier gas: Nitrogen	Montilla et al. (2006)
		Diafiltration of FOS	HT5 capillary column (12 m × 0.32 mm ID × 0.1 μm)	T programme: 130–440°C; Carrier gas: Nitrogen	
Soybean oligosaccharides (raffinose, stachyose, verbascose and maltooligosaccharides)	Methyl alditols	Water extraction and clean up with chloroform/ethanol	GC: HT SE-54 capillary column (25 m × 0.32 mm ID × 0.05 μm)	T program (GC): 40–400°C; T program (GC-MS): 70–390°C	Carlsson et al. (1992)
			GC-MS: PS264 (10 m × 0.25 mm ID × 0.02 μm)	Carrier gas: Helium	

13

Table 13.3

Carbohydrates	Derivatives	Pretreatment	Column and dimensions	Cromatographic conditions	References
Pea oligosaccharides (raffinose, stachyose, verbascose)	Trifluoroacetates and TMS	80% aqueous methanol and membrane filtration	DB5–60W capillary column (10 m × 0.32 mm ID × 0.25 μm)	T programme: 188–316°C and 80–250°C; Split ratio: 1:50,1:100 and 1:150; Carrier gas: Helium	Karoutis et al. (1992)
Polydextrose	Methyl alditols	Ultrafiltration	SA-5 capillary column coated with 5% diphenyl-95% dimethylpolysiloxane (30 m × 0.25 mm ID)	EIMS: to prove the identity of the carbohydrate derivatives; NH_3-CIMS: determination of the molecular weight	Stumm and Baltes (1997)
D-Fructofuranosyl residues of D-Fructans of different sources	Permethylated derivatives		10% SP2401 (1.83 m × 3.18 mm); (3.66 m × 3.18 mm); 3% OV-225 (2.68 m × 3.18 mm); SE-30 (1.83 m × 3.18 mm)	Different temperatures	Rolf and Gray (1984)
Honey oligosaccharides	TMSO	Nanofiltration, yeast (Saccharomyces cerevisiae) treatment, and Adsorption onto activated charcoal	DB1 capillary column (25 m × 0.25 mm ID × 0.25 μm)	T programme: 200–300°C; Carrier gas: Nitrogen	Sanz et al. (2005)

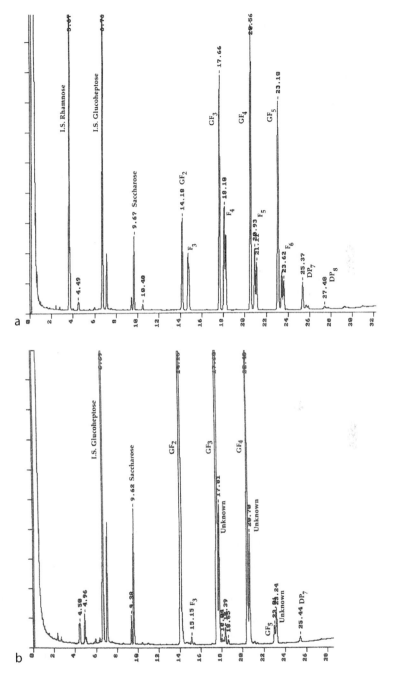

◘ Figure 13.3 (Continued)

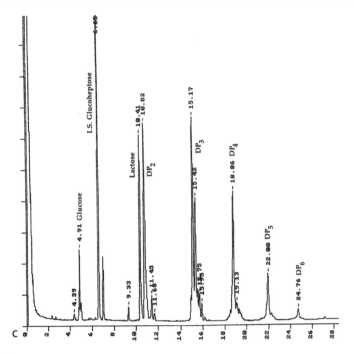

■ Figure 13.3

GC analysis of FOS and GOS. GC profile of Raftilose P95 X (a), Actilight (b) and galacto-oligosaccharides (c). From Joye and Hoebregs (2000) with permission from AOAC International.

Oligosaccharides up to DP7 were determined in foods and in various pure mixtures using conventional GC (Montilla et al., 2006). The accuracy, repeatability and reproducibility of the method were similar to the results obtained with HT-GC method from Joye and Hoebregs (2000).

Carlsson et al. (1992) developed an HT-GC methodology for the quantitative analysis of oligosaccharides in foods, diets and intestinal contents. Methylation was also performed for the identification of these oligosaccharides by GC-MS, which was able to analyze sugars up to 12 sugar units. Red lentils, soybeans, rapeseed, mung beans and chickpeas were found to contain considerable amounts of the raffinose family of oligosaccharides.

Multiple applications can be found for the structural analysis of oligosaccharides. As an example, the studies about structural determination of D-fructans (Rolf and Gray, 1984) and different oligosaccharides produced from alternansucrases (Côté and Sheng, 2006) can be pointed out. The potential prebiotic properties of these last oligosaccharides are still under study (Sanz et al., 2006).

13.2.5   Capillary Electrophoresis

The high speed of analysis, the minute amounts of analyte required and the high resolution make CE an attractive and powerful microanalytical technique to separate a wide range of charged and uncharged compounds. It is a suitable analytical tool for the analysis of foods and beverages and also has been successfully applied in other fields, such as biochemistry, biotechnology and clinical chemistry (Soga and Serwe, 2000). The advantages of CE over other traditional chromatographic methods include the extremely simple operation and the low consumption of sample and buffers (Bao and Newburg, 2008). The main drawback is the lack of sensitivity when low concentration levels are present.

13.2.5.1   Operation Modes

CE instrumentation includes a high voltage power supply (5–30 kV), buffer reservoirs, a narrow-diameter (50–100 μm) capillary, an automated sampler and detector. Separation is based on migration of compounds in narrow capillaries (length of 0.5–1.5 m) made of fused silica. The two ends of the capillary are immersed in two separated electrolyte reservoirs containing a high voltage electrode.

The mobility of analytes under an electric field depends on several factors, including the analyte charge (neutral, positive or negative), charge to mass ratio, buffer system (pH and ionic strength), presence of buffer additives (surfactants, ion-pairing agents, complexing agents), voltage applied, temperature inside of capillary, length and diameter of the capillary, and nature of the capillary wall (Bao and Newburg, 2008).

Samples can be directly analyzed with minimal sample preparation without a loss of separation performance. After detection of peaks, the hollow capillary is flushed with fresh buffer and is ready for the next injection (Soga and Serwe, 2000).

There are different operation modes of CE: capillary zone electrophoresis (CZE), capillary gel electrophoresis (CGE), micellar electrokinetic chromatography (MEKC), capillary isolectric focusing (CIEF) and capillary isotachophoresis (CITP). In all cases, the separation is achieved due to differences in migration of different solutes, on chosen electrolyte and capillary tube, under an applied electric field.

The most commonly used modes in the analysis of carbohydrates are CZE and MEKC. CZE is also called free solution electrophoresis, and the separation is based on the differences in the charge-to-mass ratio. MEKC is a pseudochromatographic mode of CE whereby separation and analysis of neutral molecules occurs through inclusion into micelle-forming detergents added to the electrophoretic medium (Cheung et al., 2007).

CE has emerged as an alternative to current analytical techniques for carbohydrates. However, carbohydrates analysis can present difficulties such as lack of electric charge and absence of chromophoric/fluorophoric groups in the analyte molecules. To overcome these limitations different procedures have been developed.

Taking into account that sugars are very weak acids (have high ionization constant, pK_a values of 12), different methodologies have been established to ionise them. Electrolyte systems based on borate complexation, metal complexation or highly alkaline pH (e.g., NaOH) have been used. The complexation of carbohydrates with borate-based electrolytes to impart the necessary charge for electrophoresis is the most widely used approach for CE separation of derivatized and underivatized sugars (E-l Rassi, 1999).

13.2.5.2   Detection

Derivatized Carbohydrates

Carbohydrates have high ionization constants (pK_a values of 12 or higher), and therefore they do not carry electrical charges at neutral pH. This, along with the fact that carbohydrates do not absorb UV light above 200 nm hinders its analysis by CE. To overcome both problems different procedures have been developed, such as derivatization with direct detection, using chromophore and fluorescent probes carriers of electrical charges, to facilitate detection via UV-VIS absorbance or laser-induced fluorescence (LIF). However, although derivatization methods lead to improve sensitivity and resolution, several drawbacks are often encountered, such as control problems due to a different reactivity of derivatizing reagents for analytes, formation of several adducts, etc. (Lee and Lin, 1996).

Similarly to HPLC, derivatization of reducing carbohydrates in CE is often performed by reductive amination, between the reducing end and an amino group of the tag reagent, using amines with strong chromophores or fluorophores such as 4-amino benzoic acid and its ethyl ester, 2-aminobenzoic acid, 4-aminobenzonitrile, 2-aminopyridine, etc. (Andersen et al., 2003). A large variety of derivatization reagents have been suggested for carbohydrate analysis (Campa et al., 2006; Cortacero-Ramirez et al., 2004).

Underivatized Carbohydrates

Indirect Detection An alternative methodology that allows CE analysis of underivatized carbohydrates includes the use of highly alkaline electrolytes, to ionize and ensure an electrophoretic mobility of the saccharides towards the anode and make them suitable for indirect UV detection using chromophore compounds and modifiers (background electrolyte; BGE) to reverse the direction of electroosmotic flow inside the capillary and promote the co-migration of the analytes (Jager et al., 2007; Soga and Serwe, 2000).

Electrochemical Detection Ionization of carbohydrates at high pH values also allows CE analysis with electrochemical detection using copper or gold electrodes. It is an interesting approach because the hydroxyl groups can be partially ionized, which in turn permits their effective separation in CZE mode. Cao et al. (2004) described a simple, reliable and reproducible CE method using NaOH (50 mmol/L) as running buffer and amperometric detection to quantify mono- and disaccharides in rice flour. The detector is composed by an electrode cell system consisting of copper working electrodes, platinum auxiliary and reference electrodes. Likewise, Chu et al. (2005) developed a miniaturized CE method and amperometric detection, with an electrode of copper, to quantify carbohydrates in soft drinks.

Complexation of alternate hydroxyl groups with borate and electrochemical and amperometric detection is another alternative for CE analysis of carbohydrates without derivatization (Cheung et al., 2007).

Refractive Index Detection Refractive index detection has been successfully used in CE to separate carbohydrates considering that they do not possess chromophore groups in their structures. Although the detection limits are relatively low, this type of detector could be used as universal detector in CE of carbohydrates (El Rassi, 1999).

13.2.5.3 Coupling with MS

The coupling of CE–MS can provide important advantages in food analysis because of the combination of the high separation capabilities of CE and the power of MS as identification and confirmation method (Simó et al., 2005). A detailed review has been written by Campa et al. (2006) which describes the advances in CE-MS of carbohydrates.

13.2.5.4   Applications

Besides the difficulties discussed above with the analysis of carbohydrates by CE, it is also necessary to consider the different structures resulting from heterogeneity in primary sequence, branching and the variety of structural isoforms of oligosaccharides. In order to address these challenges several studies have been focused on finding different separation methods. In this section a description of different CE methodologies found in the literature to analyze oligosaccharides is presented.

Human Milk Oligosaccharides (HMOS)

The difficulty of HMOS analysis is not only due to the low and variable (depending on lactational stage) content but also to the complexity of their structures. Therefore, it is necessary to arrange appropriate and powerful analytical methods to achieve efficient separation and quantification of oligosaccharides and their structural isomers.

Acidic oligosaccharides are not intrinsically strong chromophores, however they can absorb in the low UV range due to the aminoacyl moieties and sialic acid present in the structures of many HMOS (Bao and Newburg, 2008).

Underivatized HMOS

Shen et al. (2000) developed a very reproducible and sensitive CE method (fmol level) for underivatized acidic oligosaccharides with detection by UV absorbance at 205 nm. Eleven oligosaccharides of human milk, ranging from tri- to nonasaccharide (3'-sialyllactose, 6'-sialyllactose, 3'-sialyllactosamine, 6'-sialyllactosamine, disialyltetraose, 3'-sialyl-3-fucosyllactose, etc.) were resolved by MEKC. These oligosaccharides were detected in pooled human milk samples, from different donors, and comparison of oligosaccharides profiles revealed an extensive variation in the structural isomers of sialyllacto-N-tetraose. The running conditions were selected as the best compromise between resolution and running time.

The resolution of structurally similar oligosaccharides, especially those containing chemically labile sialic acid residues is a challenging problem. Shen et al. (2001) employed a CE method and UV detection (205 nm) to separate three sets of structural isomers of sialylated oligosaccharides in human milk and bovine colostrum. They developed conditions for baseline resolution of specific sets of isomers within a 35 min run. Each set of structural isomers of sialylated oligosaccharides, 3'-silayllactose/6'sialyllactose, sialyllacto-N-tetraose-a (linear), -b (branched) and -c (linear), required a unique running buffer with respect to buffer type, concentration, pH, presence of organic modifiers, and surfactants.

Likewise, Bao et al. (2007) developed a novel method to quantify sialyloligosaccharides from human milk by MEKC and UV detection at 205 nm. As running buffer, they used aqueous 200 mM sodium phosphate (pH 7.05) containing 100 mM sodium dodecyl sulfate (SDS) mixed with 4% (v/v) methanol. The method describes new CE conditions that simultaneously resolve not only separation between pairs of structural isomers of HMOS, 3'-sialyllactose/6'sialyllactose and sialyllacto-N-tetraose-a, -b and -c, but also quantification of the 12 major sialyloligosaccharides of human milk in a single 35 min run. ◉ *Figure 13.4* shows a CE separation of sialyloligosaccharide standards and HMOS found in colostrum (b) and in human milk (c). The method allowed finding differences in sialyloligosaccharide concentrations between less and more mature milk from same donors. It is possible to define acidic oligosaccharide expression in milk as function of stage of lactation, genetic variation among lactating mothers, diet, diurnal variation, stress, disease, and geographic origins of a population.

Derivatized HMOS

Major neutral oligosaccharides from human milk, such as 2'-fucosyllactose (2'FL), 3'-fucosyllactose (3'FL), lacto-N-tetraose (LNT), lacto-N-fucopentaose I (LNFP I), lacto-N-fucopentaose II (LNFP II) and fucose, have been quantified by CE using LIF as detection system (λ_{exc} = 488 nm; λ_{em} = 520 nm). Oligosaccharides were derivatized via reductive amination with 2-aminoacridone (AMAC). The CE method allowed to resolve two sets of structural isomers, 2'FL/3'FL and LNFP I/LNFP II. This was rapid, sensitive (2 fmol) and reproducible, and required a simple sample preparation (Song et al., 2002).

Schmid et al. (2002) analyzed using MEKC free oligosaccharides from human milk. They used as derivatization agent various esters of aminobenzoic acid and sodium phosphate (20 mM), pH 7.0, and sodium dodecyl sulfate (SDS) (50 mM) as buffer. Oligosaccharide was detected by UV absorbance at 285 and 310 nm. Previous to their analysis, samples were submitted to a simple clean up of deproteination and defatting before derivatization of oligosaccharides. Major human milk oligosaccharides were detected (FL, LNT, LNFP, lacto-N-difucohexaose).

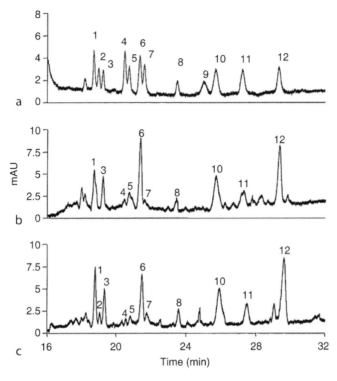

■ Figure 13.4

CE analysis of sialyloligosaccharides, colostrum sample and mature milk sample. Electro-phoregrams of sialyloligosaccharides from a mixture of 12 standard sialyloligosaccharides (a), a colostrum sample (b) and a mature milk sample (c) from a mother in the Boston area. (1) MSMFLNnH (monosialyl, monofucosyllacto-N-neohexaose; (2) MSLNnH I (mono-sialyllacto-N-neohexaose); (3) MFMSLNH I (monofucosyl, monosialyllactose-N-hexaose); (4) SLNFP II (sialyllacto-N-fucopentaose); (5) SLNT b (sialyllacto-N-tetraose); (6) SLNT c; (7)SLNT a; (8) DSMFLNH (disialyl, monofucosyllacto-N-hexaose); (9) 3′-S-3FL (3′ sialyl-3-fucosyllactose); (10) -6′-SL (6′ sialyllactose); (11) 3′-SL (3′ sialyllactose); (12), DSLNT (disialyllacto-N-tetraose). From Bao et al. (2007) with permission from Elsevier.

Galactooligosaccharides

There is scarce literature on CE methods to analyze GOS. Petzelbauer et al. (2006) separated and quantified the major GOS obtained during lactose conversion at 70°C, catalyzed by β-galactosidases from the archea *Sulfolobus solfataricus* and *Pyrocccus furiosus*. Carbohydrates were analyzed using as running buffer phosphate pH 2.5, derivatized using an aminopyridine solution and detected by UV (240 nm). The authors identified two disaccharides β-D-Galp-(1→3)-D-Glc and β-D-Galp-(1→6)-D-Glc (allolactose); and two trisaccharides β-D-Gal*p*-(1→3)-lactose and β-D-Gal*p*-(1→6)-D-lactose. As minor compound, β-D-Gal*p*-(1→6)-D-Gal was also identified.

Total GOS and di-, tri-, and tetrasaccharides, derived from lactose hydrolysis with β-galactosidases of *Lactobacillus reuteri* L103 and L461, have been also quantified by CE (Splechtna et al., 2006). The method includes a pre-column derivatization with 2-aminopyridine and detection of derivatives by UV-diode-array detector (DAD). The running buffer was 100 mM phosphoric acid, pH 2.5. The sugars eluted in groups depending on their degree of polymerization and the identified GOS were the same found by Petzelbauer et al. (2006); ◗ *Figure 13.5*

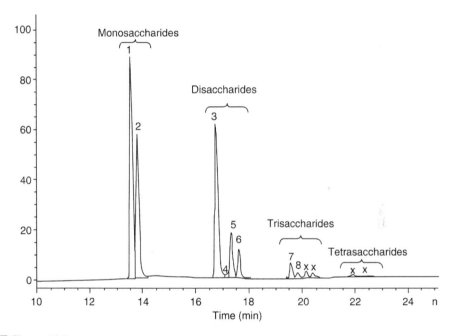

◘ Figure 13.5

CE analysis of GOS. Separation and quantification by capillary electrophoresis of individual GOS produced during the lactose conversion catalyzed by L103 or L461 β-galactosidase. The sample presents a mixture of sugars obtained after the reaction of L103 β-Gal with 205 g/L lactose. The extent of substrate conversion is approximately 67%. The identified compounds are indicated: (1) glucose, (2) galactose, (3) lactose, (4) D-Gal*p*-(1→3)-D-Glc, (5) D-Gal*p*-(1→6)-D-Glc (allolactose) with D-Gal*p* D-Gal*p*-(1→3)-D-Gal, (6) D-Gal*p*-(1→6)-D-Gal, (7) D-Gal*p*-(1→6)-Lac, and (8) D-Gal*p*-(1→3)-D-Lac. Products marked with an x are minor components and were not identified. Peaks appearing at ∼ 22 min are tetrasaccharides. From Splechtna et al. (2006) with permission from American Chemical Society.

shows the CE separation attained. Likewise, a quantification of the resulting GOS mixtures obtained from recombinant β-galactosidase of *Lactobacillus reuteri* on lactose hydrolysis has been performed by CE (Maischberger et al., 2007). A comparative study on GOS production was carried out using lactose solutions and whey permeate; and although the initial reaction rate was higher for the latter GOS, the yield was slightly lower.

Lactulose

CE can be used to separate lactulose from mixtures of carbohydrates (Vorndran et al., 1992). This allows a rapid and sensitive analysis of underivatized carbohydrates with indirect UV detection using 6 mM sorbic acid (pH 12.1) as both carrier electrolyte and chromophore. CE analysis has been used to detect carbohydrates in their original form by means of indirect photometry (Oefner et al., 1992). However, direct UV detection of saccharides derivatized pre-column with 2-aminopyridine, ethyl *p*-aminobenzoate or *p*-aminobenzoic acid, allowed a reproducible determination of aldoses and ketoses in fmol range (Oefner et al., 1992).

Underivatized lactulose along with other carbohydrates has been analyzed by CE and amperometric detection using a copper microelectrode. The separation of sugars has been performed in strongly alkaline solutions (LiOH, KOH and NaOH) at pH 13. Among the three studied reagents, the NaOH solution offered good resolution with a suitable time of analysis and it was employed as separation electrolyte. The method is simple, sensitive, and, relatively easy to implement (Colon et al., 1993).

Different CE methods have been used to determine lactulose content in heated milks; thus Guingamp et al. (1999), using an indirect UV detection at 254 nm and sorbate pH 12.04 as running buffer, evaluated heat load of commercial milks. Afterwards, Humbert et al. (2007), by improving milk sample preparation, they determined the lactulose content in pasteurized, indirect and direct UHT, and in-bottle sterilised milk by CE. Lactulose in milk samples was also measured by HPLC and an enzymatic method. The authors found a good correlation between the three methods.

Determination of lactulose along with mannitol is a highly sensitive test for the screening of the diseases that affect intestinal permeability. Paroni et al. (2006) set up a method by CE with indirect UV detection (254 nm) and sorbate, cetyltrimethylammonium bromide and LiOH as background electrolyte to estimate the lactulose-mannitol intestinal permeability in a cohort of patients with type I diabetes.

Fructooligosaccharides (FOS)

FOS are probably the most commonly used prebiotic fibers in the production of functional foods. An evaluation of prebiotic character of FOS has been carried out by Corradini et al. (2004) through short chain fatty acid (SCFA) measurement using CE.

α-Galactosides

The raffinose family of oligosaccharides (ROS) are composed of α-(1→6) galactosides bound to sucrose at C-6 of the glucose. By successive binding of one, two and three additional α-galactoside units to C-6 of the terminating galactose unit, the compounds stachyose, verbascose and ajugose are formed (Andersen et al., 2003). These oligosaccharides are decomposed in the large intestine causing unpleasant effects. However, these effects have been counterbalanced by an increasing interest in non digestible oligosaccharides as functional food ingredients.

Since α-galactosides are non-reducing oligosaccharides, borate complex formation seems to be a promising analytical methodology to analyze them. Arentoft et al. (1993) optimized a high performance capillary electrophoresis (HPCE) method to quantify ROS (raffinose, stachyose and verbascose) based on the formation of borate-carbohydrate complexes and UV detection at 195 nm. Pea seed samples were submitted to a simple extraction procedure and a purification step prior to the determination of individual oligosaccharides. This method could be adapted for the determination of other low-molecular-mass carbohydrates.

Frias et al. (1996) also quantified the ROS family by CZE using disodium tetraborate as running buffer; high-quality electrophoregrams were obtained due to a purification step of pea seeds samples, using Sep-Pak C_{18} cartridges. The ROS family was also quantified by HPAEC-PAD; both methods showed a good linearity and reproducibility and did not show significant differences.

Using indirect UV detection, Andersen et al. (2003) also quantified the ROS family by HPCE. The signal wavelength was set at 350 nm with a reference at 275 nm. As background electrolyte they used pyridine-2,6-dicarboxylic acids, sodium borate decahydrate ($Na_2B_4O_7 10H_2O$) and hexadecyltrimethylammonium bromide, adjusted to varying pH values (8.0–10.0). The method was applied for the quantification of α-galactosides in a lupine seed sample (*Lupinus angustifolius*) after extraction and purification.

Other Nondigestible Oligosaccharides

There are other oligosaccharides, the so-called "second generation of prebiotics" which could present new physical and chemical properties and different and more specific bioactivities (Joucla et al., 2004).

Homologous of glycoglucans of isomaltose and *Laminaria* have been analyzed by CZE using as running buffer 200 mM borate buffer (pH 9.5) and UV detection at 245 nm. Oligosaccharides were pre-column derivatized with 3-methyl-1-phenyl-2-pyrazolin-5-one (MPP). A good separation was achieved for each oligosaccharide series, which have various types of glycosidic linkages (Honda et al., 1991).

Oligosaccharides derived from partial hydrolysis of dextran were satisfactorily separated by HPCE using coated capillary with a copolymer of hydroxypropylcellulose and hydroxyethyl metacrylate. The running buffer was 100 mM tris-borate buffer, pH 8.8. Derivatization of oligosaccharides was performed using N-(4-aminobenzoyl)-L-glutamic acid and UV detection (Plocek and Novotny, 1997).

Also, efficient electrolyte systems for underivatized carbohydrates based on co-electroosmotic CE can be useful for the separation of derivatized counterparts. A selective separation of derivatized (reductive amination) carbohydrates (xylose, cellobiose, melibiose, maltotetraose, gentiobiose, etc.) using ethyl *p*-aminobenzoate or ethyl *p*-aminobenzonitrile can be obtained using as running buffer electrolyte borate and an organic solvent. Co-directional migration of the anionic analytes with the electroosmotic flow (EOF) was achieved by adding a cationic polymer (hexadimethrine) bromide (HDB). The carbohydrates were detected by UV (280 nm) and the method was applied to the analysis of carbohydrates of plant hydrolyzates (Nguyen et al., 1997).

A comparative study of chromophore response (CZE-UV) and electrochemical signal (HPAEC-PAD) of some model gluco-oligosaccharides (dextrans) with different DP has been carried out. UV detection was performed using 8-aminonaphtalene-1,3,6,-trisulphonic acid (ANTS) as chromophoric dye. Both methods provided similar response for DP 1,000 and 5,000 dextrans (Abballe et al., 2007).

CE with LIF (excitation at 488 nm and emission at 520 nm) and ESI-MS detection has been used to characterise gluco-oligosaccharide regioisomers synthesised by *Leuconostoc mesenteroides* NRRL B-512F with a DP ranging from 2 to 9 (Joucla et al., 2004). Resolution of APTS (9-aminopyrene-1,4,6-trisulfonate) derivatives of gluco-oligosaccharide regioisomers over a wide DP range is more appropriately performed with borate buffer systems (Joucla et al., 2004). The use of combined methods looks promising for profiling mixtures of gluco-oligosaccharides synthesised by glucansucrases.

Xyloglucans belong to the groups of hemicelluloses and are constituted by a $\beta(1\rightarrow4)$ linked glucan chain to which different short side chains are attached.

An unambiguous letter code is used for the nomenclature of each segment depending on the side chain, thus they can be classified in few types of structure: XXXG (X = xylose; G = glucose); XXGG and XXXGG. HPAEC-PAD, RP-HPLC and CE (LIF and ESI-MS detection) methods have been compared to determine xyloglucan structures in blackcurrants (Hilz et al., 2006). For CE xyloglucan oligosaccharides were labeled with APTS, separated on a polyvinyl alcohol (N-CHO) coated capillary and detected by LIF (λ_{exc} = 488 nm-λ_{em} = 520 nm). Before analysis, samples were submitted to different extraction steps and xyloglucan material was hydrolyzed with a specific endo-glucanase. The method allowed the identification the structures of xyloglucans as well as the quantification of the main oligomer as XLFG (L = galactose; F = fucose-galactose) present in black currant.

Structural isomers of short oligosaccharides have been also analyzed by CE. In this case, separation of oligosaccharides derived from maltose, cellobiose, xylobiose, and isomaltose has been performed using lithium acetate (pH 5) as running buffer. Oligosaccharides were derivatized using APTS and detected by LIF using wavelengths of excitation and emission of 488 and 520 nm, respectively. The method was applied to the analysis of structural isomers of short oligosaccharides in various plant substrates, and a baseline resolution of three different galactobioses isoforms β (1→4), α (1→4) and α (1→3) was obtained (Khandurina and Guttman, 2005).

13.2.6   Mass Spectrometry

The analysis of prebiotic carbohydrates by different analytical techniques coupled to MS has been reviewed in previous sections. However, a specific mention to this technique has to be done, considering the large number of reported applications where carbohydrates are analyzed directly by MS. Direct infusion ESI and MALDI are the most common ionization sources employed for this purpose. These techniques in combination with tandem MS analyzers have been used to solve structural problems of carbohydrates (Harvey, 1999). Although separation of isomeric oligosaccharides is not possible by MS, identification of their structures based on their different fragmentation patterns has been achieved. Nevertheless, it is not feasible yet to determine the structure of an oligosaccharide just by the study of these patterns. Comparisons to several reference carbohydrates is necessary. Moreover, isomeric oligosaccharides give rise to fragments at the same m/z values, and differences can be observed only in their abundances. Kurimoto et al. (2006) developed a quantitative procedure by quadrupole ion trap (QIT) to solve these problems. Detailed information about

MS of oligosaccharides can be obtained from an exhaustive review written by Zaia (2004), which includes the ionization methods and the mass analyzers generally used for oligosaccharide analysis.

13.2.6.1   Electrospray Ionization

Samples are introduced by direct infusion into the ESI ion source. As indicated above, different analyzers can be coupled to ESI, Q and IT being the most common ones. Nevertheless, in the last years tandem analyzers such as QToF, QIT or triple quadrupole (QqQ) are gaining a significant acceptance and becoming more widespread. While ESI produces a soft ionization the quasi-molecular ion which allows to determine the molecular weight of the analyte, its coupling to IT, QIT, QToF, QqQ or Fourier-transform ion cyclotron resonance (FT-ICR), provides higher structural information by the generation of MS/MS and MS^n. Collision-Induced Dissociation (CID) is the most common method of fragmentation, although other methods such as electron transfer dissociation (ETD), electron capture dissociation (ECD) and infrared multiphoton dissociation (IRMPD), can be applied. In CID the precursor ion is submitted to repeated collisions with a gas and product ions are formed, whereas with IRMPD, photon energy is imparted on both precursor and product ions, resulting in a higher fragmentation (Seipert et al., 2008). ECD and ETD induce fragmentation of positive ions by electron transfer. IT and FT-ICR allow a higher control over CID operation, the possibility of obtaining MS^n and low energy reactions, while QqQ and Q-ToF produce more fragmentation from CID and less operator control, and only MS/MS can be performed.

◉ Table 13.4 shows some recent applications of ESI MS to the analysis of oligosaccharides.

XOS have been analyzed by ESI MS both on positive and negative modes. Whereas the positive ESI MS allowed the identification of neutral and acidic XOS, the negative mode results in simpler MS (Reis et al., 2003a) since there is a lower adduct formation and only acidic XOS ions appear. Isomeric structures of a mixture of arabinoxylooligosaccharides (AXOS) have been also differentiated by analysing their permethylated derivatives by ESI-IT MS upon CID (Matamoros-Fernández et al., 2003); however, the direct analysis of these oligosaccharides using a Q-TOF or IT did not allow the distinction between linear and branched structures.

◻ Table 13.4

Some direct infusion ESI-MS applications for the analysis of prebiotics

Oligosaccharides	Treatment of the sample	Analyzer	Ionization method	Reference
AXOS	Permethylation	IT	CID	Matamoros-Fernández et al. (2003, 2004)
CEOS, MOS, XOS	Addition of ammonium acetate or alkali metal salts	Q-IT FT-ICR	CID	Pasanen et al. (2007)
MOS, CEOS, IMOS	Pyridylamination	Q-IT	CID	Kurimoto et al. (2006)
Sulfated human milk oligosaccharides	Pyridylamination	QqQ	–	Guerardel et al. (1999)
Human milk oligosaccharides	–	Q-ToF	CID	Kogelberg et al. (2004)
XOS		Q-ToF	CID	Reis et al. (2003a)

Fragmentation of cello- (CEOS), malto- (MOS) and xylooligosaccharides (XOS) has been recently studied by ESI MS coupled to QIT and FT-ICR (Pasanen et al., 2007). The effect of different precursor ion types (deprotonated, protonated, ammoniated and alkali metal cationized precursors) and carbohydrate structure (α or β configuration and presence of hexose or pentose units) on the fragmentation of these carbohydrates in CID was evaluated. As an example, ◉ *Figure 13.6* shows the scheme of fragmentation of a trisaccharide (◉ *Figure 13.6a*) and the different spectra obtained for deprotonated cellopentaose (◉ *Figure 13.6b*), maltopentaose (◉ *Figure 13.6c*) and xylopentaose (◉ *Figure 13.6d*). Both CEOS and MOS showed similar fragments (A, B and C), however A fragments were more abundant in CEOS. In contrast, the behavior of XOS was completely different; A fragments were the most abundant, and the intensity ratios of C fragments were clearly different from those of CEOS and MOS. In this example, the different fragmentation observed depended both on the anomeric configurations of the glycosidic linkage and on the presence of monosaccharide units. Results of this work also confirmed that the structural information obtained from the CID of oligosaccharides was dependent on the precursor ion type.

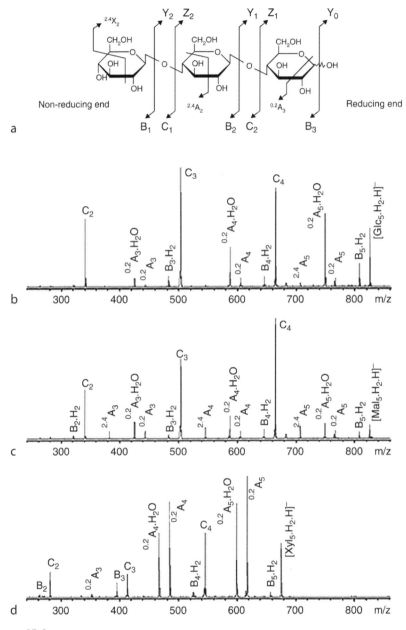

■ Figure 13.6

CID ESI MS analysis of cellopentaose, maltopentaose, and xylopentaose. Nomenclature for oligosaccharide fragments illustrated for cellotriose (a). CID spectra for deprotonated (b) cellopentaose, (c) maltopentaose, and (d) xylopentaose at a fragmentation amplitude of 0.4 V. From Pasanen et al. (2007) with permission from Elsevier.

13.2.6.2   Matrix-Assisted Laser Desorption/Ionization

An extensive and well written review about the application of MALDI MS to carbohydrate analysis which covers the period 1991–1998 has been reported by Harvey (1999). Although in this section some basic points will be referred to the mentioned study, most recent applications of prebiotic analysis will be discussed.

In MALDI, the sample is mixed with a matrix, allowed to crystallize by evaporation of the solvent and submitted to the laser whose energy is absorbed by the matrix and transferred to the carbohydrate which is ionized.

Nitrogen lasers emitting in the UV at 337 nm are those most commonly used for MALDI analysis, although other lasers that emit in the infrared (IR) have been also assayed.

It has been observed that ionization efficiency by MALDI is similar for neutral carbohydrates of different molecular weights, while efficiency of ESI decreases for carbohydrates of higher DPs (Harvey, 1999). The high sensitivity of MALDI allows the detection of oligosaccharides at picomole levels (Morelle and Michalski, 2005).

MALDI is generally coupled to a time of flight (ToF) analyzer resulting in high sensitivity, because most ions generated by the laser are recorded by the detector (Harvey, 1999), although couplings to other analyzers such as IT (Qin et al., 1996), FT-ICR (Carroll et al., 1996) ToF/ToF (Spina et al., 2004) or Q-ToF (Morelle and Michalski, 2005) have been also described. These couplings provide high-mass accuracy, high resolution, and the possibility of performing multiple methods of tandem MS which can be used to obtain higher structural and complementary compositional information. MALDI-FT-ICR has been recently used to evaluate the consumption of human milk oligosaccharides (Ninonuevo et al., 2007) and FOS (Seipert et al., 2008) by intestinal bacteria.

Harvey (1999) in his review exhaustively explained the different matrices that can be used for the MALDI analysis of free neutral carbohydrates, free acidic carbohydrates, sulfated carbohydrates and glycoproteins. Most recognized prebiotics are neutral carbohydrates, which are commonly analyzed by MALDI using 2,5-dihydroxybenzoic acid (DHB), although other matrices such as $2',4',6'$-trihydroxyacetophenone (THAP) or mixtures of DHB with different compounds have been also proposed by several authors to obtain finer crystals. One of the problems of DHB matrix is the appearance of multiple matrix peaks at low masses which could interfere with low molecular weight carbohydrates. Therefore, molecular weight of target oligosaccharides should be considered when selecting a matrix. A good repeatability of spot-to-spot or sample-to-sample, a good crystallization and the production of a spectrum with good signal to noise ratio and good resolution should be also be taken into account for matrix selection (Wang et al., 1999).

Qualitative analyses of prebiotic carbohydrates by MALDI-ToF have been widely carried out (Huisman et al., 2001; Lopez et al., 2003; Reis et al., 2003b; Sanz et al., 2006), however quantitative applications are less common (Wang et al., 1999) since they present difficulties associated to a poor shot-to-shot repeatability and to the crystal's lack of homogeneity. Quantification is normally carried out using an internal standard calibration method and it is desirable in order to obtain single alkali ion adduct peaks to gain peak intensity. Wang et al. (1999) quantitatively analyzed FOS using γ-cyclodextrin as internal standard and KCl to obtain single potassium adduct peaks. However, one of the drawbacks of this technique is the variability of ionization regarding the sample. The medium can show a significant effect on MALDI analysis (Reiffová et al., 2007). A previously proposed method to analyze FOS was used by Wang et al. (1999) to determine FOS content in food extract. Nevertheless, these extracts (i.e., from red onions) suppressed the ions produced from the internal standard although FOS could be detected. Therefore, internal standard should be selected depending on the sample to be analyzed, however, commercial standards are scarce and most of them (such as maltodextrins) posses similar molecular weight to the analytes giving the same response. Seipert et al. (2008) used deuterated reduced malto-heptaose to distinguish its masses from those of FOS with the same molecular weight.

Different comparative studies of prebiotic analyses by HPAEC-PAD and MALDI have been carried out. HPAEC-PAD has been found to be more sensitive in terms of detection limits than MALDI for the analysis of FOS and allowed the separation of linear and branched oligosaccharides. However, MALDI was a faster method, more tolerant to impurities and produced more correct molecular assignments, and was probably in general more accurate for quantitative determinations (Wang et al., 1999).

⊙ Table 13.5 shows as an example of some MALDI qualitative and quantitative studies of prebiotic oligosaccharides described in the literature.

13.2.7   Nuclear Magnetic Resonance Spectroscopy

For a long time, NMR has significantly contributed to the knowledge of the structure and conformation of carbohydrates. This technique is especially

Table 13.5

Some examples of MALDI analysis of prebiotics

Prebiotic	Analyzer	Selected matrix	Salts	Analyzes	Calibration	Reference
FOS	ToF	THAP with acetone	0.01 M KCl	Qualitative and quantitative	i.s.	Wang et al. (1999)
FOS and inulin	FT-ICR	0.4 M DHB	0.01 mM NaCl	Qualitative and quantitative	i.s.	Seipert et al. (2008)
Human milk oligosaccharides	FT-ICR	0.4 M DHB	0.01 M NaCl	Qualitative and quantitative	e.s. and i.s.	Ninonuevo et al. (2007)
Human milk oligosaccharides	ToF/ToF	10 mg mL^{-1} DHB	–	Qualitative (CID)	–	Spina et al. (2004)
FOS	ToF	10 mg mL^{-1} DHB	1 mg mL^{-1} CH$_3$COONa (only for standards)	Semi-quantitative	e.s.	Reiffová et al. (2007)
Fructans	ToF	DHB	–	Qualitative	–	Lopez et al. (2003)
Arabinogalactans	ToF	9 mg mL^{-1} DHB + 3 mg mL^{-1} 1-hydroxy-isoquinoline in water: acetonitrile (70:30)	–	Qualitative	–	Huisman et al. (2001)
XOS				Qualitative	–	Reis et al. (2003b)
XOS	ToF	1% DHB in methanol	–	Qualitative	–	Cano and Palet (2007)
GEOS	ToF	DHB in acetonitrile	–	Qualitative	–	Sanz et al. (2006)
Alternansucrase acceptor products	ToF	DHB in acetonitrile	–	Qualitative	–	Côté and Sheng (2006)

useful for detailed structural analysis of pure products, either in solution or in solid state, but it has been also applied to the study of mixtures of oligosaccharides.

NMR relies on the magnetic properties of the atomic nuclei (spin). The most used nuclei in carbohydrate chemistry are ^1H and ^{13}C. NMR operates on a time-scale slower than other spectroscopic techniques (such as IR or UV). This is not a big drawback, since NMR is usually used to elucidate structures, and not as a routine control technique.

13.2.7.1   Sample Preparation

This subject has been described by Bock and Pedersen (1983) and the main features are summarized below.

Samples can be dissolved in deuterium oxide (D_2O) and deuterated solvents, such as dimethylsulfoxide (Me$_2$SO-d_6), pyridine (Py-d_6) or chloroform (DCl$_3$). The less polar solvents are usually selected for low-molecular weight sugars, whereas water is necessary for oligosaccharides. It should be taken into account that solvent-induced shifts are low for neutral oligosaccharides when working with ^{13}C-NMR, whereas the effect on ^1H-NMR spectra is high. The effect of solvent is always important in acidic or basic carbohydrates. Concentration should be adjusted to avoid high-viscosity solutions, which can cause signal broadenings. Sample clean-up is recommended to suppress soluble paramagnetic impurities.

13.2.7.2   Reference Compounds

The most used reference signals for chemical shifts measurements of carbohydrates are probably tetramethylsilane and acetone; nevertheless, other compounds as DSS (sodium salt of 2,2-dimethyl-2-silapentane-5-sulfonic acid) and TSP ([2,2,3,3-d_4]-3-(trimethylsilyl)-propanoic acid sodium salt) are also used.

13.2.7.3   Methodology

Classic studies about NMR of carbohydrates were published by Vliegenthart et al. (1983) and Rathbone (1985). Whereas the assignment of ^1H signals is relatively easy in pure products (specially the anomeric proton), the high amount of protons when several tautomeric forms and various oligosaccharides are present makes difficult the assignment of all signals. Thus, it becomes necessary to use ^{13}C spectra and different methods including 1D and 2D homo- and heteronuclear experiments, correlation spectroscopy (COSY) including total correlation spectroscopy (TOCSY) and nuclear Overhauser effect spectroscopy (NOESY), etc. The number of different NMR experiments which have been described to elucidate oligosaccharide structures is really high. Experiments such as HSQC (heteronuclear single quantum correlation) and HMQC (heteronuclear multiple quantum correlation) which correlate the chemical shift of proton with the chemical shift of the directly bonded carbon, and HMBC (heteronuclear multiple bond correlation) which uses two or three bonds couplings, are frequently used to assign signals from complex oligosaccharides. A recent review describes several NMR methods for analyzing the structure of oligosaccharides, including assignment of all H-1 NMR signals, NOE experiments, and modification of pulse programs (Kajihara and Sato, 2003).

13.2.7.4   Applications

A series of interesting applications, covering different oligosaccharides with prebiotic properties have been summarized in ◉ *Table 13.6*. Saccharides from DP2 (lactulose) and DP3 (kestoses) to higher DP have been included.

Although basic NMR data of many pure sugars, including mono- di- and trisaccharides have been published in the 1980s, fully assigned highly resolved spectra of some prebiotic sugars have been recently reported: lactulose in solution (Mayer et al., 2004) and in crystalline state (Jeffrey et al., 1992); the three natural kestoses (Calub et al., 1999; Liu et al., 1991). α-D-Galp-(1→6)-β-D-Galp-(1→4)-β-D-fructose (three tautomers) and β-D- Galp-(1→4)-D-fructose-(1→1)-β-D-Galp (three tautomers) resulting from enzymatic transgalactosylation during lactulose hydrolysis by the galactosidase from *K. lactis* have also been characterized (Martínez-Villaluenga et al., 2008b). Twelve novel non-reducing oligosaccharides from DP3 to DP6, namely [β-D-Galp-(1→4)]$_n$-α-D-Glcp-(1→1)-β-D-Galp[-(4→1)- β-D-Galp]$_m$, with n, m = (1, 2, 3, or 4) and β-D-Galp-(1→2)-α-D-Glcp-(1→1)-β-D-Galp were characterized in a mixture produced by β-galactosidase using lactose as a substrate (Fransen et al., 1998). Although the mixture was fractionated, several oligosaccharides were found in the same fraction and several experiments were necessary to achieve complete characterization. A detailed description of signal assignments of two trisaccharides and four tetrasaccharides

◘ Table 13.6

NMR applications for the analysis of prebiotics *(Cont'd p. 519)*

Analyte	Experiments	Solvent	Reference compound	Reference
Related-kestose oligosaccharides in plants	^1H, ^{13}C	D_3HCl	Tetramethylsilane	Forshyte et al. (1990)
1-Kestose	2D homonuclear and heteronuclear	D_2O	Acetone	Calub et al. (1990)
6-Kestose, neokestose	2D homonuclear and heteronuclear	D_2O	Acetone	Liu et al. (1991)
Crystalline lactulose	^{13}C CPMAS	–	Adamantane	Jeffrey et al. (1992)
5 Trisaccharides in goat colostrum	^1H, ^{13}C, several 2D experiments	D_2O	Expressed relative to DSS, but actually measured by reference to acetone	Urashima et al. (1994)
12 non-reducing oligosaccharides	^1H, ^{13}C, several 2D experiments	D_2O	Acetone or acetate for ^1H	Fransen et al. (1998)
			Ext. glucose for ^{13}C	
XOS	^1H, ^{13}C, several 2D experiments	D_2O	Acetone	Nishimura et al. (1998)
3 Sulfated OS in human milk	^{13}C and ^1H	D_2O	Expressed by reference to DSS, but actually measured by reference to acetone	Guerardel et al. (1999)
5 FOS from *Asparagus*	1D ^1H and ^{13}C, several 2D experiments	D_2O	TSP	Fukusi et al. (2000)
Neo-FOS produced by a *Penicillium citrinum*	^{13}C	D_2O	Tetramethylsilane	Hayashi et al. (2000)
Kojioligosaccharides	^{13}C	D_2O	TSP	Chaen et al. (2001)
Lactulose	^1H, several 1D and 2D experiments		Tetramethylsilane	Mayer et al. (2004)
Inulin-type FOS from *Matricaria maritima*	2D ^1H, ^1H DQF-COSY/TOCSY and ^1H, ^{13}C HMQC/HMBC	D_2O	TSP	Cerantola et al. (2004)
Neutral oligosaccharides from human milk	^1H, several 2D experiments	D_2O	Acetone	Kogelberg et al. (2004)

⬛ Table 13.6

Analyte	Experiments	Solvent	Reference compound	Reference
Novel oligosaccharides from raffinose and stachyose	^{1}H and ^{13}C 2D-NMR including COSY, HSQC, HSQCTOCSY, HMBC and other	D_2O	TSP and 1,4-dioxane	Takahashi et al. (2005)
Oligosaccharides produced by alternansucrase	^{1}H, ^{13}C, several 2D experiments	D_2O	Acetone	Coté and Sheng (2006)
Cyclic isomaltooligosaccharides	^{13}C	D_2O	DSS	Funane et al. (2007)
IMOS	^{1}H	D_2O		Ao et al. (2007)
GOS from cyanobacterium *Nostoc commune*	^{1}H	D_2O	Acetone	Wienecke et al. (2007)
2 Trisaccharides derived from lactulose	^{1}H, ^{13}C, several 1D and 2D experiments	D_2O	Tetramethylsilane	Martinez-Villaluenga et al. (2008b)
FOS produced by *Aspergillus*	^{1}H, ^{13}C and 2D HMQCT	D_2O	Tetramethylsilane	Mabel et al. (2008)

isolated from *Asparagus* through several NMR experiments has been published by Fukushi et al. (2000).

Among the oligosaccharides with prebiotic properties, natural fructans (FOS) with the two series of inulin and levan with linkages 2→1 and 2→6 respectively, have probably been the most studied (Cerantola et al., 2004; Hayashi et al., 2000).

Most oligosaccharides obtained by enzymatic synthesis such as kojioligosaccharides (Chaen et al., 2001), cycloisomaltooligosaccharides (Funane et al., 2007) are mixtures of similar saccharides with a definite glycosidic linkage and different DP, thus the NMR signal assignment can be performed by comparison with published data.

Three xylooligosaccharides with general structure $[O\text{-}\alpha\text{-D-Glc}p\text{-}(1\rightarrow2)]_n\text{-}O\text{-}\alpha\text{-D-Xyl}p\text{-}(1\rightarrow2)\text{-}\beta\text{-D-Fru}f$ (n = 1,2,3) required several techniques for structural characterisation. The ^{1}H and ^{13}C NMR signals of each saccharide were assigned using two dimension (2D)-NMR including COSY, HSQC, HSQC-TOCSY and HMBC (Takahashi et al., 2007). Similar techniques were applied to the structural

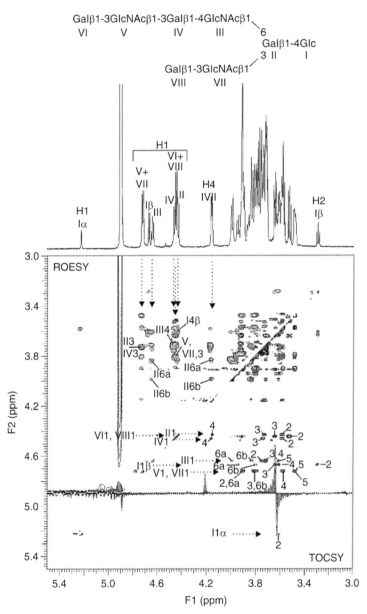

Figure 13.7

^{1}D and ^{2}D ^{1}H NMR spectra of iso-lacto-N-octaose. ^{1}D and ^{2}D ^{1}H NMR spectra (800 MHz) of iso-lacto-N-octaose, region 5.5–3.0 ppm at 15°C. Upper trace, ^{1}H NMR spectrum; top-left half, 300-ms ROESY spectrum and bottom-right half, 140-ms TOCSY spectrum. The structure is shown at the top, depicting the residue labeling. From Kogelberg et al. (2004) with permission from FEBs.

characterization of six novel oligosaccharides (one tetra-, two penta-, two hexa- and one hepta-saccharide) synthesized by glucosyl transfer from β-D-glucose-1-phosphate to raffinose or stachyose by the action of *Thermoanaerobacter brockii* kojibiose phosphorylase (Takahashi et al., 2005).

A special case is that of alternansucrase, which produces oligosaccharides with alternating α-(1→3) and α-(1→6) linkages. When this enzyme was incubated with maltose, one pentasaccharide, two hexasaccharides and one heptasaccharide were isolated as main products, with general structure [α-D-Glc$_p$-(1→6)-α-D-Glc$_p$-(1→3)]$_x$-α-D-Glc (Coté and Sheng, 2006). Experiments (gradient-enhanced band-selective HSQC and HSQC–TOCSY and gradient-enhanced band-selective HMBC) were performed at 27°C for the lower DP products and at 50°C for the higher DP ones.

Milk oligosaccharides are extremely complex: about 150–200 oligosaccharides (neutral and acidic) have been described in human milk, whereas a smaller number exist in ruminants' milk. As an example of NMR application to oligosaccharides in goat's milk, four neutral trisaccharides were characterized in goat colostrum: α-L-Fuc$p$-(1→2)-β-D-Gal$p$-(1→4)-Glc, α-D-Gal$p$-(1→3)-β-D-Gal$p$-(1→4)-Glc, β-D-Gal$p$-(1→3)-β-D-Gal$p$-(1→4)-Glc, and β-D-Gal$p$-(1→6)-β-D-Gal$p$-(1→4)-D-Glc (Urashima et al., 1994); NMR spectra were assigned by comparison with those of other previously described oligosaccharides and by several experiments, such as double quantum filtered correlation (DQF-COSY), 1D TOCSY and (^{1}H, ^{13}C) shift correlation.

Human milk neutral oligosaccharides contain galactose, *N*-acetylglucosamine, fucose, and lactose; several anionic oligosaccharides containing *N*-acetylneuraminic acid also exist; a lactose unit at the reducing end is also frequently found (Mehra and Kelly, 2006) but other substituents are possible. As an example, three sulfated oligosaccharides have been analyzed using 2D homonuclear COSY and HMQC (Guerardel et al., 1999). The structural elucidation of very complex milk oligosaccharides has been undertaken by combining two techniques: ESI-MS for determining the branching pattern and ^{1}H NMR for sequence assignment (Kogelberg et al., 2004); ⊙ *Figure 13.7* shows some spectra of *iso*-lacto-*N*-octaose.

13.3   Summary

⊙ Despite advances in analytical techniques in recent years there is still a lack of accurate and precise methods to characterize and quantify prebiotic oligosaccharides which frequently consist of complex mixtures with similar structural characteristics.

- Traditional analyses such as methylation, TLC or open chromatographic columns with RI detectors are still commonly used. The lack of standards and the similar structure of oligosaccharides are the main drawbacks to achieve a truthful qualitative and quantitative result.
- The selection of the most appropriate methodology for the analysis of prebiotics mainly depends on the nature of the carbohydrate mixture: the complexity of the sample, range of expected molecular weight, etc. No one protocol is able to cover all the possible cases.
- While GC has commonly been used to determine the composition of low molecular weight carbohydrates, oligosaccharides with high DPs are mainly characterised by HPLC.
- Considering the complexity of prebiotic samples their analyses require the use of different techniques, which are combined to obtain useful information.
- HPLC, CE or GC are used for the separation and isolation of the different constituents; methylation analysis and NMR to determine their structures; MS for studying their molecular weight and/or tandem MS systems to complement the structural information.

List of Abbreviations

2'FL	2'-fucosyllactose
2D	two dimensions
3'FL	3'-fucosyllactose
AMAC	2-aminoacridone
AMD	automatic mode development
ANTS	8-aminonaphtalene-1,3,6,-trisulphonic acid
APTS	9-aminopyrene-1,4,6-trisulfonate
ASE	accelerated solvent extraction
AXOS	arabinoxylooligosaccharides
BGE	background electrolyte
CE	capillary electrophoresis
CEOS	cellooligosaccharides
CGE	capillary gel electrophoresis
CI	chemical ionization
CID	collision-induced dissociation
CIEF	capillary isolectric focusing
CITP	capillary isotachophoresis

COSY	correlation spectroscopy (COSY)
CZE	capillary zone electrophoresis
DAD	diode-array detector
DHB	2,5-dihydroxybenzoic acid
DP	degree of polymerization
DQF-COSY	double-quantum-filtered correlation spectroscopy
DSS	2,2-dimethyl-2-silapentane-5-sulfonic acid sodium salt
ECD	electron capture dissociation
EI	electronic impact
ELSD	evaporative light scattering detectors
EOF	electroosmotic flow
ESI	electrospray ionization
ETD	electron transfer dissociation
FAB	fast atom bombardment
FID	flame ionization detector
FOS	fructooligosaccharides
FT-ICR	fourier-transform ion cyclotron resonance
GC	gas chromatography
GCC	graphitized carbon columns
GOS	galactooligosaccharides
HDB	hexadimethrine bromide
HILIC	hydrophilic interaction chromatographic
HMBC	heteronuclear multiple bond correlation
HMOS	human milk oligosaccharides
HMQC	heteronuclear multiple quantum correlation
HPAEC	high performance anion exchange chromatography
HPCE	high performance capillary electrophoresis
HPLC	high performance liquid chromatography
HPSEC	high performance size exclusion chromatography
HPTLC	high performance TLC
HSQC	heteronuclear single quantum correlation
HT-GC	high temperature gas chromatography
IMOS	isomaltooligosaccharides
IR	infrared
IRMPD	infrared multiphoton dissociation
IT	ion trap
LALLS	low angle laser light scattering
LC	liquid chromatography

LIF	laser-induced fluorescence
LNFP I	lacto-N-fucopentaose I
LNFP II	lacto-N-fucopentaose II
LNT	lacto-N-tetraose
LSI	liquid secondary ion
MALDI	matrix assisted laser desorption/ionization
MALLS	multiple angle laser light scattering
MEKC	micellar electrokinetic chromatography
MOS	maltooligosaccharides
MPP	3-methyl-1-phenyl-2-pyrazolin-5-one
MS	mass spectrometry
NMR	nuclear magnetic resonance
NOESY	nuclear overhauser effect spectroscopy
NPLC	normal phase liquid chromatography
OPTLC	over pressured TLC
o-ToF	orthogonal ToF
PAD	pulse amperometric detector
PC	paper chromatography
Q	quadrupole
QIT	quadrupole ion trap
QqQ	triple quadrupole
RI	refractive index
ROS	raffinose oligosaccharides
RP	reverse phase
SALDI	surface assisted laser desorption/ionization
SCFA	short chain fatty acid
SDS	sodium dodecyl sulfate
SEC	size exclusion chromatography
SFE	supercritical fluid extraction
THAP	2',4',6'-trihydroxyacetophenone
TLC	thin layer chromatography
TOCSY	total correlation spectroscopy
ToF	time of flight
TSP	$[2,2,3,3-d_4]$-3-(trimethylsilyl)-propanoic acid sodium salt
UPLC	ultra performance liquid chromatography
UTLC	ultra TLC
UV	ultraviolet
XOS	xylooligosaccharides

References

Abazia C, Ferrara R, Corsaro MM, Barone G, Coccoli P, Parrilli G (2003) Simultaneous gas-chromatographic measurement of rhamnose, lactulose and sucrose and their application in the testing gastrointestinal permeability. Clin Chim Acta 338:25–32

Abballe F, Toppazzini M, Campa C, Uggeri F, Paoletti S (2007) Study of molar response of dextrans in electrochemical detection. J Chromatogr A 1149: 38–45

Andersen KE, Bjergegaard CH, Moller P, Sorensen JC, Sorensen H (2003) High-performance capillary electrophoresis with indirect UV detection for determination of a-galactosides in leguminosae and brassicaceae. J Agric Food Chem 51: 6391–6397

Ao Z, Simsek S, Zhang G, Venkatachalam M, Reuhs BL, Hamaker BR (2007) Starch with a slow digestion property produced by altering its chain length, branch density, and crystalline structure. J Agric Food Chem 55:4540–4547

Arentoft AM, Michaelsen S, Sorensen H (1993) Determination of oligosaccharides by capillary zone electrophoresis. J Chromatogr A 652:517–524

Bao Y, Zhu L, Newburg D (2007) Simultaneous quantification of sialyloligosaccahrides from human milk by capillary electrophoresis. Anal Biochem 370:206–214

Bao Y, Newbrug D (2008) Capillary electrophoresis of acidic oligosaccharides from human milk. Electrophoresis 29:2508–251

Bock K, Pedersen C (1983) C-13 nuclear magnetic-resonance spectroscopy of monosaccharides. Adv Carbohydr Chem Biochem 41:27–66

Bonnett GD, Sims IM, Simpson RJ, Cairns AJ (1997) Structural diversity of fructan in relation to the taxonomy of the poaceae. New Phytologist 136:11–17

Bosch-Reig F, Marcotte MJ, Minana MD, Cabello ML (1992) Separation and identification of sugars and maltodextrines by thin-layer chromatography - application to biological-fluids and human-milk. Talanta 39:1493

Bounias M (1980) "http://sauwok.fecyt.es/apps/full_record.do?product=WOS&search_mode=GeneralSearch&qid=6&SID=S169KgcE5m9eDjdPfmg&page=1&doc=4" N-(1-naphthyl)ethylenediamine dihydrochloride as a new reagent for nanomole quantification of sugars on thin-layer plates by a mathematical calibration process. Anal Biochem 106:291–295

Brandolini V, Menziani E, Mazzotta D, Cabras P, Tosi B, Lodi G (1995) Use of AMD-HPTLC for carbohydrate monitoring in beers. J Food Comp Anal 8:336–343

Brobst KM, Lott CE (1966) Determination of some components in corn syrup by gas-liquid chromatography of trimethylsilyl derivatives. Cereal Chem 43:35–43

Bruggink C, Maurer R, Herrmann H, Cavalli S, Hoefler F (2005b) Analysis of carbohydrates by anion exchange chromatography and mass spectrometry. J Chromatogr A 1085:104–109

Bruggink C, Wuhrer M, Koeleman CAM, Barreto V, Liu Y, Pohl C, Ingendoh A, Hokke CH, Deelder AM (2005a) Oligosaccharide analysis by capillary-scale high-pH anion-exchange chromatography with on-line ion-trap mass spectrometry. J Chromatogr B 829:136–143

Calub TM, Waterhouse AL, Chatterton NJ (1999) Proton and carbon chemical-shift assignments for 1-kestose, from two-dimensional n.m.r.-spectral measurements. Carbohydr Res 199:11–17

Campa C, Coslovi A, Flamigni A, Rossi M (2006) Overview on advances in capillary electrophoresis-mass spectrometry of carbohydrates: a tabulated review. Electrophoresis 27:2027–2050

Cano A, Palet C (2007) Xylooligosaccharide recovery from agricultural biomass waste treatment with enzymatic polymeric membranes and characterization of products with MALDI-TOF-MS. J Membr Sci 291:96–105

Cao Y, Wang Y, Chen X, Ye J (2004) Study on sugar profile of rice during ageing by capillary electrophoresis with electrochemical detection. Food Chem 86:131–136

Cardelle-Cobas A, Villamiel M, Olano A, Corzo N (2008) Study of galacto-oligosaccharide formation from lactose using Pectinex Ultra SP-L. J Sci Food Agric 88:954–961

Carlsson NG, Karlsson H, Sandberg AS (1992) Determination of oligosaccharides in foods, diets, and intestinal contents by high-temperature gas chromatography and gas chromatography/mass spectrometry. J Agric Food Chem 40(12):2404–2412

Carroll JA, Penn SG, Fannin ST, Wu J, Cancilla MT, Green MK, Lebrilla CB (1996) A dual vacuum chamber Fourier transform mass spectrometer with rapidly interchangeable LSIMS, MALDI, and ESI sources: Initial results with LSIMS and MALDI. Anal Chem 68:1798–1804

Cerantola S, Kervarec N, Pichon R, Magné C, Bessierers MA, Deslandes E (2004) NMR characterisation of inulin-type fructooligosaccharides as the major water-soluble carbohydrates from Matricaria maritima (L.). Carbohydr Res 339: 2445–2449

Chaen H, Nishimoto T, Nakada T, Fukuda S, Kurimoto M, Tsujisaka Y (2001) Enzymatic synthesis of kojioligosaccharides using kojibiose phosphorylase. J Biosci Bioeng 92:177–182

Cheung RHF, Marriot PJ, Small DM (2007) CE methods applied to the analysis of micronutrients in foods. Electrophoresis 28: 3390–3413

Chu Q, Fu L, Guan Y, Ye J (2005) Fast determination of sugars in Coke and Diet Coke by miniaturized capillary electrophoresis with amperometric detection. J Sep Sci 28:234–238

Churms SC (1996) Recent progress in carbohydrate separation by high performance liquid chromatography based on size exclusion. J Chromatogr A 720:151–166

Colon LA, Dadoo R, Zare N (1993) Determination of carbohydrates by capillary zone electrophoresis with amperometric detection at a copper microelectrode. Anal Chem 85:476–481

Corradini C (2002) Characterization of nutraceutical and functional foods by innovative HPLC methods. Annali di Chimica 92:387–396

Corradini C, Bianchia F, Matteuzzi D, Amoretti A, Rossi M, Zanoni S (2004) High-performance anion-exchange chromatography coupled with pulsed amperometric detection and capillary zone electrophoresis with indirect ultra violet detection as powerful tools to evaluate prebiotic properties of fructooligosaccharides and inulin. J Chromatogr A 1054:165–173

Cortacero-Ramirez A, Segura-Carretero A, Cruces-Blanco C, Hernáinz-Bermúdez de Castro M, Fernández-Gutiérrez A (2004) Analysis of carbohydrates in beverages by capillary electrophoresis with precolumn derivatization and UV detection. Food Chem 87:471–476

Côté GL, Sheng S (2006) Penta-, hexa-, and heptasaccharide acceptor products of alternansucrase. Carbohydr Res 341: 2066–2072

Côté GL, Leathers TD (2005) A method for surveying and classifying Leuconostoc spp. glucansucrases according to strain-dependent acceptor product patterns. J Ind Microbiol Biotechnol 32:53–60

Davies MJ, Hounsell EF (1996) Carbohydrate chromatography: Towards yoctomole sensitivity. Biomed Chromatogr 10:285–289

Di X, Chan KC, Leung HW, Huie CW (2003) Fingerprint profiling of acid hydrolyzates of polysaccharides extracted from the fruiting bodies and spores of Lingzhi by high-performance thin-layer chromatography. J Chromatogr A 1018:85–95

Dreisewerd K, Kolb S, Peter-Katalinic J, Berkenkamp S, Pohlentz G (2006) Analysis of native milk oligosaccharides directly from thin-layer chromatography plates by matrix-assisted laser desorption/ionization orthogonal-time-of-flight mass spectrometry with a glycerol matrix. J Am Soc Mass Spectrom 17:139

El-Rassi Z (1999) Recent developments in capillary electrophoresis and capillary electrochromatography of carbohydrate species. Electrophoresis 20:3134–3144

El-Rassi Z (2002) Reverse-phase and ahydrophobic interaction chromatography of carbohydrates and glycoconjugates. J Chromatogr Lib Elsevier 66:41–94

Farhadi A, Keshavarzian A, Holmes EW, Fields J, Zhanga L, Banan A (2003) Gas chromatographic method for detection of urinary sucralose: application to the assessment of intestinal permeability. J Chromatogr B 784:145–154

Farhadi A, Keshavarzian A, Fields JZ, Sheikh M, Banan A (2006) Resolution of common dietary sugars from probe sugars for test of intestinal permeability using capillary column gas chromatography. J Chromatogr B 836:63–68

Fox A, Morgan SL, Gilbart J (1989) Preparation of alditol acetates and their analysis by gas chromatography (GC) and mass spectrometry (MS). In: Analysis of carbohydrates by GLC and MS. CRC Press, Florida, pp. 87–118

Fransen CTM, Van Laere KMJ, van Wijk AAC, Brull LP, Dignum M, Thomas-Oates JE, Haverkamp J, Schols HA, Voragen AGJ, Kamerling JP, Vliegenthart JFG (1998) α-D-Glcp-(1→1)- β-D-Galp-containing oligosaccharides, novel products from lactose by the action of β-galactosidase. Carbohydr Res 314:101–114

Frias J, Price KR, Fenwick GR, Hedlwy CL, Sorensen H, Vidal-Valverde C (1996) Improved method for the analysis of a-galactosides in pea seeds by capillary zone electrophoresis. Comparison with high-performance liquid chromatography-triple-pulsed amperometric detection. J Chromatogr A 719:213–219

Fukushi E, Onodera S, Yamamori A, Shiomi N, Kawabata J (2000) NMR analysis of tri- and tetrasaccharides from asparagus. Magnetic Resonance Chem 38:1005–1011

Funane K, Terasawa K, Mizuno Y, Ono H, Miyagi T, Gibu S, Tokashiki T, Kawabata Y, Kim Y-M, Kimura A, Kobayashi M (2007) A novel cyclic isomaltooligosaccharide (cycloisomaltodecaose, CI-10) produced by Bacillus circulans T-3040 displays remarkable inclusion ability compared with cyclodextrins. J Biotechnol 130:188–192

Giannoccaro E, Wang YJ, Chen P (2008) Comparison of two HPLC systems and an enzymatic method for quantification of soybean sugars. Food Chem 106:324–330

Gohlke M, Blanchard V (2008) Separation of N-glycans by HPLC. In: Kannicht (ed) Methods in molecular biology, vol 446. Humana Press, Totowa, NJ, pp. 239–254

Goulas A, Tzortzis G, Gibson GR (2007) Development of a process for the production and purification of alpha- and beta-galactooligosaccharides from Bifobacterium bifidum NCIMB 41171. Int Dairy J 17:648–656

Grizard D, Barthomeuf C (1999) Enzymatic synthesis and structure determination of neo-FOS. Food Biotechnol 13:93–105

Guérardel Y, Morelle W, Plancke Y, Lemoine J, Strecker G (1999) Structural analysis of three sulfated oligosaccharides isolated from human milk. Carbohydr Res 320: 230–238

Guingamp MF, Humbert G, Midon P, Nicolas M, Linden G (1999) Screening procedure for evaluating heat load in commercial milks. Lait 79:457–463

Haeuwfievre S, Wieruszeski JM, Plancke Y, Michalski JC, Montreuil J, Strecker G

(1993) Primary structure of human-milk octasaccharides, dodecasaccharides and tridecasaccharides determined by a combination of H-1-NMR spectroscopy and fast-atom-bombardment mass-spectrometry - Evidence for a new core structure, the para-lacto-n-octaose. Eur J Biochem 215:361–371

Hanashiro I, Abe J, Hizukuri S (1996) A periodic distribution of the chain length of amylopectin as revealed by high-performance anion-exchange chromatography. Carbohydr Res 283:151–159

Hartemink R, Van Laere KMJ, Rombouts FM (1997) Comparison of two HPLC systems and an enzymatic method for quantification of soybean sugars. J App Microbiol 83:367–374

Harvey DJ (1999) Matrix-assisted laser desorption/ionization mass spectrometry of carbohydrates. Mass Spectrom. Mass Spectrom Rev 18:349–451

Hase S (2002) Pre- and post-column detection-oriented derivatization techniques in HPLC of carbohydrates. In: El-Rassi Z (ed) Carbohydrate analysis by modern chromatography and electrophoresis. J Chromatogr Lib Elsevier 66:1043–1070

Hauck HE, Schulz (2003) Ultra thin-layer chromatography Chromatohraphia. Chromatohraphia 57S:S313–S315

Hayashi S, Yoshiyama T, Fuji N, Shinohara S (2000) Production of a novel syrup containing neofructo-oligosaccharides by the cells of Penicillium citrinum. Biotechnol Lett 22:1465–1469

Herbreteau B (1992) Review and state of sugar analysis by high performance liquid chromatography. Analysis 20:355–374

Hilz H, de Jong LE, Kabel MA, Schols HA, Voragen AGJ (2006) A comparison of liquid chromatography, capillary eletrophoresis, and mas spectrometry methods to determine xyloglucan structures in black currants. J Chromatogr A 1133:275–286

Honda S, Suzuki S, Nose A, Yamamoto K, Kakehi K (1991) Capillary zone electrophoresis of reducing mono- and oligo-saccharides as the borate complexes of their 3-methyl-1-phenyl-2-pyrazolin-5-one derivatives. Carbohydr Res 215:193–198

Huisman MMH, Brüll LP, Thomas-Oates JE, Haverkamp J, Schols HA, Voragen AGJ (2001) The occurrence of internal (1–5)-linked arabinofuranose and arabinopyranose residues in arabinogalactan side chains from soybean pectic substances. Carbohydr Res 330:103–114

Humbert G, Guimgamp MF, Gaillard JL (2007) Improvement of the ammonia measurement using a clarifying reagent and application to evaluate heat damage in commercial milk samples. Int Dairy J 17:902–906

Ikegami T, Fujita H, Horie K, Hosoya K, Tanaka N (2006) HILIC mode separation of polar compounds by monolithic silica capillary columns coated with polyacrylamide. Anal Bioanal Chem 386:578–585

Ikegami T, Tomomatsu K, Takubo H, Horie K, Tanaka N (2008) Separation efficiencies in hydrophilic interaction chromatography. J Chromatogr A 1184:474–503

Jager AV, Tonin FG, Tavares MFM (2007) Comparative evaluation of extraction procedures and method validation for determination of carbohydrates in cereals and dairy products by capillary electrophoresis. J Sep Sci 30:586–594

Jeffrey GA, Huang D, Pfeffer PE, Dudley RL, Hicks KB, Nitsch E (1992) Crystal structure and n.m.r. analysis of lactulose trihydrate. Carbohydr Res 226:29–42

Joucla G, Brando T, Remaud-Simeon M, Monsan P, Puzo G (2004) Capillary electrophoresis analysis of gluco-oligosaccharide regioisomers. Electrophoresis 25:861–869

Joye D, Hoebregs H (2000) Determination of oligofructose, a soluble dietary fiber, by high-temperature capillary gas chromatography. JAOAC Int 83(4):1020–1025

Kaack K, Christensen LP, Hansen SL, Grevsen K (2004) Non-structural carbohydrates in processed soft fried onion (Allium cepa L.). Eur Food Res Technol 218:372–379

Kajihara Y, Sato H (2003) Structural analysis of oligosaccharides by nuclear magnetic resonance method. Trends Glycosci Glycotechnol 15:197–220

Karoutis AI, Tyler RT, Slater GP (1992) Analysis of legume oligosaccharides by high-resolution gas chromatography. J Chromatogr 623(1):186–190

Katapodis P, Vrsanska M, Kekos D, Nerinckx W, Biely P, Claeyssens M, Macris BJ, Christakopoulos P (2003) Biochemical and catalytic properties of an endoxylanase purified from the culture filtrate of Sporotrichum thermophile. Carbohydr Res 338:1881–1890

Kennedy JF, Pagliuca G (1994) Oligosaccharides In: Chaplin MF and Kennedy JF (eds) Carbohydrate analysis. A practical approach, The practical approach series IRL Press, New York, pp 43–72

Kennedy JF, Stevenson DL, White CA, Viikari L (1989) The chromatographic behaviour of a series of fructooligosaccharides derived from levan produced by the fermentation of sucrose by Zymomonas mobilis. Carbohydr Polym 10:103–113

Khandurina J, Guttman A (2005) High resolution capillary electrophoresis of oligosaccharide structural isomers. Chromatographia 62:S37–S41

Knapp DR (1979) Introduction. Handbook of analytical derivatization reactions. Wiley, New York.

Kogelberg H, Piskarev VE, Zhang Y, Lawson AM, Chai W (2004) Determination by electrospray mass spectrometry and 1H-NMR spectroscopy of primary structures of variously fucosylated neutral oligosaccharides based on the iso-lacto-N-octaose core. Eur J Biochem 271:1172–1186

Koizumi K (2002) Introduction In: El-Rassi Z (ed) HPLC of carbohydrates on graphitized carbon columns. J Chromatogr Lib Elsevier 66:103–119

Kunz C, Rudloff S, Hintelmann A, Pohlentz G, Egge H (1996) High-pH anion-exchange chromatography with pulsed amperometric detection and molar response factors of human milk oligosaccharides. J Chromatogr B 685:211–221

Kurimoto A, Daikoku S, Mutsuga S, Kanie O (2006) Analysis of energy-resolved mass spectra at MSn in a pursuit to characterize structural isomers of oligosaccharides. Anal Chem 78:3461–3466

Laine RG, Sweeley CC (1973) O-Methyl oximes of sugars. Analysis as O-trimethylsilyl derivatives by gas-liquid chromatography and mass spectrometry. Carbohydr Res 27:199–213

Lamari FN, Kuhn R, Karamanos NK (2003) Derivatization of carbohydrates for chromatographic, electrophoretic and mass spectrometric structure analysis. J Chromatogr B-Anal Technol Biomed Life 793:15–36

Langer SH, Pantages P, Wender I (1958) Gas-liquid chromatographic separation of phenols as trimethylsilyl ethers. Chem Ind 50:1664–1665

Lee YH, Lin TI (1996) Determination of carbohydrates by high-performance capillary electrophoresis with indirect absorbance detection. J Chormatogr B 681:87–97

Liu J, Waterhouse AL, Chatterton NJ (1991) Proton and carbon chemical-shift assignments for 6-kestose and neokestose from two-dimensional NMR measurements. Carbohydr Res 217:43–49

Lopez MG, Mancilla-Margalli NA, Mendoza-Diaz G (2003) Molecular structures of fructans from Agave tequilana Weber var. azul. J Agr Food Chem 51:7835–7840

Lopez-Molina D, Navarro-Martínez MD, Rojas-Melgarejo F, Hiner ANP, Chazarra S, Rodríguez-López JN (2005) Molecular properties and prebiotic effect of inulin obtained from artichoke (Cynara scolymus L.) Phytochem 66:1476–1484

Mabel MJ, Sangeetha PT, Latel K, Srinivasan K, Prapulla SG (2008) Physicochemical characterization of fructooligosaccharides and evaluation of their suitability as a potential sweetener for diabetics. Carbohydr Res 343:56–66

Maischberger T, Nguyen T, Splechtna B, Peterbauer C, Lettner HP, Lorenz W, Haltrich D (2007) Cloning and expression of b-galactosidase genes from Lactobacillus reuteri in Escherichia coli and the production of prebiotic galacto-oligosaccharides. J Biotechnol 131 S223

Martínez-Castro I, Calvo MM, Olano A (1987) Chromatographic determination of lactulose. Chromatographia 23:132–136

Martínez-Castro I, Olano A (1981) Ready detection of small amounts of lactulose in dairy-products by thin-layer chromatography. Chromatographia 14:621

Martinez-Villaluenga C, Cardelle-Cobas A, Corzo N, Olano A, Villamiel M (2008) Optimization of conditions for galactooligosaccharide synthesis during lactose hydrolysis by b-galactosidase from Kluyveromyces lactis (Lactozym 3000 L HP G). Food Chem 107:258–264

Martínez-Villaluenga C, Cardelle-Cobas A, Olano A, Corzo N, Villamiel M, Jimeno ML (2008) Enzymatic synthesis and identification of two trisaccharides produced from lactulose by transgalactosylation. J Agric Food Chem 56:557–563

Matamoros-Fernández LE, Obel N, Vibe Scheller H, Roepstorff P (2003) Characterization of plant oligosaccharides by matrix-assisted laser desorption/ionization and electrospray mass spectrometry. J Mass Spectrom 38:427–437

Matamoros-Fernández LE, Obel N, Vibe Scheller H, Roepstorff P (2004) Differentiation of isomeric oligosaccharide structures by ESI tandem MS and GC-MS. Carbohydr Res 339:655–664

Mayer J, Conrad J Klaiber I, Lutz-Wahl S, Beifuss U, Fischer L (2004) Enzymatic production and complete nuclear magnetic resonance assignment of the sugar lactulose. J Agric Food Chem 52:6983–6990

McCleary BV, Rossiter P (2004) Measurement of novel dietary fibers. J.A.O.A.C. 87:707–717

McGinnis GD, Biermann CJ (1989) Analysis of monosaccharides as per-O-acetylated aldononitrile (PAAN) derivatives by gas-liquid chromatography (GLC). In: Analysis of carbohydrates by GLC and MS. CRC Press, Florida, pp. 119–126

Mehra R, Kelly P (2006) Milk oligosaccharides: Structural and technological aspects. Int Dairy J 16:1334–1340

Molnárl-Perl I, Horvath K (1997) Simultaneous quantitation of mono-, di and trisaccharides as their TMS ether oxime derivatives by GC-MS: I. In model solutions. Chromatographia 45:321–327

Montañés F, Fornari T, Martín-Álvarez PJ, Montilla A, Corzo N, Olano A, Ibáñez E (2007) Selective fractionation of disaccharide mixtures by supercritical CO_2 with ethanol as co-solvent. J Supercrit Fluids 41:61–67

Montilla A, Moreno FJ, Olano A (2005) A reliable gas capillary chromatographic determination of lactulose in dairy samples. Chromatographia 62(5):311–314

Montilla A, Van de Lagemaat J, Olano A, del Castillo MD (2006) Determination of oligosaccharides by conventional high-resolution gas chromatography. Chromatographia 63:453–458

Morales V, Sanz ML, Olano A, Corzo N (2006) Rapid separation on activated charcoal of high oligosaccharides in honey. Chromatrographia 64:233–238

Morelle W, Michalski JC (2005) Glycomics and mass spectrometry. Curr Pharm Des 11:2615–1645

Moreno FJ, Quintanilla-López JE, Lebrón-Aguilar R, Olano A, Sanz ML (2008) Mass spectrometric characterization of glycated beta-lactoglobulin peptides derived from galacto-oligosaccharides surviving the in vitro gastrointestinal digestion. J Am Soc Mass Spectrom. 19,927–937

Moura P, Cabanas S, Lourenco P, Girio F, Loureiro-Dias MC, Esteves MP (2008) In vitro fermentation of selected xylo-oligosaccharides by piglet intestinal microbiota LWT. Food Sci Technol 1–10

Mujoo R, Ng PKW (2003) Physicochemical properties of bread baked from flour blended with immature wheat meal rich in fructooligosaccharides. J Food Sci 68:2448–2452

Nguyen DT, Lerch H, Zemann A, Bonn G (1997) Separation of derivatized carbohydrates by co-electroosmotic capillary electrophoresis. Chromatographia 46:113–121

Ninonuevo MR, An H, Yin H, Killeen K, Grimm R, Ward R, German B, Lebrilla C (2005) Nanoliquid chromatography-mass spectrometry of oligosaccharides employing graphitized carbon chromatography on microchip with a high-accuracy mass analyzer. Electrophoresis 26:3641–3649

Ninonuevo MR, Ward RE, LoCascio RG, German JB, Freeman SL, Barboza M, Mills DA, Lebrilla CB (2007) Methods for the quantitation of human milk oligosaccharides in bacterial fermentation by mass spectrometry. Anal Biochem 361:15–23

Nishimura T, Ishihara M, Ishii T, Kato A (1998) Structure of neutral branched xylooligosaccharides produced by xylanase from in situ reduced hardwood xylan. Carbohydr Res 308:117–122

Oefner PJ, Vorndran AE, Grill E, Huber C, Bonn GK (1992) Capillary zone electrophoretic analysis of carbohydrates by direct and indirect UV detection. Chromatographia 34:308–316

Ohara H, Owaki M, Sonomoto K (2006) Xylooligosaccharide fermentation with Leuconostoc lactis. J Biosci Bioeng 101:415–420

Okada H, Fukushi E, Onodera S, Nishimoto T, Kawabata J, Kikuchi M, Shiomi N (2003) Synthesis and structural analysis of five novel oligosaccharides prepared by glucosyltransfer from β-D-glucose 1-phosphate to isokestose and nystose using Thermoanaerobacter brockii kojibiose phosphorylase. Carbohydr Res 338 (9):879–885

Park NY, Baek NI, Cha J, Lee SB, Auh JH, Park CS (2005) Production of a new sucrose derivative by transglycosylation

of recombinant Sulfolobus shibatae beta-glycosidase. Carbohydr Res 340: 1089–1096

Paroni R, Fermo I, Molteni L, Folini L, Pastore MR, Mosca A, Bosi E (2006) Lactulose and mannitol intestinal permeability detected by capillary electrophoresis. J Chromatogr B 834:183–187

Pasanen S, Jänis J, Vainiotalo P (2007) Cello-, malto- and xylooligosaccharide fragmentation by collision-induced dissociation using QIT and FT-ICR mass spectrometry: A systematic study. Int J Mass Spectrom 263:22–29

Paseephol T, Small DM, Sherkat F (2008) Lactulose production from milk concentration permeate using calcium carbonate-based catalysts. Food Chem 111: 283–290

Petzelbauer I, Zeleny R, Reiter A, Kulbe KD, Nidetzky B (2006) Development of an Ultra-High-Temperature process for the enzymatic hydrolysis of lactose: II. Oligosaccharide formation by two thermostable b-glycosidases. Biotechnol Bioeng 69:140–149

Plocek J, Novotny MV (1997) Capillary zone electrophoresis of oligosaccharides derivatized with N-(4-aminobenzoyl)–glutamic acid for ultraviolet absorbance detection. J Chromatogr A 757:215–223.

Qin J, Steenvoorden RJJM, Chait BT (1996) A practical ion trap mass spectrometer for the analysis of peptides by matrix assisted laser desorption-ionization. Anal Chem 68:1784–1791

Rantanen H, Virkki L, Tuomainen P, Kabel M, Schols Hb, Tenkanen M (2007) Preparation of arabinoxylobiose from rye xylan using family 10 Aspergillus aculeatus endo-1,4-β-D-xylanase. Carbohydr Polym 68:350–359

Rathbone EB (1985) Nuclear magnetic resonance spectroscopy in the structural analysis of food-related carbohydrates. In: Birch GG (eds) Analysis of food carbohydrate. Elsevier, London, pp. 149–224

Reiffova K, Nemcova R (2006) Thin-layer chromatography analysis of fructooligo-saccharides in biological samples. J Chromatogr A 1110:214–221

Reiffová K, Podolonovicová J, Onofrejová L, Preisler J, Nemcová R (2007) Thin-layer chromatography and matrix-assisted laser desorption/ionization mass spectrometric analysis of oligosaccharides in biological samples. J Planar Chromatogr 20:19–25

Reis A, Domingues MRM, Domínguez P, Ferrer-Correia AJ, Coimbra MA (2003a) Positive and negative electrospray ionisation tandem mass spectrometry as a tool for structural characterisation of acid released oligosaccharides from olive pulp glucuronoxylans. Carbohydr Res 338:1497–1505

Reis A, Domingues MRM, Ferrer-Correia AJ, Coimbra MA (2003b) Structural characterisation by MALDI-MS of olive xylo-oligosaccharides obtained by partial acid hydrolysis. Carbohydr Polym 53:101–107

Rocklin RD, Pohl CA (1983) Determination of carbohydrates by anion exchange chromatography with pulsed amperometric detection. J Liquid Chromatogr 6:1577–1590

Rohrer JS (2003) High performance anion-exchange chromatography with pulse amperometric detection for the determination of oligosaccharides in foods and agricultural products. In: Eggleston G and Coté GL (eds) Oligosaccharides in food and agriculture. ACS, Washington, DC, pp. 16–31

Rolf D, Gray GR (1984) Analysis of the linkage positions in D-fructofuranosyl residues by the reductive-cleavage method. Carbohydr Res 131:17–28

Ronkart SN, Blecker CS, Fourmanoir H, Fougnies C, Deroanne C, Van Herck JC, Paquot M (2007) Isolation and identification of inulooligosaccharides resulting from inulin hydrolysis. Anal Chim Acta 604:81–87

Rousseaua V, Lepargneurb JP, Roquesc C, Remaud-Simeond M, Paul F (2005) Prebiotic effects of oligosaccharides on selected vaginal lactobacilli and pathogenic microorganisms. Anaerobe 11:145–153

Ruiz-Matute AI, Sanz ML, Corzo N, Martín-Álvarez PJ, Ibáñez E, Martínez-Castro I, Olano A (2007) Purification of lactulose from mixtures with lactose using pressurized liquid extraction with ethanol-water at different temperatures. J Agric Food Chem 55:3346–3350

Sangeetha PT, Ramesh MN, Prapulla SG (2005) Recent trends in the microbial production, analysis and application of fructooligosaccharides. Trends Food Sci Technol 16:442–457

Sanz ML, Corzo-Martínez M, Rastall RA, Olano A, Moreno FJ (2007) Characterization and in vitro digestibility of bovine β-lactoglobulin glycated with galacto-oligosaccharides. J Agric Food Chem 55:7916–7925

Sanz ML, Côté GL, Gibson GR, Rastall RA (2006) Selective fermentation of gentio-biose derived oligosaccharides by human gut bacteria. Influence of molecular weight. FEMS Microbiol Ecol 56:383–388

Sanz ML, Martinez-Castro I (2007) Recent developments in sample preparation for chromatographic analysis of carbohydrates. J Chromatogr A 1153:74–89

Sanz ML, Polemis N, Morales V, Corzo N, Drakoularakou A, Gibson GR, Rastall RA (2005) In vitro investigation into the potential prebiotic activity of honey oligosaccharides. J Agric Food Chem 53:2914–2921

Sanz ML, Sanz J, Martinez-Castro I (2002) Characterization of O-trimethylsilyl oximes of disaccharides by gas chromatography-mass spectrometry. Chromatographia 56:617–622

Schmid D, Behnke B, Metzger J, Kuhn R (2002) Nano-HPLC-mass spectrometry and MEKC for the analyis of oligosaccharides from human milk. Biomed Chromatogr 16:151–156

Schütz K, Muks E, Carle R, Schieber A (2006) Separation and quantification of inulin in selected artichoke (Cynara scolymus L.) cultivars and dandelion (Taraxacum officinale WEB ex. WIGG.) roots by high-performance anion exchange chromatography with pulse amperometric detection. Biomed Chromatogr 20:1295–1303

Seipert RR, Barboza M, Ninonuevo MR, LoCascio RG, Mills DA, Freeman SL, German JB, Lebrilla CB (2008) Analysis and quantitation of fructooligosaccharides using Matrix-Assisted Laser Desorption/Ionization Fourier Transform Ion Cyclotron Resonance Mass Spectrometry. Anal Chem 80:159–165

Shen Z, Warren CHD, Newburg DS (2000) High-performance capillary electrophoresis of sialylated oligosaccharides of human milk. Anal Biochem 279:37–45

Shen Z, Warren CHD, Newburg DS (2001) Resolution of structural isomers of sialylated oligosaccharides by capillary electrophoresis. J Chromatogr 921: 315–321

Simó C, Barbas C, Cifuentes A (2005) Capillary electrophoresis-mass spectrometry in food analysis. Electrophoresis 26:1306–1318

Soga T, Serwe M (2000) Determination of carbohydrates in food samples by capillary electrophoresis with indirect UV detection. Food Chem 69:339–344

Song JF, Weng MQ, Wu SM, Xia QCh (2002) Analysis of neutral saccharides in human milk derivatized with 2-aminoacridone by capillary electrophoresis with laser-induced fluorescence detection. Anal Biochem 304:126–129

Sosulski FW, Elkowicz L, Reichert RD (1982) Oligosaccharides in eleven legumes and their air-classified protein and starch fractions. J Food Sci 47:498–502

Spina E, Sturiale L, Romeo L, Impallomeni G, Garozzo D, Waidelich D, Glueckmann M (2004) New fragmentation mechanisms in matrix-assisted laser desorption/ionization time-of-flight/time-of-flight tandem mass spectrometry of carbohydrates. Rapid Commun Mass Spectrom 18:392–398

Splechtna B, Nguyen TH, Steinböck M, Kulbe KD, Lorenz W, Haltrich D (2006) Production of prebiotic galactooligosaccharides from lactose using b-galactosidases from Lactobacillus reuteri. J Agric Food Chem 54:4999–5006

St Hilaire PM, Cipolla L, Tedebark U, Meldal M (1998) Analysis of organic reactions by thin layer chromatography combined with matrix-assisted laser desorption/ionization time-of-flight mass spectrometry. Rapid Commun Mass Spectrom 12: 1475–1484

Stumm I, Baltes W (1997) Analysis of the linkage positions in polydextrose by the reductive cleavage method. Food Chem 59:291–297

Sweeley CC, Bentley R, Makita M, Wells WW (1963) Gas-liquid chromatography of trimethylsilyl derivatives of sugars and related substances. J Am Chem Soc.85:2497–2507

Takahashi N, Fukushi E, Onodera S, Benkeblia N, Nishimoto T, Kawabata J Shiomi N (2007) Three novel oligosaccharides synthesized using Thermoanaerobacter brockii kojibiose phosphorylase. Chem Central J 1:18. doi:10.1186/1752–153X-1–18

Takahashi N, Okada H, Fukushi E, Onodera S, Nishimoto T, Kawabata J, Shiomi N (2005) Structural analysis of six novel oligosaccharides synthesized by glucosyl transfer from β-D-glucose 1-phosphate to raffinose and stachyose using Thermoanaerobacter brockii kojibiose phosphorylase. Tetrahedron: Asymmetry 16:57–63

Urashima T, Bubb WA, Messer M, Tsuji Y, Taneda Y (1994) Studies of the neutral trisaccharides of goat (Capra hircus) colostrum and of the one- and two-dimensional 1H and 13C NMR spectra of 6'-N-acetylglucosaminyllactose. Carbohydr Res 262:173–184

Vaccari G, Lodi G, Tamburini E, Bernardi T, Tosi S (2001) Detection of oligosaccharides in sugar products using planar chromatography. Food Chem 74:99–110

Vliegenthart JFG, Dorland L, van Halbeek H (1983) 1H-nuclear magnetic resonance spectroscopy as a tool in the structural analysis of carbohydrates related to glycoproteins. Adv Carbohydr Chem Biochem 41:209–374

Vorndran AE, Oefner PJ, Scherz H, Bonn GK (1992) Indirect UV detection of carbohydrates in capillary zone electrophoresis. Chromatogr 33:163–168

Wang J, Sporns P, Low NL (1999) Analysis of food oligosaccharides using MALDI-MS: Quantification of fructooligosaccharides. J Agric Food Chem 47:1549–1557

Wienecke R, Klein S, Geyer A, Loos E (2007) Structural and functional characterization of galactooligosaccharides in Nostoc commune: β-D-galactofuranosyl-(1→6)-[β-D-galactofuranosyl-(1→6)]-2-β-D-1,4-anhydrogalactitol and β-(1→6)-galactofuranosylated homologues. Carbohydr Res 342:2757–2765

Wilson ID (1999) The state of the art in thin-layer chromatography-mass spectrometry: a critical appraisal. J Chromatogr A 856:429–442

Wuhrer M, Deelder AM, Hokke CH (2005) Protein glycosylation analysis by liquid chromatography–mass spectrometry. J Chromatogr B 825:124–133

Yamamori A, Onodera S, Kikuchi M, Shiomi N (2002) Two novel oligosaccharides formed by 1F-fructosyltransferase purified from roots of Asparagus (Asparagus officinalis L.) Biosci Biotechnol Biochem 66:1419–1422

Ye F, Yan X, Xu J, Chen H (2006) Determination of aldoses and ketoses by GC-MS using differential derivatisation. Phytochem Anal 17:379–383

Yin JF, Yang GL, Wang SM, Chen Y (2006) Purification and determination of stachyose in Chinese artichoke (Stachys Sieboldii Miq.) by high-performance liquid chromatography with evaporative light scattering detection. Talanta 70:208–212

Zaia J (2004) Mass spectrometry of oligosaccharides. Mass Spectrom Rev 23:161–227

14 Manufacture of Prebiotics from Biomass Sources

Patricia Gullón · Beatriz Gullón · Andrés Moure · José Luis Alonso ·
Herminia Domínguez · Juan Carlos Parajó

14.1   Introduction

Biomass from plant material is the most abundant and widespread renewable raw material for sustainable development, and can be employed as a source of polymeric and oligomeric carbohydrates. When ingested as a part of the diet, some biomass polysaccharides and/or their oligomeric hydrolysis products are selectively fermented in the colon, causing prebiotic effects.

This work deals with the chemical structure, manufacture, purification, properties and applications of biomass-derived saccharides (xylans, mannans, arabinogalactans, pectins and/or their respective oligomeric products) which can be employed as food ingredients to achieve prebiotic effects closely related to the ones caused by well known prebiotic oligosaccharides, such as fructooligosaccharides, galactooligosaccharides, and isomaltooligosaccharides.

The colonic fermentation of carbohydrates is a complex problem, because the end metabolic products of given bacterial species can be used as a substrates by others, and some microorganisms may grow upon substrates that they are not able to ferment. The prebiotic behavior of dietary components is mainly related to their ability to modulate the colonic microbiota, enhancing the growth of beneficial bacteria (particularly, bifidobacteria and lactobacilli) selectively.

Prebiotic effects can be achieved by both polymeric and oligomeric carbohydrates. In the chemical processing of biomass polysaccharides, the treatments needed for separation, purification and product tailoring may result in the production of fractions with decreased degrees of polymerization (DP) in respect to the native polymers, making the distinction between "dietary fiber" (with polymeric nature) and "oligosaccharides" difficult. Usually, the term "oligosaccharide" is reserved for DP in the range 3–10 (Tungland and Meyer, 2002), whereas fractions of higher molecular weight are considered as "dietary fiber." However, the definition of this latter term has seen several revisions since it was proposed in 1953, and a variety of definitions are available based either on analytical determinations by specific methods or on their biological effects. In fact, no international consensus has been reached on a definition yet. The following definition has been proposed by the AACC Dietary Fiber Definition Committee (2001): "Dietary fiber is the edible parts of plants or analogous carbohydrates that are resistant to digestion and absorption in the human small intestine with complete or partial fermentation in the large intestine. Dietary fiber includes polysaccharides, oligosaccharides, lignin, and associated plant substances. Dietary fibers promote beneficial physiological effects, including laxation, and/or blood cholesterol attenuation, and/or blood glucose attenuation."

On the other hand, the situation has become more complicated by the fact that some "functional fibers," such as inulin-derived fructooligosaccharides, polydextrose, some polyols, and resistant starch, cannot be determined as dietary fiber by the standard AOAC methods, and they could not be confirmed as dietary fiber for labeling purposes. Finally, it has to be taken into account that in many studies, particularly in these where the distribution of molecular weights was unknown, the whole set of soluble fragments from polysaccharide degradation have been considered as "oligosaccharides" (no matter of the DP); and that some DP2 saccharides (such as xylobiose) have been considered as oligosaccharides in food studies (Vázquez et al., 2000). Based on the above ideas, the term "dietary fiber" is employed in this work to denote high molecular weight, polymeric fractions; and the term "oligosaccharides" is reserved for compounds of lower molecular weight, even if they correspond to a broad DP range.

14.2   Xylans and Xylan-Derived Products

14.2.1 Structure of Xylans

Xylans represent an immense resource of biopolymers for practical applications, accounting for 25–35% of the dry biomass of woody tissues of dicots and lignified tissues of monocots, and occur up to 50% in some grasses and tissues of cereal grains. The structure of xylans depends on the source considered. The most common xylans are made up of a main backbone of xylose linked by β-$(1{\rightarrow}4)$ bonds, where the structural units are often substituted at positions C2 or C3 with arabinofuranosyl, 4-O-methylglucuronic acid, acetyl or phenolic substituents (Ebringerova et al., 2005). Xylans are usually named according to the most abundant substituents (arabinoxylans, glucuronoxylans, arabinoglucuronoxylans, etc.). ◉ *Figure 14.1* presents structures of typical xylans.

14.2.2 Manufacture of Xylooligosaccharides

The partial hydrolysis of xylans leads to xylooligosaccharides or substituted oligosaccharides (here denoted XO), whose structures depend on the features of native xylans and on the processing conditions employed in their manufacture. Recent studies have been reported on the production of XO by chemical processing of a variety of raw materials, including crop residues, sugarcane bagasse, hardwoods, corncobs, barley and rice hulls, brewery spent grains, almond shells, corn stover and corn fiber, flax, straws, flours, bamboo and fruit wastes.

The breakdown of the xylan chains can be accomplished by different methods (Moure et al., 2006), such as:

- Direct enzymatic hydrolysis of susceptible substrates (for example, isolated xylan or glucuronoxylan, xylan-containing cellulose pulps, fruit wastes, or wheat flour arabinoxylan).
- Chemical processing (for example, by aqueous treatments with water or steam, or in media containing externally-added mineral acids) of native feedstocks or xylan-containing pre-processed solids (for example, sulphite cellulose).
- Combined chemical and enzymatic treatments, where the chemical treatments can be employed either to make the raw material accessible to enzymes or to yield soluble xylan fragments from a native feedstock (for example, alkaline processing of xylan-rich, native substrates followed by xylanolytic hydrolysis).

Our group has paid special attention to the production of XO by treatments with hot, compressed water (also called hydrothermal treatments, autohydrolysis, hydrothermolysis or water prehydrolysis), which cause the acid-catalyzed degradation of xylan. When the aqueous processing of xylan-containing substrates is carried out under suitable operational conditions, the hemicellulosic chains are β-(1→ 4) xylan

Arabinoxylan

4-O-methyl glucuronoxylan

Arabino-4-O-methyl glucuronoxylan

◻ Figure 14.1
Structure of typical xylans.

progressively broken down by the hydrolytic action of hydronium ions (generated from water autoionization and from *in situ* generated organic acids), yielding soluble products (mainly oligosaccharides), and leaving both cellulose and lignin in solid phase with little chemical alteration.

Autohydrolysis presents several advantages, such as:

- Environmentally friendly operation (no chemicals except water and lignocellulosic feedstock are needed, and no sludges are generated in the neutralization stages).
- Compatibility with further fractionation and/or hydrolysis processing of the spent solids to separate cellulose and lignin, enabling the possibility of an integral utilization of the raw material. The autohydrolysis liquors contain XO and soluble reaction byproducts, and the exhausted solids from autohydrolysis are suitable substrates for the enzymatic hydrolysis of cellulose (with or without a previous delignification step), or for applications as feed, fuel, or construction materials. This operational mode follows the "biorefinery approach," which is expected to decrease greenhouse gas emissions and to be compatible with a sustainable development.
- XO generated by autohydrolysis treatments present a rich substitution pattern, conserving the major structural features of the native xylan (Kabel et al., 2002a). For example, only a part of the ester groups are split in hydrothermal treatments, in comparison with the saponification caused by processes involving alkaline stages.

In autohydrolysis treatments, XO behave as typical reaction intermediates, and their maximal concentrations are achieved under medium-severity conditions. Optimization studies for XO production can be carried out using two different approaches:

- Pseudohomogeneous kinetic modeling.
- Empirical assessment based on the severity factor.

The kinetic modeling of lignocellulose autohydrolysis is a complex problem involving the catalytic, heterogeneous reaction of a solid substrate by means of a number of different reactions. In order to simplify the problem, the kinetic modeling of xylan solubilization has been carried out assuming irreversible, pseudohomogeneous, first-order kinetics, of several individual reactions, which are governed by kinetic coefficients following the Arrhenius equation that are assumed to be independent from time, particle size and acidity. This approach has been employed in studies involving a variety of substrates, including corncobs, brewery spent grains, rice and barley husks, hardwoods and almond shells. In some cases, the progressive breakdown of xylan chains has been interpreted through some or all of the following hypotheses:

- Xylan can be made up of two fractions with different susceptibility to hydrolytic reactions
- One or both xylan fractions are decomposed to give high-molecular weight oligomers

- High molecular weight oligomers decompose to give low-molecular weight oligomers
- Xylooligomers are decomposed to give xylose
- Xylose and terminal groups of xylooligomers can be dehydrated to furfural
- Furfural can be converted into degradation products

Additionally, some studies consider the simultaneous breakdown of arabinosyl chains to give arabinose, and the generation of acetic acid from acetyl groups.

Based on experimental data, some studies have reported on the values of preexponential factors, activation energies and mass fractions of susceptible xylan, enabling the calculation of concentration of the various chemical species under defined experimental conditions.

Alternatively to the modeling by pseudohomogeneous kinetics, the severity factor R_0 (a parameter that includes the effects of both time and temperature) has been employed to give a simplified, empirical interpretation of the measured effects (Abatzoglou et al., 1992).

R_0 can be calculated using the equation:

$$R_o = \int_0^t \exp\left(\frac{T - T_r}{\omega}\right) dt$$

where T is temperature (°C), t is time (min), and T_r and ω are parameters with values of 100 and 14.75°C, respectively. In first-order reaction systems, the measured variable and R_0 are expected to have an exponential interrelationship under either isothermal or non-isothermal operation.

R_0 was introduced to measure the hemicellulose conversion, but its ability for comparing on a quantitative basis the effects caused by hydrothermal treatments carried out under different conditions of time and temperature allowed its utilization for assessing other reaction effects (Garrote et al., 2003), including:

- Generation of XO, monosaccharides and furfural
- DP distribution of XO
- Degree of deacetylation and acidity of the reaction media
- Amount, composition, degree of polymerization and enzymatic digestibility of the exhausted solids

- Release of low molecular weight phenolics
- Antioxidant activity of byproducts present in autohydrolysis media

14.2.3 Purification of Xylooligosaccharides

In order to produce food-grade XO (whose usual commercial purity lies in the range 75–95%), the liquors from xylan conversion have to be refined. Refining involves the selective removal of undesired compounds (usually, monosaccharides and non-saccharide compounds) to obtain a concentrate with a XO content as high as possible.

Purification of XO obtained by enzymatic processing of substrates containing susceptible xylan is facilitated by the previous chemical processing of the raw material, as well as by the specific action of xylanases. On the other hand, the purification of XO present in crude autohydrolysis liquors is a complex problem, because the degradation of xylan progresses simultaneously with several side-processes (including extractive removal, solubilization of acid-soluble lignin, acetic acid generation from acetyl groups, sugar degradation to furfural or hydroxymethylfurfural, neutralization of ashes, and participation of Maillard reactions involving either the protein fraction of the feedstock or its hydrolysis products). All of these effects result in the presence of undesired, non-saccharide compounds in liquors from hydrothermal processing, which have to be removed (at least in part) before utilization.

Several strategies have been proposed for refining crude liquors. For example, in single-stage autohydrolysis reactions, a significant part of the dissolved feedstock may correspond to easily extractable compounds (for example, waxes, low molecular weight phenolics and soluble inorganic components). This kind of compound can be removed by a previous extraction stage with solvents or water. When the raw material is pretreated with hot water under mild conditions, the hemicellulosic polymers remain in solid phase almost untouched and ready for further hydrolytic conversion under harsher conditions, whereas the easily extractable compounds are separated in liquors. Ethanol extraction has been carried out before autohydrolysis processing to achieve these objectives. Alternatively, two consecutive aqueous treatments of increasing severity (the first one for removing extractives, and the second one for generating XO from xylan) can be performed. This strategy has resulted in cleaner liquors and/or in XO-containing solutions with enhanced susceptibility to further refining treatments.

However, in most cases, a sequence of different physicochemical treatments may be necessary to achieve XO concentrates of commercial purity. Focusing on the refining of autohydrolysis liquors, the mass ratio XO/total solutes can be improved by means of several separation methods (which can be assayed individually or in combination), including:

- Solvent extraction
- Vacuum evaporation
- Solvent precipitation
- Freeze-drying followed by solvent extraction
- Chromatographic separation
- Adsorption
- Ion exchange
- Membrane technologies

The corresponding objectives and effects are considered in the following paragraphs.

Ethyl acetate extraction of autohydrolysis liquors enables the integrated benefit of the several fractions from autohydrolysis of lignocellulosics, yielding:

- A refined aqueous phase having an increased mass fraction of oligomers
- An organic phase containing antioxidant compounds

Processing of lignocellulosics by sequential stages of autohydrolysis and ethyl acetate extraction leads to:

- An oligosaccharide-containing, refined aqueous phase
- An organic phase, mainly made up of phenolics and extractive-derived compounds, for which antioxidant applications have been proposed
- A solid phase enriched in cellulose, which also contains lignin, that can be fractionated by further treatments, or employed directly as a substrate for the enzymatic hydrolysis of cellulose

One of the most remarkable advantages of solvent extraction (and particularly, of ethyl acetate extraction) is the high selectivity of this method: for example, ethyl acetate extraction of rice husks autohydrolysis liquors resulted in the removal of 38% of the non-saccharide components, whereas just 5% of the initial XO were lost (Vegas et al., 2004). When the same operational procedure was applied to barley husks autohydrolysis liquors, 33% of the non-saccharide compounds was removed, in comparison with just 14% XO loss (Vegas et al., 2005).

However, the purification effects achieved by solvent extraction are usually too weak for practical purposes, and further physicochemical processing is needed to achieve products fulfilling the market specifications.

Gas chromatography coupled with mass spectrometry (GC-MS) analysis of the acetate-soluble fraction of autohydrolysis liquors from *Eucalyptus* wood showed the presence in extracts of hemicellulose-derived products (sugars and furfural) and non-saccharide components; the latter corresponding mainly to lipophilic extractives. Among them, stearic acid and palmitic acid were the main fatty acids, with minor amounts of oleic acid, 9,12 octadecanedienoic acid and tetradecanoic acid. Dehydroabietic acid was the major resin acid, whereas the phenolic compounds identified were gallic acid, vanillin, 1,2,3 trihydroxyben-zene, syringaldehyde, syringic acid and 3,4 dihydroxybenzoic acid (Vázquez et al., 2005). GC-MS analysis of the dichloromethane-soluble fraction of corncob autohydrolysis liquors allowed the identification of sugar-derived compo-unds (accounting for 35% of the extracted material), lignin-derived compounds (accounting for 59.2% of the extracted material), nitrogen-containing com-pounds (accounting for 1.1% of the extracted material) and fatty acids (account-ing for 4.7% of the extracted material). Furfural was the most abundant compound, followed by vanillin, 4-methylphenol and 4-vinylguaiacol (Garrote et al., 2007).

Vacuum evaporation has been assayed for increasing the XO concentration of autohydrolysis liquors, with simultaneous removal of volatile components (particularly acetic acid). Some flavors and/or their precursors can be also eliminated by evaporation. Depending on the case considered, vacuum evaporation can be the first step of a multistage process, or an intermediate one.

Ethanol precipitation has been assayed for isolating arabinoxylooligosacchar-ides from enzymatic hydrolysis of wheat flour, whereas precipitation of XO from autohydrolysis liquors has been assayed using ethanol, acetone or 2-propanol. The purification effects achieved and the recovery yields depended on the solvent and on the raw material employed, which controls the XO substitution pattern and the possible presence of stabilizing, non-saccharide components. Solvent precipitation of raw *Eucalyptus globulus* wood autohydrolysis liquors was strongly affected by the presence of even minimal amounts of water, which decreased the precipitation of hemicellulose-derived products (Vázquez et al., 2005). This behavior was ascribed to the XO substitution pattern (with comparatively high proportion of uronic groups) and/or to the presence of stabilizing non-saccharide components in *Eucalyptus* autohydrolysis liquors. Better results have been reported for the processing of rice husk autohydrolysis liquors (Vegas et al., 2004): ethanol precipitation of ethyl-acetate extracted liquors led to refined products with contents of non-saccharide compounds as low as 9 wt.%, but at a limited recovery yield. Comparatively, 2-propanol resulted in limited removal of non-saccharide compounds (which accounted for 12 wt.% of the concentrate), but the XO recovery yield increased up to 75.8%. Acetone presented a behavior related to the one observed for 2-propanol, with slightly lower recovery yield. Ethanol precipitation was not suitable for the processing of barley husks autohydrolysis liquors (owing to operational problems), whereas acetone and propanol achieved similar effects, with some advantage for acetone, which led to slightly higher recovery yield and provided XO concentrates containing 19 wt.% of non-saccharide products (Vegas et al., 2005).

As the presence of even minimal amounts of water affected the precipitation of hemicellulose-derived products negatively, the possibility of subjecting the autohydrolysis liquors to freeze-drying and extracting the resulting solids with organic solvents has been explored. The freeze drying-extraction of *Eucalyptus* wood autohydrolysis liquors, previously extracted with ethyl acetate and concentrated, has been assessed for operation with ethanol, 2-propanol and acetone (Vázquez et al., 2005). Extraction with 2-propanol resulted in negligible purification effects, but better results were achieved operating with acetone, which led to a concentrate containing 70.6 wt.% of substituted oligosaccharides. These latter compounds presented increased contents of uronic and acetyl substituents compared to the average values determined for the products present in the crude liquors. Ethanol extraction of the freeze-dried solids resulted in moderate recovery yields, and in enhanced removal of non-saccharide compounds (about 50% of the initial amount). The overall purification effect obtained in the sequential processing of autohydrolysis liquors by ethyl acetate extraction, vacuum concentration, freeze drying and solvent precipitation were in the range reported for more sophisticated operational schemes involving two-stage extraction, chromatographic separation and ion exchange. In a related study, the freeze drying-extraction processing of ethyl acetate-treated autohydrolysis liquors from barley husks resulted in high percentages of solute recovery, but limited purification effects were achieved (Vegas et al., 2005).

Chromatographic separation of XO leading to high purity fractions has been carried out at an analytical scale. Samples from hydrothermally treated lignocellulosics have been fractionated by anion-exchange chromatography and size-exclusion chromatography, whereas chromatographic techniques have been employed for refining samples before structural characterization of XO by ^{13}C NMR, MALDI-TOF or nanospray mass spectrometry. In the same field, simulated moving bed chromatography and size-exclusion chromatography have been employed for purification of substituted XO. Other studies reported on chromatographic separation as a part of multistage processes.

Adsorption (using adsorbents such as activated charcoal, acid clay, bentonite, diatomaceous earth, aluminum hydroxide or oxide, titanium, silica and porous synthetic materials) has been used for purification of XO, following two alternative operational modes:

- XO-containing solutions (for example, obtained by enzymatic hydrolysis of either ammonia-pretreated or native corncobs or corn stover) have been adsorbed onto charcoal, and further eluted with ethanol. Interestingly, the DP range of the fractions coming from ethanol elution depends on the alcohol concentration, allowing the fractionation of XO on the basis of their molecular weight.
- Liquors containing XO can be contacted with charcoal to remove non-saccharide compounds. Under selected conditions, adsorption was higher for lignin-related products than for XO. The separation selectivity was dependent on the properties of the adsorbent (distribution of pore sizes and basic or acidic surface groups).

Ion-exchange resins have been employed for refining XO obtained by different reaction technologies (autohydrolysis or enzymatic processing) from a variety of raw materials (including rice and barley husks, crop residues and cellulose pulps). In several cases, the target products have been acidic XO. Ion exchange enables a variety of refining effects, including:

- Removal of charged, inorganic components.
- Removal of organic compounds, either by ion exchange or by sorption in the polymeric matrix.

Ion exchange results in efficient desalination, but its performance for removing colored compounds is limited. The literature has reported on the individual utilization of either cation- or anion-exchange resins, as well as on operation with mixed resin beds. Usually, ion exchange is employed for XO purification in combination with other separation technologies. A simple refining process (ethyl acetate extraction followed by anion exchange) has been reported for XO manufacture from barley husk autohydrolysis liquors (Vegas et al., 2005). Ethyl acetate extraction was carried out in three sequential stages, and the resulting aqueous phase was treated with Amberlite IRA 400 (a strong anion exchange, quaternary ammonium, gel-type resin) or Amberlite IRA 96 (a weak anion exchange, polyamine, macroreticular resin). This operational mode resulted in remarkable results: for example, treatment with IRA 96 led to a final concentrate containing solutes with the following composition: 11.3 wt.% monosaccharides, 70.8 wt.% of substituted XO, and 17.9 wt.% of non-saccharide compounds. Spectral data confirmed that this latter fraction was mainly made up of phenolic compounds and melanoidins. Processing of rice husk autohydrolysis liquors by evaporation, ethyl acetate extraction and anion exchange with IRA 400, resulted in a final product containing 9.1 wt.% of monosaccharides, 82.1 wt.% of substituted oligosaccharides and 8.8 wt.% of non-saccharide compounds (Vegas et al., 2004). Acid hydrolysis of the concentrate showed the presence of 3.6 wt.% of phenolic compounds linked to the oligosaccharide chains (determined as acid-soluble lignin), which are types of compounds that possess antioxidant activity and that could contribute to the health benefits of the product. Determination of the nitrogen content proved the presence of melanoidins in the final isolate, which were generated in the autohydrolysis media by reaction of proteins and sugars. As melanoidins also present antioxidant activity, they could contribute to the functional properties of the target compounds.

Membrane technologies are gaining importance for XO manufacture, as they can be used for a variety of purposes, including:

- Production of XO from pure xylan or xylan-containing substrates of lignocellulosic nature, operating in enzyme-containing membrane bioreactors
- Simultaneous DP reduction and fractionation of soluble xylan-derived products from autohydrolysis treatments, operating as in the previous point
- Removal of suspended solids from reaction media by ultrafiltration
- Refining of soluble xylan-derived products present in reaction media by ultrafiltration or nanofiltration (depending on the DP of the target compounds)
- Separation of undesired, low-molecular weight components (including monosaccharides and non-saccharide compounds) from XO- containing media by diafiltration through membranes with a suitable cut-off

- Fractionation of XO on the basis of their molecular weight
- Concentration of XO by solvent removal

In this field, ultrafiltration has been employed for refining almond shell autohydrolysis liquors, as an alternative to ethanol precipitation for isolation of arabinoxylooligosaccharides from wheat bran hydrolyzed with enzymes (Swennen et al., 2005), as well as to discard the high-molecular weight fractions coming either from enzymatic treatments of oxygen-delignified pulp slurry in the manufacture of acidic oligosaccharides (Izumi et al., 2004), or from auto-hydrolysis media of *Eucalyptus* wood (Gullón et al., 2008).

In studies dealing with the ultrafiltration and nanofiltration processing of XO-containing media, several trade-offs have been identified while selecting the most appropriated membrane and when choosing operating conditions (Vegas et al., 2008a). For a feasible commercial application of this technology, an optimization based on economical profit would be needed, since (as it often happens in separation processes) the trade-offs are ultimately between recovery and purity. Yuan et al. (2004) employed nanofiltration membranes for concentrating XO obtained at the pilot plant scale by enzymatic hydrolysis of xylan from steamed corncobs in a multistage process, which led to concentrated solutions containing xylobiose and xylotriose as major products. In the framework of an alternative multistage process, nanofiltration of pretreated rice husk autohydrolysis liquors, concentrated up to a mass ratio of 4.89 kg feed/kg concentrate, resulted in the removal of 20.9–46.9% of total monosaccharides and in the decrease in the mass ratio of non-saccharide compounds compared to the total non-volatile solutes, from 0.3181 to 0.2378 kg/kg, under conditions enabling the recovery of 67.2–92% of XO. Under the same conditions, almost complete recovery of acetylated oligosaccharides was achieved. These data confirm that the beneficial effects of nanofiltration include both the concentration and purification of XO; the latter effect being achieved through the preferential removal of both monosaccharides and non-saccharide compounds in permeate (Vegas et al., 2005). In the processing of rice husk autohydrolysis liquors, low molecular weight cut-off ultrafiltration and nanofiltration ceramic membranes are especially interesting for the processing of XO-containing liquors, as they allow higher flux values than polymeric membranes (Vegas et al., 2008a). Operation with a ceramic membrane of 50 kDa cut-off was successful for retaining fractions with DP > 7, and provided a method for removing both high-DP XO and suspended solids (present in the autohydrolysis liquors or formed during membrane processing) (Gullón et al., 2008). Finally, low cut-off membranes have been employed for concentrating XO-containing media by reverse osmosis (Izumi et al., 2004).

As stated before, obtaining commercial-grade XO involves multistage processing of liquors from the hydrolytic breakdown of the xylan-containing substrate. ◉ *Figure 14.2* presents simplified diagrams proposed for the manufacture of refined XO concentrates.

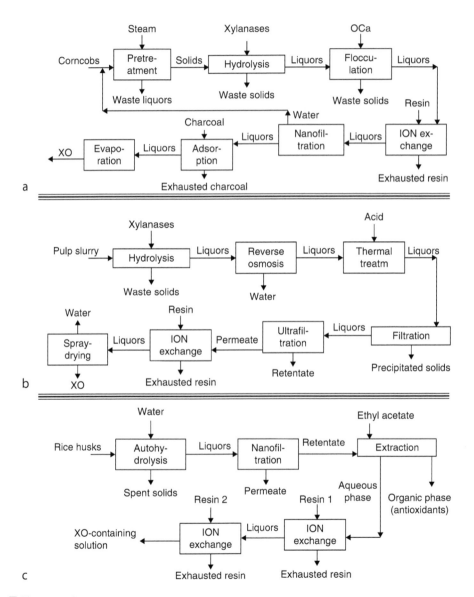

◉ Figure 14.2
Processes for XO manufacture and refining reported by: (a) Yuan et al., 2004; (b) Izumi et al., 2004; (c) Vegas et al., 2006.

14.2.4 DP Tailoring and Structure of Xylooligosaccharides

The biological activity of xylan-derived products depends on the molecular weight distribution. Even if low-DP oligosaccharides (in the range 2–4) have been preferred for food applications, it must be highlighted that compounds with higher molecular weights can also present valuable properties. In this field, experiments in rats have proved that long-chain xylooligosaccharides (DP in the range 5–20, with an average value of 12.3) can improve the intestinal function and present hypolipemic activities (Izumi and Kojo, 2003).

The DP distribution of xylan-hydrolysis products depends on the operational conditions employed in the overall processing. In the case of hydrothermal treatments, where the polymeric xylan contained in a given feedstock is converted into soluble products (which undergo further hydrolytic degradation in the reaction media), the DP distribution is strongly affected by the severity of the operational conditions: longer reaction time and/or higher temperatures lead to reaction products of decreased molecular weight. To assess this point, the influence of the reaction conditions on the DP distribution of the reaction products obtained by autohydrolysis of rice husks was studied. Rice husk samples were treated at 172 or 188°C for selected reaction times, in order to cover the severity range (measured in terms of the severity factor R_0 defined in Section 2.2) of practical interest. The DP distribution was measured by High Performance Size Exclusion Chromatograph (HPSEC) under reported conditions (Kabel et al., 2002a), and the chromatograms were integrated on the basis of the elution times determined for standard compounds, in order to estimate the relative proportions of soluble xylan-derived products having DP > 25, DP in the range 9–25 and DP < 9. The corresponding results are shown in ◉ *Figure 14.3*, which presents the relative abundance of the considered fractions as a function of the logarithm of the severity factor. The severity dependence of the proportions of low, medium and high molecular weight fractions presented a similar variation pattern for the two temperatures considered, confirming the validity of the R_0 parameter for assessing the modifications in molecular weight distribution. Operating at any of the considered temperatures, the proportions of compounds with DP > 25 accounted for a limited part of the total xylan-derived products even in treatments at low severities, and decreased slightly with severity. Under the severest conditions assayed, the high-molecular weight fractions accounted for about 2% of total xylan-derived products, no matter of the temperature considered. The low molecular weight fraction (DP < 9) was the predominant one, with proportions that increased slightly with severity owing to the increased hydrolysis. The relative amounts of medium molecular weight compounds

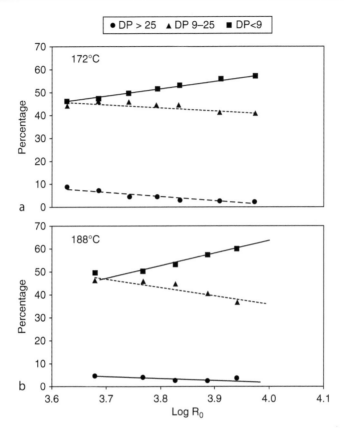

Relative proportions of XO with different DP range obtained in hydrothermal treatments of rice husks carried out at different temperatures and severities.

decreased continuously with severity (owing to the higher participation of depolymerization reactions compared to the ones causing the hydrolysis of high molecular weight fractions), to reach values around 40% under the severest conditions assayed. The main conclusion from this study was that 90–94% of the total rice husk xylan-derived products corresponded to oligosaccharides with DP < 25 operating at low severities, and that operating under harsh conditions (R_0 in the vicinity of 10^4 min), about 60% of the total xylan-derived compounds corresponded to fractions with DP < 9.

When low-DP XO are the target products, it must be taken into account that direct autohydrolysis under harsh conditions can be not suitable for operation: according to the kinetic models available, this operational strategy could result in increased generation of undesired compounds (particularly, monosaccharides and monosaccharide-degradation products), which should be removed in order to obtain purified XO, and could limit the yield in XO.

Alternatively, low DP-XO can be obtained by enzymatic DP reduction of autohydrolyis reaction products using endo-1,4-β-xylanases (also named 1,4-β-xylan xylanhydrolases, E. C. 3.2.1.8). This strategy allows the production of low DP-XO at high yields.

The effects of xylanases depend on the structures of both substrate and enzymes. Based on hydrophobic cluster analysis and similarities in amino acid sequences, xylanases have been classified into families 5, 8, 10, 11 and 43 of glycoside hydrolases. Most xylanases belong to the glycosyl hydrolase families 10 and 11 (denoted GH-10 and GH-11), which present different physicochemical properties (including structure, molecular weight, pI and thermal stability), and exert different catalytic activity in the presence of substituted xylopyranosil units. GH-10 xylanases are characterized by high molecular weights and low pI values, while the low molecular mass endoxylanases with high pI values are classified as glycanase family 11.

In order to avoid the production of xylose from XO, the enzymes selected for DP reduction should be free from xylobiase activity. For this reason, cloned endo-1, 4-β-xylanases are preferable to raw xylanase preparations from wild strains of xylanolytic microorganisms. Recent results dealing with the DP reduction of xylan-derived compounds without causing the degradation of xylobiose and low-molecular weight oligomers to xylose have been reported for commercial endo-xylanase preparations from cloned microorganisms. Shearzyme 2x (a GH-10 xylanase produced using recombinant DNA techniques from a strain of *Aspergillus oryzae* carrying the gene coding for the enzyme from *Aspergillus aculeatus*), Pentopan Mono BG (a GH-11 xylanase, obtained by heterologously expressing the *Thermomyces lanuginosus* in *Aspergillus oryzae*), and Pulpzyme HC (a GH-11 xylanase produced by submerged fermentation of a genetically modified *Bacillus* microorganism) have been employed for DP tailoring of xylan-derived products (Vegas et al., 2008b). The differences in structure and mode of action of enzymes resulted in different DP distribution of the hydrolysis products. In experiments with raw autohydrolysis liquors, Shearzyme 2x led to the highest proportion of oligomers with DP in the range 2–3, and to slightly higher amounts of compounds with DP > 7 than the GH-11 enzymes. Pentopan Mono BG and Pulpzyme HC gave oligomers with DP in the range 4–5 or 6–7 as major reaction products. Experiments with refined autohydrolysis liquors confirmed the superior ability of Shearzyme for obtaining compounds with DP in the range 2–3, and the comparative advantage of Pulpzyme for hydrolyzing XO chains to compounds with DP < 7.

Depending on the biomass feedstock, a variety of technologies have been employed for manufacturing refined, low-DP oligosaccharides. For example, an opposite philosophy to the one described above (enzymatic hydrolysis and physicochemical processing, instead of autohydrolysis and xylanase hydrolysis) has been suggested. The operational mode selected (xylanase processing of bleached kraft pulp, and further heating under acidic conditions) enabled the production of a mixture of xylose and XO having xylotriose, xylotetraose and xylopentaose as major components (Izumi et al., 2000).

The composition of XO depends on the nature of the native-xylan containing feedstocks employed as a raw material, on the possible pretreatments suffered by the raw material before performing the hydrolytic breakdown of xylan, and on the operational conditions employed in this latter stage. For example, when using hardwoods as feedstocks, preliminary processing under alkaline conditions to isolate xylan results in the saponification of acetyl groups, and the subsequent xylan hydrolysis results in non-acetylated XO. Oppositely, when native raw materials are subjected to autohydrolysis for XO production, the reaction products keep the major structural properties of the xylan present in the feedstock, presenting a rich substitution pattern.

Experimental studies have been carried out to assess the structure of XO from a variety of sources, including wheat bran, brewery's spent grains, corncobs and *Eucalyptus* wood (Kabel et al., 2002a). Under autohydrolysis conditions leading to the maximum XO yield, the xylose-degradation reactions were negligible, whereas arabinose was preferentially splitted off from the xylose-containing chains, and partially degraded to furfural and levulinic acid. Again, the relative amounts of arabinose and xylose in oligomers from autohydrolysis treatments depend on the feedstock considered: under intermediate severity conditions, XO from wheat bran autohydrolysis are essentially free from arabinose, whereas the production of XO branched with arabinose (with an arabinose to xylose ratio of 0.3) has been reported for the autohydrolysis products of brewery's spent grains. For corncobs and *Eucalyptus* wood hydrolysis products, the reported ratios arabinose/xylose were 0.04 and 0, respectively; and uronic substituents were bound in both cases. Acetylated XO containing uronic substituents are abundant in *Eucalyptus globulus* wood autohydrolyis products (Kabel et al., 2002a).

14.2.5 Biological Properties of Xylooligosaccharides

As cited before, XO cause health effects when ingested as part of the diet (for example, as active ingredients of functional foods) through the modulation of colonic microbiota. From a nutritional point of view, XO behave as nondigestible oligosaccharides (NDO's), which are not degraded by the low-pH gastric fluid and digestive enzymes, but are metabolized in the large bowel, and show prebiotic effects (discussed in detail elsewhere in this volume).

A recent review (Moure et al., 2006) identified (among others) the following potential XO biological effects:

- Antioxidant activity (with DPPH-radical scavenging and inhibition of erithrocyte hemolysis activities)
- Protection against low-density lipid peroxidation
- Prevention of atherosclerosis
- Prevention and treatment of oxidative stress, anemia and arteriosclerosis
- Treatment of vaginal and urogenital infections
- Low-glycemic index carbohydrate substitute, with applications in the prevention of type II diabetes
- Cosmetics and skin-related pharmaceuticals (for example, in the treatment of atopic dermatitis), collagen production enhancer, active components of moisturizing preparations, and treatment of epithelial covering tissue)
- Antihyperlipidemic agents, with activity against cholesterol, phospholipid and triglycerides
- Antihyperlipidemic activity
- Hair growth stimulant
- Inhibition of melanin and inhibition of melanoma cell proliferation
- Active agents against inflammation
- Therapeutic agents for osteoporosis treatment
- Hyaluronic acid-formation promoters
- Histamine-release inhibitors
- Enhancers of metal ion absorption
- Cytotoxic, antimicrobial and bacteriostatic agents
- Cancer cell apoptosis inducers
- Antiallergy agents
- Prevention and treatment of immune disorders

14.2.6 Technological Properties, Commercial Applications and Complementary Aspects

Considered as ingredients for functional foods, XO present remarkable technological properties. The sweetness of xylobiose is equivalent to 30% that of sucrose; the sweetness of other XO is moderate and possess no off-taste. XO are stable over a wide range of pH (particularly in the acidic range, even at the relatively low pH value of the gastric juice) and temperature (up to 100°C). Water activity of xylobiose is higher than that of xylose, but almost the same as glucose. Antifreezing activity of xylobiose on water at temperatures higher than $-10°C$ is the same as that of xylose, but greater than that of glucose, sucrose and maltose. Considered as food ingredients, XO have a good fragrance, and their non-cariogenic and low calorie characteristics allow their utilization in anti-obesity diets. In food processing, XO show advantages over inulin in terms of resistance to both acids and heat, allowing their utilization in low-pH juices and carbonated drinks. Moreover, the antioxidant activity of xylan-derived fractions (sometimes called "antioxidant fiber") is well known. It is usually ascribed to the presence of phenolic groups esterified to the polysaccharide chains and/or antioxidant fractions with different chemical nature in XO concentrates (for example, melanoidins coming from Maillard reactions or free low molecular weight phenolics coming from acid-soluble lignin). Recently, the ability of XO with acidic substituents for reducing iron ions has been reported, suggesting new application fields.

XO show potential in pharmaceutical and feed formulations, as well as in agricultural uses, but their most important market seems to correspond to food applications. In this field, XO present advantages compared to other oligosaccharides in terms of healthy effects and concentration thresholds, but their comparatively high production costs are hindering a wider and faster market development. To accomplish this goal, further improvements in processing technology would be necessary.

The Japanese market pioneered the incorporation of XO into foods, and today they are used in a wide variety of commercial products, including combinations with soya milk, soft drinks, tea or cocoa drinks, nutritive preparations, dairy products (milk, milk powder and yoghurts), candies, cakes, biscuits, pastries, puddings, jellies, jam and honey products, and special preparations for health food for elder people and children. Other application areas of XO include the synbiotic foods (made up of a prebiotic and a probiotic). The first synbiotic of this type was "Bikkle" from Suntory, successfully marketed since 1993, which contains bifidobacteria, XO, whey minerals, and Oolong tea extract. Additionally, the potential role of XO in the reduction of lifestyle-related diseases as well as the maintenance and improvement of human health is also a potential development factor. With the increasing health consciousness among consumers and the rapid progress of physiologically active functional foods, the future profile of products containing oligosaccharides with biological activities seems to be greatly promising (Nakakuki, 2005).

The bright future of XO for food applications is demonstrated by the dramatic increase in demand observed since 1994. For example, the production of Suntory increased from 70 tones in 1994 to 300 tones in 1996, and more recent data (Taniguchi, 2004) estimate the XO consumption in 650 ton of XO per year in Japan, which accounts for about one-half of the world market. Interestingly, the selling price reported for XO was the highest one among 13 different types of oligosaccharides (Taniguchi, 2004).

14.3   Mannans and Mannan-Derived Products

14.3.1 Structure, Occurrence, Hydrolytic Degradation and Processing

Mannans are linear or branched polymers made up of mannose, galactose, and glucose as structural monomer components. Mannan-type polysaccharides include galactomannans, glucomannans and galactoglucomannans. ❍ *Figure 14.4* shows representative mannan structures. Whereas the backbone of galactomannan is made up exclusively of β-(1→4)-linked D-mannopyranose residues in linear chains, glucomannan has both β-(1→4)-linked D-mannopyranose and β-(1→4)-linked D-glucopyranose- residues in the main chain (Ebringerova and Heinze, 2005).

Mannans are present in softwoods, hardwoods, ramie, bulbs, tubers, seeds, roots, and leaves of some non-gramineous monocotyl plants as structural, hemicellulosic polymers, as well as in gums. Galactomannans from different sources show differences in the distribution of D-galactosyl units along the polymer structure. Some of the structural units of mannans can be substituted by O-acetyl groups.

Water soluble galactomannans heavily substituted with galactopyranosyl residues (30–96%) are abundant in the cell walls of storage tissues (endosperm, cotyledons, perisperm) of seeds. Those from the endosperm of some leguminous seeds, such as guar (*Cyanopsis tetragonoloba*), locust bean or carob (*Ceratonia siliqua*), and tara gum (*Caesalpinia spinosa*), are widely used commercially.

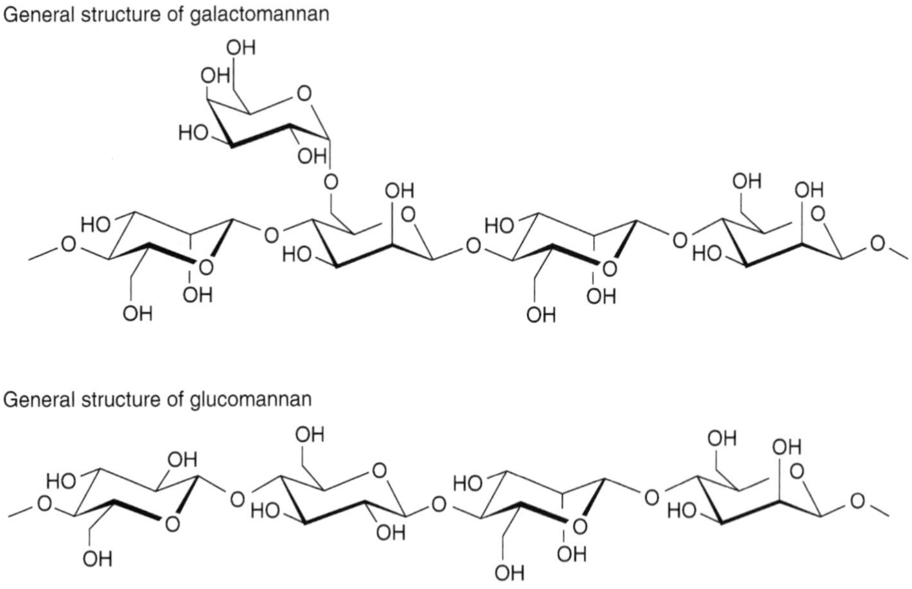

General structure of galactomannan

General structure of glucomannan

▶ Figure 14.4
Structure of representative mannans.

The most commonly used galactomannans in the food industry are guar and locust bean gums. Guar gum is composed of a mannose backbone with galactose side chains, and is used primarily as a thickening and water binding. Locust bean gum has a reduced number of galactose residues attached to the mannose backbone compared to guar gum, and is well known for its interaction with other gums such as carrageenan and xanthan gum, giving gels with particular texture characteristics.

Details on the structure and the pseudoplastic rheological behavior of locust bean galactomannans can be found in the article by Lazaridou et al. (2001). The water-soluble galactomannan present in the seeds of the Chinese traditional medicine plant fenugreek (*Trigonella foenum-graecum*) contain mannopyranose and galactopyranose residues with a molar ratio of 1.2:1.0, where the major chain is made up of β-(1,4)-linked D-mannopyranose residues, and 83.3% of the corresponding structural units are substituted at C-6 with a single residue of α-(1,6)-D-galactopyranose. Fenugreek mannan is a highly efficient water-thickening agent with applications in the food industry.

The tubers of konjac (*Amorphophallus konjac*, a potato-like plant belonging to the *Araceae* genus employed in Asia as gelling agent, thickener, emulsifier and rheological modifier) contain glucomannan of high molecular weight, which consists of sequences of mannose separated by glucose units in the backbone. The glucose to mannose ratio is 1 to 1.6, with one acetyl group per six glucose residues (Ebringerova and Heinze, 2005).

The chemical methods of mannan hydrolysis are based in the same concepts already described for XO. Both solubilization and DP reduction of the mannan-containing hemicellulosic polymers present in spruce wood have been achieved in a single step by steam processing and microwave irradiation (Palm and Zacchi, 2003). Steam treatments at 200°C led to products with a mean molecular weight of 3,400 g/mol and a maximum molecular weight of 12,000 g/mol. Operation under harsher conditions led to products of lower average molecular weight.

Manooligosaccharides have been prepared by thermal hydrolysis of spent coffee grounds (Asano et al., 2003) and by high-temperature hydrolysis of coffee bean and waste coffee (Nakamura et al., 2002). In the latter case, the main products were oligosaccharides containing up to 10 monosaccharide units.

Aqueous processing (with or without acidification by CO_2) has been employed for the manufacture of mannan oligosaccharides with molecular weights in the range 2,000–6,000 g/mol, starting from natural mannan polysaccharides (Murota and Yamanoi, 2003), whereas acid hydrolysis of coffee bean dregs and lees resulted in oligosaccharides containing up to 10 mannose residues (Fujii et al., 2001).

Partial hydrolysis of konjac glucomannan in media catalyzed by acids has been assayed for manufacturing edible films and for obtaining low-molecular weight compounds for which new application areas are expected. Except for the molar mass, no structural differences have been found between the polymeric substrate and its hydrolysis products.

As an alternative to acid or aqueous chemical treatments, the hydrolytic depolymerization and debranching of mannan can be carried out by enzymatic hydrolysis. For this purpose, both endohydrolases and exohydrolases are necessary. The main-chain mannan-degrading enzymes include β-mannanase, β-glucosidase, and β-mannosidase, but additional enzymes (such as acetyl mannan esterase and α-galactosidase) are required to remove side-chain substituents that are attached at various points on mannan.

Enzymatic hydrolysis of coffee mannan (isolated from green defatted *Arabica* beans by delignification, acid washing and subsequent alkali extraction) with *Sclerotium rolfsii* mannanases resulted in products with lower viscosity and enhanced processability. Upon enzymatic processing, various mannooligosaccharides (including mannotetraose, mannotriose, and mannobiose) were released (Sachslehner et al., 2000).

Guar gum galactomannan has been depolymerized by combined or sequential treatments with α-galactosidase and β-mannanase. Hydrolysis of guar and locust bean gums with β-mannanases from *Penicillium oxalicum* resulted in the production of low molecular-weight oligomers, which accounted for 92% of the total released saccharides. The DP range of oligomers was 2–7 for guar gum and 2–6 for locust bean gum (Kurakake et al., 2006). Enzymatic processing of konjac glucomannan with a purified β-mannanase from *Trichoderma* spp. resulted in the formation of oligosaccharides identified as M-M, G-M, M-G-M, M-G-M-M, and M-G-G-M, where G- and M- represent β-1,4-D-glucopyranosidic and β-1,4-D-mannopyranosidic linkages, respectively (Park, 2006).

Mannans can also be hydrolyzed by enzymes having pectinases and cellulases as major activities. Pectinases from *Aspergillus niger* with polygalacturonase activity have been employed for debranching and depolymerization of guar galactomannan (Shobha et al., 2005), leading to products of commercial interest as functional food ingredients. In the same way, processing of konjac with commercial cellulase preparations has been proposed for the manufacture of glucomannan hydrolysates (Al-Ghazzewi et al., 2007).

Multistage processes have been reported for obtaining either refined mannan-containing materials suitable as hydrolysis substrates or for mannooligosaccharide refining. Nunes et al. (2006) purified soluble galactomannans from roasted coffee infusions by precipitation with 50% ethanol, followed by anion exchange chromatograhy and further separation in Sepharose; whereas Wu and He (2003) started from mannan-containing gum, which was solvent-extracted at pH < 5.5, processed for solvent removal, diluted with water, treated with mannanases at controlled pH, heated, filtered and spray-dried.

14.3.2 Biological Properties of Mannans and Mannan-Derived Products

A number of health effects have been reported for polymeric and oligomeric mannans. Concerning prebiotic potential and other related effects, reported studies deal with:

- The ability of konjac mannan enzymatic hydrolysates to stimulate the growth of single strains of lactobacilli and bifidobacteria (*Lactobacillus acidophilus*, *Lactobacillus casei* or *Bifidobacterium adolescentis*) in single or mixed cultures. The growth of lactobacilli inhibited the growth of pathogens, Escherichia coli and Listeria monocytogenes.

Based on this finding it has been claimed that the hydrolysate can be used as a potential prebiotic and applied to a wide range of foods, feeds and pharmaceutical products has been claimed (Al-Ghazzewi et al., 2007).

- The beneficial health effects achieved by konjac glucomannan on the colon and bowel habits, including relief of constipation (which has been ascribed to increased stool bulk and improved colonic ecology). Konjac glucomannan is degraded almost 100% by the simultaneous action of enzymes and intestinal anaerobic bacteria present in human feces, producing formic acid, acetic acid, propionic acid, and 1-butyric acid. The proportions of these fatty acids were different among test subjects, and their total amounts ranged from 17.1 to 48.8% of the initial konjac glucomannan (Matsuura, 1998). The mechanisms by means of which konjac glucomannan modulates the bowel habit in healthy adults have been considered by Chen et al. (2006), who assessed the effects of konjac glucomannan ingestion at a dose of 4.5 g/day on the colon health. The beneficial effects observed were enhanced bowel movement, improved stool bulk and improved colonic ecology. Glucomannan supplementation increased significantly the concentrations of fecal lactobacilli, total bacteria, and daily fecal output of bifidobacteria, lactobacilli, and total bacteria. In addition, it increased the proportions of fecal bifidobacteria and lactobacilli with respect to the total fecal bacteria, whereas they decreased the relative proportions of clostridia compared to the placebo group. The suppression of clostridia growth in the human colon was achieved indirectly (through the action of bifidobacteria) or directly (through the physicochemical characteristics of glucomannan). Specific increases in the colonic bifidobacteria and lactobacilli caused by glucomannan might promote the bowel movement. The anti-constipation effects of glucomannan are well established. Treating patients with chronic idiopathic constipation with 1 g glucomannan/day resulted in statistically significant modification of the mouth-to-cecum transit time, compared with control groups, and a return to the normal range after the 10-day treatment, suggesting that chronic idiopathic constipation is a disease that involves the whole gut. Modifications of the intestinal habit and stool characteristics caused by glucomannan ingestion were assessed in a random, parallel, double blind, cross over trial study versus placebo involving 60 patients treated at two doses (3 and 4 g glucomannan/day), which led to the conclusion that the 4 g/day treatment resulted in the highest improvement of the assessed parameters. Studies in children showed that glucomannan is also suitable for pediatric applications.

- The suitability of mannan oligomers (β-1,4-D-mannobiose, β-1,4-D-mannotriose, β-1,4-D-mannotetraose, and β-1,4-D-mannopentose), manufactured from coffee mannan by enzymatic hydrolysis followed by purification by active carbon chromatography, in supporting the growth of enterobacterial strains. All mannooligosaccharides were utilised by *B. adolescentis*, *Lactobacillus acidophilus*, and *Lactobacillus gasseri*. Oppositely, detrimental bacteria such as *Clostridium perfringens* and *Escherichia coli* could not use mannooligosaccharides, affirming the prebiotic properties of the studied compounds (Asano et al., 2001). Mannooligosaccharides with DP up to 10 prepared from coffee bean dregs and lees by acid hydrolysis have been claimed as growth promoters for probiotic bacteria (Fujii et al., 2001). Mannooligosaccharides from thermally hydrolyzed spent coffee grounds, administered to volunteers at 1.0 g/day or 3.0 g/day, resulted in significantly increased contents of *Bifidobacterium* (suggesting that the ingestion of mannooligosaccharides caused *Bifidobacterium* to be the predominant bacteria in the intestine) and in improved defecation and defecating conditions. The resistance of mannooligosaccharides obtained from thermal hydrolysis of spent coffee grounds to human salivary α-amylase, artificial gastric juice, porcine pancreatic enzymes and rat intestinal mucous enzymes, together with their fermentability by human fecal bacteria leading to the formation of SCFA (mainly acetic, propionic and n-butyric acids), confirmed that mannooligosaccharides are indigestible saccharides and are potential agents for improving the large intestinal environment (Asano et al., 2003).

- The ability of hydrolyzed guar gum in promoting colonic health through the production of short-chain fatty acids at comparatively high concentrations with respect to alternative oligosaccharides and commercial dietary fiber.
- The activity of partially hydrolyzed guar gum (together with a recommended oral rehydratation solution) as an active agent for treating acute diarrhea in children, through the action of short-chain fatty acids released during its fermentation by colonic bacteria. Partially hydrolyzed guar gum has been reported to have a different *in vitro* fermentation pattern than native guar gum, whereas the concentration profiles of the acids formed were dependent on the molecular weight. The molar ratios of acetate increased with the molecular weight, while molar ratios of propionate decreased. The molecular weight of substrates leading to optimal production of short-chain fatty acids has been characterized on the basis of experimental data.
- The ability of partially hydrolyzed guar gum to regulate the bowel habit, with an increase in defecating frequency and softer stools in people with constipation, but also significantly improvement of diarrhea in patients with gastrointestinal intolerance. These effects are in relation with the improvement of the intestinal microbiota balance, which also resulted in prevention from infection and colonization by *Salmonella enteritidis*. Additionally, therapeutic benefits derived from the utilization of partially hydrolyzed guar gum have been observed in the treatment of the irritable bowel syndrome, with decrease in constipation-predominant and diarrhea-predominant forms of the disease and decreased abdominal pain. The health

effects have been ascribed to the higher colonic contents of short-chain fatty acids, and higher lactobacilli, and bifidobacteria concentrations resulting from the selective fermentation of hydrolysates. The guar gum hydrolysate has been shown to enhance the fecal lactobacilli and bifidobacteria concentrations in healthy men. In the treatment of the irritable bowel syndrome, partially hydrolyzed guar gum was well tolerated and preferred over wheat bran by patients, limiting the probability of patients abandoning the prescribed regimen, and suggesting that these products are a valid option for high-fiber diet supplementation. Based on a number of technical properties and health effects (suitability for being used in enteral products and beverages, reduction of laxative dependence in a nursing home population, reduction of the incidence of diarrhea in septic patients receiving total enteral nutrition, reduction in symptoms of irritable bowel syndrome, increased presence of *Bifidobacterium* spp. in the gut), partially hydrolyzed guar gum has been recommended for clinical nutrition applications.

• The suitability of mannans for low-calorie diets, based on their indigestible nature (Fujii et al., 2001). Obesity is a well-established risk factor for cardiovascular disease, diabetes, hyperlipidemia, hypertension, osteoarthritis, and stroke. With 50% of Europeans and 62% of Americans classed as overweight, the food industry is waking up to the potential of products for weight loss and management. Konjac glucomannan is highly effective in the treatment of obesity due to its ability to induce a satiety feeling, and daily doses in the range 2–3 g/day have been recommended as a part of anti-obesity diets. In a study of obese women observing a hypocaloric diet, administration of galactomannan preparations resulted in accelerated weight loss, decreased desire to eat, and hunger sensations before meals. A galactomannan containing, high fiber diet caused beneficial health effects in obese patients, as it induced satiety and had beneficial effects on some cardiovascular risk factors, principally decrease in plasma LDL-cholesterol concentrations. Mannooligosaccharides obtained from coffee administered to humans resulted in abdominal fat reduction. A dietary supplement containing two sources of mannan (among other components) has been reported to be suitable for increasing body weight loss, reducing the percentage of body fat and absolute fat mass, and reducing the circumferences of upper abdomen, waist and hip. Konjac glucomannan was well-tolerated at doses of 2–4 g/day, and resulted in significant weight loss in overweight and obese individuals, a fact ascribed to its beneficial effects of promoting satiety and fecal energy loss. A comparative analysis of three fiber supplements (glucomannan, guar gum and alginate) plus a balanced 1,200 kcal diet showed a comparative advantage of glucomannan for achieving weight reduction in healthy overweight subjects.

- The cardioprotective effects of mannans, for which hypocholesterolemic effects derived from interactions with the bile acids have been reported, probably in relation with their ability to form gels. The bile acids interact with the gel, loosing their ability to carry cholesterol to the mucose surface, and allowing the excretion of cholesterol in feces. On the other hand, the hepatic cells compensate the loss of bile acids by synthesis from cholesterol, decreasing cholesterolemia also by this way. Similarly, fenugreek galactomannan has been reported to inhibit bile acid absorption, leading to lower cholesterol levels. Protective effects have been observed for coffee mannan-derived oligosaccharides. The administration of a coffee drink containing mannooligosaccharides (3.0 g/day) to a group of healthy adults increased in the amount of excreted fat, and decreased the fat utilization. Ingestion of partially hydrolyzed guar gum lowered the serum cholesterol level in humans by improving lipid metabolism without reduction of protein utilization. Guar gum has been successfully used for the treatment of hypercholesterolemia, even though clinical application has not become widespread because a large amount of guar gum is unpalatable. Partially hydrolyzed guar gum ingested in a yoghurt drink (which also contained sunflower oil and egg yolk) has shown ability to lower the intestinal uptake of fat and cholesterol, reducing the risk of vascular disease. The results indicated that the hydrolyzed guar gum decreased the bioaccessibility of both fat and cholesterol in a dose-dependent manner, through a depletion-flocculation mechanism, which antagonized the emulsification by bile salts and resulted in decreased lipolytic activity, and so in lower bioaccessibility of fat and cholesterol. The effects of diet supplementation with glucomannan combined with total-body exercise program has been studied in overweight (body mass index >25 kg/m^2) sedentary men and women. The combination of resistance and endurance exercise training program with glucomannan diet regimen improved both HDL-cholesterol and total cholesterol/HDL-cholesterol ratio. The hypocholesterolemic effects obtained by supplementing the diet of normocholesterolemic subjects with 2.4 g of chitosan and glucomannan/day were confirmed by the observed significant decreases of serum total, HDL- and LDL-cholesterol concentrations, these effects being likely mediated by increased fecal steroid excretion. Administration of konjac mannan fiber to hyperlipidemic, hypertensive type 2, high-risk type 2 diabetic patients, improved blood lipid profile and systolic blood pressure, possibly enhancing the effectiveness of conventional treatment. Glucomannan has been recommended as a rationale adjunct to diet therapy in primary prevention in high risk hypercholesterolemic children, whereas a combination of glucomannan and plant sterols substantially improved plasma LDL-cholesterol concentrations.

- The effects on glycemia observed in studies with different types of mannans and mannan-containing dietary supplements. Ingestion of fenugreek with water was shown to lower postprandial blood glucose levels in type II diabetic subjects. Administration slows gastric emptying and delays the absorption of glucose in intestines, decreasing the rise in blood sugar following a meal. Delayed nutrient absorption, particularly of glucose, can reduce the glycemic index of the consumed food. Another favorable property is the ability of galactomannan to form a gel layer on the surface of stomach cells that protects against damaging compounds. High-molecular weight galactomannans are more resistant to enzymatic degradation, suggesting an ability for exerting a more sustained water binding capacity. Supplementing a high-carbohydrate diet with konjac glucomannan improved the glycemic metabolic control and the lipid profile of subjects with the Insulin Resistance Syndrome, suggesting a potential for therapeutic treatment. Partially hydrolyzed guar gum significantly reduced the level of plasma glucose, improving the acute postprandial response and the insulin response. Several clinical trials with healthy subjects and diabetic patients showed that guar gum can reduce postprandial glucose excursions. In healthy subjects, guar gum reduced the maximal rise in blood glucose by 60%, and the area under the blood glucose curve by 68%. Guar gum therapy has favorable long-term effects on glycemic control and lipid levels in non-insulin dependent diabetes mellitus subjects, and could reduce the postprandial glucose in insulin-dependent diabetes mellitus patients. The ability of glucomannan for lowering postprandial insulin levels has been confirmed in trials with healthy subjects, and its utilization has been recommended for treating patients with previous gastric surgery suffering from postprandial hypoglycaemia. In a test performed with healthy nondiabetic subjects, a novel low-viscosity beverage containing guar gum was suitable for reducing the postprandial glycemic response to an oral glucose challenge, providing a means to stabilize blood glucose levels by reducing the early phase excursion, and then by appropriately maintaining the later phase excursion in healthy nondiabetic humans.

- The antioxidative capacity of the products resulting from the *in vitro* fermentation of unhydrolyzed and two types of hydrolyzed konjac mannan with different DP. The experimental determination of antioxidant activities by a variety of experimental methods, proved that the fermentation of unhydrolyzed mannan by selected strains of bifidobacteria and lactobacilli produced antioxidative capacity mainly by preventing the initiation of ferrous ion-induced peroxidation, whereas the fermentation of mannooligosaccharides increased principally the radical-scavenging ability and hindered lipid peroxide formation.

- The mineral absorption-promoting activity reported for partially hydrolyzed guar gum.
- The immunostimulatory properties reported for mannans, including acetylated mannan from *Aloe* (acemannan) and fenugreek galactomannan. Activation of immune responses by acemannan resulted in antiviral and antitumoral activities, and its action has been partially ascribed to the ability to promote differentiation of immature dendritic cells. The immunostimulatory properties of mannan-derived saccharides have been claimed in a recent patent, whereas in experiments with human immune cells, the production of reactive oxidizing species was decreased by mannan-oligosaccharides with DP in the range 6–7.
- The suitability of low molecular-weight products obtained by enzymatic hydrolysis of natural products for ameliorating sepsis, inflammatory diseases (such as rheumatoid arthritis) and allergies. These effects have been claimed in recent patents.
- The applications of mannan and galactomannan saccharides (including gums and their partially hydrolyzed reaction products) as active agents for reducing colonization of the oral cavity by plaque and disease-causing microorganisms.

14.4 Arabinogalactans

14.4.1 Structure, Occurrence, and Technological Properties of Arabinogalactans

The term arabinogalactan refers to a kind of water-soluble polysaccharide, long, densely branched, made up of galactose and arabinose moieties (Hori et al., 2007). Arabinogalactans are widely spread throughout the plant kingdom, and appear in edible and inedible plants, including leek seeds, carrots, radishes, black gram beans, pears, maize, wheat, red wine, italian ryegrass, tomatoes, ragweed, sorghum, bamboo grass and coconut meat and milk.

Arabinogalactans occur in two structurally different forms described as type I and type II:

- Type I arabinogalactans have a linear $(1{\rightarrow}4)$-β-D-Galp backbone, bearing 20–40% of α-l-Araf residues $(1{\rightarrow}5)$-linked in short chains, in general at position 3. It is commonly found in pectins from citrus, apple and potato.
- Type II arabinogalactans, known as arabino-3,6-galactan, have a $(1{\rightarrow}3)$-β-D-Galp backbone, heavily substituted at position 6 by mono- and oligosaccharide side

chains composed of arabinosyl and galactosyl units, and occur in the cell walls of dicots and cereals often linked to proteins (known as arabinogalactan proteins). The acidic type II arabinogalactan has uronic acid residues incorporated in its side chains and belongs to the groups of gum exudates (Ebringerova et al., 2005). Type II arabinogalactan is most abundant in the heartwood of the genus *Larix* (Western larch/*Larix occidentalis* or Mongolian larch/*Larix dahurica*) and occurs as minor, water-soluble component in softwoods. Other sources of type II arabinogalactan are coffee beans and arabic gum.

Most commercially-available arabinogalactan is produced from larch trees. Certain tree parts of western larch were reported to contain up to 35% arabinogalactan, whereas high-grade products from larch are composed of more than 90% arabinogalactan. All arabinogalactans isolated from *Larix* spp. are water-soluble, nitrogen-free polysaccharides of the 3,6-β-D-galactan type. The side chains consist of combinations of single galactose sugars, as well as longer side-chains containing galactose and arabinose residues. The galactose and arabinose units (consisting of β-galactopyranose, β-arabinofuranose, and β-arabinopyranose) are in a molar ratio of approximately 6:1, and comprise more than 99% of the total glycosyl content. A trace amount of glucuronic acid is generally also found (Odonmazig et al., 1994). ◉ *Figure 14.5* shows a tentative structure proposed for larch arabinogalactan. A curdlan-like triple helical structure, based on preliminary X-ray fiber diffraction data, has been reported for this polymer (Chandrasekaran and Janaswamy, 2002). The composition of arabinogalactan from siberian larch wood and the influence of the purification conditions on the properties of the final product have been reported recently.

Alkaline treatments of coffee bean cell walls followed by cellulase digestion allowed the identification of four different structures. The outer part of cell walls presented arabinogalactan protein-rich layers. The arabinogalactan-protein

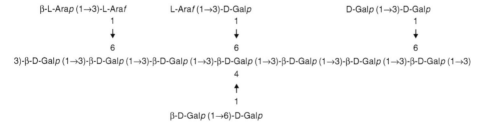

⬛ Figure 14.5
Tentative structure for larch arabinogalactan (Kelly, 1999).

(AGP) fraction of green coffee beans accounts for about 15% of the dry bean. Type II arabinogalactans are present in coffee brews, and their amount and composition depend on factors such as type of coffee, roasting and grinding degree, and brewing.

Acacia gum (gum arabic) is a natural dietary fiber produced by *Acacia* trees, principally composed of polysaccharides and proteoglycans, the latter being AGP. Gum arabic also contains complex, ramified arabinogalactans, as well as some lipids (Yadav et al., 2007). Acacia tree exudates obtained without any chemical or enzymic processing contained more than 90% soluble fiber on dry weight (Fremont, 2007). Studies on the composition of crude *Acacia* gums revealed the presence of arabinogalactan, AGP and glyco-proteins. Details on the main chemical and physical features of *Acacia* gum and its fractions isolated by hydrophobic interaction chromatography (arabinogalactan-peptide, AGP and glycoprotein fractions), have been reported.

Some technological properties of selected arabinogalactans are:

- Larch arabinogalactan is a dry, free-flowing powder, with a very slight pine-like odor and sweetish taste. Because of its excellent solubility, low viscosity, mild taste and easy dissolution in water and juices, it is easily administered. Other favorable technological properties are high water solubility and narrow molar mass distribution. According to the Generally Recognized as Safe (GRAS) Notice No. GRN 000047 (FDA, Center for Food Safety & Applied Nutrition, Office of Premarket Approval), the properties of larch arabinogalactan permit its use as a film-former, foam adhesive, additive, thickener, bulking agent, emulsifier, and as a therapeutic agent. Based on its food grade status and numerous studies supporting the safety of larch arabinogalactan, it is considered to be extremely safe with minimum to no toxicity. Larch arabinogalactan's properties make it a relatively easy therapeutic tool to employ in pediatric populations.

- Arabic gum has a unique combination of excellent emulsifying properties, low viscosity when dissolved, and is free from taste and odor, making it suitable for use in the food industry, as it can be incorporated in large amounts to foodstuffs without disturbance of their organoleptic properties. *Acacia* gums show unusual rheological properties, which have been ascribed to their surface properties, and especially to its ability to form elastic films at air/water or oil/water interfaces.

- Coffee AGP shows interesting rheological features, which suggest that coffee beans could be used as an alternative source of surface-active polymers for many commercial applications.

Reported processes for manufacturing arabinogalactan polysaccharides from different sources (including *Portulaca oleracea* and ground green or roasted coffee beans) are based on sequential stages of processing, including mechanical treatments, chemical reaction (by enzymatic hydrolysis), extraction and refining of extracts (Curti et al., 2005). The resulting products have been proposed for applications such as manufacture of functional foods, cosmetic compositions with physiological activity, formulation of glassy matrices, and beverages. *Acacia* gum commercial preparations have been proposed for the development of innovative functional foods with favorable nutritional and health claims (Meance, 2004).

14.4.2 Prebiotic Properties of Arabinogalactans

The prebiotic effects of arabinogalactans are based on their resistance to human digestive enzymes and to their fermentability by human intestinal bacteria (Crociani et al., 1994; Vince et al., 1990). Intestinal fermentation of arabinogalactan leads to the stimulation of the endogenous microbiota and to the production of short-chain fatty acids (Englyst et al., 1987; Vince et al., 1990), as well as to increased growth of beneficial anaerobes, such as bifidobacteria and lactobacilli, using culture-based methods (Vince et al., 1990).

During the last decade, larch arabinogalactan, known as larch gum in food applications, has produced emerging commercial and scientific interest, which follows closely upon recent reports related to the beneficial physiological effects of commercial arabinogalactan and its immunomodulatory properties (Ebringerova et al., 2005). As a dietary fiber supplement, larch arabinogalactan has several beneficial properties, including the abilities to promote the growth of friendly bacteria (specifically, increasing anaerobes such as bifidobacteria and lactobacilli), and to increase the production of SCFA, as well as to decrease ammonia generation. Larch arabinogalactan administration to human subjects increased the levels of beneficial intestinal anaerobes, particularly *Bifidobacterium longum*, via their fermentation specificity for arabinogalactan compared to other complex carbohydrates (Crociani et al., 1994). Arabinogalactan is fermented by mixed populations of human fecal bacteria more slowly than starch and pectin, and the fermentation occurred without accumulation of free sugars or oligosaccharides. Time-course measurements of the polysaccharide remaining in the fermentation media showed that the arabinose side chains of arabinogalactan were co-utilized with the backbone sugars, and that the polysaccharide-degrading activity was mainly cell-associated, but extracellular polysaccharidase activity increased as the fermentations progressed. In experiments, multicomponent substrate utilization occurred, leading to molar ratios of acetate, propionate, and butyrate of 50:42:8 (Englyst et al., 1987).

Prebiotic activity has also been reported for *Acacia* gums, based on human studies that showed high fermentability with no side effects even at high supplementation level. *Acacia* gum stimulates the development of beneficial bacteria and the production of SCFA, helping to regulate different functions such as gut transit or lipids metabolism (Fremont, 2007). However, thanks to its high molecular weight and to its complicated molecular structure, *Acacia* gum does not exhibit undesired nutritional side-effects such as laxative effect and/or flatulence. Its fermentation pattern is specific, leading to the preferent formation of propionate and butyrate through a progressive fermentation that could permit these beneficial end-products to reach the distal part of the colon (Meance, 2004). Ingestion of 10 or 15 g/day of a commercial *Acacia* gum concentrate for 10 days increased the stool weight and the total lactic acid-producing bacteria and bifidobacteria counts in stools, without affecting total anaerobe and aerobe counts. The magnitude of this selective effect was greater in subjects with a low initial fecal concentration of bifidobacteria. Fecal digestibility was around 95%, and its caloric value was 5.5–7.7 kJ/g. The concentrate showed good digestive tolerance, even if daily ingestions above 30 g caused excessive flatulence. Based on these data, the commercial product has been considered as a very well tolerated dietary fiber with bifidogenic properties believed to benefit intestinal health.

Most studies dealing with arabinogalactan have been conducted *in vitro*. When considering the reported results, it is important to take into account that the human colon is a complex environment, and that *in vitro* studies may not accurately represent bacterial activities within the human colon. Daily diet supplementation with 15 or 30 g arabinogalactan resulted in significant increases of total fecal anaerobes and *Lactobacillus* spp., this latter being believed to maintain and restore normal intestinal balance; but no significant changes in parameters such as fecal enzyme activity, transit time, frequency, fecal weight, fecal pH and short-chain fatty acids were observed (Robinson et al., 2001). These results have been ascribed to the fact that SCFA are believed to be quickly absorbed following their production; therefore, it is difficult to determine the total amount produced in human subjects. Additionally, it is noteworthy that arabinogalactans have been claimed as active principles of fluid diets for preventing or curing enteropathy.

14.4.3 Other Biological Properties of Arabinogalactans

Larch arabinogalactan (AG) has been reported to show immunomodulating properties (Ebringerova et al., 2005). AG from larch wood possesses a wide spectrum of physiological activities and due to that it can be applied as a medical product and biologically active food additive. Cholesterol-lowering and immunomodulating properties have been reported for arabinogalactan from several plant sources. Many herbs with well established immune-enhancing properties, such as *Echinacea purpurea, Baptisia tinctoria, Thuja occidentalis, Angelica acutiloba* and *Curcuma longa* also contain significant amounts of arabinogalactans. A purified, highly branched arabinogalactan polymer induced the proliferation of mouse lymphocytes in a dose-dependent manner. Response of mice to arabinogalactan from coffee beans showed proliferation of macrophage and splenocyte, evidencing that arabinogalactan can stimulate immunocytes and enhance immune responses. *In vitro* and animal studies suggest that the effects of larch arabinogalactan on the immune system could be related to the activation of "natural killer cells," and perhaps other white blood cells as well, and also possibly alter levels of immune-related substances such as interleukins, interferon, and properdin. Thus, arabinogalactan could be suitable for treating a number of chronic diseases characterized by decreased natural killer cell activity, including chronic fatigue syndrome, viral hepatitis, HIV/AIDS, and autoimmune diseases, such as multiple sclerosis. Stimulation of natural killer cell activity by larch arabinogalactan has been associated with recovery in certain cases of chronic fatigue syndrome. By virtue of its immune-stimulating properties, larch arabinogalactan has been shown to affect a slight increase in CD4 cell counts, in addition to a decreasing susceptibility to opportunistic pathogens. On the other hand, the enhancement of the immune response may decrease the incidence of bacterial infections, particularly those caused by Gram negative microorganisms, such as *Escherichia coli* and *Klebsiella* spp. Prophylactic applications of larch arabinogalactan might include use as an immune-building agent for individuals with a propensity to ear infections and other upper respiratory infections. Arabinogalactans have been reported to act as activators of lymphocytes and macrophage, a valuable property that could be useful in the immunoprevention of cancer.

Experiments with animals showed several physiological properties of *Acacia* gum, including increases in glucose and bile acid absorption. *Acacia* gum has been proposed as a hypoglycemic agent, based on the postprandial glucose response of type II diabetic subjects to the administration of a commercial concentrate. Other effects observed include the decrease the glycemic index of food products, making them healthier and more suitable for diabetics. However, the human response to diet supplementation with larch or tamarack arabinogalactan resulted in poor long-term results in terms of serum lipids and glucose concentrations; and administration of a commercial arabinogalactan from western larch at doses of 15 or 30 g did not result in significant changes in blood lipids or blood insulin (Robinson et al., 2001). However, significant decreases in fat consumption were observed when subjects consumed the 30 g dose of AG, a fact that could be related to increased reports of bloatedness. Thus, the sensation of fullness may have led subjects to avoid high fat foods.

The studies of Robinson et al. (2001) and Vince et al. (1990) coincided in the finding that fecal ammonia levels decreased significantly upon ingestion of arabinogalactan. This finding may be related to with the significant increases in total anaerobes, as some anaerobic colonic bacteria prefer to utilize ammonia as a nitrogen source rather than amino acids or peptides when fermenting carbohydrates. High colonic ammonia levels may have detrimental health implications, including cytopathic effects on colonic epithelial cells, effects on the intermediary metabolism and DNA synthesis of mucosal cells, and toxicity against epithelial cells. Based on these finding, arabinogalactan may be of clinical value in the treatment of portal-systemic encephalopathy, a disease characterized by ammonia build-up in the liver (Vince et al., 1990).

Additionally, a variety of health benefits, including mitogenic, antimutagenic, gastroprotective, and antimicrobial effects have been reported for arabinogalactans, which also show ability to block metastasis of tumor cells to the liver. Its utilization as a cancer protocol adjunct causing potential therapeutic benefits has been proposed.

The whole set of biological properties of arabinogalactan suggest an array of clinical uses, both in preventive medicine and in clinical medicine, as a therapeutic agent in conditions associated with different pathologies.

14.5   Pectins and Pectin-Derived Products

14.5.1 Occurrence, Structure and Applications of Pectins

Pectins are complex, acidic heteropolysaccharides with gelling, thickening and emulsifying properties, which are recognized to promote human health as dietary fiber (Gulfi et al., 2007), and are widely used in the food and pharmaceutical industries for their technological properties and health effects. The food applications of pectins, mainly derived from its ability to form gels in the presence of Ca^{2+} ions or a solute at low pH, including their utilization in jams and jellies, fruit products, dairy products, desserts, soft drinks, frozen foods, and more recently in low-calorie foods as a fat and/or sugar replacer. The food applications of pectins have been revised exhaustively by Voragen et al. (1995). In the pharmaceutical industry, they are used to achieve a number of health effects (including decrease of blood cholesterol levels and treatment of gastrointestinal disorders, as explained below). Other applications of pectin include use in edible films, paper substitute, foams and plasticizers.

Almost all higher plants contain pectins, which have a lubricating and cementing function. Pectins are major components of the primary wall and the middle lamella of plant cells, where they represent around 40 wt.% (on dry matter basis) of the cell walls of fruits and vegetables. In citrus fruits, they are present at a cellular level (membranes, juice, vesicules and core) in different quantities, depending on the fruit variety and maturity stage. Several byproducts of the food industries are used for pectin extraction, including citrus peels, apple pomace, sugar beet pulp and (in a minor extent) potato fibers, sunflower heads and onions (May, 1990).

The structure of pectins is complex. Polymers are made up of different monomers, galacturonic acid being the predominant one. Two types of "regions" have been identified in pectins: the "smooth" regions, made up of galacturonic acid, and the ramified "hairy" regions, where most of neutral sugars are located.

Pectins are believed to have structural features that are common to a variety of fruit and vegetable tissues, even if the structural elements may vary depending on the source. The following pectin structural elements can be identified:

- Homogalacturonan, composed of a α-(1→4)-linked D-galacturonic acid (GalA) backbone which forms the "smooth region," accounts usually for about 60% of the total pectin, and has a length chain of 72–100 GalA residues. The structural units can be esterified at C-6, and/or O-acetylated at O-2 or O-3. The type and degree of substitution determine the properties of pectins.
- Xylogalacturonan, composed of homogalacturonan substituted with xylosyl moieties. The degree of xylosidation is between 25 and 75% (Schols et al., 1995).
- Rhamnogalacturonan I, whose backbone is composed of repeated subunits of α-(1→2) -L-rhamnose-α-(1→4) -D-galacturonic acid. The rhamnosyl residues of rhamnogalacturonan I can be substituted at O-4 with neutral sugar side chains,

mainly composed of galactosyl and/or arabinosyl residues. These side chains are substituted by either both GalA or polymeric chains, such as arabinogalactan I or arabinan. The proportion of branched rhamnosyl residues varies from about 20 up to 80% depending on the polysaccharide source.

- Rhamnogalacturonan II, which forms a distinct region within homogalacturonan. It contains clusters of four different side chains with a variety of sugar residues.
- Arabinan, consisting of α-(1→5)-linked arabinosyl backbone, which usually is substituted with α-(1→2) or α-(1→3) linked arabinosyl side chains.
- Arabinogalactan I and arabinogalactan II, whose structures have been described above.

◗ *Figure 14.6* shows the structural units of pectins. The relative proportions of the various structural elements may vary significantly among different plant tissues (Voragen et al., 1995). Recent studies have reported data on pectins' structure and discussed their macromolecular structure, as well as additional ideas on how pectins are integrated into the plant cell wall.

Pectins are usually extracted from suitable substrates by mild acidic treatments (typically in media with pH in the range 1–3, obtained by externally added mineral acids, operating at 50–90°C for 3–12 h), which lead to products essentially constituted of homogalacturonans, with limited amounts of neutral sugars (Guillotin et al., 2005; Kravchenko et al., 1992). Pectins can be further precipitated by adding alcohol (isopropanol, methanol or ethanol), and the gelatinous mass is pressed, washed, dried and ground. Depending on the source, GalA residues in homogalacturonans can be present as free carboxyl groups, methylesterified, or acetylated; further alkaline or acid processing allows the reduction of the ester group content. Acidic processing of pectin-containing native substrates may also result in the solubilization of phenolics with antioxidant activity, which can be recovered from the liquors for further utilization (Berardini et al., 2005).

Commercial pectins are mainly classified as a function of their methylesterification degree (measured as the amount of moles of methanol per 100 moles of GalA), since this is the main parameter influencing the physical properties. The commercial products can be classified as:

- High methyl-esterified pectins, which are extracted from pomace or peels of apples under acidic conditions, and precipitated with alcohol to yield products in which 50% or more of the GalA moities are methyl-esterified. This kind of pectins can be classified according to their setting time in ultra rapid set, rapid set, medium rapid set and slow set pectins (May, 1990).

Hexoses

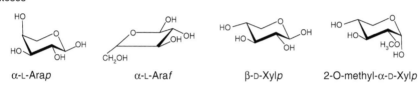

β-D-Gal*p*   α-L-Gal*p*   β-D-Glc*p*   β-D-Man*p*

Pentoses

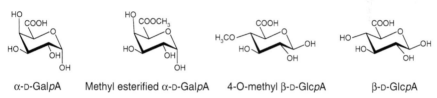

α-L-Ara*p*   α-L-Ara*f*   β-D-Xyl*p*   2-O-methyl-α-D-Xyl*p*

Hexuronic acids

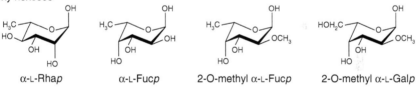

α-D-Gal*p*A   Methyl esterified α-D-Gal*p*A   4-O-methyl β-D-Glc*p*A   β-D-Glc*p*A

6-deoxy hexoses

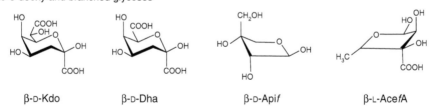

α-L-Rha*p*   α-L-Fuc*p*   2-O-methyl α-L-Fuc*p*   2-O-methyl α-L-Gal*p*

2-Keto-3-deoxy and branched glycoses

β-D-Kdo   β-D-Dha   β-D-Api*f*   β-L-Ace*f*A

□ Figure 14.6
Structural elements of pectins.

● Low methyl-esterified, non-amidated pectins, which are obtained by controlled de-
  esterification of high methyl-esterified pectins under acidic or alkaline conditions.
  Processing under alkaline conditions has to be performed under low-severity conditions to avoid polymer degradation.

● Low methyl-esterified, amidated pectins, which are obtained by chemical reaction of the methyl-ester groups in the presence of ammonia and alcohol, resulting in the presence of amide groups at the C-6 position of the GalA residue. Amidation results in improved gelling properties. Only 25% substitution of GalA by amide groups is permitted in food products.

Commercial pectins also include the "acetylated pectins," which are extracted from sugar beet residues as a low-quality material in terms of gelation due to their limited molecular weight and low proportion of GalA (derived from a higher content of neutral sugars). Acetylated pectins are used for their emulsifying properties.

Depolymerization of native pectins is carried out during the extraction process, which is usually catalyzed by chemical, enzymatic or combined methods. The final DP distribution depends on the nature of the substrate and on the operational conditions used for processing.

Partial acid hydrolysis of pectins leads to depolymerization, but also to structural modification due to the different susceptibilities of the various glycosidic bonds towards hydrolysis. Mild acid hydrolysis of galacturonans (in media containing 0.2–2 M sulfuric acid for 72 h at 80°C) resulted in the cleavage of the galacturonic acid chains into oligomeric forms without any degradation (Garna et al., 2006). Pectin fractionation may take place by β-elimination (which splits specifically the glycosidic linkages next to methoxylated galacturonic acid units without steric limitation, and is the predominant mechanism under neutral or mild acidic conditions), a reaction competitive with de-esterification. This latter reaction is promoted by cold, alkaline conditions, and results in the removal of acetyl groups and methyl esters (Kravchenko et al., 1992).

Reactions of pectins with water or steam at high temperatures and pressures have been proposed as alternative methods for chemical modification. Mitchell and Mitchell (1996) claimed the production of oligomers with DP in the range 1–20, at 25% yield, by depolymerization of polyuronic acid in aqueous media, whereas Miyazawa et al. (2008) reported a severity analysis on the hydrothermal processing of pectin using a combination of semi-batch and plug-flow reactors, in which the production of oligomers, monomers, and degradation products was assessed. In a related study, galacturonic acid and its oligomers were produced by hydrolysis of polygalacturonic acid by hydrothermal treatments with hot, compressed water at 453–533 K, which led to the formation of monomeric and oligomeric reaction products with DP < 11 at yields in the range 22–33.4% of the polymeric material. Further enzymatic hydrolysis with pectinases resulted in increased formation of monogalacturonic acid and oligomers with DP 2 and 3 (Miyazawa and Funazukuri, 2004).

Fast extraction of pectin from orange albedo was accomplished with steam processing at 15 psi for 2–6 min ("flash extraction"). The molecular weights of pectin isolates obtained under selected conditions were higher than the ones of commercial citrus pectin. Flash extraction of pectin from orange and lime peel led to premium pectin together with an oligosaccharide fraction with prebiotic activity (Hotchkiss et al., 2005). The effects of the more influential variables (processing pressure, moisture content of the feedstock treatment and processing time) on pectin extraction from orange peels have been assessed by response-surface methodology, whereas steam treatments have been coupled with hydrogen peroxide treatments for refining purposes.

Other technologies employed for the physicochemical extraction of pectins include microwave heating of orange peels in acidic media, whereas depolymerization has been carried out by ultrasound processing, extrusion at alkaline pH, and gamma ray irradiation.

Enzymes can be used for processing both pectin-containing raw materials and purified pectins (sometimes in combination with chemical processing) to accomplish a variety of objectives, including extraction, depolymerization, purification, removal of selected substituents or side chain and structural elucidation. Pectic enzymes are classified according to the mode of attack on their specific structural element of the pectin molecule. The most important enzymes employed to reduce the DP and/or to modify the structure of polymeric pectin fragments are:

- Endo-polygalacturonase (EC 3.2.1.15), which cleaves the α-($1\rightarrow4$)-D galacturonan linkages in homogalaturonan segments randomly, leading to the formation of oligosaccharides.
- Exopolygalacturonase (EC 3.2.1.67 and EC 3.2.1.82), which attack homogalacturonan chains from the non-reducing end, splitting GalA units.
- Rhamnogalacturonan hydrolase (EC 3.2.1.-), which hydrolyses the α-D-($1\rightarrow4$)-Gal$p$A-α-L-($1\rightarrow2$)-Rha$p$ linkage in the rhamnogalacturonan I backbone, leaving Rha$p$ at the non-reducing side.
- Rhamnogalacturonan lyase (EC 4.2.2.-), which enables the eliminative cleavage of α-L-($1\rightarrow2$)-Rha$p$-α-D-($1\rightarrow4$)-Gal$p$A backbone of rhamnogalacturonan I, leaving a 4-deoxy-β-L-threo-hex-4-enepyranosyluronic acid (unsaturated GalA) group at the non-reducing end.
- Endoxylogalacturonan hydrolase (EC 3.2.1.-), which hydrolyses the α($1\rightarrow4$)-D linkages of xylose-substituted galacturonan moieties in xylogalacturonan.

- Rhamnogalacturonan rhamnohydrolase, an exo-acting pectinase, which possesses a specificity to release terminal rhamnosyl residues (1→4)-linked to α-galacturonosyl groups.

- Rhamnogalacturonan galacturonohydrolase, which is able to release a GalA moiety connected to a rhamnose residue from the non-reducing side of rhamnogalacturonan I chains, but is unable to liberate GalA from homogalacturonan.

Additional effects (for example, removal of methylester or acetyl groups, and arabinan side chains) can be achieved by enzymes. Almost total solubilization ("liquefaction") of pectin-containing raw material can be accomplished by using enzyme cocktails containing cellulases, hemicellulases and pectic esterases. For example, liquefaction of apple cell walls may result in complete pectin removal, leading to a high-molecular weight pectic fraction, termed "modified hairy region," made up of highly branched rhamnogalacturonan. Liquefaction has been applied to other feedstocks, such as carrots, potato fiber and tissues from red beets, pear, carrot, leek, and onion. In the case of red beets, the poor enzymic conversion by pectolytic and cellulolytic enzymes was ascribed to the high degree of acetylation of pectins.

Partial enzymatic hydrolysis leads to the formation of oligosaccharides, an objective that can be accomplished by using free or immobilized enzymes. The latter method allowed the production of dimers to pentamers from citrus pectin using pectinases immobilized on affinity gel. Enzymatic processing may enable the recovery of valuable byproducts: for example, processing of bergamot peel with pectinolytic and cellulolytic enzymes led to the simultaneous solubilization of carbohydrates (with the formation of monosaccharides and oligosaccharides in relative amounts dependent on the reaction time) and to low molecular weight flavonoids (Mandalari et al., 2006).

As an example of mixed chemical-enzymatic treatments, the sequence of steps reported for the structural elucidation of pectins from sugar beet pulp includes extraction with water, oxalate, hot acid, and cold alkali, together with base-catalyzed β-elimination, de-esterification, and treatments with four different enzymes; the procedure allowed the identification of sequences of galacturonic acid residues with relatively little neutral sugar attached ("smooth" fragments), and "hairy" fragments containing feruloyl groups. For structural determination, the hairy- and alkali soluble- fractions of beet pectins have been subjected to the action of four different enzymes prior to mild hydrolysis by 0.05 M trifluoroacetic acid. Also, in structural studies, citrus peel pectins have been solubilized with oxalate or dilute hydrochloric acid, and the isolated products have been treated with selected enzymes. Alternatively, enzymatic processing can be employed for modifying the structure of pectins obtained by chemical methods (for example, to cause demethoxylation of fractions obtained by acid hydrolysis of citrus peels, in order to obtain products with improved emulsifying properties); or for modifying the properties of the resulting solid residues. For the latter purpose, treatments with 0.06–0.5% dilute sulfuric acid at 100–140°C had a positive effect on the rate of subsequent enzymic hydrolysis of orange peel by a mixture of cellulolytic and pectinolytic enzymes.

As previously explained in the case of xylooligosaccharides, membrane technologies are playing an increasing role in the production and refining of pectin oligosaccharides. The sequential processing of enzymically hydrolyzed pectate by ultrafiltration and nanofiltration membranes enabled the separation of di- to pentamers, whereas ultrafiltration followed by diafiltration has been proposed for making a pectin concentrate from beet pulp. The same combination has been studied as an alternative purification method to alcohol precipitation, whereas enzymatic membrane reactors have been employed for the generation of pectin oligosaccharides from orange peel albedo (Hotchkiss et al., 2003), and from methylated pectins.

14.5.2 Prebiotic Potential of Pectins and Pectin-Derived Products

The prebiotic potential of pectins and pectin-derived oligosaccharides depend on the physicochemical characteristics of the considered substrates, and particularly on their molecular weight and degree of esterification.

Concerning the comparative *in vitro* degradability of pectins by human fecal flora, their relative fermentability (as measured by total production of acetate, propionate and butyrate after 24 h of fermentation) was lower than the ones of starches, locust bean gum, arabic gum and guar gum. The whole degradation process has been reported to occur in two sequential stages: in the first one, the polymeric substrates are fragmented into unsaturated oligogalacturonans (which behave as intermediate reaction products) by the action of bacterial enzymes; and in the second one, the oligogalacturonans disappear as a result of their further fermentation by the gastrointestinal microbiota, leading to the formation of short-chain fatty acids. Low-esterified pectins were depolymerized and fermented *in vitro* by human fecal flora faster than the highly esterified one.

Fermentation with individual or mixed cultures of selected intestinal bacteria (*Bacteroides thetaiotaomicron* and *Escherichia coli*) was slower than the one carried out with a complete fecal flora, owing to the lower pectin-degrading activity. No pectin-degrading activity was found in pure cultures of *Escherichia coli*, whereas the disappearance of oligogalacturonic acids in the later stages of fermentation by intestinal bacteria resulted in an increased formation of short-chain fatty acids. *In vitro* fermentation of media derived from the enzymatic liquefaction of apple parenchyma with fecal inocula resulted in the formation of a spectrum of short-chain fatty acids, suggesting their possible application as dietary fiber.

Prebiotic effects have been suggested for highly branched, arabinan containing "hairy" regions of pectins with complex molecular structure, extracted under alkaline conditions, by means of *in vitro* fermentation assays with human fecal bacteria. Comparatively, higher amounts of short-chain fatty acids were produced from the pectin hairy regions than from commercial pectins, with particularly remarkable differences in propionate concentrations. Concomitantly, pH decreased to a larger extent in the fermentation of hairy substrates. These findings are of interest owing to their beneficial influence into colon health, suggesting that hairy regions of pectins might be an interesting dietary fiber with enhanced prebiotic properties compared to commercial pectins (Gulfi et al., 2007).

Intestinal fermentation of pectic substrates depends on the molecular weight. Pectic oligosaccharides, which can be produced from waste biomass or low-cost byproducts (such as apple pomace, orange peels or sugar beet pulp) have been considered as novel prebiotics, which may have better functionality than those currently established on the market, enabling a selective composition shift in the gut microbiota by enhancing the growth of beneficial, health-promoting bacteria. The prebiotic potential of pectic oligosaccharides has been reported to be higher than that of the pectins they are derived from, owing to the more selective fermentation. This behavior has been observed, for example, in pectic oligosaccharides prepared by enzymic hydrolysis of commercial citrus pectin and apple pectin, which enabled higher growth rates of bifidobacteria compared to the original pectins. Oligosaccharides derived from pectins by either enzymic hydrolysis or by flash extraction displayed potential prebiotic properties, as they selectively increased the populations of beneficial bacteria such as bifidobacteria and lactobacilli and decreased undesirable bacteria such as clostridia (Rastall et al., 2005). In contrast, oligosaccharides derived from high-methoxy citrus pectin, low-methoxy apple pectin and orange peel enhanced the growth of bifidobacteria and lactobacilli while limiting the growth of pathogens in mixed fecal batch cultures (Hotchkiss et al., 2004). *In vitro* fermentation of pectin oligosaccharides derived from Valencia oranges, mainly consisting of arabinogalactan pectic side chains and xyloglucan, showed bifidogenic effects, as revealed by the increases in acetate, butyrate and propionate concentrations upon fermentation (Hotchkiss et al., 2007). However, fermentation of oligosaccharides derived from orange peel pectin in mixed fecal bacterial cultures (containing glucose and pectic oligosaccharides from rhamnogalacturonan and xylogalacturonan as potential carbon sources) were suitable for increasing the concentrations of bifidobacteria and *Eubacterium rectale*, as well as the production of butyrate, but no prebiotic effect was found. The growth rates of bifidobacteria depend on the structural characteristics of pectin oligosaccharides (and particularly, on their degree of methylation). *Bifidobacterium angulatum, B. infantis* and *B. adolescentis* had higher growth rates on pectic oligosaccharides derived from pectins than on the original, polymeric substrates. In general, greater fermentation selectivity was obtained with substrates of lower degrees of methylation.

Based on the above ideas, it can be inferred that non-selectively fermented polysaccharides like pectin can have their bifidogenic properties improved by partial hydrolysis, enabling a selective increase in the populations of beneficial bacteria, such as bifidobacteria and lactobacilli, as well as a decrease in undesirable bacteria, such as clostridia. These effects are based on the more selective fermentation of oligomeric compounds in comparison with the parent polysaccharides (Rastall et al., 2005).

Looking for enhanced prebiotic activity and high tolerance, pectin oligosaccharides have been used in combination with neutral prebiotics as components of infant fourmulae. Administration of mixtures of pectin and neutral oligosaccharides to infants resulted in increased concentrations of bifidobacteria and lactobacilli in stools, softer stool consistency, and decreased fecal pH. In a study dealing with the evaluation of mixtures of galacto-, fructo- and pectin oligosaccharides carried out on a group of healthy, full-term, partially breast-fed children, ingestion of prebiotics resulted in an increase of the *Bifidobacterium* counts and a decrease in the proportions of the *Bacteroides* and *Clostridium coccoides* groups, suggesting that the proposed combination was clinically safe and effective on infant microbiota, as it minimized the alteration of fecal microbiota after cessation of breast-feeding and enhanced the proportions of bifidobacteria (Magne et al., 2008).

14.5.3 Other Biological Effects of Pectins and Pectin-Derived Products

Beyond prebiotic characteristics, pectins and pectin oligosaccharides have been reported to cause a number of healthy effects, including regulation of lipid and glucose metabolism, decreased glycemic response and blood cholesterol levels, anti-cancer and immunological properties, reduction of damage by heavy metals, anti-obesity effects, dermatological applications, anti-toxic, anti-infection, antibacterial and antioxidant properties. Owing to their ability for exerting multiple health-promoting effects (see below), pectic oligosaccharides have been proposed as a new class of prebiotics (Hotchkiss et al., 2004) which however have not been yet evaluated as such in humans.

Experimental evidence suggests that the application of pectin and pectin-derived products on health care seems to be dependent on the fine carbohydrate structures, bringing many possibilities of benefits for human being. Even if in some cases natural pectin had no activity, chemical and enzymic modifications may provide useful health care products. Selected health effects of pectins and pectin-derived products are discussed in the following paragraphs.

Epidemiological evidence exists that a diet high in water-soluble fiber is inversely associated with the risk of cardiovascular diseases. One of the ways of preventing the cardiovascular disease is to lower serum LDL cholesterol levels. This objective can be reached by pectin ingestion without concomitant effects on HDL cholesterol or triacylglycerol concentrations (Theuwissen and Mensink, 2008). Hypocholesterolemic action has been reported for a variety of pectin-containing feedstocks, whereas blood cholesterol-lowering activity has been also claimed for pectin hydrolysates, and low-molecular weight citrus pectin (prepared by hydrolysis with a crude enzyme preparation from *Kluyveromyces fragilis*) showed a repressive effect on liver lipid accumulation. Recent studies have been devoted to assess the mechanisms of pectin action on cholesterol, particularly through experiments with animals. Feeding rats with pectin-containing products from the enzymic liquefaction of apples resulted in physiological benefits derived from an increased excretion of bile acids and neutral sterols. Upon pectin ingestion, hepatic conversion of cholesterol into bile acids increases, which will ultimately lead to increased LDL uptake by the liver (Theuwissen and Mensink, 2008). Pectins with high methoxyl content show ability to bind lipids and bile acids. Experiments using rats fed with hydrophobic amidated pectins showed significantly altered cholesterol homeostasis, and the results suggested that this substrate might be considered as a clinically effective hypocholesterolemic agent. Several hypotheses have been proposed to explain the hypocholesterolemic effects, including binding of bile acids by fiber, interference of lipid absorption and reduced hepatic cholesterol synthesis by propionate. Their major hypocholesterolemic effects have been reported to be related with bile acids. Increased fecal excretion of bile acids and neutral sterols has been reported in the experimental assessment of various pectic substrates, whereas ingestion of dietary fibers (including pectin) has been correlated with decreased plasma cholesterol concentrations. The decrease of cholesterol levels in liver and blood serum and its increase in feces could explain the beneficial effects of these dietary fibers in disease prevention. The interference with bile acid absorption, and hence, reduction of cholesterol absorption, seems to be the major mechanism for the observed effects (Ikeda and Sugano, 2005). Pectins, by enhancing fecal bile acid excretion, may cause increased hepatic synthesis of bile acids and liver depletion of cholesterol, resulting in a higher rate of cholesterol synthesis and reduced serum cholesterol concentrations. Increased fecal excretion of bile acids has been reported to be negatively correlated with the serum cholesterol level. Hypolipidemic activity has been ascribed to both the lower rate of absorption and higher rate of degradation and elimination of lipids. On the other hand, the hypolypidemic action of pectins depends on the raw material and on the processing conditions: some isolates presented highly significant hypolipidemic activity, while others were less significant, or even insignificant, in their action. Hypolipidemic activity has been reported for high-methoxyl pectins through experiments with rats: pectins of increased methylation degrees enabled higher concentrations of free and secondary bile acids in the cecum and colon of conventional rats. With increasing degree of methylation, more bile acids were transported into lower parts of the intestinal tract and excreted, whereas the proportion of secondary bile acids decreased. In a related study, rats fed with dietary highly methylated apple pectin resulted in decreased body weight gains and total cholesterol and triglyceride levels compared to specimens fed with an standard diet, suggesting that high methoxyl fractions from apple could be used as a functional ingredient to decrease cardiometabolic risk factors. The structure of pectins has been found significant for the physiological effects caused in hamsters fed with either lemon pectin or "smooth regions" of lemon pectin made up of polygalacturonic acid regions; the latter fraction showed comparatively better cholesterol-lowering properties. Lemon peels and lemon peel wastes have been reported to be as effective in lowering plasma and liver cholesterol in hamsters as the pectin extracted from the peels.

Concerning the effects of pectins on glucose metabolism, these compounds (alone or in combination with other dietary fibers), have been reported to play a constructive role by releasing sugars and absorbing sugars slowly in the intestinal tract; this would be of interest, for example, for reducing the severity of diabetes mellitus (Butt et al., 2007). Pioneering studies emphasized the ability of pectins for decreasing the post-prandial hyperglycaemia, thus improving the control of blood glucose concentration. The ability of pectins to reduce the blood glucose levels (Butt et al., 2007) and the postprandial rise in blood glucose and serum insulin in patients with type-II diabetes, have been claimed in more recent studies. Pectins have been reported to be beneficial for fish oil-treated diabetic patients, whereas oligogalacturonic acids (with DP up to 20, obtained by enzymatic processing of pectin and further fractionation), and/or their physiologically acceptable salts, have been claimed as hyperglycaemia inhibitors, suitable for diabetes treatment. However, other studies did not found significant effects of pectin administration on the postprandial glucose response of rats and humans, and their potentially valuable effects could be more in relation with satiety feeling associated with sustained late blood glucose levels.

Pectins and pectin oligosaccharides may provide protection against some types of cancer. Oligosaccharides derived from pectins were able to stimulate apoptosis in these cells (Hotchkiss et al., 2004; Rastall et al., 2005), and cause the growth inhibition of colon cancer cells; this suggests that there is a possibility for developing dietary protection against colon cancer (Hotchkiss et al., 2007). The ability for inhibiting the growth of cancer cells has also been reported for citrus pectin subjected to irradiation, as well as for the dialyzed products from irradiated samples (with molecular weight <10 kDa). Some commercial pectins have also been shown to have anti-cancer activity in rats and mice: fractionated pectin significantly induced apoptosis in both androgen-responsive and androgen-independent prostate cancer cells. Lower molecular weight compounds obtained by endopolygalacturonase treatment retained the apoptosis activity, which was lost when the polymeric substrate was subjected to mild base treatment. However, pectin did not inhibit intestinal carcinogenesis in mutated mice with predisposition for intestinal tumorigenesis.

Immunomodulatory effects have been reported for pectins, which have sometimes been attributed to microbiota-dependent effects. This topic has been covered in a recent review (Vos et al., 2007) for various non-digestible-carbohydrates, including pectins, and the immunological properties of structurally characterized pectic polymers from defined sources have been assessed recently.

The ability of some pectins to bind heavy metals make them potentially suitable for treatment of poisoning by heavy metals. Study of the effects of low-esterified pectin in toxic damage to the liver caused to animals by enteral treatment with lead acetate showed that pectin treatment promoted the decrease in lead content in the liver, and the recovery of parameters of lipid metabolism; a study on the toxic injury induced in rat liver by enteral administration of lead acetate showed that pectin administration enabled a fast reduction of lead concentration in liver, resulting in decreased lipid peroxidation, as well as the normalization of total cholesterol and triglyceride levels in blood serum and liver.

Pectin oligosaccharides have been reported to provide protection against *Escherichia coli* toxins (Hotchkiss et al., 2007; Rastall et al., 2005).

Oligo-galacturonides show ability for influencing cell-cell and cell-matrix interactions, displaying activity against the adhesion of pathogenic bacteria such as enteropathogenic and verotoxigenic strains of *E. coli* (Rastall et al., 2005).

Antibacterial activity has been reported for enzymic hydrolysates of cold water-extracted pectin from *Crataegus sanguinea*, consisting of oligomers with DP > 5–6. Antibacterial activity against *Escherichia coli* has been reported for lemon pectin hydrolysates (obtained using H_2SO_4 or pectinase), with increased effects for the latter type, these were attributed to the differences in free undissociated carboxyl groups, methoxyl groups, and DP of oligogalacturonides.

Finally, it can be cited that pectins or pectin oligomers have been employed for other applications, including prevention of infection, prevention or treatment of skin inflammation, cosmetic or dermatological compositions, anti-obesity formulae, antioxidant activity, and as active principles of mineral absorption-promoter preparations.

14.6   Summary

- When ingested as a part of the diet, some biomass polysaccharides and/or their oligomeric hydrolysis products are fermented in the colon, potentially producing prebiotic effects. Xylans, mannans, arabinogalactans and pectins are suitable substrates for this purpose.
- Hydrolysis of the polysaccharides cited above can be achieved by chemical processing, enzymatic reaction, or combined chemical-enzymatic processes, resulting in products of lower DP. In many cases, refining of the reaction media is needed to obtain food-grade products, whose biological properties depend on their composition and DP distribution.
- Refining of hydrolysis media can be achieved by a number of separation methods, but multistage processing is frequently required to achieve purities in the desired range.

- Products of tailored DP (usually of oligomeric nature) may present advantages with respect to the polysaccharides they are derived from, particularly in terms of biological activity.
- The potentially prebiotic properties of biomass-derived prebiotics are related to their non-digestible nature and to their ability for being utilized by lactobacilli and bifidobacteria as specific substrates, leading to formation of SCFA (which are believed to contribute positively to colon health).
- Besides potentially prebiotic properties, the biomass-derived products considered in this study may cause a variety of favorable health effects (for example, limiting the cardiovascular risk, affecting the glucose metabolism, causing apoptosis of cancer cells or exerting immunomodulatory effects). Because of this, they can be considered as multipurpose, healthy food ingredients.
- Based on the consumers' awareness for healthy foods, the future of biomass-based prebiotics seems promisings. However, the future commercial developments must be based on sound scientific evidences of their biological properties, which need further assessment.

Acknowledgments

Authors are grateful to the Spanish Ministry of Education and Science for supporting this study, in the framework of the research Project "Properties of new prebiotic food ingredients derived from hemicelluloses" (reference AGL2008–02072), which was partially funded by the FEDER Program of the European Union.

List of Abbreviations

AG	Arabinogalactan
AGP	Arabinogalactan-Protein
AIDS	Acquired Immune Deficiency Syndrome
DP	Degree of Polymerization
DPPH	1,1-diphenyl-2-picrylhydrazyl
FISH	Fluorescent *in situ* Hybridization
GalA	Galacturonic Acid

G	β-1,4-D-glucopyranosil
GC-MS	Gas Chromatography-Mass Spectrometry
GH	Glycosyl Hydrolase
GRAS	Generally Recognized as Safe
HIV	Human Immunodeficiency Virus
HPSEC	High Performance Size Exclusión Cromatography
M	β-1,4-D-mannopyranosil
NDO's	Non Digestible Oligosaccharides
SCFA	Short Chain Fatty Acids
XO	Xylooligosaccharides
X2	Xylobiose

References

Abatzoglou N, Chornet E, Belkacemi K, Overend RP (1992) Phenomenological kinetics of complex systems: the development of a generalized severity parameter and its application to lignocellulosics fractionation. Chem Eng Sci 47:1109–1112

Al-Ghazzewi FH, Khanna S, Tester RF, Piggott J (2007) The potential use of hydrolysed konjac glucomannan as a prebiotic. J Sci Food Agric 87:1758–1766

Asano I, Hamaguchi K, Fujii S, Iino H (2003) In vitro digestibility and fermentation of mannooligosaccharides from coffee mannan. Food Sci Technol Res 9:62–66

Asano I, Nakamura Y, Hoshino H, Aoki K, Fujii S, Imura N, Iino H (2001) Use of mannooligosaccharides from coffee mannan by intestinal bacteria. Nippon Nogei Kagaku Kaishi 75(10):1077–1083

Berardini N, Knoedler M, Schieber A, Carle R (2005) Utilization of mango peels as a source of pectin and polyphenolics. Innovat Food Sci Emerg Tech 6(4):442–452

Butt MS, Ahmad A, Sharif MK (2007) Influence of pectin and guar gum composite flour on plasma biochemical profile of streptozotocin-induced diabetic male albino rats. Int J Food Prop 10(2):345–361

Chandrasekaran R, Janaswamy S (2002) Morphology of western larch arabinogalactan. Carbohydr Res 337:2211–2222

Chen HL, Cheng HC, Liu YJ, Liu SY, Wu WT (2006) Konjac acts as a natural laxative by increasing stool bulk and improving colonic ecology in healthy adults. Nutrition 22(11/12):1112–1119

Crociani F, Alessandrini A, Mucci MM, Biavati B (1994) Degradation of complex carbohydrates by Bifidobacterium spp. Int J Food Microbiol 24:199–210

Curti DG, Gretsch C, Labbe DP, Redgwell RJ, Schoonman JH, Ubbink JB (2005) Arabinogalactan isolate from green and roasted coffee for food and delivery applications and process for its production. Eur. Pat. Appl. 18 pp. EP 1600461 A1 20051130

Ebringerova A, Heinze T (2000) Xylan and xylan derivatives - biopolymers with valuable properties. 1. Naturally occurring xylans structures, isolation procedures and properties. Macromol Rapid Comm 21:542–556

Ebringerova A, Hromadkova Z, Heinze T (2005) Hemicellulose. Adv Polym Sci 186:1–67

Englyst HN, Hay S, Macfarlane GT (1987) Polysaccharide breakdown by mixed populations of human fecal bacteria. FEMS Microbiol Ecol 95:163–172

Fremont G (2007) Acacia gum, the natural multifunctional fibre. Paper Presented at the "Dietary Fibre 2006-Multifunctional Complex of Components" Conference, 3rd, Helsinki, Finland, pp 271–281

Fujii S, Aoki T, Hoshino H, Nakamura Y, Hamaguchi K, Asano I, Imura N, Umemura M (2001) Mannooligosaccharide for manufacturing probiotic bacteria growth promoter and anticariogenic food. Jpn. Kokai Tokkyo Koho (2001) 9 pp. JP 2001149041 A 20010605

Garna H, Mabon N, Nott K, Wathelet B, Paquot M (2006) Kinetic of the hydrolysis of pectin galacturonic acid chains and quantification by ionic chromatography. Food Chem 96(3):477–484

Garrote G, Cruz JM, Domínguez H, Parajó JC (2003) Valorisation of waste fractions from autohydrolysis of selected lignocellulosic materials. J Chem Technol Biotechnol 78:392–398

Garrote G, Falqué E, Domínguez H, Parajó JC (2007) Autohydrolysis of agricultural residues: Study of reaction byproducts. Biores Technol 98:1951–1957

Guillotin SE, Bakx EJ, Boulenguer P, Mazoyer J, Schols HA, Voragen AGJ (2005) Populations having different GalA blocks characteristics are present in commercial pectins which are chemically similar but have different functionalities. Carbohydr Polym 60(3):391–398

Gulfi M, Arrigoni E, Amado R (2007) In vitro fermentability of a pectin fraction rich in hairy regions. Carbohydr Polym 67(3):410–416

Gullón P, González-Muñoz MJ, Domínguez H, Parajó JC (2008) Membrane processing of liquors from Eucalyptus globulus autohydrolysis. J Food Eng 87:257–265

Hori M, Iwai K, Kimura R, Nakagiri O, Takagi M (2007) Utilization by intestinal bacteria and digestibility of arabinogalactan from coffee bean in vitro. Nihon Shokuhin Biseibutsu Gakkai Zasshi 24(4):163–170

Hotchkiss AT, Manderson K, Olano-Martin E, Grace WE, Gibson GR, Rastall RA (2004) Orange peel pectic oligosaccharide prebiotics with food and feed applications. Abstracts of Papers, 228th ACS National Meeting, Philadelphia, PA

Hotchkiss AT, Manderson K, Tuohy KM, Widmer WW, Nunez A, Gibson GR, Rastall RA (2007) Bioactive properties of pectic oligosaccharides from sugar beet and Valencia oranges. Abstracts of Papers, 233rd ACS National Meeting, Chicago, IL

Hotchkiss AT Jr, Olano-Martin E, Grace WE, Gibson GR, Rastall RA (2003) Pectic oligosaccharides as prebiotics. ACS Symposium Series 9 (Oligosaccharides in Food and Agriculture), pp 54–62

Hotchkiss AT, Widmer WW, Fishman ML (2005) Flash extraction of pectin. Abstracts of Papers, 229th ACS National Meeting, San Diego, CA

Hughes SA, Shewry PR, Li L, Gibson GR, Sanz ML, Rastall RA (2007) In vitro fermentation by human fecal microflora of wheat arabinoxylans. J Agric Food Chem 55:4589–4595

Ikeda I, Sugano M (2005) Dietary fiber and lipid metabolism: with special emphasis on dietary fibers in food for specified health uses in Japan. Foods Food Ingred J Japan 210(10): 901–908.

Izumi Y, Azumi N, Kido Y, Nakabo Y (2004) Oral preparations for atopic dermatitis containing acidic xylooligosaccharides Jpn. Kokai Tokkyo Koho 9 pp. JP 2004210666 Appl JP 2002-379881

Izumi Y, Kojo A (2003) Long-chain xylooligosaccharide compositions with intestinal function-improving and hypolipemic activities, and their manufacture Jpn. Kokai Tokkyo Koho 10 pp. JP 2003 048901 Appl JP 2001-242906

Izumi Y, Sugiura J, Kagawa H, Azumi N (2000) Alkali-oxygen bleaching of lignocellulosic pulp Eur. Pat. 32 pp. EP 1039020 A1 Appl EP 2000-302354

Kabel MA, Carvalheiro F, Garrote G, Avgerinos E, Koukios E, Parajó JC, Gírio FM, Schols HA, Voragen AGJ (2002a) Hydrothermally treated xylan rich by-products yield different classes of xylo-oligosaccharides. Carbohydr Polym 50:47–52

Kabel MA, Kortenoeven L, Schols HA, Voragen AGJ (2002b) In vitro fermentability of differently substituted xylo-oligosaccharides. J Agric Food Chem 50:6205–6210

Kelly GS (1999) Larch arabinogalactan : clinical relevance of a novel immune-enhancing polysaccharide. Alternat Medic Rev 4(2): 96–103.

Kravchenko TP, Arnould I, Voragen AGJ, Pilnik W (1992) Improvement of the selective depolymerization of pectic substances by chemical β-elimination in aqueous solution. Carbohydr Polym 19(4):237–242

Kurakake M, Sumida T, Masuda D, Oonishi S, Komaki T (2006) Production of galacto-manno-oligosaccharides from guar gum by β-mannanase from Penicillium oxalicum SO. J Agric Food Chem 54: 7885–7889

Lazaridou A, Biliaderis CG, Izydorczyk MS (2001) Structural characteristics and rheological properties of locust bean galactomannans: a comparison of samples from different carob tree populations. J Sci Food Agric 81:68–75

Magne F, Hachelaf W, Suau A, Boudraa G, Bouziane-Nedjadi K, Rigottier-Gois L, Touhami M, Desjeux JF, Pochart P (2008) Effects on faecal microbiota of dietary and acidic oligosaccharides in children during partial formula feeding. J Pediatr Gastroenterol Nutr 46(5):580–588

Mandalari G, Bennett RN, Kirby AR, Lo Curto RB, Bisignano G, Waldron KW, Faulds CB (2006) Enzymatic hydrolysis of flavonoids and pectic oligosaccharides from bergamot (Citrus bergamia Risso) peel. J Agr Food Chem 54(21): 8307–8313

Matsuura Y (1998) Degradation of konjac glucomannan by enzymes in human feces and formation of short-chain fatty acids by intestinal anaerobic bacteria. J Nutr Sci Vitaminol 44 (3):423–436

May CD (1990) Industrial pectins: source, production and applications. Carbohydr Polym 12(1):79–99

Meance S (2004) Acacia gum (FIBREGUM), a very well tolerated specific natural prebiotic having a wide range of food applications - Part 1. Agro Food Ind Hi Tec 15(1):24–28

Mitchell CR, Mitchell PR (1996) Process for manufacture of treated pectinic acid or polyuronic acid. US Pat 9 pp. US 5498702 Appl US 93-169377

Miyazawa T, Funazukuri T (2004) Hydrothermal Production of mono(galacturonic acid) and the oligomers from poly(galacturonic acid) with water under pressures. Ind Eng Chem Res 43(10):2310–2314

Miyazawa T, Ohtsu S, Funazukuri T (2008) Hydrothermal degradation of polysaccharides in a semi-batch reactor: product distribution as a function of severity parameter. J Mater Sci 43(7):2447–2451

Moura P, Barata R, Carvalheiro F, Girio F, Loureiro-Dias MC, Esteves MP (2007) In vitro fermentation of xylo-oligosaccharides from corn cobs autohydrolysis by Bifidobacterium and Lactobacillus strains. LWT 40:963–972

Moure A, Gullón P, Domínguez H, Parajó JC (2006) Advances in the manufacture, purification and applications of xylo-oligosaccharides as food additives and nutraceuticals. Process Biochem 41: 1913–1923

Murota A, Yamanoi T (2003) Maltooligosaccharides and process for manufacturing them. Jpn. Kokai Tokkyo Koho 3 pp JP 2003261589 Appl JP 2002-61415

Nakakuki T (2005) Present status and future prospects of functional oligosaccharide development in Japan. J Appl Glycosci 52:267–271

Nakamura Y, Hoshino H, Fujii S (2002) Mannooligosaccharides for control of peroxidized fat level. Jpn. Kokai Tokkyo Koho 6 pp. JP 2002262828 Appl JP 2001-70410

Nunes FM, Reis A, Domingues MRM, Coimbra MA (2006) Characterization of Galactomannan Derivatives in Roasted Coffee Beverages J Agric Food Chem 54: 3428–3439

Odonmazig P, Ebringerova A, Machova E, Alfoldi J (1994) Structural and molecular properties of the arabinogalactan isolated from Mongolian larchwood (Larix dahurica L.). Carbohydr Res 252:317–324

Palm M, Zacchi G (2003) Extraction of hemicellulosic oligosaccharides from spruce using microwave oven or steam treatment. Biomacromolecules 4:617–623

Park GG (2006) Specificity of β-mannanase from Trichoderma sp. for Amorphophallus konjac glucomannan. Food Sci Biotechnol 15:820–823

Rastall RA, Manderson K, Hotchkiss AT, Gibson GR (2005) Investigation of the biological activities of pectic oligosaccharides using in vitro models of the human colon. Abstracts of Papers, 229th ACS National Meeting, San Diego, CA

Robinson RR, Feirtag J, Slavin JL (2001) Effects of dietary arabinogalactan on gastrointestinal and blood parameters in healthy human subjects. J Am Coll Nutr 20 (4):279–285

Sachslehner A, Foidl G, Foidl N, Gubitz G, Haltrich D (2000) Hydrolysis of isolated coffee mannan and coffee extract by mannanases of Sclerotium rolfsii. J Biotechnol 80:127–134

Schols HA, Bakx EJ, Schipper D, Voragen AGJ (1995) Hairy (ramified) regions of pectins. Part VII. A xylogalacturonan subunit present in the modified hairy regions of apple pectin. Carbohydr Res 279:265–279

Shobha MS, Kumar ABV, Tharanathan RN, Koka R, Gaonkar AK (2005) Modification of guar galactomannan with the aid of Aspergillus niger pectinase. Carbohydr Polym 62:267–273

Swennen K, Courtin CM, Delcour JA (2006) Non-digestible oligosaccharides with probiotic properties. Crit Rev Food Sci Nutr 46:459–471

Swennen K, Courtin CM, Van der Bruggen B, Vandecasteele C, Delcour JA (2005) Ultrafiltration and ethanol precipitation for isolation of arabinoxylooligosaccharides with different structures. Carbohydr Polym 62:283–292

Taniguchi H (2004) Carbohydrate research and industry in Japan and the Japanese society of applied glycoscience. Starch/Staerke 56(1):1–5

Theuwissen E, Mensink RP (2008) Water-soluble dietary fibers and cardiovascular disease. Physiol Behav 94(2):285–292

Tungland BC, Meyer D (2002) Nondigestible oligo- and polysaccharides (dietary fiber): their physiology and role in human health and food. Comp Rev Food Sci Food Safety 1(3):73–92

Tuohy KM, Kolida S, Lustenberger AM, Gibson GR (2001) The prebiotic effects of biscuits containing partially hydrolysed guar gum and fructo-oligosaccharides - a human volunteer study. Br J Nutr 86:341–348

Váquez MJ, Alonso JL, Domíguez H, Parajó JC (2001) Xylooligosaccharides. Manufacture and applications. Trends Food Sci Technol 11:387–393

Váquez MJ, Garrote G, Alonso JL, Domíguez H, Parajó JC (2005) Refining of autohydrolysis liquors for manufacturing xylooligosaccharides: evaluation of operational strategies. Biores Technol 96:889–896

Vegas R, Alonso JL, Domínguez H, Parajó JC (2004) Processing of rice husk autohydrolysis liquors for obtaining food ingredients. J Agric Food Chem 52:7311–7317

Vegas R, Alonso JL, Domínguez H, Parajó JC (2005) Manufacture and refining of oligosaccharides from industrial solid wastes. Ind Eng Chem Res 44:614–620

Vegas R, Luque S, Alvarez JR, Alonso JL, Dominguez H, Parajó JC (2006) Membrane-assisted processing of xylooligosaccharide-containing liquors. J Agric Food Chem 54:5430–5436

Vegas R, Alonso JL, Domínguez H, Parajó JC (2008b) Enzymatic processing of rice husk autohydrolysis products for obtaining low molecular weight oligosaccharides. Food Biotechnol 22:31–46

Vegas R, Moure A, Dom'nguez H, Parajo' JC, Alvarez JR, Luque S (2008a) Evaluation of ultra- and nanofiltration for refining soluble products from rice husk xylan. Biores Technol 99:5341–5351

Vince AJ, McNeil NI, Wagner JD, Wrong OM (1990) The effect of lactulose, pectin, arabinogalactan and cellulose on the production of organic acids and metabolism of ammonia by intestinal bacteria in a fecal incubation system. Br J Nutr 63:17–26

Voragen AGJ, Pilnik W, Thibault JF, Axelos MAV, Renard CMGC (1995) Pectins. Food Sci Technol 67:287–339

Vos AP, M'Rabet L, Stahl B, Boehm G, Garssen J (2007) Immune-modulatory effects and potential working mechanisms of orally applied nondigestible carbohydrates. Crit Rev Immunol 27(2):97–140

Wu Z, He J (2003) Process for production of mannooligosaccharide from mannan-containing vegetable gums. Faming Zhuanli Shenqing Gongkai Shuomingshu CN 1412194 Appl CN 2001-134073

Yadav MP, Igartuburu JM, Yan Y, Nothnagel EA (2007) Chemical investigation of the structural basis of the emulsifying activity of gum arabic. Food Hydrocol 21(2):297–308

Yuan QP, Zhang H, Qian ZM, Yang XJ (2004) Pilot-plant production of xylo-oligosaccharides from corncob by steaming, enzymatic hydrolysis and nanofiltration. J Chem Technol Biotechnol 79: 1073–1079

15 Taxonomy of Probiotic Microorganisms

Giovanna E. Felis · Franco Dellaglio · Sandra Torriani

15.1   Introduction

When referring to probiotics, one refers to probiotic strains, i.e., the microbial individuals, sub-cultures of billion of almost identical cells ideally derived from the same mother cell. Therefore, beneficial effects attributed to probiotics are ascribed in fact to specific strains. However, these strains have to be, by law, clearly identified at the species level (Pineiro and Stanton, 2007). In fact, probiotics have to be safe for consumption, and the evaluation of QPS – qualified presumption of safety – status by the European Food Safety Authority (EFSA) (Opinion, 2007) is discussed for species, not for single strains.

Also, corrected names have to be reported on products labels: failure of identification of the declared species is a commercial fraud and a consumer misleading, besides being an indication of unreliability of the product.

These two examples should clarify how important is the correct taxonomic identification of probiotic strains in the assessment of their reliability and efficacy.

The aim of the present contribution is to clarify which procedures, rules and scientific knowledge stand behind microbial names, as results of taxonomic analysis. Probiotic strains described to date fall in two different groups of microorganisms, namely bacteria and yeasts, which will be the focus of this treatment.

15.2   What is Taxonomy?

Taxonomy, from the Greek meaning categorization, can be viewed as a quest for order in nature. In other words, it is the analysis of the existing biodiversity in a systematic way, with the aim of arranging it in an ordered hierarchical scheme. This hierarchy is (should be) the result of the genealogical relationships between organisms (the *taxa*). The information used for these purposes has to be as complete as possible, encompassing morphology, physiology, ecology and genetics.

A taxonomic procedure can be viewed as an iterative process, in which (i) organisms are clustered into groups on the basis of similarity; then (ii) groups are given formal names, indicated in italics, composed by a genus name and a specific identifier (e.g., *Homo sapiens*); and finally, other *taxa* are analyzed and can be either assigned to any already existing (i) and named (ii) group, or newly identified and described as novel groups. These three steps are indicated as *classification, nomenclature,* and *identification,* respectively (Staley and Krieg, 1989). The taxonomic hierarchy is based on the unit called species (see below). Then, Species are grouped in a Genus, Genera in a Family, Families in an Order, Orders in a Class, and Classes in a Domain. Three domains of life have been described comprising all the living organisms, two for the prokaryotes, Archaea and Bacteria, and one for eukaryotes. In the Domain Eukarya, four kingdoms are also recognized, i.e., Protista, Fungi, Animalia, and Plantae, to account for the diversity of eukaryotes, while no Kingdom category has been proposed above Classes in the Domains Archaea and Bacteria (Woese et al., 1990).

It is clear that a species in the bacterial domain must be, as a category, equivalent to a eukaryotic species, otherwise the "scheme of life" looses its meaning. However, clearly, biologically, species such as *Lactobacillus casei* (a bacterium) and ourselves as *Homo sapiens* are not comparable. Thus, it is clear that taxonomy is a necessary but conventional way of describing diversity. However, taxonomy has the important practical aim of making biodiversity accessible through cataloguing, therefore the species is considered to be equivalent as a category in the three domains, but it is circumscribed differently when analyzing different organisms.

15.2.1   Concept, Delineation and Naming of Species

The basic unit of the taxonomic scheme is the species. The species problem is one of the most debated topics in science, and will not be reviewed here: probiotic microorganisms are actually strains, not species, therefore it seems much more interesting here to focus on the relationship between species and strain, and explain how a species is circumscribed, i.e., which are the criteria used to determine if an isolate belongs to a known species or deserves a novel name.

Excellent reviews on the species concept and species circumscription are available for the two domains which include the probiotic strains known to date, i.e., Bacteria and Eukarya (Kingdom Fungi), (Giraud et al., 2008;

Rosselló-Mora and Amann, 2001). To date no archaeal strains with probiotic properties have been described, therefore that domain will not be reviewed here.

Considering species delineation, the practical criteria for the description of novel species will be reminded here. As for bacteria, the gene sequence for 16S rRNA gene is determined and analyzed to obtain the phylogenetic placement for the microorganism, then the closest neighbors are considered in a comparative study: DNA-DNA hybridization technique is then applied to assess the overall genomic similarity of the strains; if it is above 70% the strains are considered to belong to the same species, otherwise they represent different *taxa*. Also, relevant phenotypic characteristics of the strains are determined, e.g., patterns of fermentation of carbohydrates, in order to define a diagnostic trait which makes the novel species recognizable from its neighbors. More recently, other techniques have been suggested which should be useful in a more precise species delineation, such as Multi Locus Sequence Typing (Stackebrandt et al., 2002).

For fungi, and yeasts in particular, the procedure is less straightforward, and a novel species is described when genetic and/or morphological and/or reproductive differences are found, but these criteria can vary depending on the organisms under study.

This makes it clear that there are no rules for classification, as only experts can evaluate if a group of strains deserves a species status, and the standard procedure for species delineation of bacteria is pragmatic and useful, but does not have to be applied strictly.

On the contrary, rules do exist in nomenclature: once a group of strains is believed to represent a novel species, it deserves a name, which has to be given following precise regulation. Nomenclature of bacteria follows the International Code of Nomenclature of Bacteria (Lapage et al., 1990), and names are listed in the "Approved List 1980" and listed as Approved Lists 1980. Names published after 1980 or changes in names, to be valid, have to be published on the International Journal of Systematic and Evolutionary Microbiology – IJSEM – (known as International Journal of Systematic Bacteriology – IJSB – before 2000). They can also be published elsewhere, but, to be officially recognized, they have to be announced in the Validation Lists on IJSEM.

Nomenclature of yeasts follows, as for fungi, the International Code of Botanical Nomenclature, the recent version of which is the so-called Vienna Code, adopted by the Seventeenth International Botanical Congress, Vienna, Austria, July 2005 (McNeill et al., 2006). Species descriptions of unicellular eukaryotes are published on IJSEM, but also on other journals, and names are listed in the Index fungorum (see below for details).

As taxonomy is always linked to technical progress, taxonomic re-examinations often results in changes in nomenclature. Rules in nomenclature are necessary to avoid ambiguities and misunderstandings when taxonomic ree-valuations are performed and changes in names occur. The availability of lists of names, with rules of priority, also allows associating to new names characteristics linked to older ones, avoiding loss of information due to changes in names.

One of the most important points of the rules of nomenclature is that, when a novel species is described, a type strain (i.e., a strain chosen as a reference point among those grouped in the new species) has to be indicated, which has to be deposited in international culture collections, and be available to the scientific community for study and comparisons. Finally, it has to be pointed out that for bacteria one single correct name exists for a described species, while for yeasts there are two names, indicating the two different living forms, i.e., the teleo-morph – the sexual reproductive stage, or the anamorph, the asexual reproductive stage.

15.2.3   Useful Links for Taxonomic Information

As anticipated, nomenclature is clearly regulated, and electronic versions of the Codes are available: the Bacteriological Code (Lapage et al., 1992) at http://www.ncbi.nlm.nih.gov/books/bv.fcgi?rid = icnb.TOC&depth = 2, and the Vienna Code (McNeill et al., 2006) at http://ibot.sav.sk/icbn/main.htm.

Also, lists of up-to-date names can be checked online, on the DSMZ (German collection of microorganisms and cell cultures, http://www.dsmz.de/bactnom/bactname.htm) or on the internet pages complied by J. P. Euzéby and updated after each issue of IJSEM (http://www.bacterio.cict.fr/). For yeasts the reference is the Index Fungorum, http://www.indexfungorum.org/.

For both bacteria and unicellular microorganisms, the journal publishing novel species descriptions is the International Journal of Systematic and Evolutionary Microbiology (www.sgmjournals.org).

Considering classification, the reference book for bacteria is the Bergey's Manual of Systematic Bacteriology, now at its second edition (2001), while, for yeasts, useful indications can be found in The Yeasts: a Taxonomic Study, fourth edition (1999) (Edited by Kurtzman and Fell), and in Yeasts: Characteristics and Identification, third edition (2000) (Cambridge: Cambridge University Press). These books provide information on morphologic, metabolic and genetic characteristics of all known taxa and keys for their identification. Concerning prokaryotes, the taxonomic outline of the last version of Bergey's Manual underlines the phylogenetic placement of genera and taxonomic level above (Garrity et al., 2007a, http://www.taxonomicoutline.org/); a comprehensive picture of the phylogenetic diversity of prokaryotic microorganisms can be visualized at http://www.arb-silva.de/fileadmin/silva_databases/living_tree/LTP_tree_s93.pdf, which can also be used to analyze specific subtrees including the species of interest (Yarza et al., 2008).

Novel species descriptions require the deposit of the type strain in at least two recognized culture collections, the only international organizations devoted to storage and preservation of biodiversity. Well known collections are, among others, ATCC (www.atcc.org), LMG (www.bespo.be/bccm), DSMZ (www.dsmz.de) and JCM (http://www.jcm.riken.jp/), while CBS (http://www.cbs.knaw.nl/), DBVPG (http://www.agr.unipg.it/dbvpg/), and NRRL (http://nrrl.ncaur.usda.gov/) are mostly devoted to yeasts. Other information can be found on the web pages of the World Federation for Culture Collections (http://www.wfcc.info/datacenter.html), for other culture collections and abbreviations.

Finally, genome sequence data are expected to significantly contribute to taxonomy, therefore an interesting genomic database is GOLD (www.genomesonline.org), where sequencing projects are identified with a label: "Gc number" for complete genomes (http://genomesonline.org/gold.cgi?want = Published + Complete + Genomes), and a "Gi number" for incomplete sequences (http://www.genomesonline.org/gold.cgi?want = Bacterial + Ongoing + Genomes and http://www.genomesonline.org/gold.cgi?want=Eukaryotic+Ongoing+Genomes, for Eukaryotes, respectively), which will be referred to in the text.

15.3   Taxonomic Placement of Probiotic Microorganisms

Members of the Domain Bacteria are unicellular prokaryotic microorganisms, characterized by different cellular shapes (cocci, rods, spiral etc.) and cell size of few micrometers. To date, more than 8,000 species of prokaryotes (Garrity et al., 2007a) are described, grouping organisms isolated from almost every habitat on earth.

Concerning the taxonomy of the Domain, it is subdivided into 24 Phyla. Gram-positive probiotic strains known to date belong to the Phyla *Firmicutes* and

Actinobacteria, with low or high genome guanine-cytosine base (GC) content, respectively. Also, probiotic properties have been attributed to *Escherichia coli*, a Gram-negative bacterium belonging to the Phylum *Proteobacteria*.

Data will be presented as follows: after some general indications concerning the bacterial and yeast genera involved, detailed information will be given on species comprising probiotic strains. Also, an updated list of species included in the respective genera will be presented, with the double aim of providing an immediate check list for correct names of strains and a reference to the original description of the species themselves. Species belonging to different genera will be treated separately, even if closely related (e.g., *Lactobacillus* and *Pediococcus*), to facilitate the consultation of data for single species and genera. Finally, all genus names will be reported *in extenso* throughout the text, to avoid confusion between names of different genera with the same capital letter (e.g., *B*. could stand for *Bifidobacterium*, *Bacillus* and *Brevibacillus*).

15.3.1   Lactic Acid Bacteria (LAB)

This group of bacteria is named after the ability of fermenting carbohydrate to lactic acid. Taxonomically, it comprises diverse genera of bacteria, which appear to be also phylogenetically unrelated. The two most important genera in the probiotic field are *Lactobacillus* and *Bifidobacterium*, but some others contain species of interest, e.g., *Pediococcus*, *Enterococcus*, and *Lactococcus*.

15.3.1.1 Genus *Lactobacillus*

Lactobacilli are Gram-positive bacteria, unable to sporulate, occurring as rods or cocco-bacilli, with a GC composition of the genome usually below 50% (low GC bacteria). They are fastidious microorganisms, requiring rich media to grow, and microaerophilic. They are catalase negative, even if pseudocatalase activity can sometimes be present in some strains and in presence of a heme group.

They are almost ubiquitous and can be found in almost all the environments where carbohydrates are available, such as food (dairy products, fermented meat, sourdoughs, vegetables, fruits, beverages), respiratory, gastro-intestinal (GI) and genital tracts of humans and animals, sewage and plant material.

The genus *Lactobacillus* belongs to the Phylum *Firmicutes*, Class *Bacilli*, Order *Lactobacillales*, Family *Lactobacillaceae* and its closest relatives, being grouped within the same Family, are the genera *Paralactobacillus* and *Pediococcus* (Garrity et al., 2007b).

At the time of writing (September 2008), the genus includes 116 species with valid names, and several others are in course of publication (● *Table 15.1*).

The taxonomy of the genus is largely unsatisfactory, as the description of many novel species in the recent years has made it clear that the genus is heterogeneous and it is phylogenetically intermixed with the other two genera of the Family, i.e., *Pediococcus* and *Paralactobacillus* (● *Figure 15.1*). Moreover, metabolic differences, in terms of type and quantity of metabolites of carbo-hydrate fermentation, do not match with phylogenetic groupings and are, there-fore, unreliable markers for their classification. Two species of the genus, namely *Lactobacillus catenaformis* and *Lactobacillus vitulinus* are only poorly related to the other species and their taxonomic status requires attention (Hammes and Vogel, 1995; Pot et al., 1994); for this reason they were omitted in ● *Figure 15.1*.

Phylogenetically, the species form a number of stable groups, at least seven, based on the analysis of 16S rRNA gene sequences, but several species form distinct lines of descent (Felis and Dellaglio, 2007) as can be noted in the ● *Figure 15.1*, depicting the updated structure of the taxonomic Family.

Lactobacillus Acidophilus

The etymology of the name, meaning acid-loving, indicates the preference of this species for acid medium for growth. Strains of this species usually display a rod shape with rounded ends and occur singly, in pairs and in short chains.

The original description was based on strains isolated from the intestinal tract of man and animals, human mouth and vagina, but are also easily found in milk and dairy products. At the time of first description, it was a heterogeneous species, as it included strains later reclassified as novel species, e.g., *Lactobacillus johnsonii*. Its metabolism is homofermentative, converting hexoses almost completely to lactic acid (both isomers D and L are produced) (Hammes and Vogel, 1995). Phylogenetically, it is placed in the *Lactobacillus delbrueckii* group (Felis and Dellaglio, 2007). The type strain is ATCC 4356^T (=LMG 9433^T = DSM 20079^T), of human origin, active for acidophilus milk when used with yeast extract. The GC content of the genome is 34–37%.

Genome sequence data are available for the non-type probiotic strain NCFM (strain Gc00252), and another sequencing project is ongoing (Gi02787), but there is no accession number indicating which particular strain is being sequenced.

⬛ Table 15.1

List of valid names in the genus *Lactobacillus* (updated at September 2008) *(Cont'd p. 599)*

	Species of the genus *Lactobacillus*
1	*Lactobacillus acetotolerans*
2	*Lactobacillus acidifarinae*
3	*Lactobacillus acidipiscis*
4	***Lactobacillus acidophilus***
5	*Lactobacillus agilis*
6	*Lactobacillus algidus*
7	*Lactobacillus alimentarius*
8	*Lactobacillus amylolyticus*
9	*Lactobacillus amylophilus*
10	*Lactobacillus amylotrophicus*
11	*Lactobacillus amylovorus*
12	*Lactobacillus animalis*
13	*Lactobacillus antri*
14	*Lactobacillus apodemi*
15	*Lactobacillus aviarius*
	Lactobacillus aviarius subsp. *araffinosus*
	Lactobacillus aviarius subsp. *aviarius*
16	*Lactobacillus bifermentans*
17	*Lactobacillus brevis*
18	*Lactobacillus buchneri*
19	*Lactobacillus camelliae*
20	***Lactobacillus casei***
21	*Lactobacillus catenaformis*
22	*Lactobacillus ceti*
23	*Lactobacillus coleohominis*
24	*Lactobacillus collinoides*
25	*Lactobacillus composti*
26	*Lactobacillus concavus*
27	*Lactobacillus coryniformis*
	Lactobacillus coryniformis subsp. *coryniformis*
	Lactobacillus coryniformis subsp. *torquens*
28	***Lactobacillus crispatus***
29	*Lactobacillus crustorum*
30	*Lactobacillus curvatus*
31	***Lactobacillus delbrueckii***

◻ Table 15.1 (*Cont'd* p. 600)

	Species of the genus *Lactobacillus*
	Lactobacillus delbrueckii subsp. *bulgaricus*
	Lactobacillus delbrueckii subsp. *delbrueckii*
	Lactobacillus delbrueckii subsp. *indicus*
	Lactobacillus delbrueckii subsp. *lactis*
32	*Lactobacillus diolivorans*
33	*Lactobacillus equi*
34	*Lactobacillus equigenerosi*
35	*Lactobacillus farciminis*
36	*Lactobacillus farraginis*
37	*Lactobacillus fermentum*
38	*Lactobacillus fornicalis*
39	*Lactobacillus fructivorans*
40	*Lactobacillus frumenti*
41	*Lactobacillus fuchuensis*
42	*Lactobacillus gallinarum*
43	*Lactobacillus gasseri*
44	*Lactobacillus gastricus*
45	*Lactobacillus ghanensis*
46	*Lactobacillus graminis*
47	*Lactobacillus hammesii*
48	*Lactobacillus hamsteri*
49	*Lactobacillus harbinensis*
50	*Lactobacillus hayakitensis*
51	*Lactobacillus helveticus*
52	*Lactobacillus hilgardii*
53	*Lactobacillus homohiochii*
54	*Lactobacillus iners*
55	*Lactobacillus ingluviei*
56	*Lactobacillus intestinalis*
57	*Lactobacillus jensenii*
58	***Lactobacillus johnsonii***
59	*Lactobacillus kalixensis*
60	*Lactobacillus kefiranofaciens*
	Lactobacillus kefiranofaciens subsp. *kefiranofaciens*
	Lactobacillus kefiranofaciens subsp. *kefirgranum*
61	*Lactobacillus kefiri*
62	*Lactobacillus kimchii*

Table 15.1 (*Cont'd* p. 601)

	Species of the genus *Lactobacillus*
63	*Lactobacillus kitasatonis*
64	*Lactobacillus kunkeei*
65	*Lactobacillus lindneri*
66	*Lactobacillus malefermentans*
67	*Lactobacillus mali*
68	*Lactobacillus manihotivorans*
69	*Lactobacillus mindensis*
70	*Lactobacillus mucosae*
71	*Lactobacillus murinus*
72	*Lactobacillus nagelii*
73	*Lactobacillus namurensis*
74	*Lactobacillus nantensis*
75	*Lactobacillus oligofermentans*
76	*Lactobacillus oris*
77	*Lactobacillus panis*
78	*Lactobacillus pantheris*
79	*Lactobacillus parabrevis*
80	*Lactobacillus parabuchneri*
81	***Lactobacillus paracasei***
	Lactobacillus paracasei subsp. *paracasei*
	Lactobacillus paracasei subsp. *tolerans*
82	*Lactobacillus paracollinoides*
83	*Lactobacillus parafarraginis*
84	*Lactobacillus parakefiri*
85	*Lactobacillus paralimentarius*
86	*Lactobacillus paraplantarum*
87	*Lactobacillus pentosus*
88	*Lactobacillus perolens*
89	***Lactobacillus plantarum***
	Lactobacillus plantarum subsp. *argentoratensis*
	Lactobacillus plantarum subsp. *plantarum*
90	*Lactobacillus pontis*
91	*Lactobacillus psittaci*
92	*Lactobacillus rennini*
93	***Lactobacillus reuteri***
94	***Lactobacillus rhamnosus***

◻ Table 15.1

	Species of the genus *Lactobacillus*
95	*Lactobacillus rogosae*
96	*Lactobacillus rossiae*
97	*Lactobacillus ruminis*
98	*Lactobacillus saerimneri*
99	*Lactobacillus sakei*
	Lactobacillus sakei subsp. *carnosus*
	Lactobacillus sakei subsp. *sakei*
100	**Lactobacillus salivarius**
101	*Lactobacillus sanfranciscensis*
102	*Lactobacillus satsumensis*
103	*Lactobacillus secaliphilus*
104	*Lactobacillus senmaizukei*
105	*Lactobacillus sharpeae*
106	*Lactobacillus siliginis*
107	*Lactobacillus spicheri*
108	*Lactobacillus suebicus*
109	*Lactobacillus thailandensis*
110	*Lactobacillus ultunensis*
111	*Lactobacillus vaccinostercus*
112	*Lactobacillus vaginalis*
113	*Lactobacillus versmoldensis*
114	*Lactobacillus vini*
115	*Lactobacillus vitulinus*
116	*Lactobacillus zymae*

Species with QPS status are underlined, principal species including probiotic strains are in bold and are described in the text

Lactobacillus Casei – Lactobacillus Paracasei

The two species *Lactobacillus casei* and *Lactobacillus paracasei* are presented together because they are closely related and their nomenclatural status has been discussed for a long time and only recently solved (Judicial Commission, 2008).

The name *casei* indicate the origin of the species in cheese and dairy products in general, but it has been isolated also from sourdough, silage, human GI tract (GIT), vagina, sewage etc. The name *paracasei* indicate the close resemblance of this *taxon* to *Lactobacillus casei*.

■ Figure 15.1
Phylogenetic tree showing the relationships among the species of the Family *Lacto-bacillaceae*, including genera *Lactobacillus* (abbreviated with "*L.*" in the tree), *Paralactobacillus* and *Pediococcus* (abbreviated with "*P.*" in the tree). The tree was calculated with a maximum-likelihood-derived distance method and clustering was performed with neighbor joining method. The bar indicates number of substitutions per site.

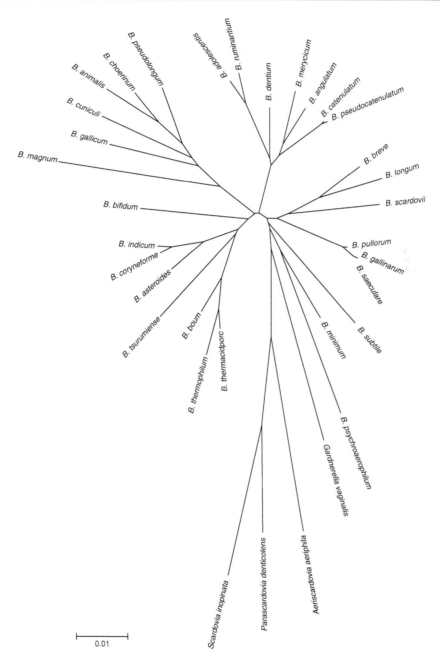

■ Figure 15.2
Phylogenetic tree depicting relationships among the species of the genus *Bifidobacterium* (abbreviated as "*B.*" in the tree) and related genera. Tree was calculated with the same procedure reported for ◉ *Figure 15.1*.

The first description of these rods with square ends, often occurring in chain, was based on few phenotypic traits, and the choice of strain ATCC 393 as the type has been considered inadequate for a long time. This has lead to the description of *Lactobacillus paracasei*, phenotypically very similar but genomically distinct. The details of the taxonomic controversy will not be explained here, but it is important to point out that the debate has been definitively concluded as follows: both species names are valid, with ATCC 393^T as the type strain of *Lactobacillus casei* and NCFB 151^T (=ATCC 25302^T) as type strain of *Lactobacillus paracasei* (Judicial Commission, 2008). This is expected to have important consequences also in marketing of probiotic strains: many of the cultures on the market called *Lactobacillus casei* are actually more similar to the type strain of *Lactobacillus paracasei* and therefore should be given that name.

As for metabolism, both species are facultatively heterofermentative microorganisms, producing lactic acid (L or both isomers, depending on the strains) from hexoses, but are also able to ferment pentoses (Hammes and Vogel, 1995). Phylogenetically, the two species together with *Lactobacillus rhamnosus* form a distinct clade in the genus *Lactobacillus* (Felis and Dellaglio, 2007), and the genome GC content for both species ranges between 45 and 47%.

Genome data have been obtained or are in progress for several strains in the two species (Gc00438, Gc00822, Gi02995, Gi03110, Gi02219, and Gi02799), including the type strain of *Lactobacillus casei* ATCC 393^T. However, the identification of the strains is not always clear, e.g., ATCC 334 (Gc00438) is indicated as *Lactobacillus casei* while it is a *Lactobacillus paracasei* strain (Judicial Commission, 2008).

Lactobacillus Crispatus

The curled morphology of the cells in liquid medium gives the name to the species, of straight or slightly curved rods, occurring singly or in short chains. Phylogenetically related to *Lactobacillus acidophilus* and *Lactobacillus delbrueckii*, its neighbor species appear to be *Lactobacillus kefiranofaciens*. Historically, the species was described based on strains previously assigned to the species *Lactobacillus acidophilus*, which showed a similar homofermentative metabolism, and also similar genome GC content (35–38% for *Lactobacillus crispatus*). Strains have been isolated from saliva, feces and urogenital tract of man and chicken, and also found in patients with different infections, but not in a causative role. The type strain is ATCC 33820^T (=LMG 9479^T = DSM 20584^T), isolated from human specimen (eye). Genome projects involving six non-type strains are ongoing (Gi02791, Gi02792, Gi03435, Gi03436, Gi03437, and Gi03438).

Lactobacillus Delbrueckii subsp. Bulgaricus

Lactobacillus delbrueckii is the type species of the genus *Lactobacillus*, and it currently includes four subspecies (❍ *Table 15.1*), three of which, i.e., subsp. *delbrueckii, bulgaricus,* and *lactis*, were first described as separate species. The unification under the same name then was performed, the name being *Lactobacillus delbrueckii,* as a rule of priority as it was described before the others. The species name is to honor Delbrück, a German bacteriologist, while *bulgaricus* indicate the geographical origin of the first isolated strains, Bulgaria. The evolution of strains of the subsp. *bulgaricus* has probably been driven by mankind: it has been used for centuries in the production of yogurt and of other dairy products; therefore its genetic make up is devoted to fermentation in milk at high temperature (Germond et al., 2003). This could also explain the unusually high genome GC content for the species, 49–51%, and the scarce ability to ferment sugars other than lactose. Also, genome reduction has been observed (van de Guchte et al., 2006), probably motivated by the proto-cooperative interaction with *Streptococcus thermophilus,* the partner for yogurt production.

Phylogenetically, the species *Lactobacillus delbrueckii* belongs to the homonymous clade in the genus *Lactobacillus,* which includes, besides the already mentioned *Lactobacillus acidophilus* and *Lactobacillus crispatus, Lactobacillus acetotolerans, Lactobacillus amylolyticus, Lactobacillus amylophilus, Lactobacillus amylotrophicus, Lactobacillus amylovorus, Lactobacillus fornicalis, Lactobacillus gallinarum, Lactobacillus gasseri, Lactobacillus hamsteri, Lactobacillus helveticus, Lactobacillus iners, Lactobacillus intestinalis, Lactobacillus jensenii, Lactobacillus johnsonii, Lactobacillus kalixensis, Lactobacillus kefiranofaciens, Lactobacillus kitasatonis, Lactobacillus psittaci,* and *Lactobacillus ultunensis.*

The type strain of the species *Lactobacillus delbrueckii* corresponds to the type strain of the homonymous subspecies and is ATCC 9649^T (=DSM 20074^T = LMG 6412^T), isolated from distillery sour grain mash incubated at 45°C, while the type strain of subsp. *bulgaricus* is ATCC 11842^T (=DSM 20081^T = LMG 6901^T), isolated from Bulgarian yogurt. Genome data are available for three strains of the species (Gc00443, Gc00394, and Gi02793), including also the type strain (Gc00394).

Lactobacillus Johnsonii

Another bacteriologist, Johnson, gives his name to the species *Lactobacillus johnsonii,* which shares with *Lactobacillus delbrueckii* not only the origin of the name, but also the phylogenetic placement and the homofermentative metabolism. Strains of this species were initially identified as *Lactobacillus acidophilus* and *Lactobacillus crispatus*: with those two species *Lactobacillus johnsonii* shares different traits, such as similar GC content (33–35% in this case), the isomer of lactic acid produced (DL) and the homofermentative metabolism. Strains of *Lactobacillus johnsonii* have been isolated from human specimens but also from feces of animals. Type strain is ATCC 33200^T (=DSM 10533^T = LMG9436^T) and genome data available at the moment are for a probiotic strain NCC533 (Gc00171), while another is in progress, but no indication of the strain is available to date (Gi02798).

Lactobacillus Plantarum

The species *Lactobacillus plantarum* is one of the first taxa described in the genus. As the name indicates, it is commonly found in plant material, but it is also commonly found in a variety of environments, i.e., dairy products, human specimens, sewage etc. This flexibility determines also the genetic and phenotypic heterogeneity of the species, in which a novel subspecies has been recently delineated (⊙ *Table 15.1*). Anyway, in general, the morphology of cells is rod-like with rounded ends, metabolism is facultatively heterofermentative, genome GC content is in the range of 44–46% and both isomers of lactic acid are produced from carbohydrates. From a phylogenetic standpoint, *Lactobacillus plantarum* closest relatives are *Lactobacillus pentosus* and *Lactobacillus paraplantarum*, and the three species form a distinct line of descent in the genus *Lactobacillus*.

The type strain is ATCC 14917^T (=DSM 20174^T = LMG 6907^T), isolated from pickled cabbage, and genome sequencing has been performed on a non-type strain with probiotic properties, WCFS1 (Gc00122), the first *Lactobacillus* strain sequenced (Kleerebezem et al., 2003), which is already becoming the model organism for the genus.

Lactobacillus Reuteri

The bent rods with rounded ends belonging to this species owe their name to the German microbiologist Reuter, who was the first to isolate and study these heterofermentative bacteria. Strains of *Lactobacillus reuteri* are isolated from intestine and feces of humans and animals, sourdough and meat. Often an antimicrobial compound, reuterin, is produced. *Lactobacillus reuteri* strains produce both isomers of lactic acid and have a genome GC content between 40 and 42%. Its closest phylogenetic relatives are *Lactobacillus antri*, *Lactobacillus coleohominis*, *Lactobacillus fermentum*, *Lactobacillus frumenti*, *Lactobacillus gastricus*, *Lactobacillus ingluviei*, *Lactobacillus mucosae*, *Lactobacillus oris, Lactobacillus panis, Lactobacillus pontis, Lactobacillus secaliphilus,* and *Lactobacillus vaginalis,* which constitute the *Lactobacillus reuteri* subclade in the genus *Lactobacillus* (Felis and Dellaglio, 2007).

The type strain of the species, isolated from human intestine is ATCC 23272^T (=DSM 20016^T = LMG 9213^T), which genome has also been sequenced by two groups (Gc00573 and Gc00786), together with other four strains (Gi01862, Gi01863, Gi00762, and Gi02800).

Lactobacillus Rhamnosus

The ability to ferment rhamnose is one of the peculiarities of these rod-shaped bacteria, phylogenetically related to *Lactobacillus casei,* which can be isolated from dairy products, sewage, and human specimens. Strains of *Lactobacillus rhamnosus* have a facultatively heterofermentative metabolism, and are, in general very similar to *Lactobacillus paracasei* strains. Also, strains of this species are among the few in the genus *Lactobacillus* being assigned to Risk group 2 and are subjected to restricted distribution (http://www.dsmz.de/microorganisms/html/bacteria.genus/lactobacillus.html) as some strains have been isolated from endocarditis and other opportunistic infections. Nevertheless, the species has been included in list for QPS status and some strains are used as probiotics. Type strain of the species is ATCC 7469^T (=DSM 20021^T = LMG 6400^T) and three genome sequencing projects are ongoing (Gi02218, Gi00316, and Gi02801).

Lactobacillus Salivarius

As the name indicates, strains of *Lactobacillus salivarius* are common in human saliva, as well as in mouth and intestinal tract of man and animals. Historically considered a homofermentative species, after a recent taxonomic reevaluation (Lee et al., 2006), it is regarded as a facultatively heterofermentative species, able to ferment both hexoses and pentoses. Many recently described species have been found to be phylogenetically related to *Lactobacillus salivarius,* which clade is at present constituted by *Lactobacillus acidipiscis, Lactobacillus agilis, Lactobacillus animalis, Lactobacillus apodemi, Lactobacillus aviarius, Lactobacillus ceti, Lactobacillus equi, Lactobacillus ghanensis, Lactobacillus hayakitensis, Lactobacillus mali, Lactobacillus murinus, Lactobacillus nagelii, Lactobacillus ruminis, Lactobacillus saerimneri, Lactobacillus satsumensis,* and *Lactobacillus vini.*

Strains of this species produce L-lactic acid, and have a genome GC content of 34–36%. Type strain is ATCCC 11742^T (=DSM 20555^T = LMG 9477^T), isolated from human saliva, while the genome sequence has been obtained for a non-type strain with probiotic properties, UCC118 (Gc00362).

Interestingly, labels of some commercial products refer to the name "*Lactobacillus sporogenes*": this name is invalid, does not indicate any *Lactobacillus* species, but the spore-forming *Bacillus coagulans* (Sanders et al., 2003) (see below).

15.3.1.2 Genus *Bifidobacterium*

The genus *Bifidobacterium*, even if traditionally listed among LAB, is only poorly phylogenetically related to genuine LAB: it belongs to the Phylum *Actinobacteria*, Class "*Actinobacteria*," Order *Bifidobacteriales*, Family *Bifidobacteriaceae*, its neighbor genera being *Aeriscardovia*, *Gardnerella*, *Parascardovia*, and *Scardovia* (Garrity et al., 2007c). Species of these genera use a metabolic pathway for the degradation of hexoses, the so-called "bifid shunt," different from that of *Lactobacillus* and related genera (Sgorbati et al., 1995; Ventura et al., 2004). The key enzyme, considered a taxonomic character for the identification of this group of bacteria, is fructose-6-phosphoketolase (EC 4.1.2.2). The genus includes, at present, 30 species (⊙ *Table 15.2*).

Bifidobacteria are Gram-positive rods, which can sometimes be branched, a characteristic which gives the name to the genus. Bifidobacteria do not form spores, are nonmotile, and anaerobic. Their genome GC content varies from 42 to 67 mol% and in fact they belong to the high GC Gram-positive bacteria. Five phylogenetic groups have been observed in the genus with also species constituting single lines of descent (Felis and Dellaglio, 2007; Matsuki et al., 2003).

Bifidobacterial strains exhibiting probiotic properties belong to the species *Bifidobacterium adolescentis*, *Bifidobacterium animalis*, *Bifidobacterium bifidum*, *Bifidobacterium breve*, and *Bifidobacterium longum*, which are not related from a phylogenetic standpoint.

Bifidobacterium Adolescentis

Bacteria belonging to this species have been isolated from feces of human adult, bovine rumen and sewage. They are phylogenetically related with *Bifidobacterium angulatum*, *Bifidobacterium catenulatum*, *Bifidobacterium dentium*, *Bifidobacterium merycicum*, *Bifidobacterium pseudocatenulatum* and *Bifidobacterium ruminantium*, and their phenotypic differentiation from *Bifidobacterium dentium* is sometimes difficult. It is characterized by genome GC content of about 58%, while the cell-wall amino acids consist of Lys (Orn) – D-Asp. The type strain, isolated from human intestine is ATCC 15703^T (=DSM 20083^T = LMG 10502^T),

☑ Table 15.2

List of valid names in the genus *Bifidobacterium* (updated at September 2008) (*Cont'd* p. 610)

	Species of the genus *Bifidobacterium*
1	***Bifidobacterium adolescentis***
2	*Bifidobacterium angulatum*
3	***Bifidobacterium animalis***
	Bifidobacterium animalis subsp. *animalis*
	Bifidobacterium animalis subsp. *Lactis*
4	*Bifidobacterium asteroids*
5	***Bifidobacterium bifidum***
6	*Bifidobacterium boum*
7	***Bifidobacterium breve***
8	*Bifidobacterium catenulatum*
9	*Bifidobacterium choerinum*
10	*Bifidobacterium coryneforme*
11	*Bifidobacterium cuniculi*
12	*Bifidobacterium dentium*
13	*Bifidobacterium gallicum*
14	*Bifidobacterium gallinarum*
15	*Bifidobacterium indicum*
16	***Bifidobacterium longum***
	Bifidobacterium longum subsp. *infantis*
	Bifidobacterium longum subsp. *longum*
	Bifidobacterium longum subsp. *Suis*
17	*Bifidobacterium magnum*
18	*Bifidobacterium merycicum*
19	*Bifidobacterium minimum*
20	*Bifidobacterium pseudocatenulatum*
21	*Bifidobacterium pseudolongum*
	Bifidobacterium pseudolongum subsp. *globosum*
	Bifidobacterium pseudolongum subsp. *Pseudolongum*
22	*Bifidobacterium psychraerophilum*
23	*Bifidobacterium pullorum*
24	*Bifidobacterium ruminantium*
25	*Bifidobacterium saeculare*
26	*Bifidobacterium scardovii*
27	*Bifidobacterium subtile*

15

◘ Table 15.2

Species of the genus *Bifidobacterium*	
28	*Bifidobacterium thermacidophilum*
	Bifidobacterium thermacidophilum subsp. *Porcinum*
	Bifidobacterium thermacidophilum subsp. *thermacidophilum*
29	*Bifidobacterium thermophilum*
30	*Bifidobacterium tsurumiense*

Species with QPS status are underlined, principal species including probiotic strains are in bold and are described in the text

which genome sequence is also available (Gc00470); a second one is ongoing (Gi01706).

Bifidobacterium Animalis

The first description of this species was based on strains isolated from feces of different animals, but different data revealed that strains described as *Bifidobacterium lactis*, isolated from dairy products, belong to the same species, but to different subspecies (Masco et al., 2004). Strains of the subsp. *lactis* are more oxygen tolerant than those of the subsp. *animalis*, and this trait is very useful for probiotic application, as it allows them to survive in high number in the non-anaerobic conditions of commercial products; another difference between the two subspecies is the ability to grow in milk. *Bifidobacterium animalis* belongs to the *Bifidobacterium pseudolongum* phylogenetic group, which includes *Bifidobacterium choerinum*, *Bifidobacterium cuniculi*, *Bifidobacterium gallicum*, and, obviously, *Bifidobacterium pseudolongum*. The DNA G + C content of the species is 61%. The type strain of *Bifidobacterium animalis* subsp. *animalis* is ATCC 25527^T (=LMG 10508^T = DSM 20104^T), isolated from rat feces, type strain of *B. animalis* subsp. *lactis* is LMG 18314^T (=DSM 10140^T), isolated from fermented milk. Genome data for three strains will be available in the future for three non-type strains of also probiotic interest (Gi01876, Gi01988, and Gi03035).

Bifidobacterium Bifidum

Bifidobacterium bifidum is the type species of the genus, strictly anaerobic. It is among the most recognizable species in the genus on the basis of the fermentation pattern, and is also clearly different from other bifidobacteria phylogenetically. Cells have, in particular conditions, a peculiar "amphora-like" shape, and often appear also as branched rods (from which the name *bifidum*). Strains of this species have been isolated from feces of humans, both adults and infants, human vagina and animal feces, but are readily used for production of fermented dairy products for probiotic purposes. The type strain is ATCC 29521^T (=DSM 20239^T = LMG 10645^T), isolated from feces of a breast-fed infant; the genome sequence for three strains, including two different subcultures of the type strain, are in progress (Gi02210, Gi02653, Gi02657).

Bifidobacterium Breve

Strains of this species are characterized by the short size, with or without bifurcation, with lysine and glycine as aminoacid in the murein, and have a genome GC content of about 58%. Strains of this species have been isolated from intestine of infants, human vagina, and sewage, and are phylogenetically related to *Bifidobacterium longum*. The type strain is ATCC 15700^T (=DSM 20213^T = LMG 11042^T) and three genome sequencing programs are ongoing, two for the different subcultures of the type strain and one for a non-type with probiotic properties (Gi00078, Gi02209, and Gi02654).

Bifidobacterium Longum

Differently from *Bifidobacterium breve*, cells of strains in this species show a very elongated cell shape with rare branching. It is anaerobic and considered the most common species of bifidobacteria, and is isolated from feces of infants and adults, human vagina, but also animal feces. Its closest phylogenetic relative is *Bifidobacterium breve*, and the two species form a couple quite unrelated to the other species in the bifidobacterial phylogenetic tree. Other relevant characteristic for strains in the species are the genome GC content of about 60%, and the presence of many plasmids, a unique feature among fecal bifidobacteria.

Two taxa formerly described as distinct species have been included in *B. longum*, namely *Bifidobacterium infantis* and *Bifidobacterium suis*. These bacteria, first recognized as biotypes of the species, have very recently been recognized as novel subspecies. The three subgroups have been isolated from slightly different niches, such as the GIT of human adults (subsp. *longum*), the GIT of infants (subsp. *infantis*), and pig's feces (subsp. *suis*).

The type strain of subsp. *longum* (and therefore of the species) is ATCC 15707^T (=DSM 20219^T = LMG 13197^T), isolated from feces of a human adult, for subsp. *infantis* is ATCC 15697 ^{T} (=DSM 20088^T = LMG 8811^T), isolated from feces of a human infant, and for subsp. *suis* is ATCC 27533^T (=DSM 20211^T = LMG 21814^T), isolated from feces of a piglet (Mattarelli et al., 2008).

Sequence data are already or will be soon available for six strains, including the type strain (Gc00811, Gc00108, Gi02659, Gi01539, and Gi02485).

15.3.1.3   Other LAB: Genera *Streptococcus, Lactococcus, Enterococcus,* and *Pediococcus*

Other species belonging to the true LAB group play an important role in food microbiology and nutrition and are, in some cases, also considered probiotics. Two taxa having an outstanding role are *Streptococcus thermophilus* and *Lactococcus lactis*. Strains of the former species are, in combination with *Lactobacillus delbrueckii* subsp. *bulgaricus*, the actors of yogurt production, while the latter is largely used for production of dairy products and, in recent years, has become the model organism for lactic acid bacteria. Other genera, which include strains attributed with probiotic properties are *Enterococcus* and *Pediococcus* and will be briefly treated below.

15.3.1.4   Genera *Streptococcus* and *Lactococcus*

The genera *Lactococcus* and *Streptococcus* form the Family *Streptococcaceae* of the Order *Lactobacillales* in the Class *Bacilli* of the Phylum *Firmicutes*.

The genus *Streptococcus* contains about 67 species of coccoid Gram positive bacteria, mostly known for the pathogenicity of some species. The species *Streptococcus thermophilus*, on the contrary, is known for its GRAS (generally recognized as safe status), due to the long history of use in food production. Also, genomic data have demonstrated that *Streptococcus thermophilus* has lost or inactivated the virulence-related genes characterized in pathogenic streptococci, during the adaptation to milk, confirming its safety (Delorme, 2008).

The genus *Lactococcus* groups coccoid bacteria formerly referred to as mesophilic lactic streptococci. It is constituted by six species and the most important for dairy fermentation is *Lactococcus lactis*, which is divided into three subspecies, namely *Lactococcus lactis* subsp. *cremoris, Lactococcus lactis* subsp. *hordniae* and *Lactococcus lactis* subsp. *lactis*, which include the biovar *diacetylactis*. The diversity within the species has recently been re-evaluated with molecular analyzes, confirming that phenotypic and genotypic diversity are not coherent (Rademaker et al., 2007).

Streptococcus Thermophilus

As the name clearly indicates, strains of this *taxon* have a preference for growth at high temperature: all of them grow at 45°C, most are able to grow up to 50°C, and some survive also after heating 60°C for 30 min. This characteristic is an indication of man-driven adaptation to yogurt production, as already explained for *Lactobacillus delbrueckii* subsp. *bulgaricus*. In the case of *Streptococcus thermophilus*, however, genome GC content is in the range 37–40%, therefore no shift towards higher percentages is evident (Hardie and Whiley, 1995). A close phylogenetic relationship with *Streptococcus salivarius* has determined some nomenclatural changes in the past, with fluctuations of *Streptococcus thermophilus* between the status of species or of subspecies ("*Streptococcus salivarius* subsp. *thermophilus*"). However at present it is fully considered in the taxonomic rank of species, although some confusion is still present in literature data.

Streptococcus thermophilus has complex nutritional requirements, in particular considering aminoacids, which are most probably the result of the adaptation to growth in a rich medium such as milk, and also a consequence of proto-cooperation with *Lactobacillus delbrueckii* subsp. *bulgaricus*. The type strain is ATCC 19258^T (=DMS 20617 = LMG 6898), isolated from milk, and genome sequence data are available or in progress for four strains (Gc00451, Gc00234, Gc00233, and Gi00621).

Lactococcus Lactis

Lactococcus lactis is the type species of the genus *Lactococcus* (❷ *Table 15.3*) and, as both genus and species names suggest, it is strictly associated with milk. However,

❏ Table 15.3
List of valid names in the genus *Lactococcus* (updated at September 2008)

	Species of the genus *Lactococcus*
1	*Lactococcus chungangensis*
2	*Lactococcus garvieae*
3	*Lactococcus lactis*
	Lactococcus lactis subsp. *cremoris*
	Lactococcus lactis subsp. *hordniae*
	Lactococcus lactis subsp. *lactis*
4	*Lactococcus piscium*
5	*Lactococcus plantarum*
6	*Lactococcus raffinolactis*

Lactococcus lactis, underlined, is the only species in the genus under consideration for QPS status one of the three subspecies it includes has been described based on strains isolated from leafhopper (subsp. *hordniae*). Moreover, *Lactococcus lactis* is considered an "old" microorganism, originally colonizing plants, and only in recent time adapted to milk. Morphologically, cells are usually spherical or ovoid, in pairs or in chains, mesophilic with a homofermentative metabolism. Genome GC content is in the range 34–36%.

Nowadays, *Lactococcus lactis* is among the lactic acid bacterial species with the biggest economic importance, as strains are largely applied as starter cultures for a variety of products; it is the model species for the genetic and metabolic study of low GC Gram-positive bacteria, second only to *Bacillus subtilis* in importance. This is also causing some confusion in nomenclature of strains in the species: it has been anticipated above that genomic groups within the species are not phenotypically homogeneous. From a taxonomic standpoint, nomenclature of subspecies depends on the results of phenotypic tests (degradation of maltose, ribose, deamination of arginine, growth in presence of sodium chloride among others) (Teuber, 1995). However, the availability of genome data and the interest in genetic traits is prevailing over classical phenotypic studies: a clear example is that of strain MG1363 (Wegmann et al., 2007): it displays many of subsp. *lactis* phenotypic traits, but it is usually referred to as subsp. *cremoris*, due to its genetic similarity to the type strain of subsp. *cremoris*. The availability of the genome sequence and the use of the strain as an exemplary for the name *Lactococcus lactis* subsp. *cremoris*, therefore the genetic-based nomenclature will probably overcome the classical and phenotype-based correct one, generating a dichotomy in taxonomic procedure of identification of strains at the subspecies level.

Type strains of the three subspecies are ATCC 19435^T (=DSM 20481^T = LMG 6898^T) for subsp. *lactis* (and therefore for the whole species), ATCC 19257^T (=DSM 20069^T = LMG 6897^T) for subsp. *cremoris*, both isolated from milk and ATCC 29071^T (=DSM 20450^T = LMG 8520^T) for subsp. *hordniae*, isolated from the insect *Hordnia circellata*.

Three genome sequences are already available (Gc00054, Gc00450, and Gc508) and other will be available soon (Gi03379).

15.3.1.5 Genus *Enterococcus*

Related to the above described two genera, the genus *Enterococcus* is also important for the field of probiotics. The first bulk of species in the genus (*Enterococcus faecalis* and *Enterococcus faecium*) were previously described as streptococci, but many others have followed, increasing the number of species in the genus to 35 (◉ *Table 15.4*).

Potential probiotic properties have been reported for strains of *Enterococcus faecium* but bacteria in the genus *Enterococcus* are amongst the leading causes of community- and nosocomial infections. Therefore the Scientific Committee of the European Authority for Food Safety (EFSA) did not propose QPS status for any species of genus *Enterococcus*.

From a taxonomic standpoint, the genus *Enterococcus* falls into the Family *Enterococcaceae*, together with *Atopobacter, Catellicoccus, Melissococcus, Pilibacter, Tetragenococcus*, and *Vagococcus*.

Enterococcus Faecium

As the name clearly indicates, source of first isolation of *Enterococcus faecium* was fecal material, but strains of this species are frequently isolated not only from the GIT of animals (mammals, birds, and reptiles), but also from raw milk and dairy products (Devriese and Pot, 1995).

Cells are ovoid, occurring in pairs and short chains, not pigmented and not motile. Genome GC content ranges from 37 to 40%. *Enterococcus faecium* gives a name also to a phylogenetic group of closely related species in the genus, which, after description of novel species (◉ *Table 15.4*), includes *Enterococcus durans, Enterococcus hirae, Enterococcus mundtii, Enterococcus ratti*, and *Enterococcus villorum*.

Type strain is ATCC 19434^T (=DSM 20477^T = LMG 11423^T), and genome sequence data are not available at the moment but are in progress for 11 strains (Gi00227, Gi00228, Gi00229, Gi02729, Gi03235, Gi03232, Gi03265, Gi03306, Gi03225, Gi02730, and Gi03362).

15.3.1.6 Genus *Pediococcus*

Another genus including coccoid bacteria, relevant for the area of probiotics, is *Pediococcus*. Its peculiarity resides in the particular type of cell division observed, in two directions of the same plain, so that cells, during division, form tetrads. Interestingly, the closest relatives to pediococci are lactobacilli, the shape and mode of division of which are different (Simpson and Taguchi, 1995).

Moreover, the description of *Pediococcus siamensis* (◉ *Table 15.5*) has determined a rearrangement of the phylogenetic structure of the genus, with the recognition of two phylogenetically distinct subpopulations, one formed by

■ Table 15.4

List of valid names in the genus *Enterococcus* (updated at September 2008)

	Species of the genus *Enterococcus*
1	*Enterococcus aquimarinus*
2	*Enterococcus asini*
3	*Enterococcus avium*
4	*Enterococcus caccae*
5	*Enterococcus camelliae*
6	*Enterococcus canintestini*
7	*Enterococcus canis*
8	*Enterococcus casseliflavus*
9	*Enterococcus cecorum*
10	*Enterococcus columbae*
11	*Enterococcus devriesei*
12	*Enterococcus dispar*
13	*Enterococcus durans*
14	*Enterococcus faecalis*
15	***Enterococcus faecium***
16	*Enterococcus gallinarum*
17	*Enterococcus gilvus*
18	*Enterococcus haemoperoxidus*
19	*Enterococcus hermanniensis*
20	*Enterococcus hirae*
21	*Enterococcus italicus*
22	*Enterococcus malodoratus*
23	*Enterococcus moraviensis*
24	*Enterococcus mundtii*
25	*Enterococcus pallens*
26	*Enterococcus phoeniculicola*
27	*Enterococcus pseudoavium*
28	*Enterococcus raffinosus*
29	*Enterococcus ratti*
30	*Enterococcus saccharolyticus*
31	*Enterococcus silesiacus*
32	*Enterococcus sulfurous*
33	*Enterococcus termitis*
34	*Enterococcus thailandicus*
35	*Enterococcus villorum*

No species in the genus *Enterococcus* is under consideration for QPS status, principal species including probiotic strains are in bold and are described in the text

◧ Table 15.5

List of valid names in the genus *Pediococcus* (updated at September 2008)

	Species of the genus *Pediococcus*
1	***Pediococcus acidilactici***
2	*Pediococcus cellicola*
3	*Pediococcus claussenii*
4	*Pediococcus damnosus*
5	*Pediococcus dextrinicus*
6	*Pediococcus ethanolidurans*
7	*Pediococcus inopinatus*
8	*Pediococcus parvulus*
9	***Pediococcus pentosaceus***
10	*Pediococcus siamensis*
11	*Pediococcus stilesii*

Species with QPS status are undelined, principal species including probiotic strains are in bold and are described in the text

Pediococcus claussenii, Pediococcus pentosaceus and *Pediococcus acidilactici*; the other including *Pediococcus damnosus, Pediococcus inopinatus, Pediococcus parvulus, Pediococcus ethanolidurans, Pediococcus stilesii, Pediococcus siamensis* and *Pediococcus cellicola. Pediococcus dextrinicus* is scarcely related with other species in the genus, and is more similar to lactobacilli (Felis and Dellaglio, 2007).

Pediococcus Acidilactici

The name of this first described species of the genus *Pediococcus* clearly indicates its ability to produce lactic acid. Strains of this species are mostly isolated from plant material (silage, cereal and potato mashes, barley, malt), but some strains have been isolated also from meat products. It produces both isomers of lactic acid from glucose and other carbohydrates, while it is usually unable to degrade maltose, a sugar usually associated with plant environment. Strains can also grow up to 50°C and in presence of sodium chloride at high concentration (10%). Genome GC content is in the range 38–44%.

The species has always shown high overall similarity with *Pediococcus pentosaceus* (see below), and this is confirmed also from their phylogenetic relatedness: they form a distinct clade in the phylogenetic tree of the genus, together with *Pediococcus claussenii,* and *Pediococcus stilesii.*

The type strain is DSM 20284^T (=LMG 11384^T), and no genome sequence data are available nor in progress (GOLD database).

Pediococcus Pentosaceus

As the name indicates, strains of this species are able to degrade pentoses (except strains of the formerly valid subsp. *intermedius*). It is very similar to *Pediococcus acidilactici* in terms of shape, culture conditions and physiological traits, and they are also isolated from almost the same plant niches. In principle the two species can be phenotypically distinguished on the basis of maltose degradation, and the slightly heat resistance (39–45°C but not 50°C) and genome GC content (35–39%), but reliable differentiation can only be obtained with molecular methods. Some strains are able to produce pediocin.

The type strain is ATCC 33316^T (=DSM 20336^T = LMG 11488^T), and genome data are available for a non-type strain (Gc00439).

15.3.2   Non LAB: Genera *Propionibacterium, Bacillus, Brevibacillus, Sporolactobacillus, Escherichia*

15.3.2.1   Genus *Propionibacterium*

The genus *Propionibacterium*, similarly to the genus *Bifidobacterium*, belongs to the Class *Actinobacteria* (Garrity et al., 2007c), which comprises high G + C content Gram-positive non sporeforming bacteria; its neighbor genera in the Family *Propionibacteriaceae* are *Brooklawnia, Jiangella, Luteococcus, Microlunatus, Propioniferax, Propionomicrobium,* and *Tessaracoccus.*

Propionibacteria take the name from the ability to produce propionic acid, acetic acid and carbon dioxide from carbohydrates and lactic acid; these characteristics are desirable for the production of some types of cheeses, e.g., the Swiss type, where propionibacteria are used as starter cultures. In general, morphology of propionibacteria might be very different: coccoid, bifid or even branched rods have been described, which vary also with the physiological state and culture conditions (Jan et al., 2007).

The genus comprises 12 species (◉ *Table 15.6*) divided in two ecologically distinct groups, in terms of habitat/source of isolation: the "*acnes* group," of human origin, and the "dairy" or "classical" propionibacteria, isolated from milk and dairy products. The latter group includes the only species proposed for QPS status, i.e., *Propionibacterium freudenreichii*. Other species have been isolated from different environments (e.g., *Propionibacterium cyclohexanicum,* from spoiled orange juice), and phylogenetically assigned to different subgroups,

◘ Table 15.6

List of valid names in the genus *Propionibacterium* (updated at September 2008)

	Species of the genus *Propionibacterium*
1	*Propionibacterium acidipropionici*
2	*Propionibacterium acnes*
3	*Propionibacterium australiense*
4	*Propionibacterium avidum*
5	*Propionibacterium cyclohexanicum*
6	**Propionibacterium freudenreichii**
	Propionibacterium freudenreichii subsp. *freudenreichii*
	Propionibacterium freudenreichii subsp. *Shermanii*
7	*Propionibacterium granulosum*
8	*Propionibacterium innocuum*
9	*Propionibacterium jensenii*
10	*Propionibacterium microaerophilum*
11	*Propionibacterium propionicum*
12	*Propionibacterium thoenii*

Species with QPS status are underlined, principal species including probiotic strains are in bold and are described in the text

demonstrating that the correlation between phylogenetic and ecological structure of the genus is not obvious.

Propionibacterium Freudenreichii

Propionibacterium freudenreichii is the type species of the genus *Propionibacterium*, and it owes its name to the microbiologist von Freudenreich, who first described, with Orla-Jensen, the "*Bacterium acidi propionici* a." It includes two subspecies, namely *Propionibacterium freudenreichii* subsp. *freudenreichii* and *Propionibacterium freudenreichii* subsp. *shermanii*. Strains of this species are commonly found in cheese, where are often used as starter cultures; to date only strains belonging to this species have probiotic properties confirmed in humans. As other propionibacteria, strains of this species can be rod-shaped or branched, can be observed as single cells or in pairs, or in groups, and grow anaerobically.

Propionibacterium freudenreichii belongs to the group of dairy species, but its closest phylogenetic relatives are two recently described species, i.e.,

Propionibacterium cyclohexanicum and *Propionibacterium australiense*, isolated from spoiled orange juice and granulomatous bovine lesions, respectively. The type strain of the subspecies is ATCC 6207^T, isolated from Swiss cheese.

Currently the genome sequencing for two strains in this species is ongoing, one for a non-type strain, not identified at the subspecies level (Gi00483), and a second one for the type strain of the subsp. *shermanii* CIP 103027^T (Gi 00772).

15.3.2.2   Spore-Forming Bacteria: Genera *Bacillus*, *Brevibacillus*, *Paenibacillus*, and *Sporolactobacillus*

The first description of genus *Bacillus* dates back to 1872 (Cohn) and it was based on two species, i.e., *Bacillus anthracis* and *Bacillus subtilis*. Nowadays, 147 species are ascribed to this genus (◉ *Table 15.7*), and many other species have also been transferred to other genera, i.e., *Alkalibacillus, Alicyclobacillus, Aneurinibacillus, Brevibacillus, Geobacillus, Gracilibacillus, Lysinibacillus, Marinibacillus, Paenibacillus, Pullulanibacillus, Salimicrobium, Sporolactobacillus, Sporosarcina, Ureibacillus, Virgibacillus,* and *Viridibacillus* (Garrity et al., 2007b).

According to a recent re-analysis of all archaea and bacteria described to date (Yarza et al., 2008) the genus *Bacillus* is sub-divided in at least eight phylogenetic groups, intermixed with other bacterial genera (see http://www.arb-silva.de/fileadmin/silva_databases/living_tree/LTP_tree_s93.pdf), in the taxonomic lineage of *Firmicutes*.

Members of the genus *Bacillus* (and related genera) are a heterogeneous group of aerobic rod shaped bacteria, producing lactic acid and other metabolites (e.g., carbon dioxide, diacetyl, bacteriocins), with a range of mole G + C genome content from 32 to 69%, with the most striking characteristic of producing endospores, a characteristic important also for probiotic application. Their heterogeneity results in difficult identification based on phenotypic but also on genetic tests (Sanders et al., 2003).

Bacillus species are commonly associated with soil, but frequent sources of isolation are also water, dust, and air, but are also easily found in the GIT of humans, mammals, aquatic animals and invertebrates, whether as contaminant or as residing *taxa*. Industrially, bacilli are employed for production of antibiotics, industrial chemicals, and enzymes. Also, the potential for food spoilage is known for some species. Furthermore, heat resistant spores are very problematic for some specific areas such as dried milk industry, and some species are known as possible agents in biological warfare in terroristic actions. However, *B. subtilis*

15

■ Table 15.7

List of valid names in the genus *Bacillus* (updated at September 2008) (*Cont'd* p. 622)

	Species of the genus *Bacillus*
1	*Bacillus acidiceler*
2	*Bacillus acidicola*
3	*Bacillus aeolius*
4	*Bacillus aerius*
5	*Bacillus aerophilus*
6	*Bacillus agaradhaerens*
7	*Bacillus akibai*
8	*Bacillus alcalophilus*
9	*Bacillus algicola*
10	*Bacillus altitudinis*
11	*Bacillus alveayuensis*
12	*Bacillus amyloliquefaciens*
13	*Bacillus anthracis*
14	*Bacillus aquimaris*
15	*Bacillus arseniciselenatis*
16	*Bacillus arsenicus* Shivaji et al. (2005)
17	*Bacillus asahii*
18	*Bacillus atrophaeus*
19	*Bacillus aurantiacus*
20	*Bacillus azotoformans*
21	*Bacillus badius*
22	*Bacillus barbaricus*
23	*Bacillus bataviensis*
24	*Bacillus benzoevorans*
25	*Bacillus bogoriensis*
26	*Bacillus boroniphilus*
27	*Bacillus butanolivorans*
28	*Bacillus carboniphilus*
29	*Bacillus cellulosilyticus*
30	*Bacillus cereus*
31	*Bacillus chagannorensis*
32	*Bacillus cibi*
33	*Bacillus circulans*
34	*Bacillus clarkii*
35	***Bacillus clausii***
36	***Bacillus coagulans***
37	*Bacillus coahuilensis*

◻ Table 15.7 (Cont'd p. 623)

	Species of the genus *Bacillus*
38	*Bacillus cohnii*
39	*Bacillus decisifrondis*
40	*Bacillus decolorationis*
41	*Bacillus drentensis*
42	*Bacillus edaphicus*
43	*Bacillus endophyticus*
44	*Bacillus farraginis*
45	*Bacillus fastidiosus*
46	*Bacillus firmus*
47	*Bacillus flexus*
48	*Bacillus foraminis*
49	*Bacillus fordii*
50	*Bacillus fortis*
51	*Bacillus fumarioli*
52	*Bacillus funiculus*
53	*Bacillus galactosidilyticus*
54	*Bacillus gelatini*
55	*Bacillus gibsonii*
56	*Bacillus ginsengihumi*
57	*Bacillus halmapalus*
58	*Bacillus halodurans*
59	*Bacillus hemicellulosilyticus*
60	*Bacillus herbersteinensis*
61	*Bacillus horikoshii*
62	*Bacillus horti*
63	*Bacillus humi*
64	*Bacillus hwajinpoensis*
65	*Bacillus idriensis*
66	*Bacillus indicus*
67	*Bacillus infantis*
68	*Bacillus infernus*
69	*Bacillus insolitus*
70	*Bacillus isabeliae*
71	*Bacillus jeotgali*
72	*Bacillus koreensis*
73	*Bacillus kribbensis*
74	*Bacillus krulwichiae*
75	*Bacillus lehensis*

Table 15.7 (Cont'd p. 624)

	Species of the genus *Bacillus*
76	*Bacillus lentus*
77	*Bacillus licheniformis*
78	*Bacillus litoralis*
79	*Bacillus luciferensis*
80	*Bacillus macauensis*
81	*Bacillus macyae*
82	*Bacillus mannanilyticus*
83	*Bacillus marisflavi*
84	*Bacillus massiliensis*
85	*Bacillus megaterium*
86	*Bacillus methanolicus*
87	*Bacillus mojavensis*
88	*Bacillus mucilaginosus*
89	*Bacillus muralis*
90	*Bacillus murimartini*
91	*Bacillus mycoides*
92	*Bacillus nealsonii*
93	*Bacillus niabensis*
94	*Bacillus niacini*
95	*Bacillus novalis*
96	*Bacillus odysseyi*
97	*Bacillus okhensis*
98	*Bacillus okuhidensis*
99	*Bacillus oleronius*
100	*Bacillus oshimensis*
101	*Bacillus panaciterrae*
102	*Bacillus patagoniensis*
103	*Bacillus plakortidis*
104	*Bacillus pocheonensis*
105	*Bacillus polygoni*
106	*Bacillus pseudalcaliphilus*
107	*Bacillus pseudofirmus*
108	*Bacillus pseudomycoides*
109	*Bacillus psychrodurans*
110	*Bacillus psychrosaccharolyticus*
111	*Bacillus psychrotolerans*
112	***Bacillus pumilus***
113	*Bacillus pycnus*

☐ Table 15.7

	Species of the genus *Bacillus*
114	*Bacillus qingdaonensis*
115	*Bacillus ruris*
116	*Bacillus safensis*
117	*Bacillus salarius*
118	*Bacillus saliphilus*
119	*Bacillus schlegelii*
120	*Bacillus selenatarsenatis*
121	*Bacillus selenitireducens*
122	*Bacillus seohaeanensis*
123	*Bacillus shackletonii*
124	*Bacillus silvestris*
125	*Bacillus simplex*
126	*Bacillus siralis*
127	*Bacillus smithii*
128	*Bacillus soli*
129	*Bacillus sonorensis*
130	*Bacillus sporothermodurans*
131	*Bacillus stratosphericus*
132	<u>*Bacillus subterraneus*</u>
133	*Bacillus subtilis*
	Bacillus subtilis subsp. *spizizenii*
	Bacillus subtilis subsp. *subtilis*
134	*Bacillus taeanensis*
135	*Bacillus tequilensis*
136	*Bacillus thermantarcticus*
137	*Bacillus thermoamylovorans*
138	*Bacillus thermocloacae*
139	*Bacillus thioparans*
140	*Bacillus thuringiensis*
141	<u>*Bacillus tusciae*</u>
142	*Bacillus vallismortis*
143	*Bacillus vedderi*
144	*Bacillus vietnamensis*
145	*Bacillus vireti*
146	*Bacillus wakoensis*
147	*Bacillus weihenstephanensis*

Species with QPS status are underlined, principal species including probiotic strains are in bold and are described in the text is traditionally used in Eastern countries to produce specific foods, which demonstrates its safe use for food application, together with other species. In fact, members of the genus *Bacillus* are used as probiotics, where one of the main advantages over classical lactobacilli is the availability of endospores, which may be stored desiccated almost indefinitely (Fritze and Claus, 1995).

Spore-forming bacteria proposed for probiotics production belong mainly to the species *Bacillus cereus, Bacillus clausii, Bacillus coagulans, Bacillus indicus, Bacillus licheniformis, Bacillus pumilus, Bacillus subtilis,* "*Bacillus laterosporus,*" "*Bacillus laevolacticus,*" "*Bacillus polymyxa*" and "*Bacillus polyfermenticus.*" Species names reported in inverted commas are not valid either because they are not updated ("*B. laterosporus*" is now *Brevibacillus laterosporus,* "*Bacillus polymyxa*" corresponds to *Paenibacillus polymyxa,* and "*Bacillus laevolacticus*" belongs to the genus *Sporolactobacillus,* with the same species name) or they have not been formally described ("*Bacillus polyfermenticus*"). Some details on the genera *Brevibacillus* and *Sporolactobacillus* are given below, before species explanations.

The genus *Brevibacillus,* the name meaning short rod, has been described in 1996 to accommodate, on the basis of phylogenetic relatedness, ten species of Gram-positive, motile, aerobic spore-forming bacteria previously belonging to the genus *Bacillus* (Shida et al., 1996). After publication of that new genus name, other four species have been described, i.e., *Brevibacillus ginsengsoli, Brevibacillus invocatus, Brevibacillus levickii,* and *Brevibacillus limnophilus* (◉ Table 15.8). Ecologically, *Brevibacillus* inhabit the same environments as *Bacillus. Brevibacillus brevis* is the type species of the genus.

Species in the genus *Paenibacillus* are facultatively anaerobic or strictly aerobic bacteria, rod shaped, generally Gram-positive but also variable, usually motile, almost all catalase positive. They have different degrading abilities, excreting diverse proteolytic and/or extracellular polysaccharide-hydrolyzing enzymes. The G + C contents range from 45 to 54 mol%. The type species is *Paenibacillus polymyxa.* The genus groups more than 90 species, none of them has been proposed for QPS status, strains considered probiotics seem to belong to *Paenibacillus polymyxa* only.

The genus *Sporolactobacillus* comprises six species of catalase-negative, facultative anaerobic or microaerophilic endosporeformers (◉ Table 15.9). Due to the metabolic similarities with genuine lactic acid bacteria, species of this genus were originally thought to belong to the genus *Lactobacillus,* but then assigned to *Bacillus,* due to their ability to form spores, and eventually described as novel genus, in the Family *Sporolactobacillaceae* (Garrity et al., 2007b). The type species for *Sporolactobacillus* is *Sporolactobacillus inulinus.* Members of the genus have been isolated from chicken feed, soil, dairy products, and pickle.

☐ Table 15.8

List of valid names in the genus *Brevibacillus* (updated at September 2008)

	Species of the genus *Brevibacillus*
1	*Brevibacillus agri*
2	*Brevibacillus borstelensis*
3	*Brevibacillus brevis*
4	*Brevibacillus centrosporus*
5	*Brevibacillus choshinensis*
6	*Brevibacillus formosus*
7	*Brevibacillus ginsengisoli*
8	*Brevibacillus invocatus*
9	***Brevibacillus laterosporus***
10	*Brevibacillus levickii*
11	*Brevibacillus limnophilus*
12	*Brevibacillus parabrevis*
13	*Brevibacillus reuszeri*
14	*Brevibacillus thermoruber*

No species has been proposed for QPS status. Species including proposed probiotic strains are in bold and are described in the text

☐ Table 15.9

List of valid names in the genus *Sporolactobacillus* (updated at September 2008)

	Species of the genus *Sporolactobacillus*
1	***Sporolactobacillus inulinus***
2	*Sporolactobacillus kofuensis*
3	*Sporolactobacillus lactosus*
4	***Sporolactobacillus laevolacticus***
5	*Sporolactobacillus nakayamae*
	Sporolactobacillus nakayamae subsp. *nakayamae*
	Sporolactobacillus nakayamae subsp. *Racemicus*
6	*Sporolactobacillus terrae*

No species has been proposed for QPS status. Species including proposed probiotic strains are in bold and are described in the text

Bacillus Cereus

The species *Bacillus cereus*, the species name meaning wax-coloured, is associated with food poisoning, due to the production of toxins, however strains belonging to the "var. *vietnami*" have been indicated as probiotics (Duc et al., 2004).

The indication of a variety (the "var.") has no meaning in bacterial taxonomy. In fact this name has been assigned to one strain exhibiting physiological traits similar to *Bacillus cereus* but apparently not phylogenetically related to it. According to Hoa et al. (2000), these data could justify the description of a novel species. For the time being, this description has not appeared, therefore the name cannot be considered valid. Also, another organism attributed with probiotic properties is "*Bacillus toyoi*" or "*Bacillus cereus* var. *toyoi*," used mostly for animals (e.g., De Cupere et al., 1992; Scharek et al., 2007), but similar considerations presented for "var. *vietnami*" apply.

In general, *Bacillus cereus* is motile, hemolytic on blood agar, penicillin resistant. Its closest relatives are *Bacillus mycoides*, *Bacillus weihenstephanensis*, *Bacillus thuringensis*, *Bacillus anthracis* and *Bacillus pseudomycoides*. According to the tree obtained by Yarza et al. (2008) this group of bacteria seems to be more closely related with the genus *Gemella* than with other *Bacilli*.

Type strain of *Bacillus cereus* is DSM 31^T (=ATCC 14579^T) which has also been sequenced (Gc00135) together with other strains, already completed (Gc00617, Gc00215, and Gc00173) or ongoing (40 more), due to the clinical interest for the species.

Bacillus Clausii

Bacillus clausii, described in 1995 with other alkaliphilic species in the genus *Bacillus*, is named after Claus, a German microbiologist who made fundamental contribution to the study of bacilli, and it groups bacteria isolated from soil. The closest phylogenetic relatives are *Bacillus oshimensis*, *Bacillus lehensis*, *Bacillus patagoniensis*, *Bacillus gibsonii*, *Bacillus murimartini*, and *Bacillus plakortidis*. As a species, it is characterized by the production of catalase and oxidase, is able to reduce nitrate and to hydrolyze starch and gelatin, finally, it grows between 30 and 50°C and with up to 10% sodium chloride. Genome GC content is around 43 mol% (Nielsen et al., 1995). The type strain, isolated from Garden soil, is ATCC 700160^T (=DSM 8716^T). Genome sequencing projects exist for strains in the species, one complete (Gc00228), and one incomplete (Gi00061), strains chosen for sequencing are not the type strain.

Bacillus Coagulans

Bacillus coagulans was first described in 1915 isolated from spoiled canned milk, in which it had caused coagulation (from which the names) is a thermotolerant and microaerophilic species, associated to spoilage of foods such as milk products, vegetables and fruits. Also *Bacillus coagulans* is exploited industrially, e.g., as a source of thermostable enzymes, and it is employed also as growth promoting additive in food and feed, often under the invalid name of "*Lactobacillus sporogenes*." Different morphologies of the cells, spore surfaces and sporangia have been reported, complicating the recognition of *Bacillus coagulans* as a single species. An emended description of the species based on a polyphasic approach and on 31 strains confirmed the heterogeneity of the species (De Clerck et al., 2004), the strains of which have been isolated from different environments, suggesting high flexibility. The closest phylogenetic relatives are *Bacillus oleronius*, *Bacillus sporothermodurans, Bacillus acidicola*, and *Bacillus ginsenghumi*, which, as a clade, appear to be quite unrelated to *Bacillus subtilis* group. The type strain is ATCC 7050^T (=LMG 6326^T = DSM 1^T) isolated from evaporated milk, but other strains have been isolated from soil. A genome project is ongoing (Gi01001) for a non-type strain.

Bacillus Licheniformis

The first description of the *taxon* called today *B. licheniformis* dates back to 1898, and it owes its name to its shape, the name meaning lichen-shaped. As other bacilli, it is Gram-positive, microaerophilic, and motile. *Bacillus licheniformis* is an industrially important strain as it produces enzymes and bacteriocins. Strains of *Bacillus licheniformis* have been isolated from soil, milk, water, but also septic wounds, and are generally considered safe. However, Salkinoja-Salonen et al. (1999) and other studies have reported on the toxigenic activity of *Bacillus licheniformis, Bacillus pumilus*, and *Bacillus subtilis*. Interestingly, these three species, proposed for probiotic application, also belong to the same phylogenetic group.

The type strain is ATCC 14580^T (=DSM 13^T = LMG 12363^T), and genome data for DSM 13^T have been obtained by two groups independently (Gc00213, and Gc00221).

Bacillus Pumilus

Strains of this species, the name of which means "small," are used for enzyme production on industrial scale, and for many other applications.

Strains in this species are aerobic and motile, positive for catalase as well as for oxidase, beta-galactosidase and amylase. Acid production is observed from glucose, arabinose, mannitol and xylose. As for pH it is a neutrophilic species, while, considering heat resistance of vegetative cells, it is mesophilic. This species is also associated to food poisoning due to toxin production. It belongs to *Bacillus subtilis* phylogenetic group (see below). Its type strain is ATCC 7061^T (=DSM 27^T = LMG 18928^T). Genome sequences for four strains are completed or in progress, including one for the type strain (Gc00656, Gi03241, Gi01901, and Gi00674).

Bacillus Subtilis

The slender shape of these rods, less than 1 μm wide, gives the names to the *taxon*, which is the type species of the genus. As anticipated, *Bacillus subtilis* is the classical model organism for genetic research in Gram-positive bacteria, but it is also widely used in traditional and industrial fermentation processes as well as in agriculture. Phylogenetically it is very closely related to species *Bacillus altitudinis, Bacillus aerius, Bacillus aerophilus, Bacillus amyloliquefaciens, Bacillus atrophaeus, Bacillus licheniformis, Bacillus mojavensis, Bacillus pumilus, Bacillus safensis, Bacillus sonorensis, Bacillus stratosphericus, Bacillus vallismortis,* and *Bacillus velezensis* but also to *Brevibacterium halotolerans*. Its sporangia are not swollen, and spores are ellipsoidal. It grows between 15–20 and 45–55°C, with an optimum at 28–30°C. It also grows in presence of 7% sodium chloride. It is a strictly aerobic and catalase positive species, able to hydrolyse starch and casein.

Recently, a second subspecies has been described, i.e., subsp. *spizizenii,* named after American bacteriologist Spizizen, on the basis of significant sexual isolation found between two genotypically distinguishable populations within the species and DNA-DNA hybridization levels in the range of 58–68%. These values are thus just below the threshold usually applied for the delineation of species (70%) but, due to the high overall similarity of this second group of bacteria with the type strain of *Bacillus subtilis* the two populations were retained in the same species (Nakamura et al., 1999). Type strains are ATCC 6051^T (=DSM 10^T = LMG 7135^T), for subsp. *subtilis,* and NRRL B-23049^T (=LMG 19156^T), for subsp. *spizizenii.* Genome sequence data are available or in progress for 10 strains (Gc00010, Gi01245, Gi03233, Gi01244, Gi01246, Gi03234, Gi03274, Gi03229, Gi03230, and Gi03239).

Brevibacillus Laterosporus (Formerly "Bacillus laterosporus")

Brevibacillus laterosporus strains take their name from the lateral position of spores. These bacteria show low level of larvicidal activity, and have also been associated with human infections. As for temperature for growth, it ranges between 15 and 50°C, with an optimum at 30°C. The species reduces nitrate and is capable of casein and gelatin hydrolysis, but not of starch.

Phylogenetically the genus appears quite homogeneous. Type strain is ATCC 4517^T (=DSM 25^T) while no sequence data are in progress at the moment, according to lists in GOLD database.

Paenibacillus Polymyxa (Formerly "Bacillus polymyxa")

As anticipated, *Paenibacillus polymyxa* is the type strain of the genus. Its closest phylogenetic relatives are *Paenibacillus kribbensis, Paenibacillus peoriae, Paenibacillus jamilae, Paenibacillus brasilensis,* and *Paenibacillus terrae.* It is able to hydrolyze pectin and xylan and fix nitrogen, and it produces acid from various sugars. Another interesting characteristic is the production of slime, which gives the origin to the name, which literally means "much slime." Its habitat is probably soil. The type strain is ATCC 842^T, and genome sequence data are available (Gi00423) for a non-type strain.

Sporolactobacillus Inulinus

Sporolactobacilli, in general, and species *inulinus* in particular, display intermediate characteristics with respect to *Bacillus* and *Lactobacillus.* In common with the former is the ability to form spores, and the presence of diaminopimelic acid as cell wall component. Similarly to the latter, they are catalase negative, microaerophilic and produce lactic acid from glucose through homolactic fermentation. The lactic acid isomer produced is D (-). Genome GC content is 38–39%. The type strain is ATCC 15538^T (=DSM 20348^T = LMG 11481^T), and no genome sequence days appear to be in progress in GOLD database.

Sporolactobacillus Laevolacticus

The taxonomic status of this motile bacterium producing D(−) – lactic acid has been questioned for a long time, but only recently (Andersch et al., 1994; Hatayama et al., 2006) its description has been validly published, including motile bacteria with diaminopimelic acid in the cell wall. The species is facultatively anaerobic, catalase positive, and mesophilic, with 40°C as the maximum temperature for growth. It hydrolyzes starch and degrades sugars through homolactic fermentation. Strains have been isolated from rhizospheres of plants. The type strain is ATCC 23492^T (=DSM 442^T = LMG 6329^T), and no sequence data are available.

Notably, *Bacillus polyfermenticus* is not a validly published name. Therefore, even if research reports on its probiotic properties have been published, no reliable characteristics can be assigned to this name, and no strain can be indicated as reference point.

15.3.2.3 Genus *Escherichia*

The genus *Escherichia* belongs to the Phylum *Proteobacteria*, Class *Gammaproteobacteria*, Order *Enterobacteriales*, Family *Enterobacteriaceae*. Genera in the same family are: *Arsenophonus, Brenneria, Buchnera, Budvicia, Buttiauxella, Cedecea, Citrobacter, Dickeya, Edwardsiella, Enterobacter, Erwinia, Ewingella, Hafnia, Klebsiella, Kluyvera, Leclercia, Leminorella, Moellerella, Morganella, Obesumbacterium, Pantoea, Pectobacterium, Phlomobacter, Photorhabdus, Plesiomonas, Pragia, Proteus, Providencia, Rahnella, Raoultella, Saccharobacter, Salmonella, Samsonia, Serratia, Shigella, Sodalis, Thorsellia, Tatumella, Trabulsiella, Wigglesworthia, Xenorhabdus, Yersinia*, and *Yokenella* (Garrity et al., 2007d).

Escherichia includes six species of Gram-negative rods, chemo-organotrophic with both oxidative and fermentative metabolism, catalase-positive which produce acid and gas from glucose.

From a taxonomic standpoint, the genus should also include the four species of the genus *Shigella*, which are retained separate only for historical and clinical reasons (Lan and Reeves, 2002).

Escherichia coli is the best known species of the genus, and it is a widespread commensal of the lower intestinal tract of humans and other vertebrates. Clones of *Escherichia coli* cause intestinal and extra-intestinal diseases with devastating effects on the host. However, as a commensal, some *Escherichia coli* strains are also used as probiotics. The type strain of *Escherichia coli* is ATCC 11775^T (=LMG 2092^T = DSM 30083^T = JCM 1649^T). It was isolated from urine but it has not been sequenced although about 71 genome sequences are reported as completed (13) or ongoing (58) projects for the species *Escherichia coli*, not considering *Shigella*.

In a recent genomic study, Willenbrock et al. (2007) analyzed the genomic content of four probiotic strains by microarray. Results showed that the probiotic strains were most similar, in terms of gene pool, with *Escherichia coli* K-12 strains, and with H10407, which is an enterotoxigenic strain, the virulence of which is plasmid-encoded. Some virulence related genes where also detected in these isolates, indicating that both pathogenic and non-pathogenic *Escherichia coli* strains use common strategies for adaptation to their niche. Finally, genetic flexibility was witnessed by the presence of strain-specific phage genes, transposases, insertion elements and mobile-elements-related genes. These data cannot be used in devising a taxonomic scheme as they are not stable elements, but are of utmost importance for characterization at the strain level.

15.3.3   Yeasts

Yeasts are unicellular eukaryotic microorganisms. The strains used as probiotics are referred to as "*Saccharomyces boulardii*" however this species name has no meaning in taxonomy as it is a synonym for *Saccharomyces cerevisiae* (Edwards-Ingram et al., 2004; Hennequin et al., 2001; McCullough et al., 1998; Mitterdorfer et al., 2002).

According to Kurtzman and Fell (1999), the genus *Saccharomyces* includes 14 species. However more recently, Kurtzman (2003), on the basis of Multigene Sequence Analysis, proposed a new *Saccharomyces* genus that includes only seven of the previous species (*Saccharomyces cariocanus, Saccharomyces cerevisiae, Saccharomyces bayanus, Saccharomyces mikatae, Saccharomyces kudriavzevii, Saccharomyces paradoxus*, and *Saccharomyces pastorianus*); the rest of the previous species are in a new genus, namely *Kazachstania*. *Saccharomyces cerevisiae* is the type species of the genus.

Considering the taxonomic lineage, the genus *Saccharomyces* is grouped in the Domain Eukaryota, Kingdom Fungi, Phylum Ascomycota, Subdivision Saccharomycotina, Class Saccharomycetes, Order Saccharomycetales, Family Saccharomycetaceae. It can be noted that names above the genus level are not written in italics, according to the rules of the International Code of Botanical Nomenclature (McNeill et al., 2006) and differently from bacterial nomenclature.

Cells of *Saccharomyces cerevisiae* reproduce vegetatively by multilateral budding (a characteristic of the Class Saccharomycetes), and are transformed directly to asci, containing ascospores, when grown on acetate agar. Also, a characteristic of the genus *Saccharomyces* is the vigorous fermentation of sugars, which is the desired characteristic of strains of *Saccharomyces cerevisiae* (and other species) used for food production, from breadmaking to production of alcoholic beverages from a wide range of vegetable raw materials. Besides its commercial importance, *Saccharomyces cerevisiae* is a very important model system for molecular biology and genetics, as the basic cellular mechanisms of this simple eukaryote are largely conserved also in higher organisms, including mammals. The name of the anamorph for *Saccharomyces cerevisiae* is *Candida robusta*, and a large number of synonyms have been determined. The type strain is CBS 1171^T (=ATCC 18824^T = DBVPG 6173^T = NRRL Y-12632^T), isolated in a Dutch brewery (Vaughan-Martini and Martini, 1999). Genome sequence for one strain has been determined (Gc00006) but 35 others are ongoing.

15.4   Summary

- Names are the results of taxonomic procedures, i.e., extensive and time consuming characterization of organisms under different aspects (genetic, physiology, etc) in order to define their diversity.
- Taxonomic studies of microorganisms have always been dependent on scientific and technological advancements and development of novel techniques of investigation, due to their extremely small size. Application of new techniques can highlight novel traits and therefore can modify the understanding of diversity of organisms, thus changing also names.
- Correct names of organisms are essential as they (i) allow retrieving all the updated information on *taxa*, (ii) constitute a standard for the unambiguous identification of bacteria, both for scientific purposes, for definition of lists of safe and pathogenic bacteria, and for reliable commercial information on microorganisms.
- Taxonomic traits, i.e., characteristics useful for characterization of genera, species and subspecies can be different in different organisms, as they depend also on ecology and evolution. Therefore, classification of different groups of bacteria could be slightly different and it has to be considered an "agreement among experts." On the other hand, nomenclature, i.e., the assignment of names to recognized taxa, is strictly regulated, to ensure clarity.
- Probiotic properties are strains specific and not species specific, but an accurate identification at species level is essential to have more information on the strains.

List of Abbreviations

EFSA European Food Safety Authority
GIT gastro-intestinal tract
LAB lactic acid bacteria
QPS qualified presumption of safety

References

Andersch I, Pianka S, Fritze D, Claus D (1984) Description of Bacillus laevolacticus (ex Nakayama and Yanoshi 1967) sp. nov., nom. rev. Int J Syst Bacteriol 44:659–664

Approved List of Bacterial Names (1980) Int J Syst Bacteriol 30:225–420

Garrity GM (2001) Bergey's manual of systematic bacteriology, 2nd edn. Springer, New York

De Clerck E, Rodriguez-Diaz M, Forsyth G, Lebbe L, Logan NA, DeVos P (2004) Polyphasic characterization of Bacillus coagulans strains, illustrating heterogeneity within this species, and emended description of the species. Syst Appl Microbiol 27:50–60

De Cupere F, Deprez P, Demeulenaere D, Muylle E (1992) Evaluation of the effect of 3 probiotics on experimental Escherichia coli enterotoxaemia in weaned piglets. Zentralbl Veterinarmed B 39:277–284

Delorme C (2008) Safety assessment of dairy microorganisms: Streptococcus thermophilus. Int J Food Microbiol 126:274–277

Devriese LA, Pot B (1995) The Genus Enterococcus. In: Wood BJB, Holzapfel WH (eds) The genera of lactic acid bacteria, vol. 2. Blackie Academic & Professional (UK), London, pp. 327–368

Duc LH, Hong HA, Barbosa TM, Henriques AO, Cutting SM (2004) Characterization of Bacillus Probiotics Available for Human Use. Appl Environ Microbiol 70:2161–2171

Edwards-Ingram LC, Gent ME, Hoyle DC, Hayes A, Stateva LI, Oliver SG (2004) Comparative genomic hybridization provides new insights into the molecular taxonomy of the Saccharomyces sensu stricto complex. Genome Res 14:1043–1051

Felis GE, Dellaglio F (2007) Taxonomy of Lactobacilli and Bifidobacteria. Curr Issues Intest Microbiol 8:44–61

Fritze D (2004) Taxonomy of the genus Bacillus and related genera: The aerobic endospore-forming bacteria. Phytopathol 94:1245–1248

Fritze D, Claus D (1995) Spore forming, lactic acid producing bacteria of the genera Bacillus and Sporolactobacillus. In: Wood BJB, Holzapfel WH (eds) The genera of lactic acid bacteria, vol. 2. Blackie Academic & Professional (UK), London, pp. 368–391

Garrity G, Lilburn T, Cole J, Harrison S, Euzeby J, Tindall B (2007a) Introduction to the taxonomic oultine of bacteria and archaea (TOBA) release 7.7. The Taxonomic Outline of Bacteria and Archaea, 7(7), from http://www.taxonomicoutline.org/index.php/toba/article/view/190/223

Garrity G, Lilburn T, Cole J, Harrison S, Euzeby J, Tindall B (2007b) Part 9 – The Bacteria: Phylum Firmicutes: Class "Bacilli. The Taxonomic Outline of Bacteria and Archaea, 7(7), from http://www.taxonomicoutline.org/index.php/toba/article/view/186/218

Garrity G, Lilburn T, Cole J, Harrison S, Euzeby J, Tindall B (2007c) Part 10 – the bacteria: phylum actinobacteria: class "actinobacteria". The Taxonomic Outline of Bacteria and Archaea, 7(7), from http://www.taxonomicoutline.org/index.php/toba/article/view/187/219

Garrity G, Lilburn T, Cole J, Harrison S, Euzeby J, Tindall B (2007d) Part 5 – the bacteria: phylum proteobacteria, class gammaproteobacteria. The Taxonomic Outline of Bacteria and Archaea, 7(7), from http://www.taxonomicoutline.org/index.php/toba/ article/view/181/214

Germond JE, Lapierre L, Delley M, Mollet B, Felis GE, Dellaglio F (2003) Evolution of the bacterial species Lactobacillus delbrueckii. Mol Biol Evol 20:93–104

Giraud T, Refrégier G, Le Gac M, de Vienne DM, Hood ME (2008) Speciation in fungi. Fungal Genet Biol 45:791–802

Hammes WP, Vogel RF (1995) The genus Lactobacillus. In: Wood BJB, Holzapfel WH (eds) The genera of lactic acid bacteria, vol. 2. Blackie Academic & Professional (UK), London, pp. 19–54

Hardie JM, Whiley RA (1995) The genus *Streptococcus*. In: Wood BJB, Holzapfel WH (eds) The genera of lactic acid bacteria, vol. 2. Blackie Academic & Professional (UK), London, pp. 5–124

Hatayama K, Shoun H, Ueda Y, Nakamura A (2006) *Tuberibacillus calidus* gen. nov., sp. nov., isolated from a compost pile and reclassification of *Bacillus naganoensis* Tomimura et al. 1990 as *Pullulanibacillus naganoensis* gen. nov., comb. nov. and *Bacillus laevolacticus* Andersch et al. 1994 as *Sporolactobacillus laevolacticus* comb. nov. Int J Syst Evol Microbiol 56: 2545–2551

Hennequin C, Thierry A, Richard GF, Lecointre G, Nguyen HV, Gaillardin C, Dujon B (2001) Microsatellite typing as a new tool for identification of *Saccharomyces cerevisiae* strains. J Clin Microbiol 2001 39:551–559

Hoa NT, Baccigalupi L, Huxham A, Smertenko A, Van PH, Ammendola S, Ricca E, Cutting AS (2000) Characterization of *Bacillus* species used for oral bacteriotherapy and bacterioprophylaxis of gastrointestinal disorders. Appl Environ Microbiol 66:5241–5247

Hong HA, Duc LH, Cutting SM (2005) The use of bacterial spore formers as probiotics. FEMS Microbiol Rev 29:813–835

Jan G, Lan A, Leverrier P (2007) Dairy Propionibacteria as probiotics. In: Saarela M (ed) Functional dairy products, vol. 2. Woodhead Publishing Limited, Abington, USA, pp. 165–194

Judicial Commission of the International Committee on Systematics of Bacteria (2008) The type strain of *Lactobacillus casei* is ATCC 393, ATCC 334 cannot serve as the type because it represents a different taxon, the name *Lactobacillus paracasei* and its subspecies names are not rejected and the revival of the name '*Lactobacillus zeae*' contravenes Rules 51b (1) and (2) of the International Code of Nomenclature of Bacteria. Opinion 82 Int J Syst Evol Microbiol 58:1764–1765

Kleerebezem M, Boekhorst J, van Kranenburg R, Molenaar D, Kuipers OP, Leer R, Tarchini R, Peters SA, Sandbrink HM, Fiers MW, Stiekema W, Lankhorst RM, Bron PA, Hoffer SM, Groot MN, Kerkhoven R, de Vries M, Ursing B, de Vos WM, Siezen RJ (2003) Complete genome sequence of *Lactobacillus plantarum* WCFS1. Proc Natl Acad Sci USA 100:1990–1995

Kurtzman CP (2003) FEMS Yeast Res 4:233–245

Kurtzman CP, Fell JW (1999) Definition, Classification and Nomenclature of the Yeasts. In: Kurtzman CP, Fell JW (eds) The yeasts: a taxonomic study, Elsevier Science BV, Amsterdam, pp. 3–5

Lan R, Reeves PR (2002) *Escherichia coli* in disguise: molecular origins of *Shigella*. Microbes Infect 4:1125–1132

Lapage SP, Sneath PHA, Lessel EF, Skerman VBD, Seelinger HPR, Clark WA (eds). (1992) International code of nomenclature of bacteria (1990 revision). Bacteriological code. American Society for Microbiology, Washington, DC

Li Y, Raftis E, Canchaya C, Fitzgerald GF, van Sinderen D, O'Toole PW (2006) Polyphasic analysis indicates that *Lactobacillus salivarius* subsp. *salivarius* and *Lactobacillus salivarius* subsp. *salicinius* do not merit separate subspecies status. Int J Syst Evol Microbiol 56:2397–2403

Masco L, Ventura M, Zink R, Huys G, Swings J (2004) Polyphasic taxonomic analysis of *Bifidobacterium animalis* and *Bifidobacterium lactis* reveals relatedness at the subspecies level: reclassification of *Bifidobacterium animalis* as *Bifidobacterium animalis* subsp. *animalis* subsp. nov. and *Bifidobacterium lactis* as *Bifidobacterium animalis* subsp. *lactis* subsp. nov. Int J Syst Evol Microbiol 54:1137–1143

Matsuki T, Watanabe K, Tanaka R (2003) Genus- and species-specific PCR primers for the detection and identification of bifidobacteria. Curr Issues Intest Microbiol 4:61–69

Mattarelli P, Bonaparte C, Pot B, Biavati B (2008) Proposal to reclassify the three biotypes of *Bifidobacterium longum* as three subspecies: *Bifidobacterium longum* subsp. *longum* subsp. nov., *Bifidobacterium longum* subsp. *infantis* comb. nov. and *Bifidobacterium longum* subsp. *suis* comb. Int J Syst Evol Microbiol 58:767–772

McCullough MJ, Clemons KV, McCusker JH, Stevens DA (1998) J Clin Microbiol 36:2613–2617

Mcneill J, Barrie FR, Burdet HM, Demoulin V, Hawksworth DL, Marhold K, Nicolson DH, Prado J, Silva PC, Skog JE, Wiersema JH (Members) (2006) Turland N.J. (Secretary of the Editorial Committee) International Code of Botanical Nomenclature (Vienna Code). Regnum Vegetabile 146. A.R.G. Gantner Verlag KG

Meile L, Le Blay G, Thierry A (2008) Safety assessment of dairy microorganisms: *Propionibacterium* and *Bifidobacterium*. Int J Food Microbiol 126:316–320

Mitterdorfer G, Mayer HK, Kneifel W, Viernstein H (2002) Protein fingerprinting of *Saccharomyces* isolates with therapeutic relevance using one- and two-dimensional electrophoresis. Proteomics 2:1532–1538

Nakamura LK, Roberts MS, Cohan FM (1999) Relationship of *Bacillus subtilis* clades associated with strains 168 and W23: a proposal for *Bacillus subtilis* subsp. *subtilis* subsp. nov, and *Bacillus subtilis* subsp. *spizizenii* subsp. nov. Int J Syst Bacteriol 49:1211–1215

Nielsen P, Fritze D, Priest FG (1995) Phenetic diversity of alkaliphilic *Bacillus* strains: proposal for nine new species. Microbiology 141:1745–1761

Opinion of the Scientific Committee on a request from EFSA on the introduction of a qualified presumption of safety (QPS) approach for assessment of selected microorganisms referred to EFSA (2007) EFSA J 587:1–16

Pineiro M, Stanton C (2007) Probiotic bacteria: legislative framework–requirements to evidence basis. J Nutr 137 (3 Suppl 2): 850S–853S

Pot B, Ludwig W, Kersters K, Schleifer KH (1994) Taxonomy of Lactic Acid Bacteria. In: de Vuyst L, Vandamme EJ (eds) Bacteriocins of lactic acid bacteria: microbiology, genetics and applications. Blackie Academic & Professional (UK), London, pp. 13–90

Rademaker JL, Herbet H, Starrenburg MJ, Naser SM, Gevers D, Kelly WJ, Hugenholtz J, Swings J, van Hylckama Vlieg JE (2007) Diversity analysis of dairy and nondairy *Lactococcus lactis* isolates, using a novel multilocus sequence analysis scheme and (GTG)5-PCR fingerprinting. Appl Environ Microbiol 73:7128–7137

Rosselló-Mora R, Amann R (2001) The species concept for prokaryotes. FEMS Microbiol Rev 25:39–67

Salkinoja-Salonen MS, Vuorio R, Andersson MA, Kämpfer P, Andersson MC, Honkanen-Buzalski T, Scoging AC (1999) Toxigenic Strains of *Bacillus licheniformis* Related to Food Poisoning. Appl Environ Microbiol 65:4637–4645

Sanders ME, Morelli L, Tompkins TA (2003) Sporeformers as Human Probiotics: *Bacillus, Sporolactobacillus,* and *Brevibacillus*. Comp Rev Food Sci Food Saf 2:101–110

Scharek L, Altherr BJ, Tölke C, Schmidt MF (2007) Influence of the probiotic *Bacillus cereus* var. toyoi on the intestinal immunity of piglets. Vet Immunol Immunopathol 120:136–147

Sgorbati B, Biavati B, Palenzona D (1995) The genus *Bifidobacterium*. In: Wood BJB, Holzapfel WH (eds) The genera of lactic acid bacteria, vol. 2. Blackie Academic & Professional (UK), London, pp. 279–306

Simpson WJ, Taguchi H (1995) The genus *Pediococcus* with notes on the genera *Tetragenococcus* and *Aerococcus*. In: Wood BJB, Holzapfel WH (eds) The genera of lactic acid bacteria, vol. 2. lackie Academic & Professional (UK), London, pp. 25–172

Stackebrandt E, Frederiksen W, Garrity GM, Grimont PA, Kämpfer P, Maiden MC, Nesme X, Rosselló-Mora R, Swings J, Trüper HG, Vauterin L, Ward AC, Whitman WB (2002) Report of the *ad hoc* committee for the re-evaluation of the species definition in bacteriology. Int J Syst Evol Microbiol 52:1043–1047

Staley JT, Krieg NR (1989) Classification of procaryotic organisms: an overview. In: Staley JT, Bryant MP, Pfennig N, Holt JG (eds). Bergey's manual of systematic bacteriology, vol. 3. Williams & Wilkins, Baltimore, pp. 1601–1603

Teuber M (1995) The genus *Lactococcus*. In: Wood BJB, Holzapfel WH (eds) The genera of lactic acid bacteria, vol. 2. Blackie Academic & Professional (UK), London, pp. 173–234

van de Guchte M, Penaud S, Grimaldi C, Barbe V, Bryson K, Nicolas P, Robert C, Oztas S, Mangenot S, Couloux A, Loux V, Dervyn R, Bossy R, Bolotin A, Batto JM, Walunas T, Gibrat JF, Bessières P, Weissenbach J, Ehrlich SD, Maguin E (2006) The complete genome sequence of *Lactobacillus bulgaricus* reveals extensive and ongoing reductive evolution. Proc Natl Acad Sci USA 103:9274–9279

Vaughan-Martini A, Martini A (1999) *Saccharomyces* Meyen ex Reess In: Kurtzman CP, Fell JW (eds) The yeasts: a taxonomic study. Elsevier Science BV, Amsterdam, pp. 358–371

Ventura M, van Sinderen D, Fitzgerald GF, Zink R (2004) Insights into the taxonomy, genetics and physiology of bifidobacteria. Antonie Van Leeuwenhoek 86:205–223

Wegmann U, O'Connell-Motherway M, Zomer A, Buist G, Shearman C, Canchaya C, Ventura M, Goesmann A, Gasson MJ, Kuipers OP, van Sinderen D, Kok J (2007) Complete genome sequence of the prototype lactic acid bacterium *Lactococcus lactis* subsp. *cremoris* MG1363. J Bacteriol 189:3256–3270

Willenbrock H, Hallin PF, Wassenaar TM, Ussery DW (2007) Characterization of probiotic *Escherichia* coli isolates with a novel pan-genome microarray. Genome Biol 8: R267. doi:10.1186/gb-2007-8-12-r267

Woese CR, Kandler O, Wheelis ML (1990) Towards a natural system of organisms: proposal for the domains Archaea, Bacteria, and Eucarya. Proc Natl Acad Sci USA 87:4576–4579

Yarza P, Richter M, Peplies J, Euzeby J, Amann R, Schleifer K-H, Ludwig W, Glockner FO, Rosselló-Mora R (2008) The All-Species Living Tree project: A 16S rRNA-based phylogenetic tree of all sequenced type strains. Syst Appl Microbiol. doi:10.1016/j.syapm.2008.07.001

Subject Index

Subject Index